Developments in Agricultural Engineering 6

Housing of Animals

OTHER TITLES IN THIS SERIES

1. Controlled Atmosphere Storage of Grains
   by J. Shejbal (Editor)
   1980 viii + 608 pp.

2. Land and Stream Salinity
   by J.W. Holmes and T. Talsma (Editors)
   1981 iv + 392 pp.

3. Vehicle Traction Mechanics
   by R.N. Yong, E.A. Fattah and N. Skiadas
   1984 xi + 307 pp.

4. Grain Handling and Storage
   by G. Boumans
   1984 xiii + 436 pp.

5. Controlled Atmosphere and Fumigation in Grain Storages
   by B.E. Ripp et al. (Editors)
   1984 xiv + 798 pp.

Developments in Agricultural Engineering 6

Housing of Animals

CONSTRUCTION AND EQUIPMENT OF ANIMAL HOUSES

A. MATON
Doctor h.c. of Agricultural Sciences (Univ. of Keszthely, Hungary)
Doctor of Science (Univ. of Lille, France)
Agricultural Engineer (Univ. of Gent, Belgium)

J. DAELEMANS
Doctor of Agricultural Sciences (Univ. of Giessen, W. Germany)
Agricultural Engineer (Univ. of Gent, Belgium)

J. LAMBRECHT
Master of Management Sciences (Univ. of Leuven, Belgium)
Industrial Engineer (Agric.) (Gent, Belgium)

Rijksstation voor Landbouwtechniek, Van Gansberghelaan 115, B-9220 Merelbeke, Belgium

English Text: **F. LUNN**, Gr.Sc.
Drawings: **A. STEVENS**

ELSEVIER

Amsterdam — Oxford — New York — Tokyo 1985

ELSEVIER SCIENCE PUBLISHERS B.V.
Molenwerf 1
P.O. Box 211, 1000 AE Amsterdam, The Netherlands

Distributors for the United States and Canada:

ELSEVIER SCIENCE PUBLISHING COMPANY INC.
52, Vanderbilt Avenue
New York, NY 10017, U.S.A.

Library of Congress Cataloging-in-Publication Data

Maton, A. (André)
Housing of animals.

(Developments in agricultural engineering ; 6)
Translation of: De huisvesting van dieren.
Includes bibliographies.
1. Livestock--Housing. I. Daelemans, J.
II. Lambrecht, J. III. Title. IV. Series.
SF91.M3713 1985 728'.92 85-16020
ISBN 0-444-42528-4

ISBN 0-444-42528-4 (Vol. 6)
ISBN 0-444-41940-3 (Series)

Printed in The Netherlands

CONTENTS

Chapter 4

Chapter 5

Chapter 6

Chapter 7

Chapter 8

Chapter 9

Chapter 10

PREFACE

Animal husbandry plays an increasingly important role throughout the world in the fulfilment of the primary need of mankind : food. The growing demand for food leads to an intensification of the production of livestock. Climatic conditions in large parts of the world require the accommodation of livestock in suitable buildings during certain periods in a year. The housing must also contribute towards creating a healthy environment for the animals and their attendants and has to make possible a rational organization of the involved labour, an economically, ecologically and ethologically justified production of livestock, conform to the latest standards of Public Health. The housing must also blend with the landscape.

Our research carried out since the Fifties in the National Institute for Agricultural Engineering in Merelbeke and a thorough study of the literature concerning the housing of animals have led successively to a first edition in the Dutch language of a book on this subject (494 pp.) in 1971, which was translated into French and published in Paris in 1972 and was later also translated into Spanish and published in Madrid in 1975. The second and third completely revised editions of this book in the Dutch language were respectively published in 1976 and 1983.

The English edition of "Housing of Animals" by Elsevier Science Publishers is in fact not merely a translation of the last edition of our book in the Dutch language but is a new text. It is true that large parts of this book were taken from the Dutch edition, but on the other hand a great deal of recent information of international dimension concerning animal husbandry, construction and equipment of animal houses has been included in this edition. Last but not least the book was updated with the latest scientific and technological findings on this subject published in 1983 and 1984.

I would certainly fail in my duty not to express very special thanks to my collaborators of the National Institute for Agricultural Engineering in Merelbeke : Dr. Ir. J. Daelemans and M. Sc. Ing. J. Lambrecht who, together with myself, wrote the original manuscript. Their knowledge and devotion were of great value in completing this book.

I am especially grateful to Gr. Sc. F. Lunn, attached to our Institute, for the elaborate work involved in writing the English text and in the editorial preparation of this book.

To Mr. A. Stevens I owe a special debt of thanks for making the many drawings with great skill.

I also want to thank M. Sc. Ing. J. Lambrecht for taking as well as developing and printing the many photographs.

I am particularly grateful to Mrs. M. Van den Hauwe for preparing and typing the manuscript in a professional way and for being so patient with all of us.

I finally wish to thank Mrs. M.C. De Ghouy-De Wandel and Mr. H. Van De Sype for the preparation of the numerous plates.

We hope that this new book, like its predecessors in the other languages will prove its services to all who are involved in or confronted with the housing of animals, such as agricultural engineers, veterinary surgeons, architects, students, progressive farmers and all who for various reasons show interest in this subject. We dare to hope that this book will facilitate a scientifically based and judicious choice between the large variety of technical means which are nowadays at the disposal of the animal keeper and that it will serve the interest of agriculture and of mankind throughout the world.

The authors would be grateful to those readers who want to make suggestions or remarks.

Merelbeke, Belgium, March 2, 1985.

A. MATON

Chapter 1

BRIEF HISTORY OF THE HOUSING OF DOMESTIC ANIMALS

Chapter 1

BRIEF HISTORY OF THE HOUSING OF DOMESTIC ANIMALS

The housing of domestic animals has through the centuries been linked to the way animal breeding was practised, so far that the history of cattle breeding and the housing of animals cannot really be separated from each other (Lindemans, 1952 ; Trefois, 1978).

In ancient days man had to provide for his own supplies of food by picking fruits and parts of plants which he found in nature, and by hunting and fishing i.e. by killing wild animals, mainly mammals, birds and fishes. Soon he found that more plentiful and regular food provision was possible by domesticating certain animals and so cattle breeding, which is older than arable farming, originated some ten thousand years ago. The keeping of animals and the cultivation of plants gave greater freedom and possibilities to mankind. They formed the basis for the development of settlements, towns, states, cultures and civilizations (Rohrs, 1974). This development took place in several steps.

At the time of the first civilization man already possessed herds to produce wool, meat and dairy products. These people were wandering herdsmen who drove their herds from one place to another. The animals found their pastures in the landscape and as soon as the surroundings of "the camp" were grazed, the herdsmen drove the herd to new pastures. Each community required an extensive wandering area in order to find pastures for their herd throughout the year. During the winter the herd was brought to sheltered places viz. under the dense cover of the forests, in gorges or in secluded valleys ; the animals survived from dry leaves, young saplings and moss. This was *the nomadic pastoral cattle husbandry*. It disappeared in Europe probably before the historical chronology, at least in the form as it still exists nowadays in some Asiatic steppes.

In a further stage each tribe or community had its own fixed location and the first rudiments of agriculture originated with produce of the field. In this first period of sedentary agriculture the land was not fertilized : the crops utilized the humus and nutrients which were abundantly present in the cultivated primitive soil. New fields were cultivated when the peasants found that the fertility of the soil diminished and the previous fields were abandoned and left fallow. Later this land became farmhold with a three course rotation (land divided in three parts where the principle of three course rotation was applied : winter-grains as food for man, summer-grains as feed for the animals and the third part was left fallow), using straw-manure and livestock droppings. The surrounding pastures became a permanent part of the settlement.

As the number of villages increased, it became necessary to demarcate and delineate the pastures of each community. In this way each community had its own land and the peasants kept it scrupulously for their own, if need be by force. All land that was not cultivated and was not appropriated by a "lord" for his own use, could be used as pasture for the herd belonging to the community. The land was and remained wild and covered with its original vegetation. It was generally denoted as pasture land. Cattle breeding was still pastoral. The herd of the community grazed for the greater part of the year under the vigilant eyes of the herdsmen. It grazed the wild herbs and low shrubs. To protect the animals against the rough climate during the winter, pounds and primitive confinements were built by the farmer.

In the period of *sedentary pastoral cattle husbandry* the herds were quite numerous, at least during the summer. Apart from the fields belonging to the village community, the landscape was covered with forests and heaths, where the herds ranged over an immense pasture. The winter was however rather perilous for the farmer : the residue of the grain harvest, which in fact was owned by the community, certainly did not suffice to bring the herd through the winter, even if hay was available. Therefore, all slaughter-ready animals and certainly a number of saleable horses were taken to the market. Slaughtering also took place in the village where the meat was salted, dried or smoked and the hides were tanned. Meat and hides together with wool were in fact the main and probably even the only products which the farmer could sell since the field crops were mainly required as nourishment for his family. There were mainly herds of pigs and flocks of sheep. Which animals overwintered and the way this happened is still a mystery. Probably the pedigree sheep and horses were kept and fed in confinements, while for the pigs the pedigree-animals and non-slaughterable remained in the forests and had to provide for themselves just like the wild boars.

The community sent its animals to a common pasture under the care of the herdsmen. Every type of animal (horses, dairy cows, sheep, pigs, geese) formed a separate herd. Early in the morning the herdsmen blew their horns to notify the villagers to open their sheds and the animals joined the herds. Along certain paths the herds were taken to the respective pastures by the herdsmen, where they remained all day. At midday the herd was assembled under the shadow of the trees for the time of rest. Before darkness the herds were brought back to the village and after the herdsmen blew their horns again, the animals found their way back to their sheds.

The domains, just like the village communities, practised pastoral cattle husbandry. Their herds of sheep, pigs, cattle and horses were driven into private wild pastures. By the standards of sedentary pastoral husbandry these herds were rather large and often each was under the care of a different herdsman. The numerous herds which were owned by the abbeys were under the care of a pastor equorum, a pastor ovium, a pastor vaccarium and a pastor porcorum.

In sedentary pastoral cattle husbandry the animals were the root of the whole active business of the farmer. The cultivated land was rather small compared to the wild heaths and forests in which the animals were driven. The cultivated land was fertilized with the droppings from the animals in the pasture. A lot of these droppings were lost at the pasture but those left behind in the sheds during the night were probably sufficient to give a certain degree of fertilizing to the cultivated land. The produced straw-manure was probably rather scarce since the straw was used more as feed for the overwintering herds and less as littering. When the animals could be driven in forests where oak and beech dominated the husbandry was mainly concentrated on pigs. Near the sea and the rivers mainly sheep were kept but also dairy cows (vaccariae). Sheep produced wool for the drapery, but sheep and cows also delivered milk to the cheese-monger.

On the heaths the sheep dominated since only they found enough food on the scanty grounds. They produced wool and sheepskin and probably mutton. There were also herds of horses found on the heaths which were bred in the wild, in large fenced heath pastures. Pastoral horse breeding already existed for many centuries on the immense heaths.

In many regions the pastoral cattle husbandry with its immense, often collective pastures evolved into a *cattle husbandry with limited and privately owned pastures*. A more intensive cattle breeding arose. One of the most important causes of this new kind of husbandry was the disintegration of large early-medieval domains and their alienation by donation or sale to the minor landed nobility and especially to religious communities. The rough grounds, which previously were pastures for extensive cattle husbandry were now cultivated and converted to arable land. The new independent farms still possessed some pastures for their cattle, although they were now necessarily restricted.

To provide a pasture for the cattle the farmers were forced to let a limited part of the land uncultivated or to create some artificial pastures in the cultivated land, covered for instance with peas or vetches and mixed with oats.

The permanent housing of dairy cattle already existed in the Middle-Ages : fodder was collected on the land and fed to the stalled animals.

Contrary to the permanent stalling as practised in some regions where the animals were kept in the stalls winter and summer, the stalling was interrupted in other regions in the summer during daytime to allow some grazing. Here the farmers possessed some permanent pastures and orchards near the farm. During late summer the remains of hay pasture was grazed and during autumn stubble supplemented for instance with turnip leaves was grazed. The cattle were stalled during the night and the hottest hours of a summer day, where they consumed clover in the stall ; the remaining hours were spent grazing.
This type of cattle keeping with alternate grazing and housing was from a hygienic and zootechnical point of view much better than permanent stalling. It aimed at the highest quantitative and qualitative yield from the cattle together with a good manure production. This method of cattle keeping superseded that with permanent stalling in most regions in the course of the 19th century. Due to the permanent or interrupted stalling of the cattle, the dairy cows

became more important and privileged. The flocks of sheep became less numerous and finally disappeared from most farms except from some privileged farms which owned the right to keep a flock of sheep on the common grounds.

The different periods in the development of cattle husbandry have not succeeded each other in Europe during the course of the centuries. They arose or were forced, due to the circumstances, earlier in one area than in another. The population, the type of landed property which was dominating in a certain region - small, medium, large or very large - and the extent of the forests in the landscape were the main factors influencing the cattle husbandry. In places where small and medium landed properties were dominant, sedentary pastoral cattle husbandry became impossible and the farmer was forced to keep the cattle on restricted, private pastures ; on large and very large domains however and also where immense forests provided unlimited pastures, it was advantageous for the farmer to practise pastoral husbandry.

As cattle husbandry with permanent or periodic stalling gained acceptance, it became necessary to pay more attention to the housing of the animals and especially of dairy cattle whereby the following description gives an idea.

We may assume that in many countries the cow-shed, where cattle were stalled during the winter, was a construction of woodwork and loam, without windows and covered with a low straw roof. A wide door formed the entrance for the dung-cart. The animals were tied to wooden poles which were driven into the soil, at a safe distance from the walls to prevent damage from the horns. The straw and eventually the hay was thrown in front of the animals. In the cow-shed the space above the animals was probably reserved as storage for straw and hay. This hay-loft had previously been in the byre. The walls supported a frame of wooden stakes which formed the hay-mow. Houses without a hay-mow are more spacious : an empty hay-mow provided a cool and airy space to the animals during the summer.

At an early stage the farmer on poor sandy soils learned to appreciate the fertilizing power of farmyard manure and knew that it formed the most important factor for his whole farm and for his existence. He kept treating the manure as in the old sheds : he left it there, piled up under the animals even if they were permanently stalled. Therefore these byres were dug out to about one metre and sometimes deeper, whereby the animals initially stood in a pit ; the straw together with the solids of the manure held by the straw, accumulated in this pit. In one or both of the opposing walls wide gates were provided allowing free entry of the dung-cart.

In the old, improved, Flemish cow-shed, the animals were placed next to each other, in one row. They were tied to horizontal stakes, connected to the poles. They were at a safe distance from the loam wall to prevent damage from the horns. In this way a passage was created in front of the animals which was useful to "serve" their fodder. This passage was floored with bricks and was connected, through a door, with the room used for the feed preparation. Such byres were already mentioned in the 16th century (figs 1.1, 1.2).

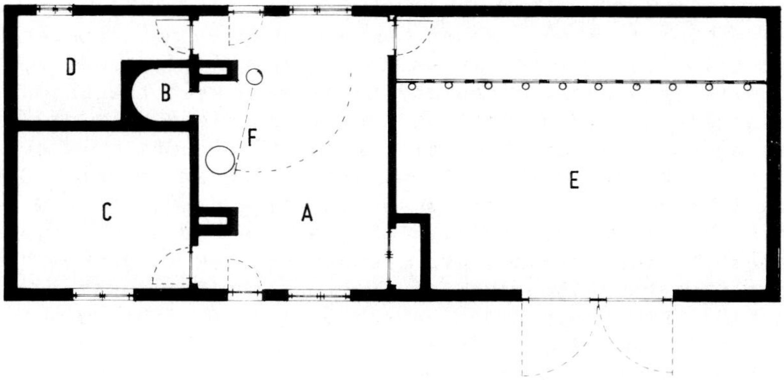

Fig. 1.1 View at and layout of a Flemish long façade farm (Belgium).

Legend : A = living quarters ; B = oven ; C = bedroom ; D = room ;
E = cow-shed ; F = swing-beam with feeding kettle.

Description : In the living quarters: projection of the swing-beam which brings the kettle from the fireplace to the entrance of the feed passage.
In the cow-shed : projection of the stakes and the wide door to evacuate the manure.

Fig. 1.2 The living-quarters with swing-beam and feed kettle (open-air museum at Bokrijk, Belgium).

As long as the walls were made from loam the danger of damage remained and it was impossible to stall animals along the outer side. The construction of byres in brickwork brought important changes to the equipment : the mangers were placed along an outer wall and the space in front of the animals disappeared. In order to "serve" the animals it was now necessary to pass between the animals. A service passage remained behind the cattle. Possibly this way of construction was aimed at a reduction of the building costs by decreasing the space.

For the same number of animals, the loam animal houses were more spacious than the later houses in brickwork : the row of animals stood in the lengthwise direction of the byre and a space was reserved adjacent to the cow-house to prepare the feed. The new type of animal house, without the space in front of the tied animals, originated, as mentioned above, when houses were built from durable materials. They first appeared in the richest and most fertile regions. They were probably earlier on the rented farms than on freehold farms and it seems that the initiative for the building of such houses was taken by the abbeys and monasteries during the prosperous years of the 17th and mainly the 18th century. This new type of housing resembled the animal houses which were already built in natural stone by the monks since the Middle-Ages. Freehold farmers built such animal houses in the good years of the 19th century (before 1880) and gradually this type of housing became a model. Animal houses in brickwork with a feeding passage (and other improvements from the hygienic point of view) have been built only for about the last half a century.

We must admit that the old animal houses left much to be desired from the hygienic point of view. One of the main faults was the restricted space which was available to the animals. Another fault was the insufficient ventilation. The only supply of fresh air was through a "half heck door" or through some holes in the wall. The latter were often filled with straw as soon as it became colder. The ceiling or hay-loft was much too low. As a result the old animal houses had a sweltering atmosphere of damp, warm and putrid air.

The animals often stood on a hard floor, paved with hard sandstone or sometimes with porphyry. It was better than nothing but far from ideal since the stones were not jointed and the liquid of the dung penetrated the soil below. Removal of the urine was also rather primitive. A furrow behind the animals carried the urine through an opening in the wall out of the house, while a second furrow carried it to the dung-pit.

Generally, the animals took their feed from a manger, except in some regions where tubs were used. The local methods of preparing the feed, the mashing and slopping, were probably the reason for it.

The *pigsty* had its own history. Some old calendars dating back to the 16th century show the hogs running loose on the farmstead. Probably the young pigs remained on the farm. Other pictures show the hogs in a hog's-cote under the pigeon platform. In autumn the hogs were driven under the oaks and beeches to be fattened. After the acorn-time and sometimes even earlier, the hogs were stied. The world famous Flemish painters Breughel and Teniers have left us some pictures of such a swine-sty : a tiny wooden pen adjacent to the byre-wall, with one opening above the trough through which the hog was fed and got fresh air. This situation remained unchanged until the 20th century and it was only after the Second World War that more attention was paid to rational and hygienic housing of pigs.

Finally, some words about the *housing of poultry*. On some old calendar pictures the chicken-ladder is often found adjacent to the house. It is certainly an old usage. In the evening the chickens entered along this ladder and were safe on the "cock-perch" or on a stake of the ceiling above the byre or floor out of reach of foxes and other chicken-thieves. It was not always very hygienic since the poultry house was sometimes above the living quarters, rather than above the byre.

From this brief survey it is obvious that the housing of domestic animals has been through a long history. For many centuries the houses have given but the most required protection against rough weather or have been used to obtain a better valorization through stall-feeding of the fodder crops. Little attention was paid to the hygiene of man and animal and the care of the animals required slavish labour in the animal houses. Only from the 20th century can we really speak of the housing of animals. Modern constructions focussed on the requirements of animal hygiene and to the rational labour organization only started to penetrate after the Second World War. This does not mean that we only find modern animal houses today, for a lot of old farm buildings are still in use.

The modern housing has come a long way since the primitive sheds that were used before and modern techniques have in barely twenty years contributed more to the progress in rational construction and equipment of animal houses than was achieved in the many centuries before. Until recently it was possible to modernize animal houses by remodelling existing houses (Goedseels and Vanhaute, 1978). The actual trend towards specialization together with an important enlargement (Verhulst and Bublot, 1980) often oblige farmers to build new houses. The aesthetical implantation of new buildings on existing farms is often, though unjustly, neglected (Goedseels and Vanhaute, 1978) and deserves our attention with regard to the protection of landscape and environment.

REFERENCES

Goedseels V. and Vanhaute L., 1978. Hoeven op land gebouwd, Belgische Boerenbond (Editor), Lannoo, Tielt, (Belgium), 237 pp.

Lindemans P., 1952. Geschiedenis van de Landbouw in België, (two parts), De Sikkel, Antwerp, (Belgium), 1013 pp.

Rohrs M., 1974. Artgemässe und verhaltensgerechte Haltung von Haustieren, Der Tierzüchter, 26 : 509-511.

Trefois C., 1978. Ontwikkelingsgeschiedenis van onze landelijke architectuur, Danthe, St. Niklaas, (Belgium), 305 pp.

Verhulst A. and Bublot G., 1980. De Belgische Land- en Tuinbouw, verleden en heden, Ministerie van Landbouw, Nationale Dienst voor Afzet van Land- en Tuinbouwprodukten (Editor), Mercatorfonds/Cultura, Brussels, (Belgium), 127 pp.

Chapter 2

THE HOUSING OF DOMESTIC ANIMALS IN RELATION TO MODERN FARMING AND PRESENT-DAY SOCIETY

Chapter 2

THE HOUSING OF DOMESTIC ANIMALS IN RELATION TO MODERN FARMING AND PRESENT-DAY SOCIETY

The housing of animals must in the first place be adapted to the requirements of the modern exploited farm but also to those set by present-day society (energy, environmental problems, protection of the animals).

Even the outsider is nowadays aware of the developments in agriculture and animal breeding which thoroughly changes the oldest activities of mankind. Up to a quarter of a century ago the farm was basically the same as many centuries before. Its appearance had of course been thoroughly changed by modern mechanization, farm building techniques, plant- and animal breeding methods etc. but the internal structure remained principally the same as it had always been : an independent, geographically isolated, family concern, run by proud, hard working people with relatively restricted material needs and with a pronounced conservative and individualistic tendency. Those characteristics of agricultural activity remain in countries where the economy is based on agriculture.

In the industrialized world an unmistakable evolution has taken place which is from our point of view an expression of the adaptation process of agriculture and farmers to the economic and social requirements of the industrialized society. The unexpected expansion of industrial activity in Western countries has established an unknown prosperity and social security amongst almost all levels of society and this despite the fact that labour time has been shortened and work has become physically easier. This remains true even in the serious economic recession which currently is taking place in the Western world.

It is a well known fact in the industrialized Western world that a disparity exists between the income of the farmer and that of the other groups of the active population and this to the disadvantage of the farmer. The cause of this inequality is, according to some experts, due to the different magnitudes of increases during the last decade in the price of agricultural products in comparison with those of the production means in agriculture. Others, however, find that besides the unfavourable trends in prices, the lag in the economic and social situation of farmers is due to the fact that a number of farms are still run according to old-fashioned methods. The search for a new equilibrium in modern agriculture through an *adaptation process* is based on this inequality.

Agricultural development is primarily expressed in the aim for *concentration*, the formation of larger units. The enlargement allows of course, certainly in the initial phase, to reduce markedly the production costs per produced unit (e.g. cow, pig, chicken...).

The number of farms shows a distinct reduction, which for five important West-European industrialized countries is represented in table 2.1 (Anon., 1983). In the period 1960-1981, the number of farms in Germany, the United Kingdom and the Netherlands has more or less been halved. In Belgium there was an even greater reduction, while for France the reduction was limited to only one third. On the contrary, the acreage of the remaining farms was increased as shown in table 2.2.

TABLE 2.1 The number of farms in five European countries (Anon., 1983).

Year	1960		1970		1981	
Country	Number	%	Number	%	Number	%
W. Germany	1,385,250	100	1,083,118	78.2	780,469	56.3
France	1,773,500	100	1,420,924	80.1	1,129,000	63.7
Netherlands	230,312	100	164,119	71.3	126,156	54.8
Belgium	198,706	100	130,397	65.6	89,131	44.9
U. Kingdom	443,100	100	311,478	70.3	244,481	55.2

TABLE 2.2 Relative % distribution of the number of farms of five European countries, according to their acreage, 1960-1981 (Anon., 1983).

	≥ 50 ha			20 - 50 ha			10 - 20 ha			1 - 10 ha		
	1960	1970	1981	1960	1970	1981	1960	1970	1981	1960	1970	1981
D	1.2	1.8	4.1	8.8	14.5	22.6	20.7	24.7	22.6	69.3	59.0	50.6
F	5.5	8.5	13.6	20.5	26.0	30.4	26.6	25.0	21.1	47.4	40.5	34.9
NL	0.9	1.5	3.2	10.9	17.0	24.5	23.4	31.7	28.6	65.1	49.8	43.7
B	1.1	2.0	4.4	6.2	13.1	21.3	17.7	25.7	26.5	75.0	59.2	47.8
UK	18.7	27.0	33.1	22.4	25.9	27.2	16.3	16.0	16.0	42.6	31.1	23.6

where D = W. Germany ; F = France ; NL = Netherlands ; B = Belgium ; UK = United Kingdom.

The number of farms with an acreage between 1 and 10 ha decreased in every country although this group is still the most important, except in the United Kingdom. In 1981 this group represented less than 25 % of all the farms in the United Kingdom, about 35 % in France and about half of all the farms in the other mentioned countries. There is a net rise in the number of farms between 20 and 50 ha and even of those larger than 50 ha, which are now very important in the United Kingdom and to some extent in France. The importance of the number of farms between 10 and 20 ha differs from country to country and has a recent tendency to decrease in W. Germany, France and the Netherlands whereas they become more important in Belgium and remain unchanged in the United Kingdom. Mainly in livestock keeping a large concentration has taken place during the last 25 years. As an example the recent increase of the number of pigs per farm in 9 E.E.C.-countries is given in table 2.3 (Verduijn, 1983).

TABLE 2.3 Increase of the number of pigs per farm in the E.E.C. (Verduijn, 1983).

Country	Degree of self-sufficiency in 1979	Pig farms in % per member country in 1979	Number of pigs per farm		
			1973	1976	1979
W. Germany	89	25.8	26	34	41
France	89	16.4	21	25	30
Italy	75	47.9	7	8	9
United Kingdom	63	1.6	142	193	225
Netherlands	225	2.2	104	159	205
Belgium/Lux.	162	2.2	67	92	113
Ireland	144	0.5	32	66	114
Denmark	368	3.4	82	99	127
Total	102	100.0	25	29	35

There are different forms of concentration. Let us first of all mention the *horizontal integration* under the form of co-operatives which, at the production level, has known some success in a few countries (e.g. France) while in other countries (e.g. Germany, Belgium) it disappeared after a short time.

The formation of *privately-owned large farms*, whether or not exploited by the proprietor, is another example of this integration. In many countries such privately-owned large farms raised their number of animals up to 100 or more cows, 200 or more sows, 1,000 or more fattening pigs, 20,000 or more layers or 40,000 or more broilers per farm. The dairy farms mostly took over small mixed farms from the vicinity.

An ultimate form of concentration is the *communist horizontal integration* whereby under pressure or by order of the Government, the previous existing farms have been assembled to large production units (Landwirtschaftliche Produktionsgenossenschaften in East-Germany, sovchoses and kolchoses of Russia and other East-European countries). The poor results of the horizontal integration in most communist countries, together with the negative remunerativeness of some large farms exploited by a paid manager for account of a third party who takes the farm for a capital investment, clearly demonstrates the dangers of concentration in agriculture, when the stimulus of self-interest in the farm is taken away.

The concentration has indeed its limits : as the farms become larger they are less surveyable and their management is more difficult. The risks are also increasing since the capital involved is larger as the size of the farm grows and because the break-out of an epidemic disease can lead to catastrophic results. It seems to us of utmost importance that the concentrated farms should also be run by farmers for their own account ; for certain enterprises such as dairy cattle husbandry where dedication plays an important role it seems to us to be indispensable.

The *vertical integration* has excited a lot of interest : many find it "the" perfect way to reform the distribution trades and to make the difference between the production-price and the consumer-price more acceptable both to the farmer and to the customer. Another typical expression of the adaptation process of agriculture to the industrialized society is the striving to *specialization* of the farms. The old days in which a small farm produced almost everything are definitely past and those adhering to this conservative principle are doomed to leave. The amalgamate of concentration and specialization allows a far advanced and justified mechanization, the building costs per unit product decrease remarkably, an efficient utilization of the committed manpower is possible and by acquiring a great experience and professional knowledge the manager is able to produce better. The many farms keeping thousands of layers or broilers form one of the typical examples of specialization coupled to concentration. The maximum advanced specialization whereby only one product is involved leads however to a greater vulnerability of the farm : the decrease of prices together with the damage caused by diseases are undeniable dangers.

It is obvious that *farm products should meet the quantitative and qualitative demands of the market*. In general it can be stated that the demands put forward by the consumer in the industrialized countries are very high both to the quality and to the presentation of food, which forces the producer to pay more attention to them at the risk of expulsion from the market. The fattening of pigs is a misnomer : the fattening ought not to be the production of fat pigs but has to be directed to the production of lean pork which exclusively fulfils the taste-demands of the inhabitants of a welfare state. The consumer is exceptionally sensitive to the use of hormones, antibiotics etc. in the production of meat whereby the economic imperatives of the production are not compatible to the requirements of public health. It is obvious that the adaptation to the market has now to be carried out internationally instead of nationally.

The *labour conditions of the farmer* and his helpers have gained a considerable interest. The question of, and indeed the demands for comfort, as are known to the inhabitants of a modern city, become more pronounced in the rural society. The young farmer wishes to work in hygienic rooms, he wants to carry out his job in rationally designed buildings and using rational methods, he wants to be productive and tends to attach a very high importance to mechanization : on one hand he wants shorter working times, lighter, more pleasant work and on the other hand he is forced to mechanization by the ever increasing shortage of labourers. Especially, weekend work is avoided or reduced to a minimum ; the young family at the farm, quite justly claims, similar to other citizens, the right to take a holiday or have entertainment. The decrease of the number of labourers in agriculture is a phenomenon which seriously confronts all industrialized countries.

The number of employed labourers in agriculture has decreased remarkably in the E.E.C.-countries, as shown, for eight countries, in table 2.4 (Anon., 1983). In W-Germany, Italy and Belgium this number was in 1982 reduced to one third of the number employed in 1960, in France the number has been more than halved, in Ireland it has been reduced by about 50 %. In Belgium and the United Kingdom less than 3 % of the active population was in 1982 employed in agriculture, in Germany and the Netherlands a mere 5 % and in France and Denmark 8 %.

TABLE 2.4 Number of employed persons in agriculture, in eight European countries (x 1,000) (Anon., 1983).

Absolute number								
	D	F	I	NL	B	UK	IRL	DK
1960	3,623	4,189	6,611	485*	300	1,134	390	362
1970	2,262	2,821	3,878	340	174	784	283	266
1982	1,382	1,758	2,545	248	107	632	196	207
Relative number in % (1960 = 100 %)								
	D	F	I	NL	B	UK	IRL	DK
1960	100	100	100	100	100	100	100	100
1970	62.4	67.3	58.7	70.1	58.0	69.1	72.6	73.5
1982	38.1	42.0	38.5	51.1	35.7	55.7	50.3	57.2
in % of the active population								
	D	F	I	NL	B	UK	IRL	DK
1960	13.8	21.4	31.7	12.1	8.4	4.7	37.0	
1970	8.5	13.5	19.6	6.8**	4.6	3.2	26.9	11.3
1982	5.4	8.2	12.1	4.9	2.9	2.7	17.1	8.4

*interpolation (1959 - 1962)

**1971

where D = W-Germany ; F = France ; I = Italy ; NL = Netherlands ; B = Belgium ; UK = United Kingdom ; IRL = Ireland ; DK = Denmark.

It is obvious that the development of agriculture and cattle husbandry towards concentrated, specialized, strong labour productive, mechanized and market-orientated farms has had radical repercussions on the ideas concerning construction and equipment of service-buildings including the animal houses and the farm- house. The buildings have to make possible the above-mentioned adaptation of agriculture and of the farmer to the changed production requirements. The buildings and the equipment have to form a rational production instrument, the farm-house has to contribute to the social integration of the farmer's family (Maton, 1968). We can notice two aspects in the issue of farm buildings : on one hand the transformation of obsolete service-buildings in view of their adaptation to modern management and on the other hand the construction of new farms. Besides these two aspects we would like to stress the importance of integrating the constructional element with the equipment and mechanization in the buildings.

The transformation of obsolete farm buildings can sometimes enable the adaptation of the farm to modern production requirements. When for instance a (too) large barn in good condition is available it can often be converted to a good cubicle house for dairy cattle with a minimum of costs, whereby under these circumstances modern equipment has to be provided (mechanical mucking-out or slatted floors, milking parlour, etc.). The conversion of existing buildings to rationally equipped houses for sows with or without piglets is also possible. These particular transformations are actually still carried out since they require markedly lower investments than new buildings. It is however not our intention to pretend that remodelling ("Althofsanierung") will solve the problems of the adaptation of our farms to the development of agriculture.

It is beyond doubt that the *erection of new farm buildings incl. animal houses* is often essential for, or will be the final phase of the adaptation of a farm to the modern requirements. This creates many problems both technically and financially and the indication of ways to solve the problems of housing animals is the main aim of this study-book. The exploitation requirements of a house for 100 cows are indeed completely different than those of a house for 10 animals.

Besides the problems in relation to the exploitation of the large scale farm there are others, namely those in relation to the construction methods of large-scale enterprises whereby the effects on the cost price are at least as important. The important question which arises in this context is undoubtedly that of the possibilities offered by the assembly of farms from prefabricated building elements : the turn-key assembly ("Schlüsselfertiges Vollmontagehof") or the construction according to the principles of children's building blocks.

There are already a number of interesting examples in Europe of assembly-farms made of prefabricated elements for which both the building industry as well as the agricultural world show a great interest.

It became clear that only the production of large series of building elements and thus the construction of farms on a large scale will enable the reduction of the cost price of a full assembly to an acceptable level. This level has not yet been reached and generally the prefab has not been widely propagated in the building of new animal houses. The housing of animals must not only fulfil the requirements of a modern exploited farm but has to be fitted in a suitable manner, in our modern society. In relation to this special care has to be taken of the energy problems, the requirements of the environment and the protection of animals.

The *energy problems* are actually at the centre of interest. We can without any doubt say that cheap energy together with cheap raw materials and rapid technological progress have been the foundation of the welfare society and of the comfortable life in the Western world in the last decades.

The Western economy (industry, agriculture, transport, tourism, etc.) has become increasingly dependent on the supply of primary energy from foreign countries. The import of petroleum products, consumed in the West is largely of Arabic origin and especially of the countries of the Middle-East. Those countries were, until World War II, controlled by Western industrialized countries (England, France) but became independent after the war. They were however still economically dominated for another two decades by the West, which is quite understandable if one thinks of the limited know-how and work capacity in those countries compared to the West. Step by step the oil-sheikhs became aware of their immense power and gathered in the club of the so-called OPEC-countries. They provoked a first oil-crisis in 1973-1974 which was followed by a second crisis in 1979-1980 resulting in economic shock waves in Western countries but, and this is often forgotten, also caused great havoc in the already precarious economy of the developing countries. Both oil-crises and especially the last one have produced price-increases as never seen before.

In addition to the direct consumption of energy, agriculture consumes indirectly large amounts of energy, mainly incorporated in feeds, fertilizers, phytosanitary products and also in agricultural machines and materials.

Both the oil-crises have resulted in a considerable increase of the production costs in agriculture but they have, until now, not led to fundamental changes in the production techniques. Also under those changed conditions biological agriculture, extensive arable farming and extensive breeding have not succeeded in a break-through. The draught-horse has not returned to our farms because it is simply impossible to put the clock a hundred years back and to return to the life of a century ago.

In order to prevent such a return alternative energy sources have to be sought for oil-products and natural gas. The application of nuclear energy is, despite the many demonstrations against it, probably the only realizable, and contrary to what has been said, a safe alternative for the production of electricity. The re-application of coal, the use of geothermal energy (warm water-sources in the ground), of solar energy (collectors, photo-voltaic cells), of wind energy (turbines) etc. are other possibilities which are being investigated for industrial and agricultural purposes, but of which, it seems to us, the application on a large scale is not yet economically justified. This last conclusion is also valid for typically agricultural alternative energy sources viz. the production of biogas and the combustion of straw, etc.

The saving of energy has of course become a necessity. In animal husbandry attention must especially be paid to good insulation of the different parts of an animal house, viz. : the walls, roof and floor which all have to be kept dry : for this topic we refer to the third chapter of this book. The ventilation shall preferably be of the natural type, certainly in cattle- and slaughter pig houses and the heating will be restricted to those rooms where it is essential and remunerative (e.g. farrowing pens). The burner has to be maintained in good working order, regularly cleaned and professionally adjusted (carbon dioxide-content in the exhaust gases 10 to 11 %, temperature of the exhaust gases : 200 - 240°C). Heat recuperation by means of a heat pump connected to the milk cooling tank, where simultaneously heat is released and then utilized for the heating of water (cleaning of milking equipment, household applications) is certainly economically justified and is widely used. Heat recuperation of the outlet air by means of a heat exchanger which warms the entering air is probably economically justified but the application of a heat pump for this purpose is, we believe, not (yet) justified.

The *environmental problems* still, and quite justly, attract interest, although one should beware of the rather frequent exaggeration. Animal breeding contributes to the embellishment and vitalization of the landscape but also produces wastes which may be annoying to the neighbourhood. Especially the practice of intensive pig keeping can, in densely populated countries, lead to problems concerning the utilization of liquid manure and odour control. This can also be the case in areas frequented by tourists. The so-called "large-scale bio-industrial plants" where the animals are practically always housed without straw, produce large amounts of liquid manure. The cheapest method to dispose of this liquid manure is by spreading on farmland. Provided that certain precautions are taken such as spreading windwardly off the habitation, spreading before or during rain and certainly by avoiding spreading with sunny weather or during the weekend, and by prompt ploughing or subsoil manuring (by means of a liquid manure injector) etc., the odour nuisance can be restricted to a minimum.

Often, the intensive slaughter pig farm has little or no ground. The liquid manure can then be seen as waste which has to be disposed of in one way or the other. Pasture-land does not tolerate large amounts of liquid manure and surpluses are mainly to be used on arable land. This requires storage and transport which involves high costs, especially for liquid manure. It can therefore be essential, together with the fact that excessive doses of liquid manure can lead to pollution of groundwater, to treat it at the farm in such a way as to make it suitable for discharge to a watercourse.

The most suitable treatment here is biological oxidation : in a basin the liquid manure is enriched with oxygen by means of a surface aerator which enables a microbial, aerobic destruction of the organic matter present in the liquid manure. The aerated liquid manure is then led over a sand-filter before being drained in surface waters. In this way it is possible to obtain an effluent which is almost conform to the legal requirements for discharge to a watercourse (Priem, 1977). The cost price of this treatment of liquid manure does not allow its application on the farm, except perhaps with subsidization by the government. It is however possible to discharge the effluent, obtained by means of biological oxidation and filtration, into a sewer connected to a local purification plant. This purification plant will be able to convert household- and industrial wastewater together with agricultural wastewater to a dischargeable effluent, against compensation. For manure with a high dry-matter content, such as chicken manure, a solution could be the drying in a drum drier, followed by an after-burning of the combustion gases (Priem, 1974). Further research is necessary to find the most suitable treatment for animal waste.

The odour nuisance is another problem for intensive stock-breeding and especially for pig-keeping. In contrast to the traditional animal breeding, which occasionally causes odour nuisance, for instance during spreading, the bio-industry is often responsible for a continuous odour nuisance to the local population. This nuisance is caused by the great concentration of animals, by the anaerobic storage of manure and by supplying the animals feeds with a high animal protein content. Two methods are actually applied for the abatement of odour nuisances. By generating electrical discharges in the air of the animal house, the oxygen of the air is partly converted into ozone which, as a powerful oxidizer, transforms the odourous components to odourless or less smelling components. Another method, mainly applied in Holland, consists of leading the foul air through a biological air scrubber, thereby bringing the foul air in contact with water enriched with active sludge, which retains most of the odour components. The addition of air-oxygen and/or chemical products to the (pig) liquid manure reduces the emanation of stenches from the manure in the air of the animal house. All installations for manure processing and odour abatement in stock-breeding require considerable investment from the animal-keepers. It seems justified that the whole community contributes to the costs of the abatement of environmental nuisance.

Besides environmental pollution, public opinion has recently become more sensitive to the way in which animals are kept. Some of the forms of housing disturb the societies for *animal protection*. Of considerable importance is the West-German "animal protection law" of July 24, 1972 which, amongst other things, determines : "Wer ein Tier hält, betreut oder zu betreuen hat muss dem Tier eine verhaltensgerechte Unterbringung gewährleisten" i.e. one who keeps, takes care or has to take care of animals must guarantee their suitable housing. The law involves severe punishment in cases where animals are badly housed. The question of what one is allowed to do with the animals and where the boundaries of economic necessities in animal breeding are situated are both ethically and scientifically becoming more relevant. A point of exaggeration in animal protection is certainly the anthropomorphical approach of the problem. It is indeed very difficult to determine exact criteria for the welfare of animals (Hofstra, 1975) which should not be directly compared to humans. The ethologist has quite a task to determine the boundaries between comfort, discomfort and vexation in relation to the different breeding- and housing methods. Additionally, an objective education of the consumer of animal food is urgently required. The consumer is not always aware of the fact that his striving towards "natural" food can encourage torment. The meat of the sheep in the Scottish Highlands for instance is certainly natural but those animals are suffering a hard time in the barren hills during the winter months whereby vexation and mortality are more substantial than with the sheep "artificially" kept in houses. Furthermore the consumer often sets quality-requirements to the meat which will inevitably lead to torment. The consumer for instance prefers white veal and therefore calves are kept on an iron-free diet whereby all iron is carefully excluded from the animal house and as a result they become anaemic ; this is indeed an example of wilful torment. The exceptionally cruel way of producing "pâté de foie gras" can be blamed on the demands of the gastronome.

Finally the consumer must understand that the "torment-free" housing of animals will often result in for instance keeping less animals per square metre (more expensive buildings) or ample littering and/or free run in yards (more labour) etc. In these cases the production costs of animal food will increase and the consumer must therefore be willing to pay considerably more for his food package than is actually the case. On the subject of the relation between housing and welfare of the animal the golden mean has to be sought. On one hand the economic imperatives remain dominant but on the other hand both vexation and exaggerative care of the animal have to be excluded.

In this connection we can quote the view expressed by the Parliamentary Select Committee on Agriculture of the House of Commons of the U.K. in 1981 : "We do not accept the contention, frequently stated or implied, that the public demand for cheap food decrees that the cheapest possible methods of production must be adapted... Society has the duty to see that undue suffering is not caused to animals and we cannot accept that that duty should be set aside in order that food may be produced more cheaply. Where unacceptable suffering can be eliminated only at extra cost, that cost should be borne or the product foregone".

The problem of the welfare of laying hens is actually being posed. The pressure of public opinion, sensitized by the societies for animal protection and the "ecologists" has grown in such a way that the E.E.C.-authorities are debating requirements applicable to the member countries concerning layer cages. This important problem will largely be covered in Chapter 6.

In accordance with the foregoing statements the requirements to be complied with by animal houses can be formulated as follows :

- the animal houses have to guarantee the health and welfare of the animals and - not at least - of the people taking care of them : special attention must be paid to the climatization and the hygiene of the animal houses ; animals of different species, age and breeding purpose will principally be kept apart ; the housing will as much as possible take into account the ethological requirements of every type of animal ;
- the animal house and its equipment have to contribute in the production of high quality animal products (dairy, meat, eggs) : hygienic milk production, production of clean eggs and lean, tender meat ;

- the animal houses must enable the rationalization of labour required for the care of the animals : the location of and the connections between the different buildings, together with the internal equipment and mechanization, have to contribute in labour saving and labour relief in the care of the animals ;
- the animal houses have to be built in such a way as to allow an eventual expansion of the stock and if necessary the conversion to another enterprise : the farmer has to adapt his enterprises quantitatively and qualitatively to the economic imperatives including the sales potentials and therefore the utilization of multipurpose hangar type buildings is recommended ;
- the animal houses have to be protected against fire ;
- the animal houses shall require only these investments which are economically justified : relatively less attention is paid to the durability since if the houses are built to withstand the ages they may form an obstacle to modernization or adaptation to new techniques ;
- the animal houses shall form an harmonious whole with the farmhouse and the other buildings, they shall give satisfaction from the aesthetical point of view and their implantation shall suit the landscape ; the animal houses and other buildings shall be built in accordance with the existing regulations of town-planning and public health and shall cause a minimum of environmental nuisance.

REFERENCES

Anon., 1983. Yearbook of Agricultural Statistics 1978-1981, Statistical Office of the European Communities - Eurostat, Brussel, 286 pp.

Hofstra S., 1975. Dierenbescherming, Tijdschrift voor Diergeneeskunde, Vol. 100, nr. 13, pp. 697-702.

Maton A., 1968. Moderne tendenzen in de konstruktie en de uitrusting van de landbouwbedrijfsgebouwen, Landbouwtijdschrift, Brussel, 21 : 3-17.

Priem R., 1974. Het drogen van kippemest, Mededelingen van het Rijksstation voor Landbouwtechniek, Merelbeke, (Belgium), nr. 55, 85 pp.

Priem R., 1977. Onderzoek betreffende de aëroob biologische afbraak van varkensmengmest door middel van een aktiefslibinstallatie, Mededelingen van het Rijksstation voor Landbouwtechniek, Merelbeke, (Belgium), nr. 70, 105 pp.

Verduijn J.J., 1983. Bedrijfsekonomische ontwikkeling in de varkenshouderij, Bedrijfsontwikkeling, 14 : 585-590.

It is our intention to demonstrate in the following chapters how the above-mentioned requirements can be fulfilled by means of a suitable construction and a judicious choice of building materials together with an efficient equipment of the animal houses intended for the main types of animals.

Chapter 3

SOME FUNDAMENTALS CONCERNING THE CONSTRUCTION OF ANIMAL HOUSES AND THE BUILDING MATERIALS TO BE USED

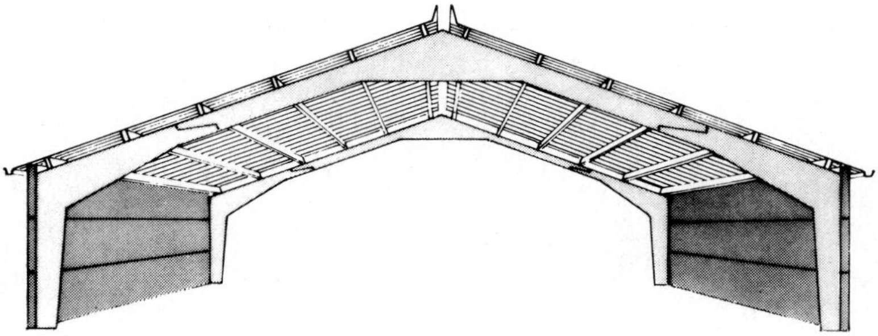

Chapter 3

SOME FUNDAMENTALS CONCERNING THE CONSTRUCTION OF ANIMAL HOUSES AND THE BUILDING MATERIALS TO BE USED

3.1 INTRODUCTION

The general concept for the construction of modern animal houses is the hangar-structure. The hangar can accommodate the animal houses with all its components such as machines together with the storage of roughage, concentrates, straw etc. Machines and fodders are however often stored in a separate hangar attached to the animal house. The farm-house is either detached from (fig. 3.1) or connected to (fig. 3.2) the hangars. In the latter case a lock (e.g. milking parlour) is required between the farm-house and the animal house.

The hangar construction allows a rational arrangement of different rooms whereby the length of the connecting passages is kept to a minimum thus resulting in an important labour saving. From the latter point of view interesting possibilities arise by limiting all constructions to ground floor level : it allows a considerable saving on building costs (few or no supporting walls, no ceilings) and is "the" modern way of building animal houses and other agricultural buildings. Vaulting is far too expensive and a storage on the first floor has its drawbacks : storing progresses less smoothly than on the ground floor and eventual free space cannot be used for other purposes (such as the accommodation of machines for example). The fact that less hay and often no straw is used anymore has made two storey-construction unnecessary.

The hangar construction offers the possibility to convert certain areas to other purposes than those for which they were originally planned. The hangar acts as a polyvalent building whereby the internal utilization is not bound to certain limited forms. The complete hangar complex can be extended by installing additional frames or spans in line with the already existing structure.

3.2 THE FUNDAMENTALS OF STABILITY FOR AGRICULTURAL BUILDINGS

For the construction of agricultural buildings more than the usual regulations for the calculation of building structures (snow and wind load) are in force. Waaijenberg (1984) of the Dutch Institute for Mechanization, Labour and Buildings (IMAG - Wageningen) has worked out a proposal for draft standards specifically meant for agricultural buildings. The formerly existing regulations contain a number of functional requirements not specific for agricultural buildings.

For the wind load for example no allowance was made for the fact that it concerns a rural building with low façades, a slight roof inclination and a large span as with a cubicle house. Specific loads for animals, manure and fodders were not included in the building regulations. For those reasons a proposal was made for a draft stan-

Fig. 3.1 Dairy cattle house in hangar-construction with detached farm-house.

Fig. 3.2 Dairy cattle house with attached farm-house.

dard specifically intended for agricultural buildings. This draft standard contains some modifications and additions to existing building regulations.

3.2.1 Snow load

In agreement with the existing standards a uniformly distributed vertical roof load of 0.5 kN/m² (50 kgf/m²) projected to ground level was used (fig. 3.3).

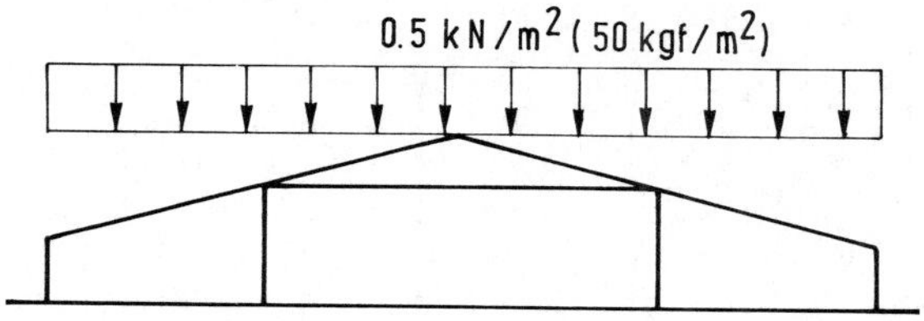

Fig. 3.3 The uniformly distributed vertical roof load (in operational condition) (Waaijenberg, 1984).

This is only valid for buildings for which the structure, location and use give no inducement to include other loads than snow load or a load caused by persons carrying tools for construction and repair. This is valid for practically all agricultural buildings. A concentrated vertical load and a perpendicular "in line" load are taken into account in accordance with the building regulations. For roofs where snowdrift is to be expected due to obstacles or valley gutters this additional load concentration has to be included. Fig. 3.4 represents a snow load for a number of roofs coupled by means of valley gutters. The basic snow load is calculated for a load S = 0.375 kN/m² (37.5 kgf/m²) multiplied by a factor Cs which depends on the roof slope. Since most agricultural buildings are situated in a rural area the combination of wind and increased local snow load can eventually determine the structural strength and stability of parts of the agricultural buildings.

Aerial photographs and the analysis of damage have shown that snow dunes can be formed on the lee side over part of the roof surface where the wind load is zero. The complete truss surface has to be checked for the simultaneous occurrence of wind and snow load according to fig. 3.5. Here the wind direction is perpendicular to the ridge.

3.2.2 Wind load

An agricultural building is often characterized by its relatively low safety, its limited lifetime and the use of lightweight materials. Therefore the wind load is here more determinant than for houses or offices. Wind suction on roof surfaces and increased local wind load at roof edges and corners often cause storm damage to farms. In fact storm damage is the most frequent type of damage to these buildings and the trussless front seems to be particularly sensitive to this

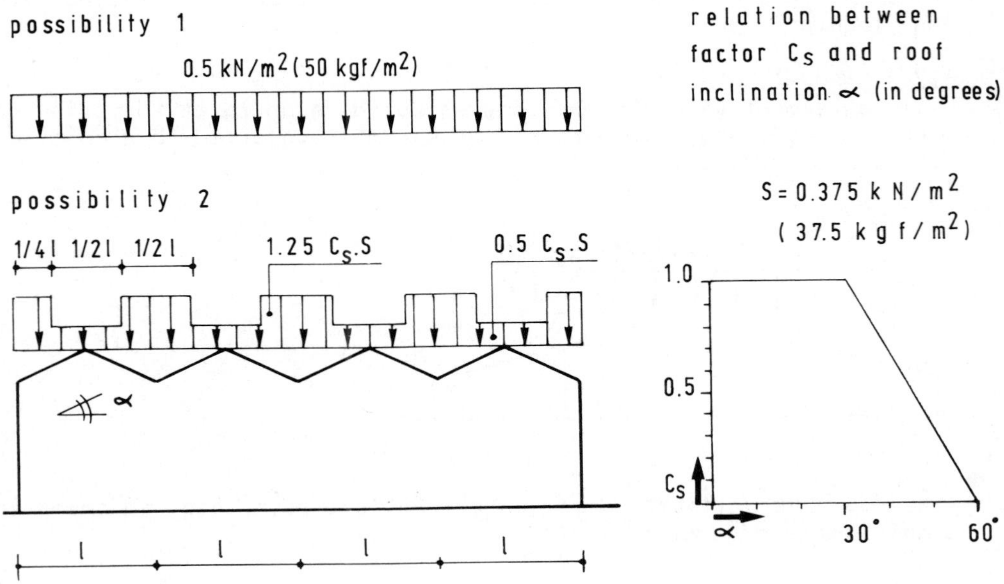

Fig. 3.4 Snow load for coupled roofs (in operational condition) (Waaijenberg, 1984).

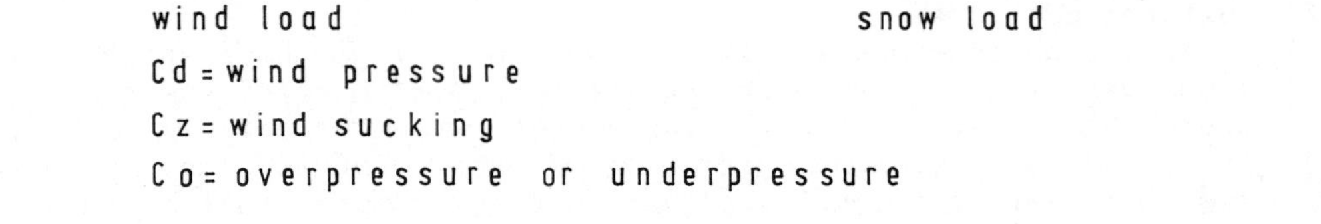

wind load

snow load

Cd = wind pressure

Cz = wind sucking

Co = overpressure or underpressure

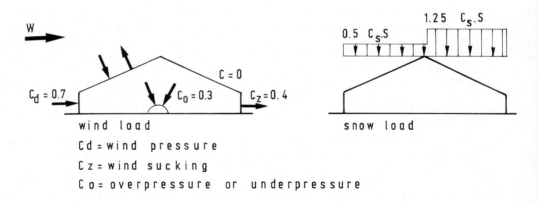

Fig. 3.5 Combination of wind and snow load (Waaijenberg, 1984).

damage.

From a research carried out in a wind tunnel it is clear that whatever the wind direction might be, all roof slopes of less than 20° will cause sucking on all roof surfaces. For roof slopes between 20° and 40°, pressure or sucking can occur on the first roof of the windward side. Both cases of load have to be checked in any design. A summary is found in fig. 3.6.

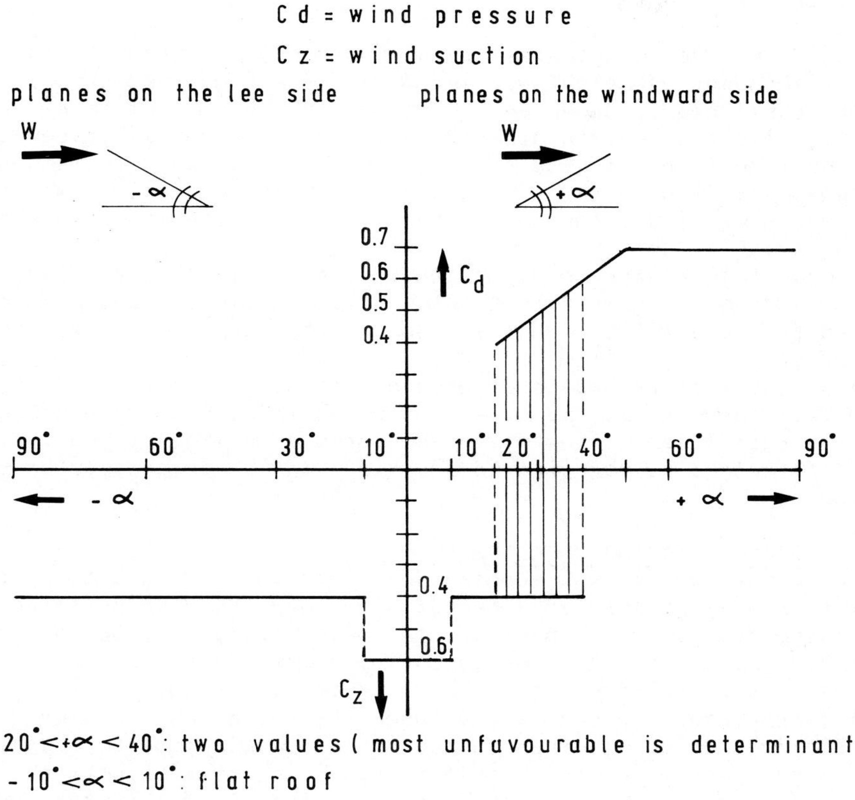

Fig. 3.6 Relation between coefficients Cd and Cz and the angle (+α and -α) which forms a plane with the land level (Waaijenberg, 1984).

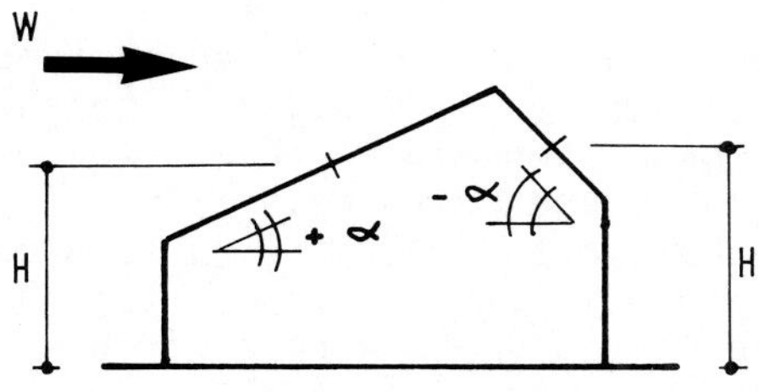

H in m	q in kN/m²	(kgf/m²)
2	0.51	(51)
3	0.55	(55)
4	0.59	(59)
5	0.63	(63)
6	0.67	(67)
7	0.71	(71)
(NEN 3850)		

Fig. 3.7 Relation between pressure thrust q and the height H (in operational condition) (Waaijenberg, 1984).

From these the real developed pressure or suction can be calculated with the formula :

$$q_w = C.q.A$$

where q_w = real developed pressure or suction ; q = pressure thrust per m² (dependent on height H, Fig. 3.7) ; A = wind load area ; C = form coefficient, dependent on the angle α with land level.

In fig. 3.6 the horizontal axis represents the angle between a plane and land level. To the right of the vertical axis we find the planes situated on the windward side, while to the left we find those on the lee side. The coefficient C can be determined on the vertical axis.

The magnitude of the load, caused by the wind, is calculated by multiplying the pressure thrust value by coefficient C which is determined by the position and the slope of the plane in respect to the wind direction.

The pressure thrust value depends on the difference in height of ground level and the middle of a roof plane (fig. 3.5). The standard gives the coefficient C for different forms of buildings and for two wind directions : a. parallel to the ridge and b. perpendicular to the ridge.

3.2.3 Overpressure and underpressure

Through the many leaks, ventilating ridges, flaps and doors the wind will cause not only an external wind load upon the building but also underpressure or overpressure in the building. The building is either being pressurized or being sucked empty. This over- or underpressure must be combined with the external wind load.

The overpressure Co can vary between the values +0.3 and +0.8 A calculation method is given in the standard allowing the determination of the overpressure coefficient. According to this norm the value depends on the ratio : permeability (leaks) of the windward side and permeability of the other surfaces at the outer side where there is sucking. When more than 20 % of the windward side is open (for example a one-sided house or hangar) and the other surfaces are relatively closed the value of overpressure Co reaches +0.8.

3.2.4 Wind on vulnerable places

The analysis of damage to agricultural buildings shows that most of the damage occurs at the roof surface near the crossing to another surface where wind turbulences are generated and which exert additional forces on the different sides of the building.
Therefore the standard prescribes that an additional wind load has to be taken into account for all parts of walls and roofs together with their fittings which are present in a strip with a width of at least 2 m on both sides of a change in slope (fig. 3.8).

These increased coefficients C have to be combined with the coefficients for respectively overpressure and underpressure.

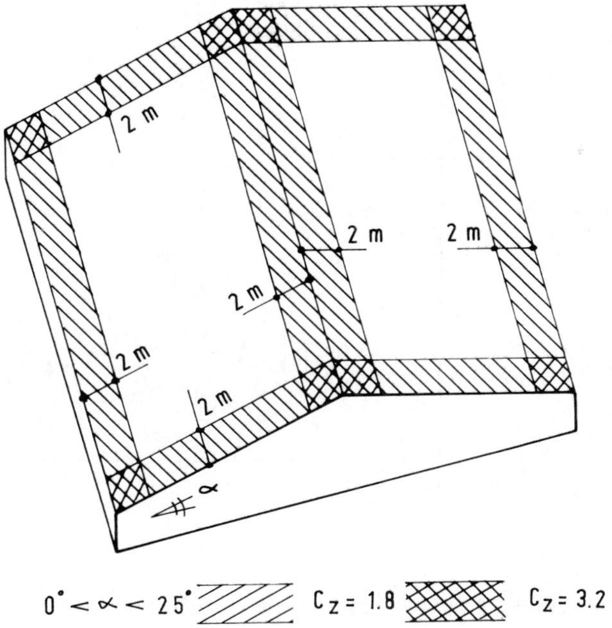

Fig. 3.8 Strips with increased wind load (Waaijenberg, 1984).

3.2.5 Consequences

What is actually the practical meaning of those - rather deviating - regulations concerning the wind load for agricultural buildings ?

Fig. 3.9 illustrates the method by means of a current type of cubicle house where the wind load expressed in kN/m² is indicated according to the standards.

Here the two prescribed wind directions have been calculated : parallel and perpendicular to the direction of the ridge. It is obvious that this load has to be combined with the dead load of the construction.

The resulting wind load on the roof surfaces is for example directed outwards. This wind sucking on the roof can increase according to an increased number of openings in the wall in the windward direction (increased overpressure).

Some points requiring additional attention are :

- the upper edge of a truss must be checked for instability by tilt since the moment is mainly negative. With slim profiles the bottom edge tends to tip over ;
- the purlins in the zones with increased wind load (fig. 3.8) have to be checked for this particular wind load. This can eventually require a reduction of the distance between the purlins for the

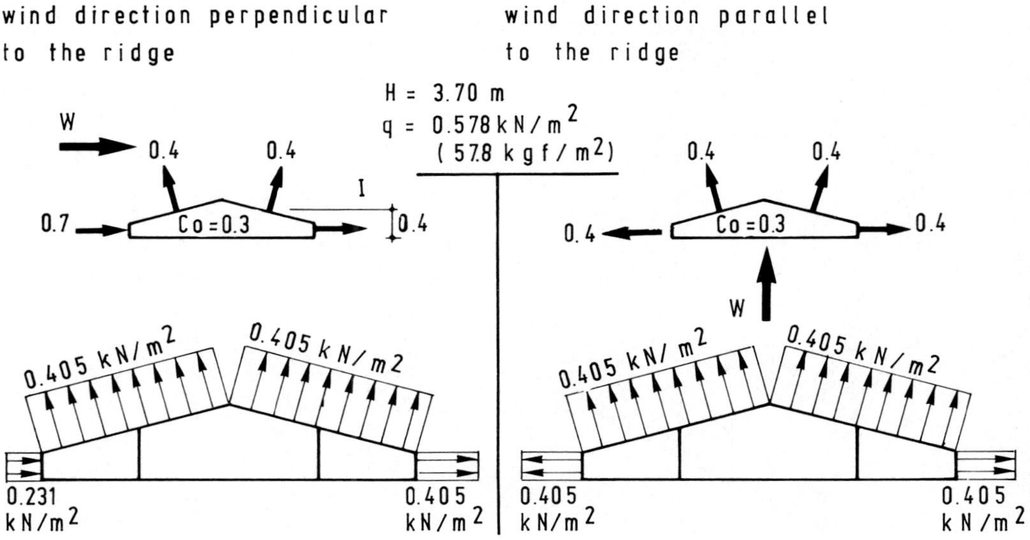

Fig. 3.9 Example of the wind load exerted on a current type of cubicle house (in operational condition) (Waaijenberg, 1984).

outer truss surfaces. The fittings of the purlins to the trusses are also important in relation to wind suction ;

- effective stability bonds in the roof and the walls have to be mounted to assure the stability in the longitudinal direction. These wind braces must be tied to the junctions of rafters with purlins and of columns with wall plates or gable rafters ;
- the fixation of the front wall to the roof must be carried out with a heavy framing girder in absence of an end truss. This is not required if the columns are directly attached to the ends of the purlins ;
- the corrugated asbestos cement roofing has to be fixed with an adequate number of wood- or coachscrews - particularly in damage-sensitive zones - and the latter have to be durably protected against rust. The holes must be drilled in the asbestos cement roofing and the wood- or coachscrews screwed in and not driven in.

When all these points are taken into account, the damage caused by storm and snow will drastically be reduced.

3.3 THERMODYNAMICAL FUNDAMENTALS

In one-storey hangars the pillars will often carry the weight of the construction. The role of the walls, which in fact are only a filling between the pillars, is primarily in contributing to a favourable climate in the animal house. The animal house has to be ventilated to reduce the relative humidity (max. 60 to 80 % R.H.) and the concentration of noxious gases (ammonia, hydrogen sulphide, carbon dioxide) below the maximum admissible level. The animals continuously produce water vapour and other gases which have to be extracted by means of ventilation. This must not lead to a too low temperature in the animal house. Since during winter-time the animals have often to provide the necessary heat production - although this is rather limited - the walls have to contribute in keeping the temperature at a suitable level in spite of the ventilation. When the animal house is heated by means of a heating installation the insulation has to restrict the heat losses in order to keep the energy consumption within reasonable limits. In summer time the ventilation will prevent a too high temperature in the animal house.

In order to obtain an insight in the insulation capacity of a wall of a building it is necessary to know some principles concerning thermal insulation.

3.3.1 Thermal conductivity

All materials will conduct a certain amount of heat when they are situated between two zones having different temperatures. This is called *thermal conductivity*. The purpose of thermal insulation is to limit the heat transport from a warmer to a colder zone. Thermal insulating materials are therefore used and those materials will seriously restrict the heat conduction. Stationary air is one of the best and certainly the cheapest insulation material. Keeping the air stationary by encapsulating it in a large number of small cavities is often applied in insulation materials (e.g. cellular concrete, wool, expanded polystyrene). The inclusion of large amounts of air make these insulation materials light-weight : 1 m^3 of cellular concrete weighs only approx. 500 kg while 1 m^3 of copper for example weighs about 8,500 kg. Polyurethane weighs only approx. 30 kg per m^3. The cubic metre weight of a building material thus gives a good indication about its insulation capacity, but a better standard is the thermal conductivity.

Thermal conductivity is the rate at which heat is transferred through a material i.e. the amount of heat which passes through a material divided by the time, the thickness and the temperature difference between both sides. The thermal conductivity is represented by the Greek letter lambda (λ) and is expressed in W/(m.K). The smaller this value the less heat is transported through the material and the better it insulates.

Since water is a good heat conductor, approx. 15 times better than air, it is necessary to keep the building materials as dry as possible. This might not always be possible and therefore an adjusted λ-value has to be used according to the considered wall : outer walls will take a lot of moisture while inner walls will stay rather dry.

The thermal conductivity of different materials is given in table 3.1.

TABLE 3.1 Some technical properties of building materials (Anon., 1967).

Material	Cubic metre weight in kg	Thermal conductivity in $\frac{W}{m.K}$		Specific heat capacity c in $\frac{kJ}{kg.K}$
	ρ	λ_i^*	λ_e^*	
aluminium	2,800	200	200	0.88
hard natural stone	2,750	2.3	2.9	0.84
concrete, compressed				
- reinforced	2,500	1.9	2.3	0.84
- unreinforced	2,400	1.75	2.2	0.84
concrete, uncompressed				
- reinforced	2,300	1.4	1.9	0.84
- unreinforced	2,200	1.3	1.75	0.84
brickwork of				
- hard bricks	1,900	0.70	1.2	0.84
- ordinary bricks	1,700	0.65	1.0	0.84
- hollow or porous br.	1,500	0.56	0.87	0.84
- hollow or porous br.	1,300	0.46	0.75	0.84
lightweight concrete	1,300	0.46	0.81	0.84
lightweight concrete	1,100	0.39	0.60	0.84
lightweight concrete	700	0.23	0.47	0.84
lightweight concrete	500	0.17	-	0.84
asbestos cement	1,600-1,900	0.35-0.70	0.93-1.2	0.84
glass	2,500	0.8	0.8	0.84
cement rendering	1,900	0.93	1.5	0.84
plastering	1,600	0.70	0.81	0.84
hard wood (oak, beech)	800	0.17	0.23	1.88
pine wood	550	0.14	0.17	1.88
plywood	700	0.17	0.23	1.88
chipboard	600	0.15	-	1.88
chipboard	450	0.10	-	1.88
cement chipboard	700	0.21	-	1.47
cement chipboard	350	0.09	-	1.47
rubber	1,200-1,500	0.17-0.29	-	1.47
cork	100-200	0.041-0.046	-	1.76
polyester	1,200	0.17	-	1.47
polyvinyl chloride	1,400	0.17	-	1.47
mineral wool	35-250	0.041	-	0.84
polystyrene foam	15-40	0.035	-	1.47
polyurethane foam	30-60	0.023-0.035	-	1.47

λ_i^* = the thermal conductivity for inner walls

λ_e^* = the thermal conductivity for outer walls

3.3.2 Surface and cavity thermal transmittances

Surfaces of materials have an additional insulation capacity due to the boundary layer of liquid or air held against it under frictional resistance. This *surface thermal transmittance* is the amount of heat which passes per time, per area of material and per temperature difference from a liquid or gaseous to a solid material or vice versa. Also the presence of any air space between materials results in an additional insulation capacity i.e. the *cavity thermal transmittance*. Both transmittances are represented by the Greek letter alpha (α) and are expressed in $W/(m^2.K)$.

The surface and cavity thermal transmittances are given in table 3.2. The outer surface thermal transmittance is influenced by the wind velocity. In table 3.2 an air speed of 4 m/s is assumed. The cavity thermal transmittance varies with the pattern of air movement which explains the differences between vertical and horizontal cavities given in table 3.2.

TABLE 3.2 Surface and cavity thermal transmittances in $W/(m^2.K)$ (Anon., 1981a).

Wall and direction of the heat flow		α_i	α_e	α_s
1. vertical wall horizontal heat flow	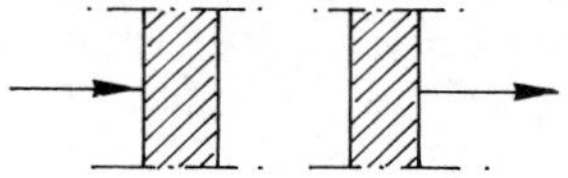	8	23	-
2. horizontal wall vertical heat flow upwards	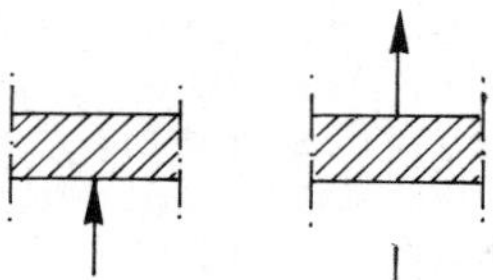	8	23	-
3. horizontal wall vertical heat flow downwards	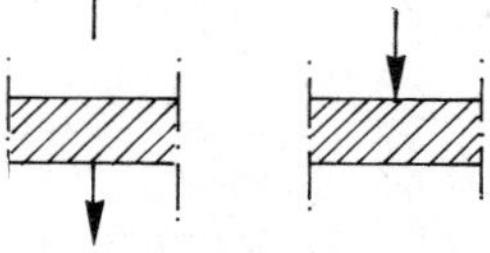	6	23	-
4. vertical air layer (0.02 m to 0.2 m) horizontal heat flow	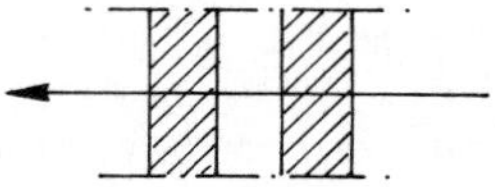	-	-	6
5. horizontal air layer (0.02 m to 0.2 m) vertical heat flow upwards	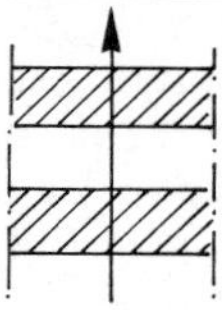	-	-	7

α_i = surface thermal transmittance for the inner side of the wall ;

α_e = surface thermal transmittance for the outer side of the wall ;

α_s = cavity thermal transmittance.

Fig. 3.10 represents the temperature variation in a cavity wall at an inside temperature of +20°C and an outside temperature of −10°C.

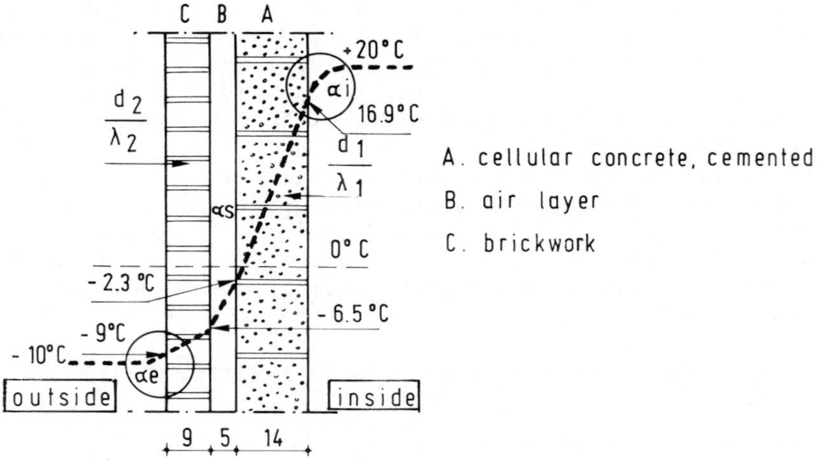

Fig. 3.10 Schematical representation of the temperature variation through a cavity wall.

The inner side of the cellular concrete brick-wall is 3.1°C colder than the inside air. The other side of the cellular concrete brick-wall, in the cavity, has a temperature of −2.3°C and is thus 19.2°C lower than the inner wall temperature. The temperature of the brick wall along the cavity measures −6.5°C while the temperature of the outer wall is −9°C i.e. 1°C above the outside temperature.

3.3.3 Thermal transmittance

When two different temperatures prevail at both sides of a wall or roof, a certain amount of heat is transmitted per time through the wall or the roof. This amount is equal to the sum of the thermal conductivity and the surface thermal transmittances. The thermal transmittance is always directed to the colder side or from the place having the highest temperature to the place having the lowest.

This thermal transmittance depends on a number of factors i.e. :

- the temperature difference between in- and outside ;
- the area of the walls ;
- the thickness of the walls ;
- the nature and composition of the walls ;
- the time ;
- the transfer losses of the inside air towards the inner side of the wall ;
- the heat losses through the eventual cavity.

The *thermal transmittance* is the amount of heat which passes through a wall per time, per wall area and per difference in temperature at both sides of the wall. This thermal transmittance, also called k-value (or U-value), is expressed in $W/(m^2 .K)$.

A wall often consists of different materials, for example : brickwork + cavity + cellular concrete (fig. 3.10). The k-value of such a wall can be calculated by the following formula :

$$k = \frac{1}{\frac{1}{\alpha_i} + \frac{d_1}{\lambda_1} + \frac{d_2}{\lambda_2} + \ldots \frac{1}{\alpha_s} + \frac{1}{\alpha_e}}$$

where k = thermal transmittance in $W/(m^2 .K)$; d_1, d_2 ... = thickness of each layer in metres ; λ_1, λ_2 = thermal conductivity of each layer in W/(m.K) ; α_i, α_e = inside and outside surface thermal transmittances in $W/(m^2 .K)$; α_s = cavity thermal transmittance in $W/(m^2 .K)$.

The average k-values of the boundary surfaces of an animal house (walls, ceilings, floors) in a temperate climate are :

$k \leq 1.2 \frac{W}{m^2 .K}$ for : cattle houses (except : veal calf houses) ;

$k \leq 0.9 \frac{W}{m^2 .K}$ for :
- slaughter pig houses ;
- houses for dry and pregnant sows ;
- cage houses for laying hens ;

$k \leq 0.7 \frac{W}{m^2 .K}$ for :
- farrowing and weaner houses ;
- deep litter houses for laying hens ;
- houses for broilers ;
- veal calf houses ;

$k \leq 0.6 \frac{W}{m^2 .K}$ for : storehouses for potatoes, beets, etc... .

Compared to the walls, the roofs need a better insulation (see later).

The total heat flow (loss or gain) through walls, roofs or floors can be calculated by the formula :

$$Q = k \times F \times (T_i - T_e)$$

where Q = total heat flow through a wall, roof or floor expressed in watt ; k = thermal transmittance in $W/(m^2 .K)$; F = total area of the wall in m^2 ; $T_i - T_e$ = the difference between inside and outside temperature in °C.

The required minimum wall thickness of some building materials to obtain a value $k \leq 0.9$ $W/(m^2 .K)$ are mentioned in table 3.3.

3.3.4 The basic outside temperature

In the planning stage of an animal house it is necessary to calculate if additional heating is required and in that case the maximum heating capacity has to be determined. The outside temperature plays herein a predominant role.

TABLE 3.3 The minimum wall thickness in metres of some simple walls in order to obtain a k $\leq$ 0.9 W/(m² .K).

Materials	λ in $\frac{W}{m.K}$	thickness in m
reinforced concrete (2,300 kg/m³)	1.90	1.79
unreinforced concrete (2,200 kg/m³)	1.75	1.65
hollow bricks (1,500 kg/m³)	0.87	0.82
lightweight concrete (1,100 kg/m³)	0.60	0.57
lightweight concrete (700 kg/m³)	0.47	0.44
hard wood (800 kg/m³)	0.23	0.22
soft wood (550 kg/m³)	0.17	0.16
polyurethane foam (50 kg/m³)	0.03	0.03

The temperature of the outside air T_e which has to be taken into account when determining heat losses depends on the climatological conditions of the area. The basic outside temperature is the normalized outside temperature whereby the heating in the animal house still provides an acceptable thermal comfort. This of course is not the lowest outside temperature which has ever been measured. The basic outside temperature is always taken into account for dwellings. The question arises if it is necessary to include the basic outside temperature for animal houses, in fact they can be regarded as service-buildings. Here, one would rather try to reach an economical thermal comfort. The basic outside temperatures assumed for dwellings are probably too rigid for animal houses. Only a slight production inhibition will result when it is temporarily impossible to maintain inside the assumed temperature by too low a capacity of the heating installation. Taking an economically justified risk is quite normal when carrying out climate calculations for animal houses. The risk taken by choosing a minimum basic temperature for animal houses higher than for dwellings is certainly economically justified when an insignificant reduction in production is covered by a relatively important saving in the costs of the climatic installation of the animal house (insulation and heating). When extremely low outside temperatures occur, other measures can be taken in the animal house to reduce the effects viz. a severe limitation of the ventilation.

The basic outside temperature accepted for an animal house has to be estimated on the basis of an analysis of the meteorological statistics and of the investment. In Flanders (N. Belgium) for example calculations concerning insulation are based on an outside temperature of -5°C. Besides the conclusions drawn from meteorological data the variations in occupancy of the house have to be taken into account. The chance that a minimal density coincides with a period of minimum outside temperatures has not to be exaggerated.

When calculating the fuel consumption it is not so much the basic outdoor temperature which is of importance but the average outside temperature during the heating period. In Flanders (Northern Belgium) for instance an average outside temperature of +6.5°C, during the heating season, from September 15 to May 15, is taken into account (Poncelet and Martin, 1947).

3.3.5 The basic inside temperature

The desirable indoor temperature in animal houses is represented in table 3.5 (Anon., 1981b).

3.3.6 Thermal resistance

The reciprocal of the amount of heat which passes through a wall is called heat resistance or thermal resistance. This is proportional to the thickness of the material and inversely proportional to the thermal conductivity :

$$R = \frac{d}{\lambda}$$

R is expressed in $\frac{m^2.K}{W}$.

Similarly the thermal transmittance can be expressed as a thermal resistance viz. surface or cavity thermal resistance :

$$r_{i,e,s} = \frac{1}{\alpha_{i,e,s}}$$

The air layers (cavities) form a special case. Heat transmission in a cavity is caused by convection and radiation as well as by conduction. A cavity is not a homogeneous material where the heat transmission is only caused by conductivity. The thermal resistance of air layers is represented in table 3.4.

TABLE 3.4 The heat resistance of air layers (Anon., 1981a).

Wall and direction of the flux	r_s
vertical air layer 2 - 20 cm 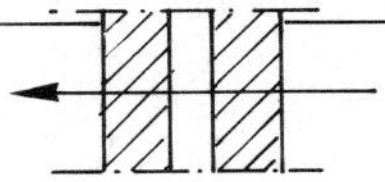	0.17
horizontal air layer 2 - 20 cm heat passage upwards 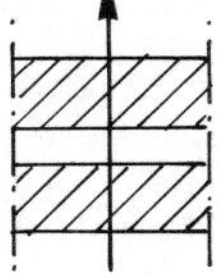	0.14
horizontal air layer 2 - 4 cm heat passage downwards 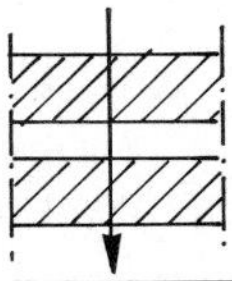	0.20

The total thermal resistance, expressed as $(m^2.K)/W$, of a wall composed of different materials and an air layer becomes :

$$R_T = \frac{1}{K} = r_i + \frac{d_1}{\lambda_1} + \frac{d_2}{\lambda_2} + \ldots + r_s + r_e$$

TABLE 3.5 The optimum environmental factors for different animals (Anon., 1981b).

Animal	Temperature (a) in °C		
Cattle	min.	opt.	max. (b)
a. Dairy cattle			
- full grown cattle	- 5	5 to 15	25
b. Beef cattle			
- heavy	- 5	5 to 15	25
- young or on slats	+ 5	5 to 15	25
c. Suckling cows	- 5	5 to 15	25
d. Rearing calves			
- less than 3 months old	10	20 to 15	20
- older than 3 months	5	10 to 15	20
e. Veal calves	10	20 to 15	20
Pigs	fully slatted	partly slatted (c)	strawed
a. Slaughter pigs			
- 20 kg	22 - 24	20 - 24	20 - 22
- 45 kg	19 - 21	17 - 21	15 - 19
- 90 kg	17 - 19	15 - 19	14 - 18
b. Breeding pigs			
- boars, young sows (from ca. 90 kg), dry and pregnant sows		12 - 16	
- suckling sows		17 - 20	
- piglets during the first days (micro-climate)		30 - 35	
- weaners (d)	fully slatted	strawed	
- ca. 3 kg	26 ± 1	24 ± 1	
- ca. 8 kg	24 ± 1	22 ± 1	
- 16 - 20 kg	22 ± 1	20 ± 1	
Poultry			
a. Rearing pullets	30 - 35 (c) to be decreased by 4°C per week to 20°C in cages and 15°C on litter		
b. Broilers	30 - 25 to be decreased by 4°C per week to 20°C from 5 weeks onwards		
c. Laying hens	20 - 22		
d. Brood hens	15		
Rabbits			
a. Rearing rabbits	16 - 18		
b. Slaughter rabbits	14 - 16		
(e), (f) : see remarks on next page			

Remarks concerning table 3.5.

(a) : these temperatures are measured with a dry thermometer, inside the house at a minimum distance of 1.50 m from any outside wall and at the location of the animals. The temperatures of the inside walls, of insulated houses, should not be more than 3°C lower than the environmental temperature ;

(b) : - minimum : the minimum temperature required for servicing (e.g. working climate) or for zootechnical reasons ;
- optimum : the temperature to which the thermostat or other climatic installation is adjusted for zootechnical reasons (thermoregulation of the animals) ;
- maximum : the maximum temperature allowed for servicing or for zootechnical reasons ;

(c) : only the practical optimum zone is given within which the measured temperature of the animal house is allowed to vary ;

(d) : sudden temperature variations have to be kept to a minimum for new weaners ;

(e) : the air velocity in winter time is not to exceed 0.25 m/s for cattle and 0.20 m/s for other animals ;

(f) : the optimum relative humidity is 60 - 80 % for cattle, pigs, laying hens and brood hens, 50 - 70 % for rearing pullets and broilers, 70 % for rabbits.

For the calculation of the thermal resistance and the thermal transmittance of cavity walls the outside-lambda (λ_e) is used for the outer wall part whereas the inside-lambda (λ_i) is taken for the inside wall part.

3.3.7 The thermal capacity of building materials

Building materials have the property to take up heat which can later be returned to the atmosphere. This in fact is heat accumulation or heat storage. Different materials however have a different heat accumulation power, or, as is normally said, a divergent *specific heat capacity*. The greater the specific heat capacity of a building material, the longer it will take to attain a higher temperature. We have to know the specific heat capacity (c) of a material before we are able to calculate the heat capacity of that material. The energy required to increase the temperature of 1 kg of that particular material with 1°C is called the specific heat. This c-value is expressed in J/(kg.K) or Ws/(kg.K).

The heat capacity per m^3 or the so-called w-value is then equal to the specific heat capacity (c), multiplied by the specific mass (ρ) and is expressed in J/(m^3.K).

The specific heat capacity (c) and the cubic metre weight ρ of some building materials are given in table 3.1.

In order to describe the heat capacity quantitatively we need to know the concept of heat content or *enthalpy* (Q). This is the amount of heat which is stored in a volume of material (V) at its temperature t, or

$$Q = \rho \cdot c \cdot V \cdot t \quad \text{or} \quad Q = w \cdot V \cdot t$$

Q is expressed in J.
Q can either be negative or positive according to t being lower or higher than 0°C.

Let us illustrate this with an example. Fig. 3.11 represents the variation of temperature in a wall made of 14 cm lightweight concrete (700 kg/m³) inside, 4 cm polyurethane (50 kg/m³), a 4 cm air layer and finally a 9 cm hollow brick (1,500 kg/m³) at the outside.

The thermal resistance R_T is equal to :

$$R_T = \frac{1}{\alpha_i} + \frac{d_1}{\lambda} + \frac{d_2}{\lambda} + \frac{1}{\alpha_s} + \frac{d_3}{\lambda} + \frac{1}{\lambda_e}$$

$$R_T = \frac{1}{8} + \frac{0.14}{0.23} + \frac{0.04}{0.03} + \frac{1}{6} + \frac{0.09}{0.87} + \frac{1}{23} = 2.38 \frac{m^2.K}{W}$$

or the k-value is : $k = \frac{1}{R_T} = 0.42 \frac{W}{m^2.K}$

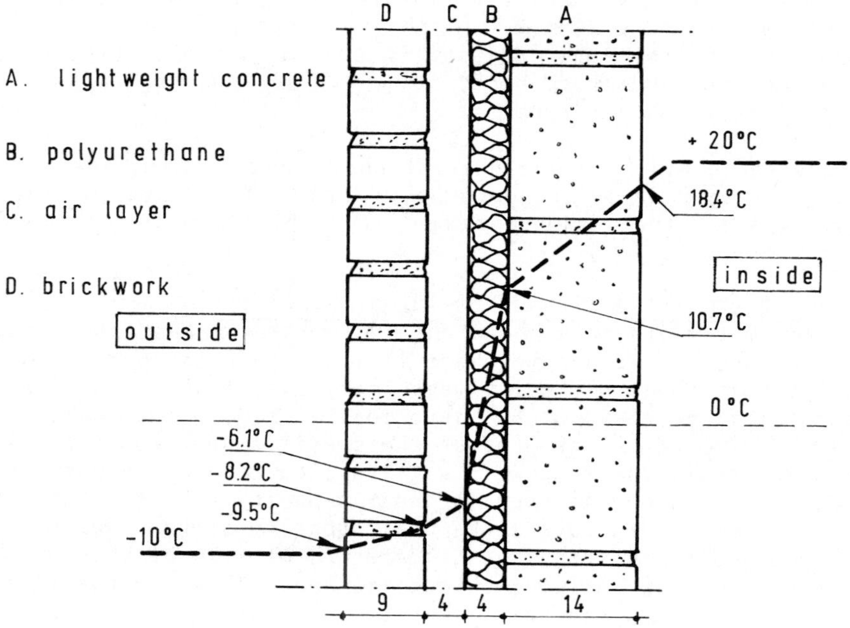

Fig. 3.11 Temperature variation through a wall-construction with an insulation-layer in the cavity, against the inner wall.

The decrease in temperature through the wall is directly proportional to the thermal resistance of each material and the temperature difference and is indirectly proportional to the total thermal resistance of the wall. It can be calculated as follows :

- inner transfer : $\frac{0.13}{2.38} \times 30°C = 1.6°C$

- material A : $\frac{0.61}{2.38} \times 30°C = 7.7°C$
- material B : $\frac{1.33}{2.38} \times 30°C = 16.8°C$
- cavity C : $\frac{0.17}{2.38} \times 30°C = 2.1°C$
- material D : $\frac{0.10}{2.38} \times 30°C = 1.3°C$
- outer transfer : $\frac{0.04}{2.38} \times 30°C = 0.5°C$

Total 30.0°C

The average temperatures of the different wall materials are :

- material A : $\frac{18.4 + 10.7}{2} = 14.55°C$
- material B : $\frac{10.7 - 6.1}{2} = 2.30°C$
- cavity C : $\frac{-6.1 - 8.2}{2} = -7.15°C$
- material D : $\frac{-8.2 - 9.5}{2} = -8.85°C$

The heat capacity for the different wall materials are :

- material A : $700 \frac{kg}{m^3} \times 0.84 \frac{kJ}{kg.K} \times 0.14\ m = 82.32 \frac{kJ}{m^2.K}$
- material B : $50 \frac{kg}{m^3} \times 1.47 \frac{kJ}{kg.K} \times 0.04\ m = 2.94 \frac{kJ}{m^2.K}$
- cavity C : $1{,}293 \frac{kg}{m^3} \times 1 \frac{kJ}{kg.K} \times 0.04\ m = 0.05 \frac{kJ}{m^2.K}$
- material D : $1{,}500 \frac{kg}{m^3} \times 0.84 \frac{kJ}{kg.K} \times 0.09\ m = 113.40 \frac{kJ}{m^2.K}$

The total heat content of a m² of this wall is :

$$Q = (14.55 \times 82.32) + (2.3 \times 2.94) + (-7.15 \times 0.05) + (-8.85 \times 113.40) = 200.56 \frac{kJ}{m^2}$$

Fig. 3.12 shows the same wall as fig. 3.11 but differs in the fact that the polyurethane insulation layer is now applied on the inside wall instead of in the air cavity.

The thermal resistance R_T is the same as that of figure 3.11 :

$$R_T = \frac{1}{8} + \frac{0.04}{0.03} + \frac{0.14}{0.23} + \frac{1}{6} + \frac{0.09}{0.87} + \frac{1}{23}$$

$$R_T = 0.13 + 1.33 + 0.61 + 0.17 + 0.10 + 0.04 = 2.38 \frac{m^2.K}{W}$$

or the value $= \frac{1}{R_T} = 0.42 \frac{W}{m^2.K}$

The k-value is naturally the same as that of fig. 3.11. The temperature decrease obtained per material is the same as in the previous

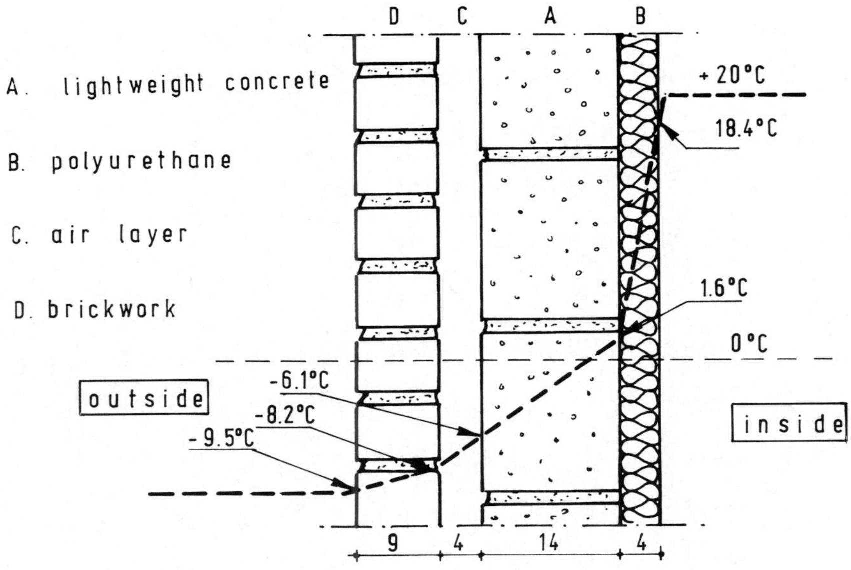

Fig. 3.12 Temperature variation through a wall construction with an insulation layer at the inside.

case. Different average temperatures of the materials are however obtained :

- material B : $\frac{18.4 + 1.6}{2} = 10.00°C$
- material A : $\frac{1.6 - 6.1}{2} = -2.25°C$
- cavity C : $\frac{-6.1 - 8.2}{2} = -7.15°C$
- material D : $\frac{-8.2 - 9.5}{2} = -8.85°C$

The heat capacity per m² of each wall material and per °C is also the same, the total heat content of the wall then becomes :

$$Q = (10.0 \times 2.94) + (-2.25 \times 82.32) + (-7.15 \times 0.05) + (-8.85 \times 113.40)$$
$$= -1,159.78 \frac{kJ}{m^2}$$

If we compare the heat contents of both wall constructions, resp. of figs 3.11 and 3.12, we find a very important difference viz. $+200 \frac{kJ}{m^2}$ and $-1,160 \frac{kJ}{m^2}$

Although both walls have the same k-value or in other words the same insulation power, they will exert a completely different influence on the internal climate.

When the temperature of the animal house starts to decrease (e.g. during the night) the wall of fig. 3.11 (with an insulation layer in the cavity) will still be able to return a large amount of heat to the inside. The wall of fig. 3.12 (with an insulation at the inside) is unable to do so. On the contrary by its negative heat content more heat will be absorbed : the temperature in the house will drop even faster.

The inverse effect takes place when the building is heated (e.g. during the morning). When the insulation layer is situated in the cavity (fig. 3.11) it will take longer to reach the required temperature since the wall will absorb an important amount of heat. Having the insulation at the inside reduces the amount of heat absorbed by the walls. The required temperature will thus be achieved sooner. Places which are only used during a short time and which must be heated fast have to be provided with a heavy insulation as close as possible to the inner wall. On the contrary houses which have to be maintained at constant temperature for a longer time such as farrowing and weaner houses can best be provided with walls having a heavy insulation layer as far away as possible from the inside wall. In this manner the heat capacity of the wall is fully used.

Moreover, it is advisable to use building materials which not only have a good heat capacity but which also form a good insulation such as cellular concrete and cellular earthenware.

In table 3.6 a number of building materials are compared as a function of their thermal capacity. This can either take place for an identical wall thickness of these materials or for a wall with an identical k-value, which obviously gives an important difference.

The thermal capacity per volume of a material will increase according to an increased cubic metre weight. The best conducting building materials, thus the least insulating, will be the best accumulators of heat. This means that heavy materials will only prove to be good insulators when a sufficient thickness is used.

Light insulation materials have nothing to offer concerning heat accumulation. A combination of both can be optimum but the economical and constructional considerations will also play an important role in the choice of those materials.

3.3.8 Temperature fluctuations

The outdoor temperature fluctuates from hour to hour as is shown in fig. 3.13. The wall forms a separation between inside and outside and will adopt a temperature which is situated between both. The temperature of the walls will fluctuate also but with a certain delay. The higher the heat capacity of the wall the greater the delay will be.

The so-called phase shift is of importance (fig. 3.13). The ideal case is a phase shift of 12 hours or in other words a wall temperature which is then the highest when the outdoor temperature is the lowest and vice versa. The phase shift can be estimated from the following formula (Anon., 1981c) :

$$F = 0.727 \,.\, d \,.\sqrt{\frac{\rho . c}{\lambda}}$$

where F = phase shift in hours ; d = thickness of the wall in metres ; ρ = cubic metre weight in kg/m^3 ; c = specific heat capacity of the

TABLE 3.6 The heat capacity kJ/(m² .K) of some building materials for either an identical wall thickness (d = 0.30 m) or the same k-value (k = 0.9 W/(m² .K)).

Material		$k = 0.9\ \frac{W}{m^2 .K}$		$d = 0.30$ m	
	kg/m³	d in m	Q in $\frac{kJ}{m^2 .K}$	k in $\frac{W}{m^2 .K}$	Q in $\frac{kJ}{m^2 .K}$
reinforced concrete	2,300	1.79	3,460	3.06	580
unreinforced concrete	2,200	1.65	3,048	2.94	554
hollow bricks	1,500	0.82	1,033	1.95	378
lightweight concrete	1,100	0.57	523	1.50	277
lightweight concrete	700	0.44	261	1.24	176
hard wood	800	0.22	326	0.68	451
soft wood	550	0.16	166	0.52	310
polyurethane foam	50	0.03	2	0.09	22

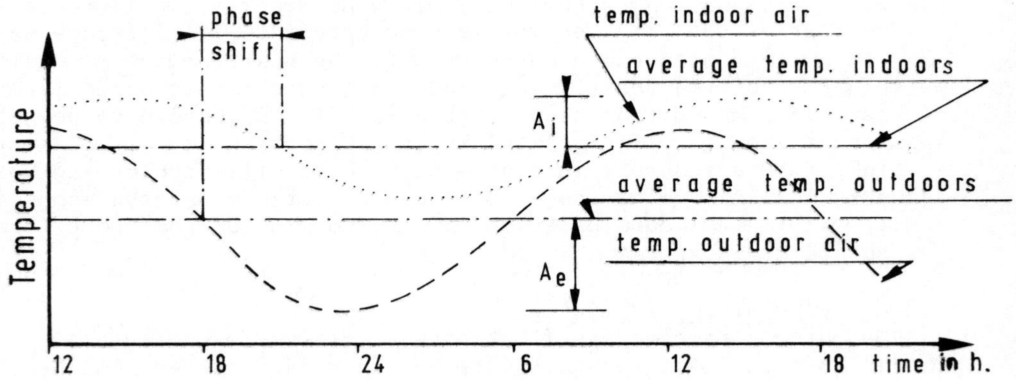

Fig. 3.13 The variation of in- and outside temperature.

wall in kJ/(kg.K) ; λ = the heat conduction coefficient in W/(m.K).

In a heavier wall with a large heat capacity there are never such large temperature variations as in the outside air. This is the so-called amplitude damping (see fig. 3.13). Lighter walls, which may insulate well but have a smaller heat capacity will follow the outside air temperature much quicker. Those walls have a small phase shift and amplitude damping and are the cause of the so-called "attic or glasshouse-effect" wherein one sweats during the day and shivers during the night.

This amplitude damping can be estimated with the following formula (Anon., 1981c) :

$$A_d = \frac{7.5 \times d^3}{3{,}162.28 \times \left(\sqrt{\frac{\lambda}{\rho . c}}\right)^3} + 1$$

This amplitude damping is in fact an abstract number and indicates how many times the fluctuation of the inside temperature is smaller than this of the outside air temperature.

If A_e is the outside air amplitude and A_i is that of the inside air then the amplitude damping can be represented by :

$$A_d = \frac{A_e}{A_i}$$

If A_d is 10 for example then the amplitude of the inside air amounts to 1/10th of that of the outside air. When $A_d = 1$ the inside air temperature fluctuates as much as the outside air.

In table 3.7 a comparison is made between the same building materials as mentioned in table 3.6 and this as a function of their phase shift and amplitude damping. Again a distinction is made between an equal wall thickness and an equal k-value of these materials.

TABLE 3.7 The phase shift F (in hours) and the amplitude damping A_d of some building materials for either the same wall thickness (d = 0.30 m) or the same k-value (k = 0.9 W/(m² .K)).

material*	$k = 0.9 \frac{W}{m^2 .K}$			d = 0.30 m		
	d in m	F in hour	A_d	k in $\frac{W}{m^2 .K}$	F in hour	A_d
reinforced concrete	1.79	41.5	442.1	3.06	7.0	3.1
unreinforced concrete	1.65	39.0	366.6	2.94	7.1	3.2
hollow bricks	0.82	22.7	73.1	1.95	8.3	4.5
lightweight concrete	0.57	16.3	27.5	1.50	8.6	4.9
lightweight concrete	0.44	11.3	9.9	1.24	7.7	3.8
hard wood	0.22	12.9	14.4	0.68	17.6	34.9
soft wood	0.16	9.1	5.6	0.52	17.0	31.4
polyurethane foam	0.03	1.1	1.0	0.09	10.8	8.8

*See cubic metre weight in table 3.6.

With identical k-values the heaviest building material will also have the largest phase shift and amplitude damping. "Strong" insulation materials will show practically no phase shift or amplitude damping. With an identical wall thickness but varying k-value smaller variations of phase shift and amplitude damping are obtained. It seems that walls, having a same k-value, cannot directly be labelled as equivalent. Therefore the k-value should not be taken as the only criterion for the evaluation or comparison of a wall construction.

Also the heat capacity is an important factor to be taken into consideration.

3.3.9 The comfort temperature or sensorial temperature

The temperature which we "feel" in a room is not only dependent on the air temperature but is also determined by the wall temperature. Several authors (Anon., 1975a ; Anon., 1981c ; Flamand, 1981) call the sensorial temperature the average between wall- and air temperature or :

$$t_v = \frac{t_w + t_i}{2} \quad \text{or} \quad t_v = 0.5\ t_w + 0.5\ t_i$$

where t_v = sensorial temperature in °C ; t_w = wall temperature in °C ; t_i = inside air temperature in °C.

With an inside air temperature of for example 20°C and a wall temperature of 14°C a temperature of 17°C is felt. This formula can only be applied for small places such as living-rooms, where one is always relatively close to a wall. In larger confinements such as pig houses the air temperature is of greater importance than the wall temperature. According to Holmes and Mount (1966) the temperature a pig "feels" decreases by 1/4°C for each °C that the wall temperature is lower than the temperature of the air inside the piggery. The formula can then be changed to :

$$t_v = 0.75\ t_i + 0.25\ t_w$$

A temperature of 23°C in a piggery and a wall temperature of only 15°C will create a sensorial temperature of 21°C (Holmes and Mount, 1966). These calculations have merely a theoretical value. The actual position of the animal in the house with respect to the walls is of greater importance.

Buildings constructed from lightweight but good insulating panels or with a great deal of glass will give a warm impression during daytime but cold during the night, even if the thermostat is kept constantly at 20°C. This is due to a too high or too low wall temperature or in fact a less optimum heat capacity of the used materials wherebythese materials show insufficient *buffering capacity*.

3.4 BUILDING MATERIALS FOR THE CONSTRUCTION OF ANIMAL HOUSES

The building materials used for the construction of animal houses can be divided in two categories according to their function. The first category involves those materials which are mainly necessary for the stability of the building while the second are mainly intended for insulation.

3.4.1 Building materials applied mainly for the structural stability of the building

Bricks : a traditional building material which is made of earthenware, put in a form and baked in a furnace. Bricks withstand pressure but are susceptible to pull and shear : the admissible tension is respectively 8 - 1.5 - 0.8 kg per cm². Bricks can be manufactured with relatively large holes, either in the vertical or horizontal direction, which are filled with air. Due to the bad heat conduction of stationary air the heat transmission is limited and the thermal insulating properties of the bricks are thereby improved. Bricks having a large number of small holes, filled with air, are called porous bricks.

Recently, a new brick with lots of small pores (cellular earthenware) has come on the market : the so-called Poroton-brick. In a Swedish patented manufacturing process the clay is thoroughly mixed with an appropriate amount of expanded polystyrene beads. During the baking process of the stones the beads are vaporized, leaving no residue, but producing a very large number of pores. A good insulation is thereby achieved. The cubic metre weight of this particular material is approx. 850 - 900 kg (compared to approx. 1,800 kg/m^3 for ordinary bricks).

Concrete : is a widely used building material composed of cement, rubble or gravel, sand and water. These materials are dosed and homogeneously mixed in a concrete mixer. It is applied as a semi-liquid mixture between metal or wooden encasements, which have the form of the future part of construction. After hardening to "concrete", the formwork is removed. Concrete has an important resistance against pressure-forces but a low resistance against pull and shear. The admissible tensions are respectively 60 - 3 - 3 kg/cm^2. In case pull and shear are to be expected - e.g. a concrete beam whose ends are supported but receives a uniformly distributed load on its full length, whereby the beam will show some bending so that the upper side is compressed and the lower side is stretched - the produced pull has to be absorbed by steel bar reinforcement included in the concrete. This is the so-called *reinforced concrete*.

A special kind of reinforced concrete is the *prestressed concrete*. Prestressed concrete is made by including steel cables in the formwork before the introduction of the concrete. The fabrication is carried out on a stretching device having a length of up to 100 metres and which has a stiff metal encasement upon which vibrators are mounted. The concrete is introduced around the steel cables prestressed by means of a hydraulic jack screw.

After the hardening of the concrete the cables are cut and the prestress is transmitted by means of "adhesion" on the concrete elements. The concrete is in fact compressed, thus increasing the resistance against external influences markedly, compared to ordinary reinforced concrete. The application of prestressed concrete results in important savings on concrete and steel bar reinforcement, while the maintenance costs are practically nil since the chances of cracks and clefts are minimal. Ordinary steel bar reinforcement has an ultimate tensile strength of 1,400 kg per cm^2 and wire for prestressed concrete approx. 9,000 kg per cm^2. Prestressed concrete is widely used for the construction of concrete beams since it avoids the following drawbacks which are inherent in ordinary concrete :

- the concrete around the steel bar reinforcement is unable to support the deformation caused by the load from pulling forces and starts to crack ;

- the concrete beams must have relatively large dimensions in view of the resistance to tensile stress which is the result of shear, they thus have a relatively large dead weight ;
- the concrete is subjected to shrinkage and can show clefts, even in the absence of external influences ;
- a complete utilization of the large natural resistance of concrete against pressure forces is not possible.

Concrete has a high λ-value, thus allowing an important transfer of heat, and has no insulating properties, but on the other hand it has a high cubic metre weight and an important resistance against pressure.

Solid concrete blocks are also manufactured, they are of course very heavy (2,200 kg per m^3) and little insulating. They are applied for foundations and as watertight blockwork for slurry tanks and channels, containing liquid manure.

Lightweight concrete : different sorts of lightweight concrete can be distinguished :

- lightweight concrete made from mineral aggregates : composed of cement mortar supplemented with pumice, natural pumice grit, expanded blast-furnace slag (Hüttenbims), expanded clay (argex, leica) cinders (ash) etc. Those sorts of concrete show a "closed" texture (the voids between the aggregates are filled with mortar) with a porous character ;
- granular concrete : composed of cement, mixed with coarse or medium sized aggregates as mentioned earlier but with the exception of sand or any other fine aggregate. This material has an "open" texture with many voids ;
- cellular concrete : (e.g. Siporex, Durox, Hebel) has an alveolar structure with many small gaseous bubbles enclosed in the material :
  - air- or gas-filled concrete : made from cement, pebble, lime and eventually fine additional materials, mixed with a gas-evolving product such as aluminium powder ;
  - foam concrete : similar to gas-filled concrete but instead of aluminium powder a foam-producing material is supplemented ;
  - lime-silica cellular blocks (Ytong) manufactured by mixing at high temperature and under high pressure, finely ground lime with particles having a high silica content e.g. : slate, sand, blast-furnace slag, fly ash, and supplemented with water and aluminium powder. These insulating blocks, originally manufactured in Scandinavia, are now widely used in Europe.

All lightweight types of concrete and similar materials are characterized by their good thermal insulating value, due to the presence of entrained air or gas in the holes or pores which minimize the heat flow. The inclusion of a gaseous component results in a low cubic metre weight and thus in a rather low resistance and bearing capacity.

Asbestos cement : (Eternit, SVK, Johns-Manville etc.) building material fabricated from cement and asbestos fibres with or without colouring agents. It is widely used in agricultural constructions :

roofing (corrugated a.c. sheets), divisions between pens (a.c. boards), feeders, air releaser caps and prefabricated wall panels ;

Aluminium : building material obtained from bauxite-winning after enrichment and chemical processing. Its weight is only 1/3rd of that of iron. The admissible compression and tensile stress are respectively 250 and 200 kg per cm² ;

Wood : the building material throughout the centuries, derived mainly from oak, spruce or pine. It is well known for its insulating and bearing properties. The admissible compression strength is 80 kg per cm² perpendicular to the fibres and 20 kg per cm² parallel with the fibres.

3.4.2 Building materials applied for the insulation of the building

Insulation materials used in combination with the above-mentioned materials are abundantly present on the market. Materials in the form of mats or boards can easily be applied in animal houses and especially under asbestos cement roof sheets.

Cork : derived from the periderm of the cork oak ; *peat* : derived from vegetable matter decomposed in the soil and partly carbonized by chemical reaction ; *glass-wool* : glass spun out to fine fibres ; *rock-wool* : is a fibrous material made from rock and spun out to fibres ; *expanded plastics* : derived from chemically manufactured plastics such as polyvinyl chloride, polystyrene, polyurethane ; *vermiculite, flax-loam etc.* : are all insulation materials, which owe their thermal insulation capacity to the air-entrained spaces and which are especially used as infill for cavities and holes in the walls ; *cement-wood fibre board* : manufactured from wood shavings which are cemented and compressed into stiff boards (e.g. Heraklith, Dhenaklith, ...) or possibly as sandwich-boards which between the two skins contain a powerful insulation material such as foam plastic (e.g. Dhenatherm).

3.5 THE PRACTICAL EXECUTION OF THE CONSTRUCTION OF ANIMAL HOUSES

Important constructional parts of livestock buildings are : the frame, the roof, the walls, the doors and windows and the floors.

3.5.1 Frames and roofs

Hangars can nowadays be built according to three methods viz. :
- construction utilizing a portal frame ;
- construction utilizing a propped portal frame ;
- construction utilizing trusses.

3.5.1.1 Portal frames

Trusses are nowadays often omitted and main rafters alone are used forming a frame which is referred to as a portal frame. The portal frame is characterized by the absence of intermediate posts in the middle of the building thus creating a completely free space. The stresses and strains are very much more complex with portal frames than with traditional trusses where the structural load is rather straightforward. Portal frames can be made of steel, timber or concrete. They are however more expensive than constructions having intermediate supports. The strength of the portal frame or the load upon it depends on several different factors such as its weight, the permanent load of roofing or flooring, the eventual operational load and the snow load. Besides the vertical loads we have to include also oblique loads such as wind loads. The portal frame consists of two columns with projections supporting an inverted V-concrete beam (fig. 3.14). The connections are accomplished with bolts at precise locations i.e. where the moment is zero under permanent load. For the commonly used values of span, free height and slope this point is located at a favourable place for the division in elements for prefabrication. Since the connection is established with several bolts, a couple of forces can arise which will allow the absorption of binding moments created by wind load.

Fig. 3.14 The concrete portal frame.

Spans up to 30 metres can, according to this system, be transported and installed in three parts. This system lends itself admirably to the realization of multi-span structures.

The manufacturers offering concrete beams deliver them with span widths of 10 to 30 metres. They can often be obtained for different roof slopes of 15 to 40 cm/m. Concrete portal frames require no maintenance whatsoever and are characterized by their long lifetime. They are fireproof, not susceptible to atmospheric or chemical corrosion or damage caused by animals.

3.5.1.2 Propped portal frames

These are an alternative to portal frames. Intermediate posts are used to support the roof. Propped portal frames can be made of timber, steel or concrete. Fig. 3.15 shows a construction made of steel I-profile beams and supported by steel tubular posts. It requires no extensive maintenance (paintwork) but should be covered with a good protective layer (a rust preventing paint or hot zinc dipping) prior to installation. Spans of up to 30 m can be obtained.

Fig. 3.15 The steel I-profile beam as used in a propped portal frame.

3.5.1.3 Trusses

This is an easier and more economic way of building a roof. Timber, steel and eventually concrete can be used.

- *traditional nailed trusses* are commonly used where intermediate supports are provided or in buildings with a restricted width. Such a timber truss tolerates spans of up to approx. 15 m. This type of truss is less applied in modern farm building ;
- *nailed and glued frame trusses* (fig. 3.16) consist of sloping frames at a mutual distance of approx. 1 m and composed of a set of planks. The upper and lower part of this frame consist of pairs of longitudinal frame planks between which bridgings are mounted. The bridgings are sawn at an angle on both sides and glued to the frame planks. The diagonal bridgings are also nailed to the planks at the places where they are glued to each other. These frames will form the truss which supports the roof. The advantage

of this construction is the saving on timber for the same bearing capacity compared to the ordinary trusses. They are mainly used in Switzerland and France. This type of truss will allow a span of up to 20 metres ;

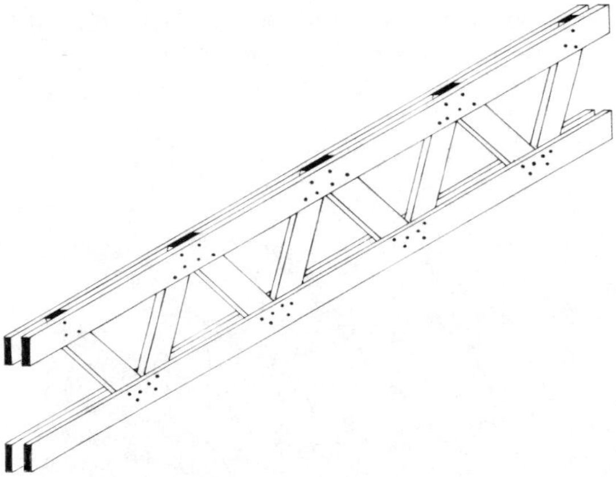

Fig. 3.16 The nailed and glued frame truss.

- *glued trusses* are always prefabricated in the factory and then transported to the construction site where straightforward erection is possible. These trusses are lightweight compared to concrete trusses. Red Norwegian pine from areas with a severe climate and slow growth are therefore used. The planks are attached to each other by means of a "finger-pressure connection" and then glued with resorcinol-phenol formaldehyde, a glue which is resistant to water and severe climatological conditions. Large spans, even up to 50 metres are possible. These glued trusses lead to an aesthetic design. They are however more expensive than ordinary trusses and are therefore less applied ;
- *steel trusses* were widely applied in earlier days. Nowadays they are rather rare. They require regular painting. The trussed steel girders allow a span of up to 20 metres.

3.5.2 Roof cladding

Frames, trusses or rafters support the *purlins* which in turn will carry the *roofing*. The *roof insulation* can be installed below the purlins. When steel profiles are used, the purlins are installed between the flanges of the profiles. When concrete beams are used, the purlins are mostly also made of concrete. The spacing between the purlins must be adapted to the roofing. The length of the purlins (= distance between the rafters) is 4 to 6 metres. Spans up to 10 metres can be achieved by using prestressed concrete purlins. In this case the rafters have to be calculated for the higher load. Some long prefabricated roof sheets can be applied without purlins. The roof sheets are then directly supported by the rafters.

The roofing of animal houses is often made of corrugated asbestos cement sheeting but aluminium sheets are also sometimes used.

Most of the roof constructions are composed of a very thin roof sheeting (6 mm thick corrugated asbestos cement sheets) under which there is a layer of air between the purlins and under this layer a high quality insulation material is eventually installed. A low k-value can be obtained with even a small thickness of the insulation layer, as shown in table 3.8.

TABLE 3.8 The necessary thickness in cm of some insulation materials in order to obtain different k-values when corrugated asbestos cement sheets are used as roofing.

Insulation material		k-value in $\frac{W}{m^2 . K} \leq$									
	λ in $\frac{W}{m . K}$	1.3	1.2	1.1	1.0	0.9	0.8	0.7	0.6	0.5	0.4
Mineral wool	0.041	1.9	2.1	2.4	2.8	3.3	3.8	4.6	5.5	6.9	9.0
Polystyrene	0.035	1.6	1.8	2.1	2.4	2.8	3.3	3.9	4.7	5.9	7.6
Polyurethane	0.03	1.4	1.6	1.8	2.1	2.4	2.8	3.3	4.1	5.1	6.6

A disadvantage however is the fact that in such a roof no heat accumulation can take place. The influence of sunshine in the summer and of cold nights in the winter will quickly be noticed in the animal house.

It is strongly recommended to insulate the roof more than the walls. On one hand warm air constantly rises so that the roof is in touch with warmer air than the walls. As a consequence more heat is lost through the roof than through the same area of wall. On the other hand low-built houses have a relatively larger roof area compared to the area of the walls.

3.5.3 Walls

A suitable wall must fulfil the following requirements :

- be sturdily built ;
- built with materials having low permeabilities to water vapour (vapour-proof barrier) ;
- be of good insulating material or in other words have a bad thermal conductivity ;
- have a good heat accumulating capacity.

Good wall insulation can be obtained by :

- a suitable choice of building materials, such as mineral building materials having a cellular structure and/or hollow bricks possibly provided with a rough surface to limit air movement ;
- the kind of construction, using a cavity and an insulating material inasmuch as this is covered with a moisture-repellent layer ;
- choosing a suitable site for the building, thereby taking into account that an isolated building loses more heat and that a high

plantation around the building will produce not only a more pleasant view but also less air flow and hence better insulation (fig. 3.17).

Fig. 3.17 A piggery surrounded with plants is more pleasant and allows some savings on the energy-bill.

Both the thermal-insulating capacity and the lifetime of the walls are markedly influenced by their moisture content. With regard to this it is necessary, during the construction of the wall, to include a *damp-proof-course* (e.g. bituminous flashing) over the entire area of the wall at a height of approx. 10 cm above the floor of the housing which in turn is about 10 cm above ground level (fig. 3.18).

3.5.3.1 The solid wall

In unheated houses the insulation requirements of walls are less stringent than those for heated houses. The latter type of housing must be adequately insulated. To achieve a sufficiently low heat transmission or a sufficient insulation, materials with a cellular or hollow structure are preferably used. Table 3.9 represents the required thickness of some building constructions in order to achieve the desired k-value. From this table we can conclude that solid outside walls are only satisfactorily insulated with a large and hence uneconomical wall thickness. For this reason a composite wall is chosen. This type of wall is either provided with a cavity or lined with a valuable insulation material (e.g. polyurethane boards, wood wool slabs etc.) which is applied against a solid building material.

No advantages however can be achieved from the heat capacity of

Fig. 3.18 The *damp-proof-course* (bituminous flashing) in the outer wall of an animal house.

TABLE 3.9 The required thickness in cm of a solid wall to achieve a desired k-value, (without cavity).

Material	k-value in $\frac{W}{m^2.K} \leq$				
	1.2	1	0.9	0.8	0.7
concrete (2,200 kg/m³)	116	146	165	189	221
ordinary bricks (1,700 kg/m³)	66	83	94	108	126
hollow bricks (1,300 kg/m³)	50	62	71	81	95
lightweight concrete (700 kg/m³)	31	39	44	51	59
hollow bricks + ordinary bricks (9 cm)	43	56	64	74	88
lightweight concrete + ordinary bricks (9 cm)	27	35	40	47	55

the walls if those insulation plates are applied at the inside. Since the animals can come in contact with the inner side of the outer walls the insulation boards have to be provided with a strong plastering. Such a construction is therefore rather seldom applied.

If the outside is lined with an insulation panel, full utilization of the heat capacity is guaranteed. In this case the outside insulation has to be protected against moisture by a hydrofuge rendering. Such an external rendering can be decorative but is too expensive for animal houses and is therefore seldom used. If no other solution is possible, one can think of the infill of the cavity with an insulation material.

3.5.3.2 The cavity wall

The cavity wall consists of an internal wall, an air layer and an external wall. If the external wall is strongly vapour-tight, such as is the case with enameled bricks or painted walls, it will be necessary to ventilate the cavity. If the outer side of the cavity is normally vapour-permeable a non-ventilated cavity can be used.

A cavity can also partly or fully be filled with a high-quality insulation material. A *fully-filled cavity* requires an exterior side which is vapour-permeable and moreover is not exposed to driving rain. Ideally the cavity should be filled with a non-capillary, vapour-permeable, water-repellent insulating material (Anon., 1980). Those requirements are fulfilled with materials such as : semi-hard, hydrophobically-treated mineral slabs, siliconized perlite and glass foam beads. Existing cavities can be injected with a loose-fill type of insulation material such as :

- hydrophobically-treated mineral wool slabs ;
- urea-formaldehyde foam ;
- siliconized perlite ;
- glass foam beads.

Cavities, fully-filled with such materials are not always problem-free, moisture problems can arise and for this reason a fully-filled cavity doesn't seem appropriate.

A great number of moisture problems can be solved by using *partially filled cavities*. In this case an insulation board is fitted to the internal cavity side (fig. 3.11) hence the cavity now extends from the insulation board to the outer brick wall. Following insulation materials, amongst others, can be taken into consideration :

- hard boards of mineral wool ;
- polystyrene foam sheets ;
- glass foam boards ;
- polyurethane foam sheets ;
- polyisocyanurate foam boards.

Normally an insulation thickness of 3 to 5 cm and a cavity of 3 to 4 cm are used. Ventilation is recommended with partly-filled cavities (a minimum opening of 6 cm^2 per running metre of wall, both at the upper and lower side). It is regularly found that partly-filled cavities are not always built to the rules of good craftsmanship. A common practice is to erect the inner and outer cavity wall simultaneously and to introduce the insulation material while the construction progresses. As a result the insulation material might get in touch with the exterior side and can then become wet. In the small air layer the possibility of mortar- and stone bridging exists and this in turn will lead to moistening of the insulation material whereby the insulation value of the wall is drastically reduced.

Table 3.10 gives the required thickness of the inner cavity side for a number of common cavity-constructions. The outer cavity side is always assumed to be half a brick (9 cm). In case of a partially-filled cavity we assume a 3 cm layer of polyurethane (λ = 0.03 W/(m.K))

TABLE 3.10 The required thickness, expressed in cm, of a building brick X at the interior side of a wall, to obtain the required k-value for a number of cavity constructions, assuming a 9 cm brick at the exterior.

	Building material and kind of cavity	kg/m³	k-value in $\frac{W}{m^2 . K} \leq$						
			1.2	1	0.9	0.8	0.7	0.6	0.5
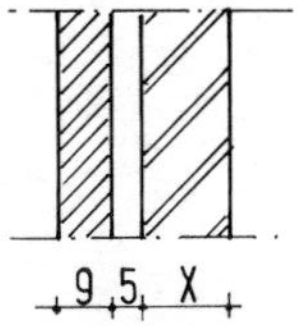	*unfilled cavity*								
	- ordinary bricks	1,700	27	37	45	54	65	87	102
	- hollow bricks	1,300	19	26	32	38	46	57	72
	- lightweight concrete	700	9	13	16	19	23	29	36
	- lightweight concrete	550	7	10	12	14	17	21	27
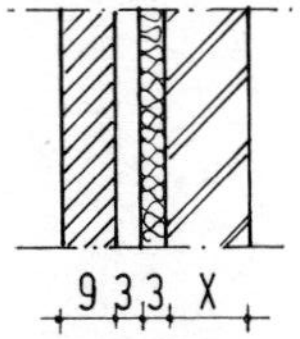	*partially-filled cavity** (3 cm layer of polyurethane)								
	- ordinary bricks	1,700	0	0	0	0	0	16	37
	- hollow bricks	1,300	0	0	0	0	0	11	26
	- lightweight concrete	700	0	0	0	0	0	6	13
	- lightweight concrete	550	0	0	0	0	0	4	10
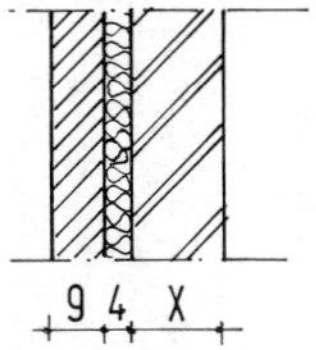	*fully-filled cavity** (4 cm layer of mineral wool)								
	- ordinary bricks	1,700	0	0	0	1	13	28	50
	- hollow bricks	1,300	0	0	0	1	9	20	35
	- lightweight concrete	700	0	0	0	0	5	10	18
	- lightweight concrete	550	0	0	0	0	3	7	13

* Moisture problems may arise and the thermal capacity might be inadequate.

while for a fully-filled cavity this is assumed to be a layer of semi-hard, hydrophobically-treated mineral wool slabs (λ = 0.041 W/(m.K)).

In case no insulation is applied in the cavity, an adequately insulated wall can only be obtained by using good insulating bricks such as cellular earthenware and cellular concrete.

From table 3.10 we can conclude that when an insulation layer is applied in the cavity we will obtain a k-value which is always inferior to 0.8 W/(m^2.K) : even a building brick of zero cm against the inner side of the wall will be sufficient. In practice, the minimum commercially obtainable size of brick will be chosen, which will not endanger the stability of the wall. To achieve a k-value inferior to 0.7 W/(m^2.K) or even 0.5 W/(m^2.K) we do not need a thick wall. Cavity insulation will always contribute in obtaining a sufficient global insulation even with a thin wall. The thermal capacity however is rather small. All these reasons and the possibility of moisture problems make us believe that a cavity insulation is not strictly necessary. A cellular concrete or earthenware wall with an empty cavity will also give sufficient insulation.

3.5.3.3 The wall-elements

Wall elements are commonly used in modern (prefabricated) constructions (fig. 3.19). These wall elements can be manufactured from wood, asbestos cement, reinforced concrete, cellular concrete, clinker concrete, etc. An insulation material (e.g. air) is sometimes introduced between both faces of the elements : the so-called sandwich-boards. The dimensions and weights of the boards vary greatly. Some of these elements require no special tools for installation and can easily be mounted by two persons. Other types of panels are heavy, even up to two tonnes or more, hence requiring a crane to load, unload and mount the elements. Most of the elements have a height of 0.60 m. The length varies between ca. 1.8 m and 6 m.

Wall elements made of cellular concrete or clinker concrete must have a minimum thickness of respectively 20 cm and 25 cm to satisfy the standard requirement of resp. $k \leq 0.9$ W/(m^2.K) and $k \leq 0.7$ W/(m^2.K).

Elements made of asbestos cement extrusion (ACE) (fig. 3.20) can be provided internally with a high quality insulation material. A disadvantage of these elements, despite its insulation, is the fact that a lot of cold bridges remain. Insulated plates of even 12 cm thick still have a k-value of 0.92 W/(m^2.K). An additional disadvantage of these asbestos cement elements is their very low heat accumulation capacity. These wall elements are therefore more suitable for animal houses requiring less stringent insulation conditions such as dairy cattle houses.

3.5.4 Doors and windows

In unheated animal houses, such as loose houses for dairy cattle, doors and windows will not pose problems. For well heated animal houses, such as farrowing houses for sows, it is senseless to optimize the insulation of the walls and waste the reclaimed heat through doors

Fig. 3.19 A cattle house constructed from wall elements of reinforced concrete.

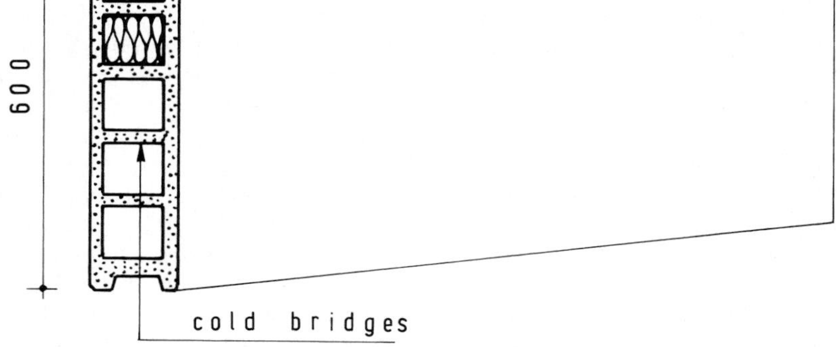

Fig. 3.20 An element in asbestos cement extrusion (ACE) with several cold bridges.

and windows. Doors and windows should therefore not be neglected from this point of view. This problem is partly solved by restricting the number of windows, especially in the outside walls. The number of outside doors can be limited by using compartments and a central passage in sow houses.

Windows installed in outside walls will lead to important heat losses. A single glazed window has a k-value of ca. 5.7 W/(m^2.K). The installation of double glazing will reduce the k-value to approx. 2.9 W/(m^2.K), which is clearly still too much. Furthermore double glazing is rather seldom applied in heated animal houses for economic reasons. The only practical alternative is the windowless construction, with the exception of places which do not require extensive heating, such as the alleys and the feed room, the houses for dry and pregnant sows, etc.

Doors can also be true cold bridges as shown in table 3.11. Home-made doors can be constructed by lining a wooden frame with 2 mm plywood on both sides and filling the space between with a 2 to 4 cm layer of mineral wool. A lightweight but relatively well-insulating door is thus obtained.

TABLE 3.11 The k-value of some door constructions for animal houses.

Material	k-value in $\frac{W}{m^2.K}$
18 mm flat asbestos cement board ($\lambda_e \approx$ 1W/(m.K))	5.36
20 mm hard wood	3.91
30 mm hard wood	3.35
40 mm hard wood	2.92
2 x 2 mm plywood + 30 mm cavity	2.84
2 x 2 mm plywood + 20 mm mineral wool in-fill	1.48
2 x 2 mm plywood + 30 mm mineral wool in-fill	1.09
2 x 2 mm plywood + 40 mm mineral wool in-fill	0.86

3.5.5 The flooring of animal houses

The floor is the only boundary surface in the housing with which the animals are in direct contact. Therefore it is necessary to keep the floor warm and dry in order to give the animals a warm and dry lying area. The floor must be calculated for the weight it has to support. The insulation values of different floorings are treated here without their eventual straw covering. If the lying area is littered the insulation value will consequently be increased considerably.

Calculations of heat losses often neglect the losses through the floors since the difference in temperature is small. In animal houses with a temperature of 20°C or more it is necessary to include the floors in the calculations of heat losses since the difference in temperature becomes relatively large. The heat losses through the floors in slaughter pig houses are also considerable since littering is here seldom applied and slaughter pigs spent 80 % of the time with 20 % of their body surface on the floor.

A draft proposal has been worked out (Anon., 1981a) for standardization of the calculation of the temperature-difference for walls in contact with the floor. Three zones can be distinguished starting from ground level (point 0 in fig. 3.21) :

- the first zone is situated over a distance of 2 m from ground level along the dug-in wall (OA) : the basic temperature of this zone is 0°C ;
- the second zone is situated over a distance of 8 m (AB) : the basic temperature of this zone is 10°C ;
- the third zone includes all surfaces in direct contact with the soil and which fall outside the first and second zone : the basic temperature is the same as the inside temperature t.

The k-value of insulated floors can be negatively influenced by damp rising from the soil. A plastic moisture-proof membrane is normally fitted between the insulation layer and the hardcore-concrete foundation. This foil which has to be applied with great care, must have a thickness of at least 0.2 mm. The overlapping is at least 20 cm and joints are preferably glued. The flooring insulation can also become wet from the top by faeces and cleaning water. A water-proof top layer must therefore be applied and finished with great care.

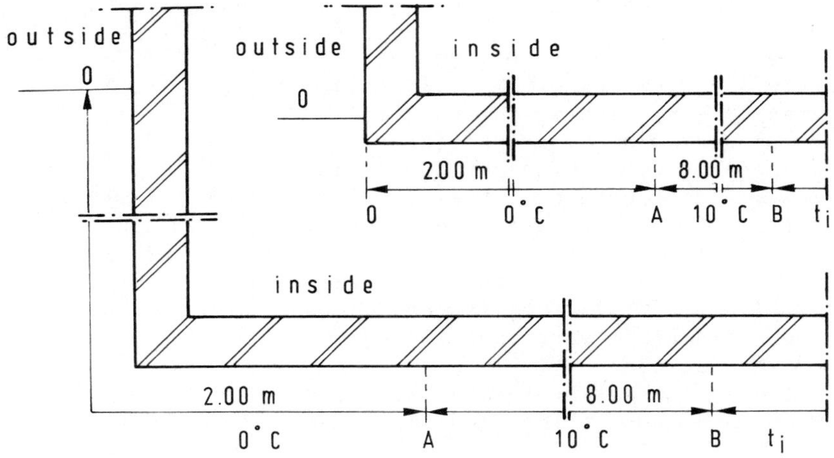

Fig. 3.21 Indication of the different zones of the basic outside temperature for walls in contact with the soil.

Lying floors for cattle houses consist mainly of concrete covered with a top layer or a mat. A mixture of 1 volume cement and 3 volumes sand is commonly employed as top layer. There are of course other jointless floor coverings (e.g. Bernit, Stallit, Steinit) which can be directly applied on the concrete surface and which have a number of advantages : they are insulating, pressure-proof, water-sealing, gas-tight and stiff. Their cost price is however rather high. As for mats, one has the choice between a number of rubber mats and plastic mats. We shall discuss these extensively in Chapter 4. *Lying-floors for pigs* have to be more insulated than those for cattle. As flooring insulation in pig houses the following materials are mainly used :

- expanded clay beads : 10 to 15 cm thick ;
- cellular concrete : 9 to 14 cm thick ;
- hard boards of high quality insulation materials (mineral wool, polystyrene, glass foam, polyurethane) : 3 to 5 cm thick.

When using hard boards of high quality insulation material it is strongly recommended to fit a waterproof plastic membrane on top of the boards to prevent the pores of being filled with cement slurry of the top layer. Table 3.12 gives the k-values of some commonly used floor constructions. Only by using high-quality insulation material (here polyurethane) the standard k $\leq$ 0.7 W/(m².K) is met. This floor construction is generally employed in farrowing houses equipped with floor heating (see Chapter 5).

TABLE 3.12 The k-value in W/(m².K) of some floor constructions.

Floor construction	k-value in $\frac{W}{m^2.K}$
15 cm concrete + 2 cm top-layer (screed)	3.76
15 cm concrete + 2 cm rubber mat (1,500 kg/m³)	3.11
7 cm concrete + 9 cm lightweight concrete (700 kg/m³) + 2 cm top-layer (screed)	2.43
7 cm concrete + 14 cm lightweight concrete (700 kg/m³) + 2 cm top-layer (screed)	1.93
5 cm concrete + 10 cm lightweight concrete (1,100 kg/m³) + 3 cm top-layer (screed)	2.62
5 cm concrete + 15 cm lightweight concrete (1,100 kg/m³) + 3 cm top-layer (screed)	2.15
7 cm concrete + 4 cm polyurethane (50 kg/m³) + 4 cm top-layer (screed)	0.64
7 cm concrete + 4 cm polyurethane + 7 cm concrete + 2 cm top-layer (floor heating)	0.63

Since the animals are directly in contact with the lying floors, not only the k-value is important but also the temperature of the floor surface. Top layers, which feel warmer are probably better for pigs. Floor heating is therefore also interesting for pigs. For cattle, the temperature requirements for floors are less stringent and consequently the k-value is of less importance.

3.6 THE PREFABRICATION OF ANIMAL HOUSES

In the last decade an interest has grown for prefabrication i.e. the mounting of building elements manufactured in a factory and their assembly in situ to a complete unit of walls, floors and roof. Both in the building of houses and animal houses, prefabrication has gained some, although rather limited, application. The following advantages are attributed to prefabrication :

- the construction time is reduced to a minimum : one to two months are ample for the construction of a prefab cattle house (excl. foundations and layout) ;
- prefab lends itself better than traditional building to the concept of turn-key housing, whereby the principal deals with only one firm and does not have to call in a range of skilled labourers (contractor, carpenter, plumber, floorer, plasterer, electrician, etc.) and where the co-ordination is not always easy to realize ;
- less skilled personnel is required, since mounting of a prefab is mainly assembly-work ;
- prefab creates the possibility to use the same standard designs over and over again ;
- prefab enables the manufacture of the building elements in a factory, thereby reducing the number of labourers on the construction site and this in turn allows better supervision and more efficient work ;
- the building price is fixed in advance and "surprises" are practically non-existent as compared to traditional building where this happens rather often ;
- the building costs can be reduced but prefabricated building manufacturers look for quantity as an essential in reducing their costs, but up to now this is rather seldom achieved.

Foundations are not often prefabricated and hence have to be constructed by a (separate) builder in advance.

Prefab assumes that the dimensions of all building elements are reducible to one and the same module or basic lattice. Internationally (ISO) it is agreed to represent M as the module dimension, whereby M = 10 cm or the so-called basic module which is, most of the time, also employed in Europe. Often a multiple module is employed i.e. 3M (= 30 cm) or 6M (= 60 cm). It is recommended to use the internal dimensions for the description of an animal house and to choose them in such a way that they are in conformity with the modular co-ordination. The height of a side wall of a building is measured along the inside wall from zero level to the underside of the roof or the upperside of the purlin. The distances of the trusses are measured between the centrelines and are dependent on the used materials and the building system. A span of up to 6 m can be achieved when using steel trusses in combination with steel purlins. This is also possible for the combination of concrete trusses and purlins although they are normally installed at a distance of only 5 or 6 m. These distances are seriously reduced for wooden trusses and purlins and this in function of the width of the housing (3 to 4 m).

Some guidelines have been worked out for dairy cattle houses (Anon., 1974 ; Anon., 1975b) and for pig houses (Anon., 1976) based on the multiple module 3M (= 30 cm). In fig. 3.22 some of those standard house profiles for cubicle houses intended for dairy cattle are illustrated.

It is clear, that even starting from the multiple module 3M a large number of construction variants are possible. The modular co-ordination has however not led to an advanced standardization in the development of building systems. No agreement is yet reached on the complete uniformity of the prefab supply. The application and the interchangeability of building elements of different manufacturers is not always possible ; the applied materials are often different for different manufacturers ; the offered prefab range differs from firm to firm : some supply both houses and service buildings, others offer only the latter, some offer walls, floors, roofs and interior equipment, others supply only skeletons and walls, etc.

Of importance in this relation would be the achievement of uniformity, not only for the prefab-agricultural-constructions but also for the manufacture of as many as possible building elements suitable for both industrial and agricultural buildings in order to enable the fabrication of much larger series which would thereby reduce the cost price. International understanding would greatly contribute to this purpose.

3.7 THE VENTILATION OF LIVESTOCK BUILDINGS

The problem of ventilation will be briefly discussed and only to the extent to which it is related with the building of animal houses. Moreover, numerous studies have been published on this particular problem (Anon., 1965 ; Anon., 1981b ; Brandsma, 1976 ; Bruce, 1975 ; Carpenter, 1974 ; Christiaens and Debruyckere, 1977 ; Debruyckere and Neukermans, 1973 ; Debruyckere et al., 1982 ; Hoorens et al., 1973 ; Karle, 1981 ; Mertens and Brandsma, 1973 ; Mitchell, 1972 ; Randall, 1977 ; Sällvik, 1979 ; Tol, 1972).

Livestock buildings have to be ventilated for several reasons. The oxygen consumed by the animals has to be replaced. The produced water vapour and noxious gases have to be eliminated. The temperature of the house has to be kept as constant as possible, whereby a high extraction rate can be required to cope with extreme summer conditions. The ventilation shall be at a minimum level in severe winter conditions in order to prevent an undesirable temperature drop.

Basically, the inlet of fresh air and the outlet of stale air are separated. The air circulation is established in a natural way (by wind force or/and gravity : open ridge and chimney outlet) or in an artificial way (by applying electric energy : fans).

3.7.1 Natural ventilation

Natural ventilation is based upon the physical principles of the stack-effect and the wind effect.

The principle of the *stack-effect* is based on the fact that the specific weight of the entering colder air is greater than that of the warmer foul air and therefore pushes the latter upwards where it must be removed. The difference in specific gravity between stale

TYPES OF HOUSING	WIDTH	STANDARD PROFILES
2 rows with outside feeding	6.90	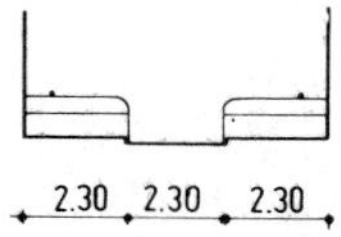
4 rows with outside feeding	13.80	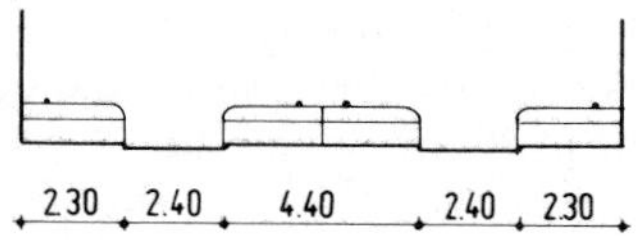
2 rows with sidelong feeding passage	14.40	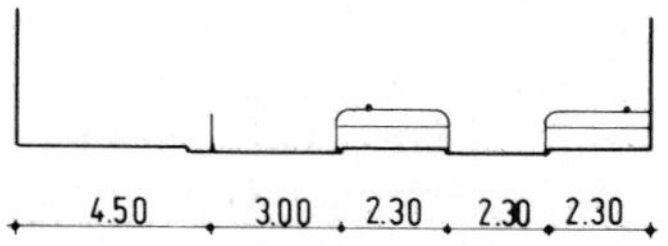
3 rows with sidelong feeding passage	16.50	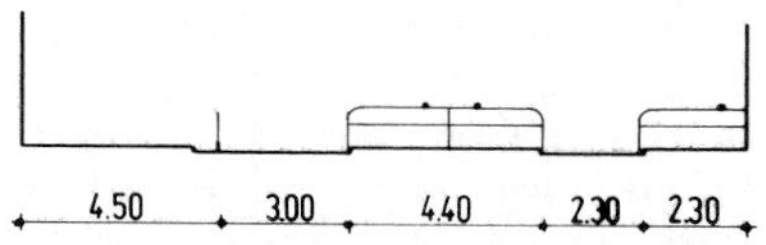
2+1 rows with central feeding passage	19.80	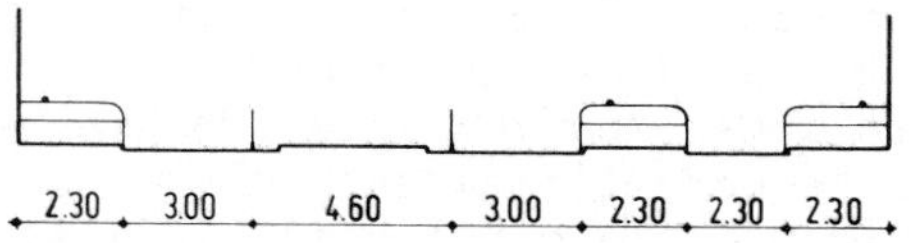
2+2 rows with belt conveyor feeding	21.00	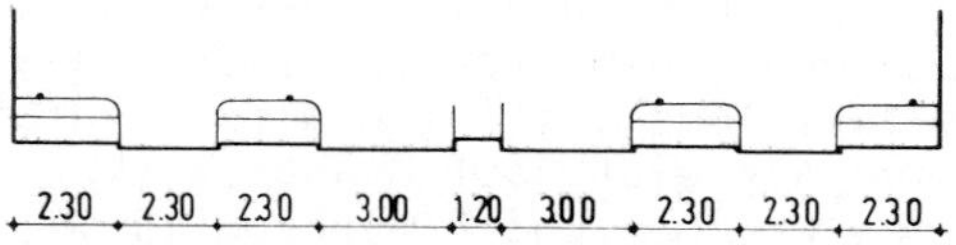
3+1 rows with central feeding passage	21.60	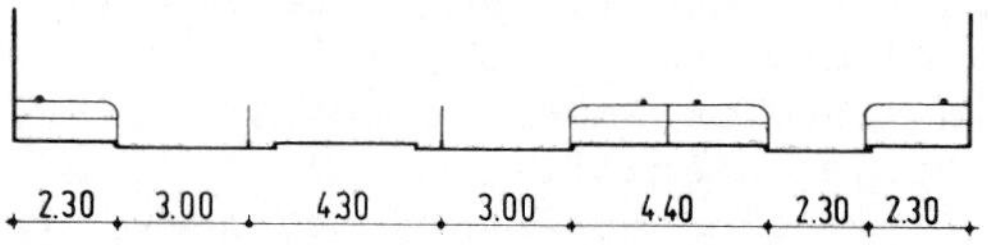
2+2 rows with central feeding passage	24.60	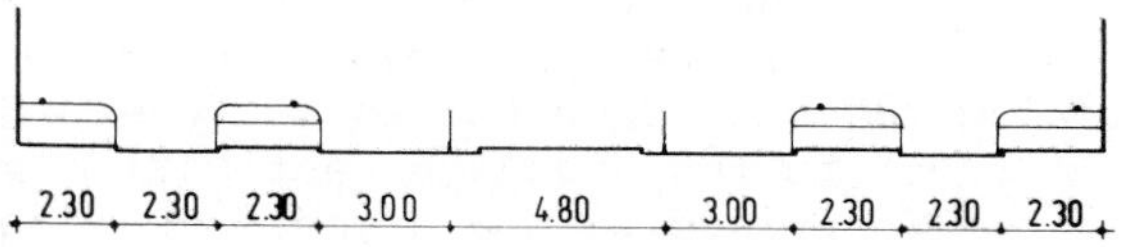
1+1 rows feeder-cubicles with central feeding passage	12.30	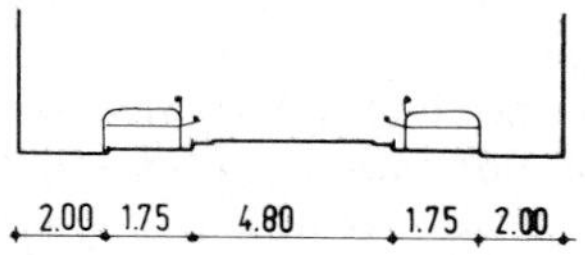

Fig. 3.22 Some standard profiles for cubicle houses intended for dairy cattle.

air and fresh air is dominant for the working of natural ventilation. A 5°C temperature rise will cause a weight decrease of approximately 24 g per cubic metre of air. Colder weather will enable a stronger ventilation while in fact the ventilation requirement is lower. In summer, on the contrary, with warmer weather, the ventilation possibilities will be restricted due to the small temperature difference while in fact the requirements will then be the greatest.

The so-called *stack-height* i.e. the height from the level of the inlets to the top of the outlet is of great importance in the natural draught system. The larger the difference in height, the better the ventilation will work. A large height difference can be obtained by installing the inlets for fresh air as low as possible and by utilizing a greater rise of the roof. The inlets should not be placed too low however, since then the likelihood of draught onto the animals becomes more important.

The *wind* will create a pressure phenomenon on and in the livestock building. The pressure difference between the outer and inner side of the house will cause ventilation. This pressure difference is mainly dependent on the wind direction, the wind intensity and the form and dimensions of the building. A greater slope of the roof will create a larger wind velocity above the open ridge, which results in a larger underpressure thereby extracting the air more quickly out of the house (Brandsma, 1976). A larger stack height is then also obtained. The influence of the wind is incalculable : the influence of the wind can sometimes increase the ventilation but a decrease is also sometimes possible (Hoorens et al., 1973). Adverse wind effects on the ventilation can be avoided or reduced by installing a wind deflector along the open ridge and a hood at the air inlet.

The *stack* functions better as the vertical temperature gradient and hence the pressure difference at the bottom and at the top is greater. This implies a good insulation of the stack, otherwise the rising air will cool drastically causing a cold plug in the upper part of the chimney which eventually will block further extraction of foul air. Insufficient insulation might lead to condensation. Often, two waterproof boards, nailed on a wooden frame are used and the void space between the two skins is filled with a high quality insulation material.

Fig. 3.23 represents the above described stack. It is also possible to use insulated plastic cylindrical chimneys. The round form is aerodynamically the most advantageous. To prevent down draughts the stack outlet has to be placed at least 50 cm above the ridge.

In order to restrict the ventilation during the winter the stack has to be provided with a damper or a flap and to avoid short-circuiting of air, the chimney has to be mounted as far away as possible from air inlets and doors.

Table 3.13 gives the required dimensions of a stack according to the German standards (DIN) (Anon., 1965). The figures in table 3.13 are only valid for chimneys with a stack height greater than 4 m. For stack heights smaller than 4 m, as often applied in modern low level livestock buildings, an empirical standard of 0.5 to 1.0 m² stack per 100 m² flooring surface, according to the density, is used.

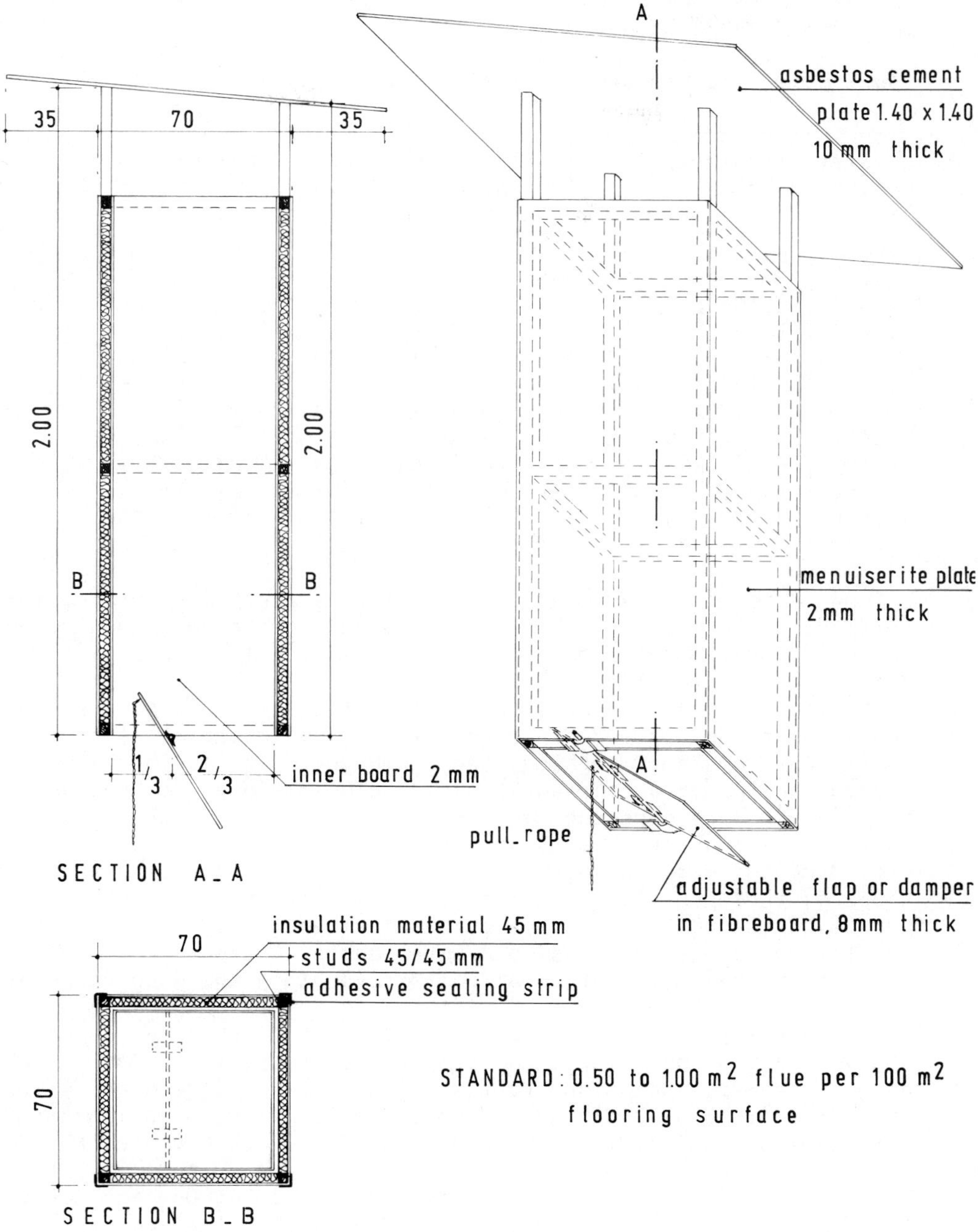

Fig. 3.23 The construction of a flue or stack.

For aerodynamic reasons, chimneys smaller than 0.25 m² or larger than 1 m² are to be avoided.

TABLE 3.13 The required dimensions for stacks (Anon., 1965).

Species of animals and weight in kg			Side or diameter of chimney in cm (square- or round shaped section) with a flue height of :				
cattle	pigs	poultry	4 m	6 m	8 m	10 m	12 m
2,000	1,500	375	52	47	44	42	40
3,000	2,250	502	63	58	54	51	49
4,000	3,000	750	74	67	62	59	56
5,000	3,750	939	82	74	69	66	63
6,000	4,500	1,125	90	81	76	72	69
7,000	5,250	1,312	97	88	82	78	74
8,000	6,000	1,500	--	94	88	83	79
9,000	6,750	1,687	--	100	93	88	84
10,000	7,500	1,875	-	-	98	93	88

The *open-ridge ventilation* offers a cheap possibility for ventilation. Fresh air is taken in via adjustable continuous openings on both longitudinal sides and escapes via openings in a part or the full length of the ridge of the roof.

Similar to the flue or stack, the difference in temperature of in- and outside air and the stack height are important factors which contribute to a good working open-ridge ventilation. The wind exerts a great influence on the working of the open ridge. It can happen that by wind pressure the wind strikes through the open ridge and that the air outlet acts as an air inlet. This can be avoided by providing two dampers in the vertical plane of the roof along both sides of the open ridge. During stormy weather the flap on the windward side is then manually or automatically closed.

Flow pattern studies using a water table (Bruce et al., 1978) have shown that the installation of deflectors along the open ridge eliminates wind-striking.

Principally, the same calculation methods of chimneys can be applied to the open ridge. In low-profile houses with small stack height the empirical standard assumes 2 to 3 cm ridge-opening per metre run of house width according to the density. A horizontal damper is also required with the open ridge. In cattle houses this damper is often neglected.

A good open-ridge design is shown in fig. 3.24. One can see that the cover plate overlaps the opening below it but lies below the top of the upstands, which are laid with the flat side to the roof so that drainage is very rapid. The air passage width is maintained (150 mm and 2 x 75 mm). This open ridge was always found to operate as an outlet even when the windward inlet-slots were closed up and no rain was entering through it (Bruce et al., 1978). The size of the opening must be designed for each animal house.

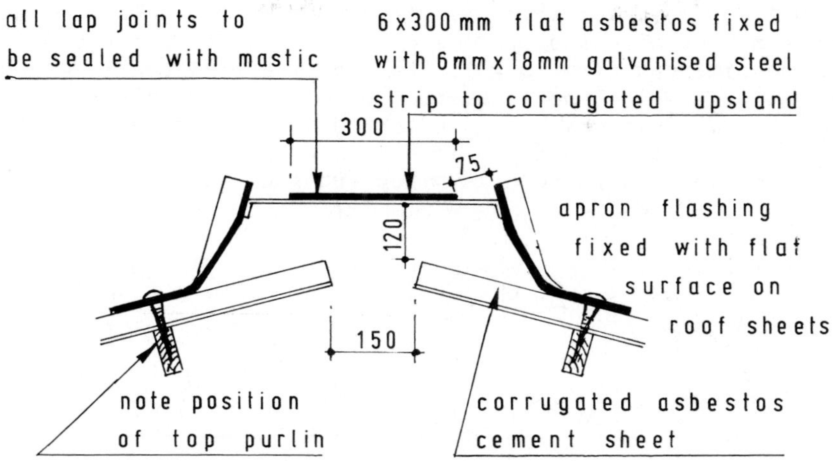

Fig. 3.24 A good open-ridge design (Bruce et al., 1978).

3.7.2 Mechanical ventilation

In contrast, mechanical ventilation is distinguished from natural ventilation by the origin of the motive power which is required to provide ventilation. Mechanical ventilation is carried out by means of electric energy and the natural force is not or only to a small extent applied.

Four systems are possible with mechanical ventilation :

- *extraction ventilation* whereby the stale air is extracted and fresh air enters through the normal inlets ;
- *pressurized ventilation* whereby outside air is forced into the building and where the stale air is discharged via pressure flaps ;
- *equilibrium ventilation* whereby fresh air is forced into the building and foul air is simultaneously extracted ;
- *hybrid recirculation ventilation* whereby an adjustable quantity of the ventilated air can be recirculated.

3.7.2.1 Extraction ventilation or conventional extraction

Extraction ventilation is the conventional method of mechanical ventilation since it is rather easy to realize and gives satisfactory results.

With extraction ventilation the foul air is extracted from the building by means of a fan whereby a slight vacuum is created in the house which will draw in fresh outside air through the inlets of the animal house. The fans can be mounted in chimney trunks placed on or just to the side of the ridge. The construction of the chimney trunk is similar to that of a stack. The trunk must be insulated, not so much for efficient working but mainly to avoid condensation.

In narrow houses (having a width of less than 6 to 8 metres) and with roof ventilation the risk always exists of air short-circuiting i.e. fresh air which is directed from the inlet to the fan without mixing with the stale air. In such cases cross ventilation is recommended (fig. 3.25). The inlets are placed in one longitudinal wall while the fans are mounted in the other longitudinal wall. In wide houses it is better to provide the extraction via the ridge and to install the inlets on both side walls. In fig. 3.26 the air flow pattern is given for roof ventilation.

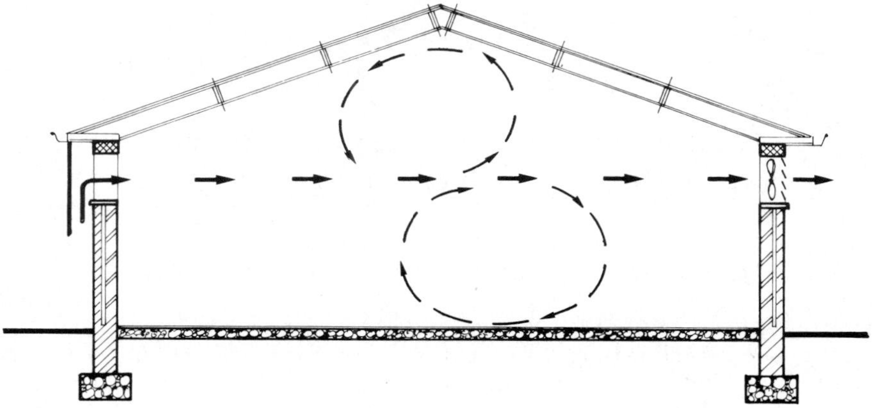

Fig. 3.25 The air movement with fans in one side wall (cross ventilation)

Since the degree of pollution of the air stream increases with the distance travelled, the distance between in- and outlet, especially in livestock units with a high density, should not be too great. The distance is not only of importance for the composition of the air but also for the air volume which has to be supplied by the inlet.

In practice, a maximum distance of 7.5 to 10 metres is acceptable (Hoorens et al., 1973) which means that the building must not be wider than 15 to 20 metres if a double sided air flow is used.

Fig. 3.25 shows the air pattern with cross ventilation. The fan is preferably placed in the wall opposite the dominating wind direction. A lower ventilation capacity is obtained if "reverse" winds occur ; in our region, the outside temperature is mostly lower and thus less ventilation is required. Cross ventilation in wide livestock buildings gives high air velocities which may result in draught.

In wide-span animal houses it is also possible to install the fans in both side walls and to supply fresh air through the ridge

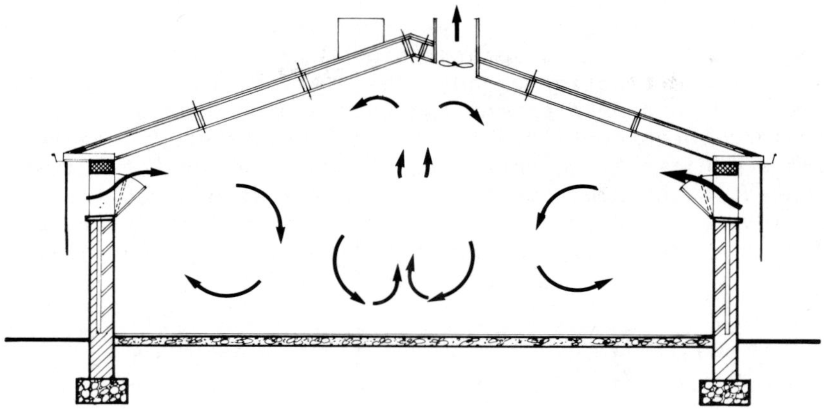

Fig. 3.26 The air-flow pattern with mechanical roof ventilation of a livestock building.

(see fig. 3.27). The unnatural air-flow, obtained by this type of ventilation requires a higher electricity consumption than with roof ventilation. A hood must be provided over the fans, when the latter are installed in side walls, to prevent direct blow-in when the fans are not in use.

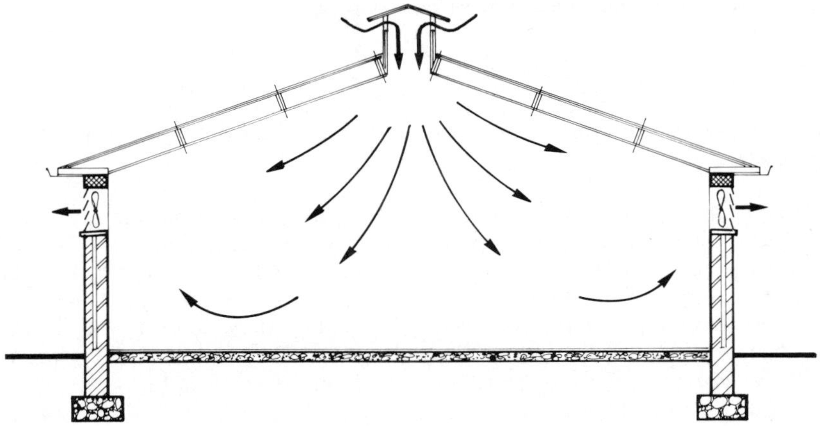

Fig. 3.27 The air movement with fans in both side walls.

A special form of extraction ventilation is the floor extraction, which takes place through a duct placed very low near the floor and, with slatted floors, even under the floor. In this way the most polluted air is pulled out. This technique offers also the advantage that the colder air is removed from the animal house thus enabling a higher average environmental temperature. In summer however it can become quite hot in the animal house if floor ventilation is used (Hoorens et al., 1973). In slaughter pig houses mainly, with their high density, the extraction through the slatted floor can produce a much lower foul air development and thus a more pleasant environment. Therefore large ventilation ducts are built beside the animal house which are directly connected to the dung cellars. A suitable fan is mounted in the upper part of the ventilation duct. Fresh air is brought into the piggery via the ridge or through the side walls. Fig. 3.28 gives a schematical representation of this technique.

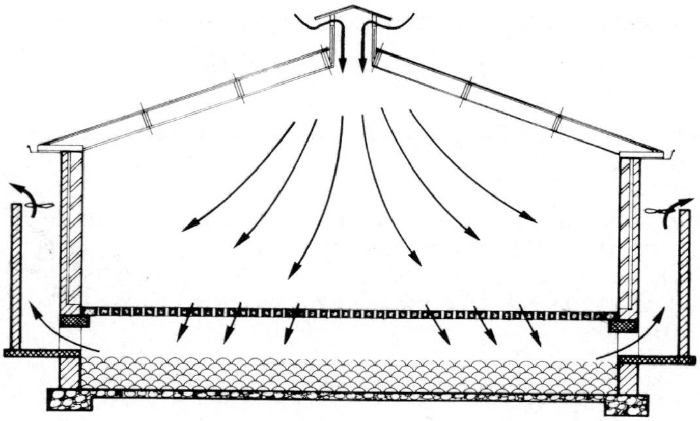

Fig. 3.28 The air-flow pattern with the floor extraction system in an animal house.

In practice, however, it is found that there is a tendency that even the heavier noxious gases in the animal house, such as carbon dioxide and hydrogen sulphide will accumulate together with lighter gases and water vapour in the upper part of the building. Therefore air inlets and extraction points must be carefully sited and adequately spread. Dutch investigations (Mertens and Brandsma, 1973) have shown that the ammonia concentration in slaughter pig houses is significantly lower with floor ventilation than with ridge extraction, in other words to obtain the same ammonia concentration during the winter, less ventilation (ca. 30 %) is required. From these investigations it seems that animals in houses with floor extraction are generally cleaner than those in houses with roof extraction.

With floor extraction there is less danger of accumulation of toxic gases in the house during mixing or pumping of the liquid manure. The construction of high ventilation trunks and the resulting high resistances make floor extraction more costly than roof extraction.

3.7.2.2 Pressurized or plenum ventilation (fig. 3.29)

Pressurized ventilation may be achieved by forcing fresh air by means of a fan directly or through an air distribution duct into the animal house thereby creating a slight pressure which drives the stale air outwards through the outlets. An advantage of the pressurized ventilation is that it enables the preheating of fresh air by means of a heater. This system is therefore more applied in broiler houses where no additional local heating is then installed. Often an oil-fired heater is used to preheat the entering air. Fresh outside air, eventually mixed with recirculated air is forced through a heater by means of a fan. Most of all, the air is forced into the longitudinal direction of a building and relatively high air velocities can be produced. No adverse effects to the animals will be noticed if the air temperature is high enough (20 to 35°C). The air volume delivered by the heating unit is often insufficient to provide the total ventilation, therefore additional possibilities for (pressure) ventilation must be present.

The draught-free supply of fresh outside air by pressurized ventilation necessitates the provision of a distribution duct under the *ridge* running over the length of the house. Choosing the dimensions of the duct and its inlet-holes in relation to the fan capacity is difficult. The pressure flaps are not always free of problems too. Furthermore the energy consumption is higher. Pressurized ventilation is therefore rather unattractive and in practice preference is given to conventional extraction.

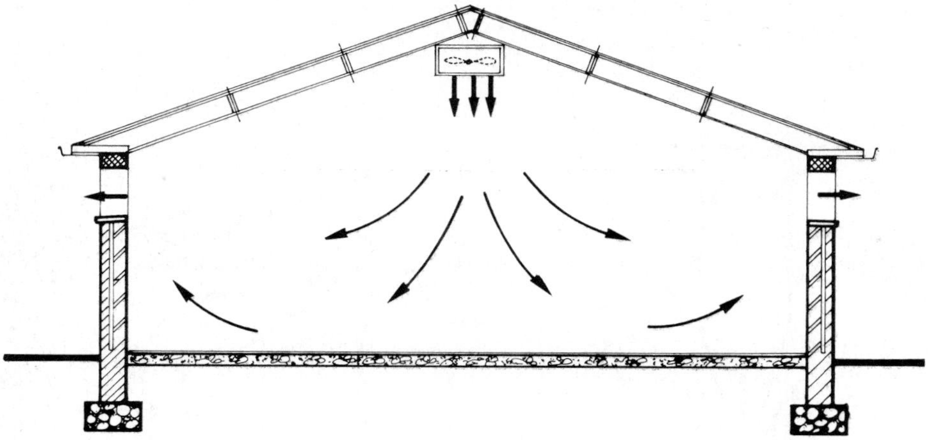

Fig. 3.29 The air movement in an animal house equipped with mechanical pressurized ventilation with an air duct.

3.7.2.3 Equilibrium ventilation

This arrangement relies on the entering of fresh air at one side of the building and extracting at the other. Therefore two "matched" fans at precise locations are required. In fact a specially developed system can be applied which employs only one fan in a double channel which simultaneously extracts and enters air. By providing an adjustable damper it is possible to use such a system for ventilation with fresh outside air, for recirculation or partial recirculation with any ratio of fresh air-recirculated air.

3.7.2.4 Hybrid-recirculation ventilation

Hybrid-recirculation ventilation systems are derived from extraction or pressurized ventilation systems by providing an adjustable recirculation device at each fan. Fig. 3.30 shows two different hybrid extraction ventilation systems. The recirculation flaps are mounted behind the fan (pressure side). Fig. 3.31 illustrates the hybrid-recirculation principle in pressurized ventilation. In that case the recirculation flaps ought to be installed in front of the fan (suction side) and an air distribution duct with air straightener is essential (Owen, 1981).

In both cases the recirculation flaps permit a continuous regulation from 100 % fresh air (100 % ventilation) to 0 % fresh air (100 % recirculation). However, the last possibility has to be eliminated by construction in order to avoid asphyxiation and idle ventilation.

Recirculation ventilation is indicated when the minimum fan capacity is still too high. It may be the only possible way to improve existing ventilation systems with too high a capacity fans, especially in winter time.

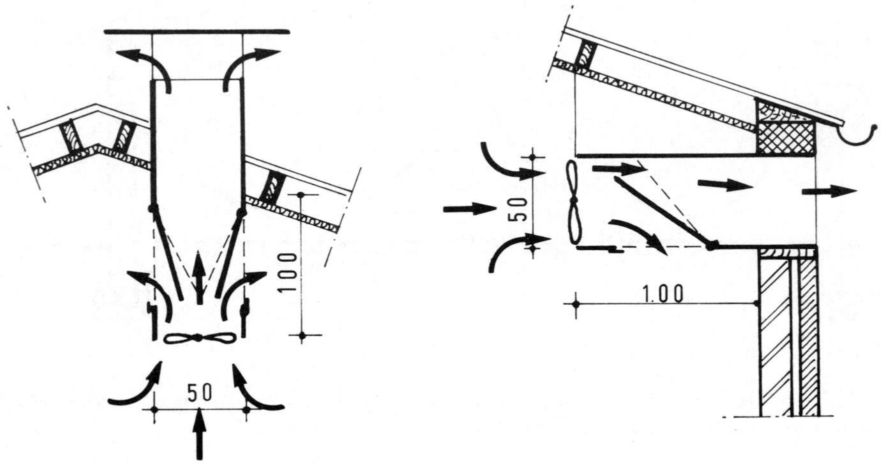

Fig. 3.30 Two hybrid-recirculation ventilation systems.

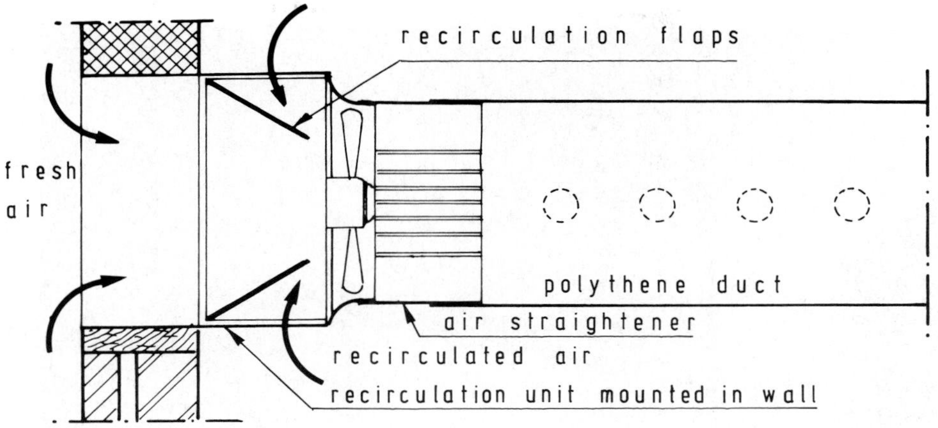

Fig. 3.31 A recirculation unit in a pressurized ventilation system (Owen, 1981).

3.7.2.5 Ventilation volumes with fan-ventilation

The following maximum ventilation volumes are taken into account with fan ventilation (expressed in m^3 per hour and per kg liveweight) (Anon., 1981b) :

- cattle : 0.50 m^3 per h and per kg ;
- pigs : 1 m^3 per h and per kg ; weaners : +20 % ; sows : -20 % ;
- chickens : 4 m^3 per h and per kg ;
- rabbits : 3 m^3 per h and per kg.

3.7.3 The air inlets

The air inlets are of great importance in an animal house. They have to realize a good mixing between entering, often colder, fresh air and the warm air inside the building. Highly placed inlets are to be preferred since they will enable a partial heating of the entering air before the air reaches the animals. A disadvantage however is the decrease of the stack height with natural ventilation.

The inlet has to be provided with an adjustable flap (fig. 3.32) which enables the regulation of the air velocity of the incoming air. Entering air with too low a velocity, which sometimes occurs with fans working too slowly and with large inlets will increase the likelihood of down draughts on the livestock, while the heating of the air by mixing with the air inside the building will become insufficient.

Until recently a maximum air entrance speed of 1 m per second was advised. Recent work proved that, with mechanical ventilation, a higher inlet speed through slot type inlets is necessary in order to keep the air-flow pattern in the animal house constant. Randall (1977) calls this high-speed jet inlets and promotes a constant inlet air speed of 5 metres per second. In that case the inlet opening has to be continuously adjusted. This can easily be done by a "Ten Elsen"

Fig. 3.32 Adjustable air inlets in a side wall of a livestock building.

balancing inlet flap as shown in fig. 3.33. The higher the fan speed the higher the pressure difference between inside and outside and thus the higher the suction will become. At higher fan speeds the inlet opens wider. In this case the hopper type deflector is important in giving the entering air the right direction (inlet flap = deflector).

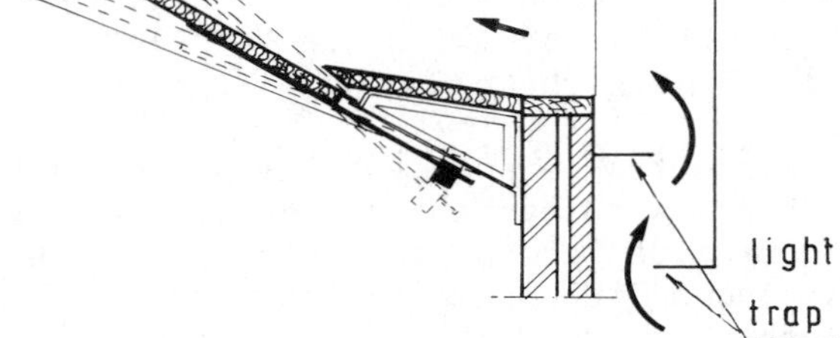

Fig. 3.33 A "Ten Elsen" balancing inlet flap.

In order to obtain a suitable ventilation the air inlets have to be evenly spread over the entire building and preferably be provided with hoods to neutralize the wind influences. In hen houses the hood may be equipped with a supplementary light trap.

The air inlets should not be placed :

- above water pipes or drinking bowls because of the danger of frost ;
- at less than 2 metres from corners of the building to avoid accumulation of cold air in the corners ;
- at less than 2 metres from the thermostat ;
- at less than 3.5 metres from a stack to avoid short-circuiting of air.

A special form of air inlet is the so-called curtain in the upper part of both longitudinal walls (fig. 3.34). This curtain is constructed from woven nylon thread and its minimum height is 2 cm per m house width. It is installed in the upper part of the outer side walls. Along the inner side of the wall a wire netting (1 cm mesh) is placed at the same height which prevents the entrance of birds.

Fig. 3.34 A curtain acting as an air inlet at the top of a longitudinal wall of an animal house.

The opening of the curtain can manually or automatically be controlled according to the outside temperature. With a manually operated system the curtain is suspended from vertical steel cables which, via rollers, are mounted on a horizontal cable, having the same length as the curtain. This horizontal cable is wound upon a winch, which can be turned in either direction by means of a handle. In this way the curtain can be lowered or raised thereby adjusting the opening.
An automatically controlled system uses an electric motor which is

coupled to the winch and is driven by a relay, controlled by a thermostat. This thermostat is installed at a suitable place in the livestock building. The automatic control is very sensitive and follows the changing weather conditions precisely, thereby keeping the house temperature within preset limits. Even with a fully-closed curtain a minimum ventilation is guaranteed through the gauze.

The use of such a woven curtain as an air inlet with natural ventilation is already widely spread in houses for laying hens and is also used to a certain extent in finishing pig and cattle houses. A disadvantage of this system is the interference with the light scheme during the laying period.

The adjustable air inlet area is :

- equivalent to the outlet of stacks with less than 4 m stack height ;
- twice to three times the outlet area of stacks with a stack height greater than 4 m ;
- a continuous inlet under each eave of 1/3rd of the ridge area, i.e. about 1 cm height per m house width and per side wall ;
- 0.28 m^2 per mechanical ventilation volume of 1,000 m^3 per hour if a normal entrance speed of max. 1 m/s is expected and 0.06 m^2 if a high-speed jet type inlet (5 m/s) is preferred.

3.7.4 Condensation of water on walls, doors and windows

All the above described measures concerning wall constructions are aimed at the limitation of thermal losses. The relative humidity of the air in the animal house can nevertheless be quite high if one knows that in a cow house about 650 g of water vapour per cow and per hour are produced (Comberg and Hinrichsen, 1974). The relative humidity in a cattle house should not exceed 80 % as mentioned in table 3.4. To prevent condensation on the walls we have to keep the temperature of the wall surface at the inside above the dew point. The following formula allows the calculation of the maximum value of the thermal transmittance (k-value) of a wall to prevent condensation on the wall :

$$k_k \leq \frac{\alpha_i . (T_i - T_k)}{T_i - T_e}$$

whereby k_k = the maximum admissible thermal transmittance to prevent condensation, in $W/(m^2.K)$; α_i = the surface thermal transmittance at the inner side of the wall in $W/(m^2.K)$; T_i = the inside temperature in °C ; T_e = the outside temperature in °C ; T_k = the condensation temperature in °C under the actual or the expected conditions.

The application of this formula is a valuable approach to condensation problems and allows actions to prevent this phenomenon. If from these calculations it seems that uneconomically thick walls have to be used one can either :

- ventilate more ;
- increase the temperature ;
- apply a vapour barrier along the inner side ;
- allow a temporary condensation (e.g. on the windows).

Furthermore, each wall, as long as no vapour barrier is employed, is able to transport some moisture.

A brick wall for example can transport with the heat stream approximately 300 g of water vapour per m^2 and per day through the wall. Even below freezing point the wall is still able to accumulate moisture.

The building material itself will influence the moisture transfer. Bricks, pumice grit concrete, light concretes and plastering are moisture permeable. With the ever increasing use of insulating blocks as inside walls it is recommended not to plaster them. Sometimes rendering of the inside walls cannot be avoided : the often-washed walls of the milking parlour and the milk house or those in contact with animals must be rendered. A smooth but less permeable rendering is therefore employed in the milk house and milking parlour (cement-sand, 2 cm thick). The walls along the dung passage of a fattening pig house can be rendered up to a height of 1 metre to avoid damage by the animals.

It is however not always possible to avoid condensation on the windows of an animal house. In any case one should prevent the condensation water from running down the walls by the use of a suitable window construction.

REFERENCES

Anon., 1965. Klima im geschlossenen Stall, Lüftung, DIN 18910, Sept. 1965, Blatt 2, 3 pp (revised in 1974).

Anon., 1966. Stallklima - A.L.B., Heft Nr. 27, 50 pp.

Anon., 1967. Eigenschappen van bouw- en isolatiematerialen, deel 9, Stichting Bouwresearch, N. Samsom, Brussel, Belgium, 33 pp.

Anon., 1974. Richtlijnen : Bindstallen voor melkkoeien, Direktie voor Landbouwtechniek, Ministerie van Landbouw, Brussel, 24 pp.

Anon., 1975a. Bouw- en isolatietechnieken, Uitgaven van Open School B.R.T. en Fonds voor Vakopleiding in de Bouwnijverheid, Brussel, 168 pp.

Anon., 1975b. Richtlijnen : Ligboxenstallen voor melkkoeien, Direktie voor Landbouwtechniek, Ministerie van Landbouw, Brussel, 34 pp.

Anon., 1976. Richtlijnen : Varkensstallen, Direktie voor Landbouwtechniek, Ministerie van Landbouw, Brussel, Belgium, 55 pp.

Anon., 1980. Isoleer uw huis, Brochure published by:Ministerie van Economische Zaken, Dienst voor Energiebehoud, Brussel, 68 pp.

Anon., 1981a. Normvoorstel van het W.T.C.B., Berekeningen der warmteverliezen van gebouwen, BIN dok. 734/152N, 28 pp.

Anon., 1981b. Berekeningsgrondslagen voor klimaatbeheersing in veestallen, Richtlijnen voor landbouwbedrijfsgebouwen, Direktie voor Landbouwtechniek, Ministerie van Landbouw, Brussel, (in preparation).

Anon., 1981c. Isoleer zonder isolatiekosten, Brochure published by: Belgisch Cellenbeton, Brussel, 33 pp.

Brandsma C., 1976. Invloed van het buitenklimaat op het ventileren van varkensstallen, Publikatie nr. 74 van het Instituut voor Mechanisatie, Arbeid en Gebouwen, Wageningen, the Netherlands, 37 pp.

Bruce J.M., 1975. The open ridge as a ventilator in livestock buildings, Farm Building R and D Studies, nr. 6, November 1975, 8 pp.

Bruce J.M., Ross P.A. and Burnett G.A., 1978. Protected open ridge design, Farm Buildings Progress, 53, July 1978, pp. 9-10.

Carpenter G.A., 1974. Ventilation of buildings for intensively housed livestock. In : J.L. Monteith and L.E. Mount (Eds), Heat Loss from Animals and Man, Butterworths, London, G. Britain, pp. 389-403.

Christiaens J. and Debruyckere M., 1977. Natuurlijke ventilatie in veestallen, Landbouwtijdschrift, 30 : 255-281.

Comberg G. and Hinrichsen K., 1974. Tierhaltungslehre, Ulmer Verlag, Stuttgart, W. Germany, 464 pp.

Debruyckere M. and Neukermans G., 1973. Algemene richtlijnen in verband met de klimaatregeling in gesloten stallen, Landbouwtijdschrift, 26 : 251-282.

Debruyckere M., Van Laken J., Christiaens J. and Van Der Biest W., 1982. Klimaatregeling in stallen, Ventilatie en Verwarming, Vereniging der Elektriciteitsbedrijven in België, Brussel, Belgium, 115 pp.

Flamand C., 1981. Thermische isolatie, akoestische isolatie, Brochure published by : Isoverbel N.V., Brussel, Belgium, 110 pp.

Holmes C.W. and Mount L.E., 1966. Effect of ambient temperature on heat loss, weight gain and water consumption of young pigs, Animal Production 8 : 363-368.

Hoorens J., Debruyckere M., De Moor A., Maton A., Oyaert W., Pensaert M., Vandeplassche M. and Vanschoubroek F., 1973. Ziekten, Voeding, Huisvesting van het varken, Story Scientia, Gent, Belgium, 496 pp.

Karle D., 1981. Kriterien der Stallklimagestaltung in Mastschweineställen, Landtechnik 36 : 336-340.

Mertens J.A.M. and Brandsma C., 1973. Het afzuigen van ventilatielucht in varkensstallen onder de roosters in de mestgang, Mededeling van het I.L.B., nr. 63, Wageningen, the Netherlands, 3 pp.

Mitchell C.D., 1972. Open ridge designs, Farm Buildings Progress, nr. 29, July 1972, pp. 15-18.

Owen J.E., 1981. The hybrid-recirculation ventilation system, Communications of C.I.G.R. Seminar Section II, Aberdeen S.F.B.I.U., Scotland, pp. 159-168.

Poncelet L. and Martin H., 1947. Hoofdtrekken van het Belgisch Klimaat, Koninklijk Meteorologisch Instituut, Brussel, Belgium, 265 pp.

Randall J.M., 1977. A handbook on the design of a ventilation system for livestock buildings using step control and automatic vents, Rep. nr. 28, Nat. Inst. Agric. Engng., Silsoe, England, 47 pp.

Sällvik K., 1979. Principles for mechanical exhaust ventilation systems in animal houses, Institutionen för lantbrukets byggnadsteknik Rapport 8, 93 pp.

Tol J.C., 1972. Lichtdoorlatende open nok-konstruktie in verbeterde uitvoering, Mededeling van het I.L.B., nr. 62, Wageningen, the Netherlands, 6 pp.

Waaijenberg D., 1984. Norm voor de constructie van agrarische bedrijfsgebouwen, Landbouwmechanisatie, 35 : 203-206.

Chapter 4

THE HOUSING OF CATTLE

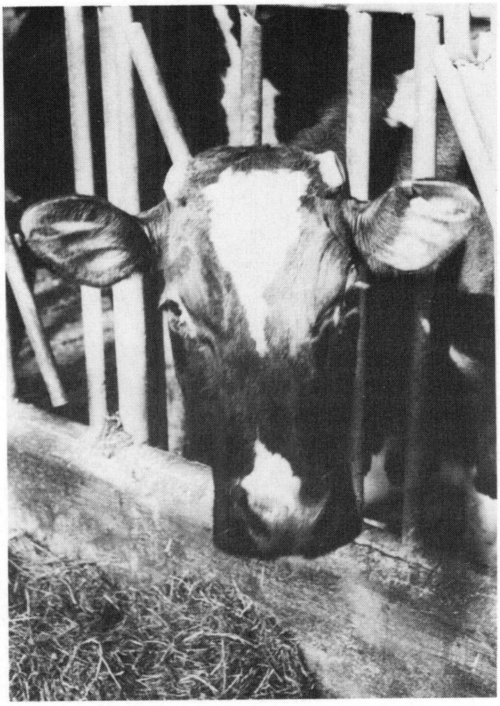

Chapter 4

THE HOUSING OF CATTLE

4.1 GENERALITIES

The construction and the equipment of cattle houses are closely related to the life-cycle of the animal and are based upon their physical characteristics. It is therefore useful to recapitulate briefly the most important biological characteristics of cattle upon which the different designs and dimensions of a cattle house and all its appurtenances are based. The characteristics mentioned are only guidelines since variability is inherent to all living species.

A newborn calf weighs 40 to 50 kg and is 1.00 to 1.20 m long. It can be kept as a future breeding animal or as a veal calf, which after ca. 18 weeks weighs ca. 200 kg, is ca. 1.50 m long and ca. 40 cm wide, or for baby-beef, which at the age of 12 months weighs ca. 480 kg, is 2.0 to 2.2 m long and ca. 60 cm wide, or for beef which after 18 months weighs ca. 500 kg, is then 2.0 to 2.2 m long and 60 cm wide.

A young animal weighing 200 - 300 kg is 1.7 m to 1.9 m long and 45 - 50 cm wide, at a weight of 300 - 400 kg it is ca. 2 m long and 50 - 60 cm wide. A heifer is either naturally or artificially inseminated when she is 1.5 to 2 years old, she delivers her first calf after 9.5 months and then starts to produce milk. She can calve each year. A cow of 500 kg is ca. 2.2 m long and 65 cm wide, at 600 kg she is ca. 2.3 m long and 65 cm wide and at 700 kg ca. 2.4 m and 70 cm wide. A cow is kept as long as possible for the production of milk and on an average she is slaughtered at 5 to 7 years because of a decreased milk production or for other reasons. With cattle breeds, such as the Holstein-Friesian, specialized dairy farms obtain an average milk yield of 6,000 l per cow and per annum, and more. A bull can normally be used for breeding after 18 months and can reach 1,000 to 1,400 kg.

Today the *low-level hangar type* of building, i.e. a hangar without an upper level, is generally used for cattle houses. For the stanchion barn this hangar can consist of a composite wall which carries the roof construction. The wall has then both a supporting as well as an insulating function. The hangar can also be constructed of trusses supported by pillars, whereby the walls are merely a filling between the pillars. In this case the pillars perform a supporting function while the walls have simply an insulating function. A suitable flooring completes the construction. Concerning the choice of building materials we refer to Chapter 3.

A loose house is generally not insulated and its walls consist of prefab panels or of blockwork up to a height of 1.80 m. The walls are built on a concrete plinth. Above the wall corrugated asbestos cement sheets or translucent corrugated plastic sheets are fitted to eaves level i.e. to a height of 2.5 metres.

4.2 THE HOUSING OF DAIRY CATTLE

4.2.1 The construction and equipment of cow houses

From the point of view of layout two main types of cow houses are found :

- the *stanchion barn* : where the animals are tied to a stall and have a limited freedom of movement ; a distinction can be made between a strawed stanchion barn and a stanchion barn with grids ;
- the *loose house* : where the animals, preferably dehorned, are allowed, within certain limits, to move around but are for certain activities (milking, sometimes feeding) restricted in their freedom of movement, a distinction can be made between the littered loose house and the cubicle house.

4.2.1.1 The strawed stanchion barn for dairy cattle

The strawed stanchion barn with either a long stall (2.0 m - 2.5 m) or a medium stall (1.7 m - 2.0 m) is still found on small, mostly older farms but is nowadays no longer built. The animals are often dirty thereby making a hygienic milk production impossible ; the straw consumption is much too high (4 kg/cow/day), as is also the labour demand, and the area per animal is too large.

New-built strawed stanchion barns are all of the short-stall type. Cow houses with one, two or more stall rows are found. New-built stanchion barns, housing at least 40 cows, are mostly of the two-row type. Fig. 4.1 shows the perspective of a double row face-to-face stanchion barn. This stanchion barn consists of :

- a *feed passage* : which is 2.7 m wide (2.4 m to 3 m) for a single row stanchion barn and 3.3 m (2.7 m to 3.9 m) for a double row stanchion barn. This allows the direct delivery of roughage by means of a tractor with forage trailer from the storage or the field (late crops) to the crib. A feed pass of 1.5 m (1.2 m to 1.8 m) is sufficient for manual feeding ;
- the *manger* has a width of 0.6 m (0.5 m to 0.7 m) and includes the kerb which is 0.2 m high and 0.1 m wide. It is placed lower than the raised feeding passage, whereby the feed distribution and the cleaning of the manger can be rationally carried out i.e. fast and without great effort ;
- the *stall* : is the place where the animals stand or lie and its dimensions are dependent on the breed. The length varies from 1.5 m to 1.6 m. The stall platform is preferably 3 cm (2 cm to 4 cm) lower than the manger in order to facilitate the intake of the feed (Anon., 1974a). According to Zeeb (1973) the stall platform can from an ethological point of view be up to 15 cm lower than the manger. This, however, leads to construction problems due to level differences with the raised feeding passage.

Between every two animals a partition made of tubes with a length of 0.80 m is provided. The stall has a width of 1.1 m per animal. From our research (Maton et al., 1978) it appears that the time the animals spend lying increases according to the increase of the width of the stall. Therefore the stall should certainly have a width of

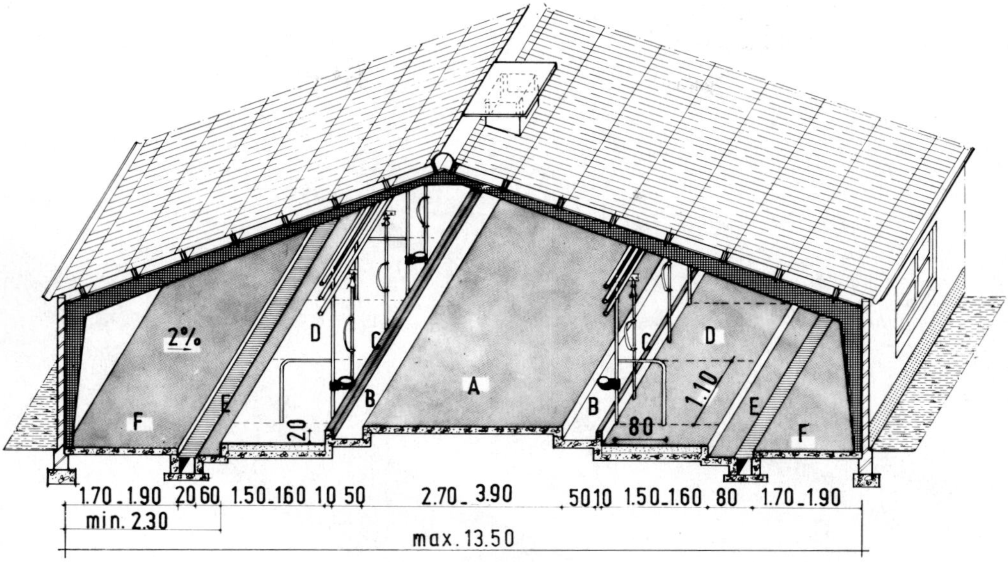

Fig. 4.1 The strawed stanchion barn.

Legend : A = feed passage ; B = manger ; C = kerb ; D = stall platform ; E = dung plate ; F = service passage.

1.10 m, whereby the cows have a satisfactory comfort and remain clean. A stall width of 1.20 m will, compared to one of 1.10 m, hardly or not improve the comfort of the animals but will increase the building costs (Maton et al., 1978). A cow which has a stall next to a wall cannot step aside and therefore this stall must have a width of 1.20 m. The stall is strawed with ca. 2 kg of straw per animal and per day.

- the *dung plate*, where faeces and urine are collected has for manual muck-out a width of 0.8 m including the 0.2 m wide gutter. The plate has a fall of 1 % towards the reception pit which, by means of a siphon, is connected to the slurry store. Its capacity is calculated to hold 0.5 m³ per cow and per month ;
- the *litter or service passage* is used for milking, littering and mucking-out. To facilitate help with calving the total width of the dung plate and litter passage must be at least 2.30 m (Anon., 1974a). The service passage has a fall of 2 cm per metre towards the dung plate.

Figure 4.1 shows the construction of a stall. On a foundation of rubble concrete, a layer of concrete or clinkers is applied (ca. 10 cm thick) which is covered with straw acting as an insulation.

For tying the cow chain and collar (fig. 4.2), the yoke tying of American design (fig. 4.3 and 4.4) and the vertical chain tying (fig. 4.5) can be applied. These types of yoke tying are often found today, which allow the tying and releasing of a batch of cows by

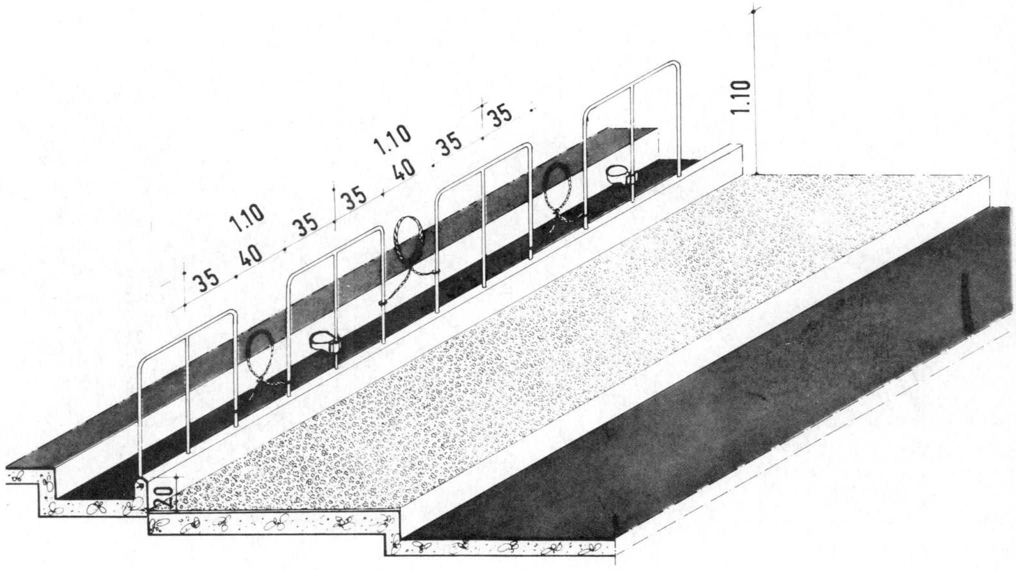

Fig. 4.2 Tying system with ordinary tie and double chain.

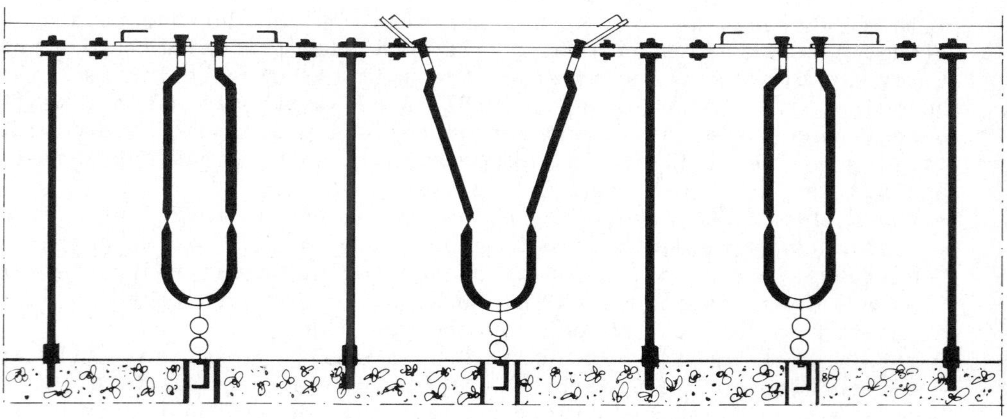

Fig. 4.3 The American yoke with articulated bars.

one simple manipulation and this by means of a mechanical transmission. It is thereby possible to keep one or more cows tied while the others are released with one movement. This group-tying is particularly applied at farms where the cows, during the summer, are brought to the

Fig. 4.4 The American yoke with articulated bars.

farm to be milked. This tying can also be done by the cows themselves. These self-catching yokes are an important labour saving but require a greater investment.

The underside of the vertical chain tie and the American yoke is attached to a hook in the floor, at 10 cm behind the kerb and recessed in the stall floor. With the vertical chain tying the chain is more and more replaced by a nylon strap which reduces the noise in the barn. The American yoke has to be of the articulated type which, unlike the original stiff yoke, allows the animals to rise and lie down smoothly (Maton and De Moor, 1975).

Slurry handling can be done mechanically by means of one of the following systems :

- the *dung scraper* (fig. 4.6) which by means of a cable and a winch driven by an electric motor is pulled over the entire length of the gutter and which is also economically justified for a small herd ;
- the *automatic dung plate cleaner* (fig. 4.7) : composed of hinged flaps, i.e. metal laths ca. 45 cm long, which are mounted on a steel pushing bar. This pushing bar with its flaps has a forward and backward movement produced for example by an electric motor equipped with an invertor. The flaps are transversely placed during the forward movement thereby pushing the manure towards the manure pit. During the idle motion the flaps are situated along the pushing bar. The dung channel, in which the flaps are moving, must be 2 cm wider than the length of the flaps ;
- the *endless chain type of automatic dung plate cleaner* is economically preferred in a barn with two or four rows instead of the push-type system. It moves over the dung plate which

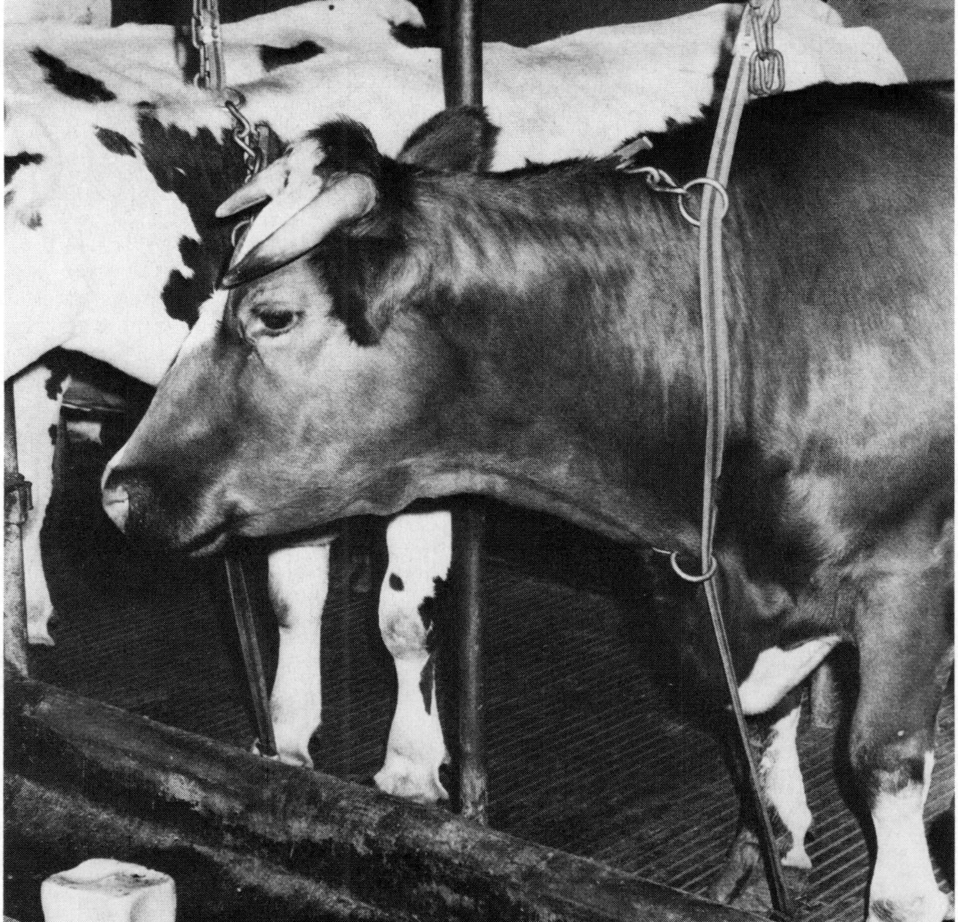

Fig. 4.5 Vertical chain tying with nylon strap and collar chain.

Fig. 4.6 A strawed stanchion barn equipped with a dung scraper.

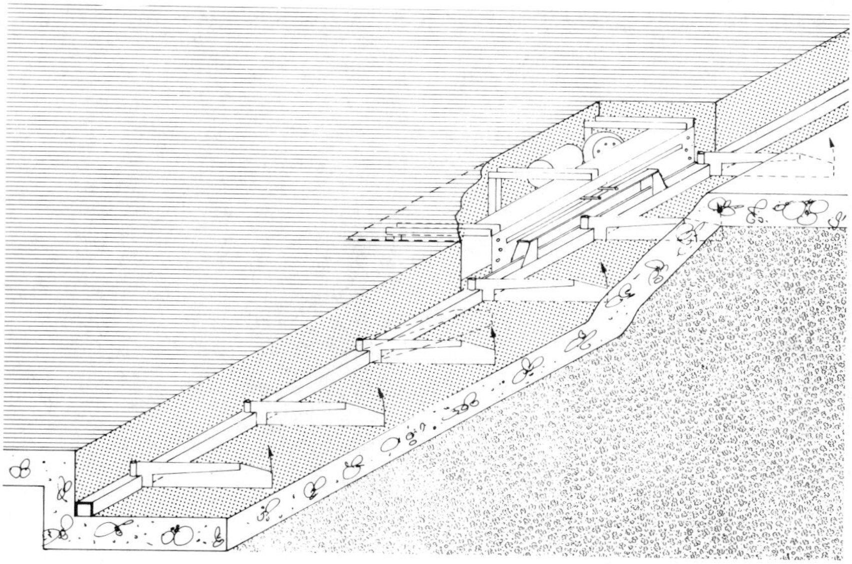

Fig. 4.7 The strawed stanchion barn equipped with an automatic dung plate cleaner.

has a width of ca. 40 cm and is provided with non-hinged flaps. Both types of automatic dung plate cleaners are equipped with an elevator gutter by means of which the manure is dumped on the dung stack outside the barn ;

- the *front loader* consists of a metal bucket mounted in front of a tractor and which is pushed over the dung plate. The dung plate and the litter passage are at the same level and have together a width of 2.3 m if this alley abuts against the wall and 2.7 m for a central service passage. The liquid dung gutter is sunk into the floor and is situated 1 m behind the stall.

The *dung stack* consists of a concrete apron, 15 to 20 cm thick, upon which the farmyard manure from the stall barn is piled up. This apron has a fall of 1 to 2 % towards the gutter which via a siphon is connected to the slurry pit. The area of the dung yard is estimated at 0.5 m^2 per cow and per month storage. If the highest groundwater level is low enough, the concrete plate is made 1 m below ground level and this will contribute to a larger storage capacity. It is advisable to provide a wall along three sides of the dung yard.

The *drinking water supply* is provided by drinking bowls (one per two animals) which are automatically filled when the animals press the lever (fig. 4.8), since they are connected by pipes to the main or to a private water distribution installation. According to the valve closure system two different types of drinking bowls can be distinguished : the valve is activated either by a counterweight (fig. 4.8) or a spring (fig. 5.19). The upperside of the drinking bowl must be at a height of 60 cm above the stall platform otherwise fouling of the water may occur ; this can quickly put cows off drinking.

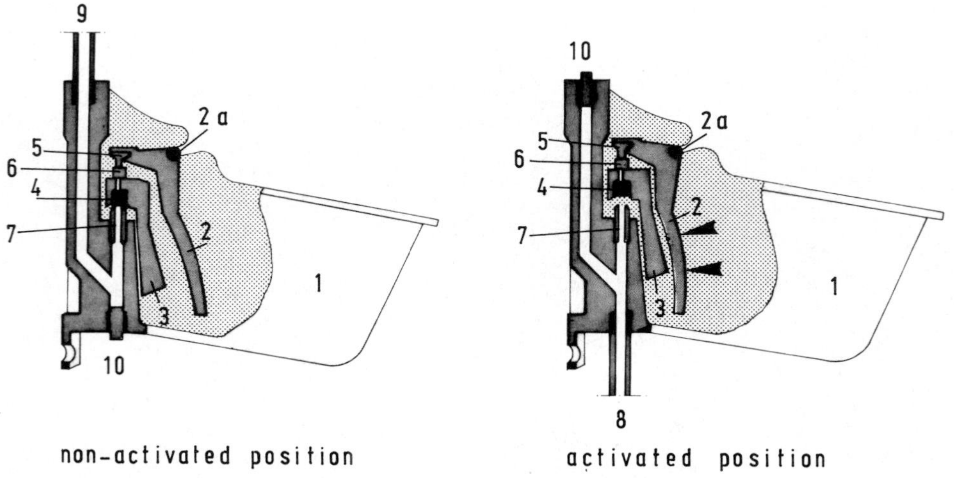

Fig. 4.8 The drinking bowl for dairy cattle.

Principle of operation : the water enters either through 8 or 9 and can only leave via a small tube provided at its end with an orifice 7. In the non-activated position (left drawing) the admission of water to the bowl 1 is blocked by a valve 4 which is pressed on the orifice 7 by means of a counterweight 3. The lever 2 is attached to the counterweight 3 by means of a rod which is provided with an adjusting screw 5 and a set screw 6. The lever 2 revolves over point 2a and when pressed by the cow it will force the counterweight upwards thereby allowing water to flow through the orifice 7 to the bowl and this as long as the lever is activated (right drawing). A cap plug 10 shuts the unused entrance of the water supply tube 8 or 9.

Milking takes place in the barn, where a vacuum pipeline (and often a milk pipeline) is provided at a height of 1.6 to 1.8 m above the kerb (figs 4.9, 4.10 and 4.11). A vacuum pump VP, driven by an electric motor M, extracts air from the pipeline in the barn (bucket milking machine : 50 l/min + 60 l/min for each cluster ; pipeline milking installation : 150 l/min + 60 l/min for each cluster Cr). This vacuum pipeline is equipped with an interceptor for condensed water I, a vacuum gauge G and a vacuum regulator R. The vacuum pipeline is connected either to the bucket B via vacuum taps T and by means of a vacuum tubing Vtu, or to the milking pipeline. The pulsator system P provides alternately vacuum (50 kPa) and atmospheric pressure, admitted via an airbleed in the claw Cw, to each teat cup shell t via the pulse tubes Ptu and this at a constant rate and ratio. The ratio can sometimes be adjusted from 1:1 to 3:1. In fig. 4.9 at the left a vacuum exists in the pulsation chamber and a flow of milk is obtained, at the right air is admitted between teat cup shell and liner and this through an orifice in the claw : massage of the teat. The pulsation rate varies between 44 and 60 and depends on the ratio suction stroke-rest stroke.

A widely used suction-rest ratio of 60:40 corresponds with a pulsation rate of ca. 52 pulses/min. One pulsator per milking cluster is provided, although during the last few years a tendency exists towards the installation of one central pulsator for all clusters in use. The milk flows through short (s) and long (l) milk tubes to standing or suspended milking buckets B or through a milking pipeline to a milk cooling tank.

This milk cooling tank, which is almost always installed in an adjacent room (the dairy), makes it possible to store the milk under favourable conditions (< 5°C) for two or even three days. The required capacity of the cooling tank can be calculated from the following formula (Anon., 1974b) :

$$C = 12 \times n \times m$$

where C = minimum content of the milk cooling tank in litres ; n = number of milk producing cows ; m = number of milkings, of which the milk has to be stored (4 to 6).

The working principle of a refrigerated farm bulk tank is represented in fig. 4.12. The refrigerant commonly used is dichlorodifluoro-

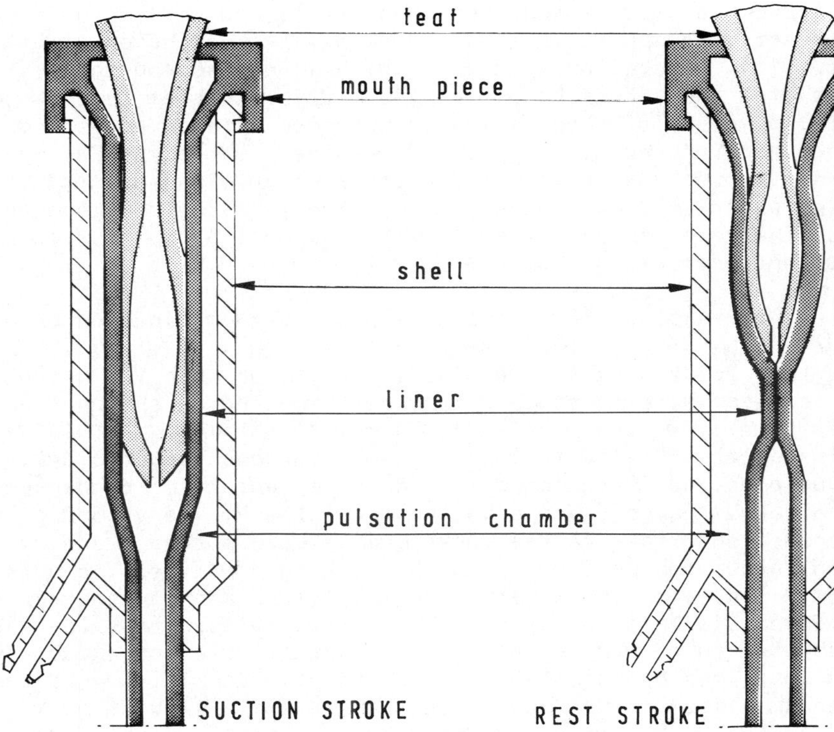

Fig. 4.9 The working principle of a milking machine with its different parts.

methane, known alternatively as Freon 12. This refrigerant flows under pressure from the condenser to the expansion valve. This valve allows the refrigerant to expand from a high pressure to a suitable low pressure whereby the refrigerant is easily evaporated. This process takes place in the evaporator and requires heat. The evaporator extracts the required heat from the immediate surroundings thereby creating the desired cold zone. The vapour, transporting the heat absorbed from the milk, is drawn into the compressor and is compressed to a high pressure consequently producing heat again. From the compressor the cycle is continued to the condenser where the heat, produced by the evaporator and the compressor, is released to the surroundings and whereby the refrigerant condenses back to liquid. The environment into which heat is released is generally, and certainly for small installations, the air, which is then drawn over the condenser by means of a small fan. Since the heat of the condenser is released into the atmosphere it will cause an increased temperature of the air and this has a double disadvantage : not only the ambient temperature in the room where the milk tank is installed rises but

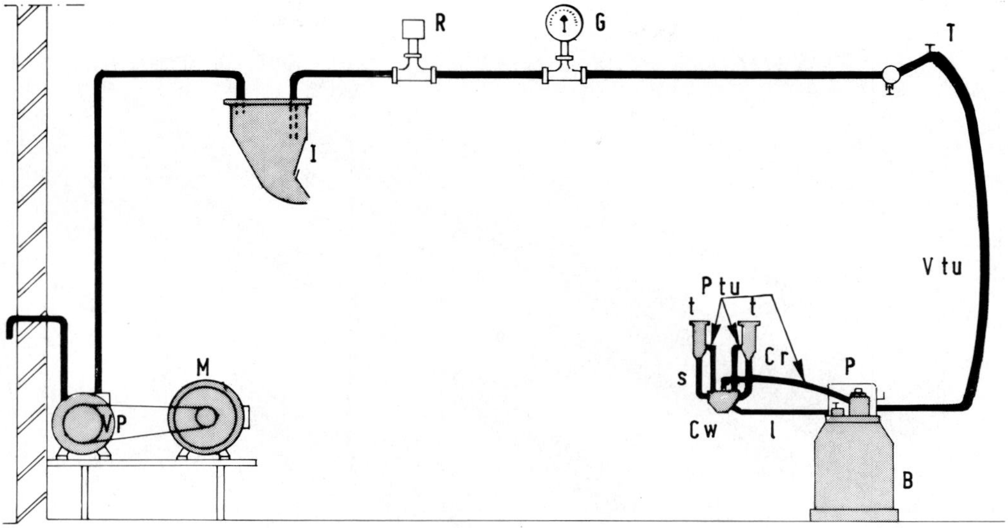

Fig. 4.10 The different parts of a milking installation used with a bucket unit.
Legend : VP = vacuum pump ; M = electric motor ; I = interceptor of condensing water ; R = vacuum regulator ; G = vacuum gauge ; T = vacuum tap ; Vtu = vacuum tubing ; B = bucket ; P = pulsator ; Ptu = pulsator tubing ; Cr = cluster ; t = teat ; Cw = claw ; s = short milking tube ; l = long milking tube.

also the temperature of the air used for cooling the condenser is increased. The efficiency of the cooling equipment depends largely on the good working (temperature) of the condenser. The condenser should preferably be supplied with air as cold as possible. It is also a good practice to install the milk cooling tank in an amply ventilated room.

Two types of refrigerated farm bulk tanks exist : the direct cooling or direct expansion type of tank and the indirect cooling or the ice building chilled water bulk tank. With direct cooling the evaporator is installed at the bottom and in a part of the sides of the cooling tank. In this way the milk comes in direct contact with the evaporator whereby a direct heat transfer is created from the milk to the cold source. The transfer of heat with an indirect cooling system takes place by means of an ice bank. The bulk tank is either immersed in an ice-water bath or ice-water is being pumped and sprayed on the walls of the vessel (sump and spray type of refrigerated milk tank).

The time required for cooling milk is less with an indirect cooling tank than with a direct expansion milk cooling system. The ice-bank allows an immediate cooling, while in a direct expansion system the

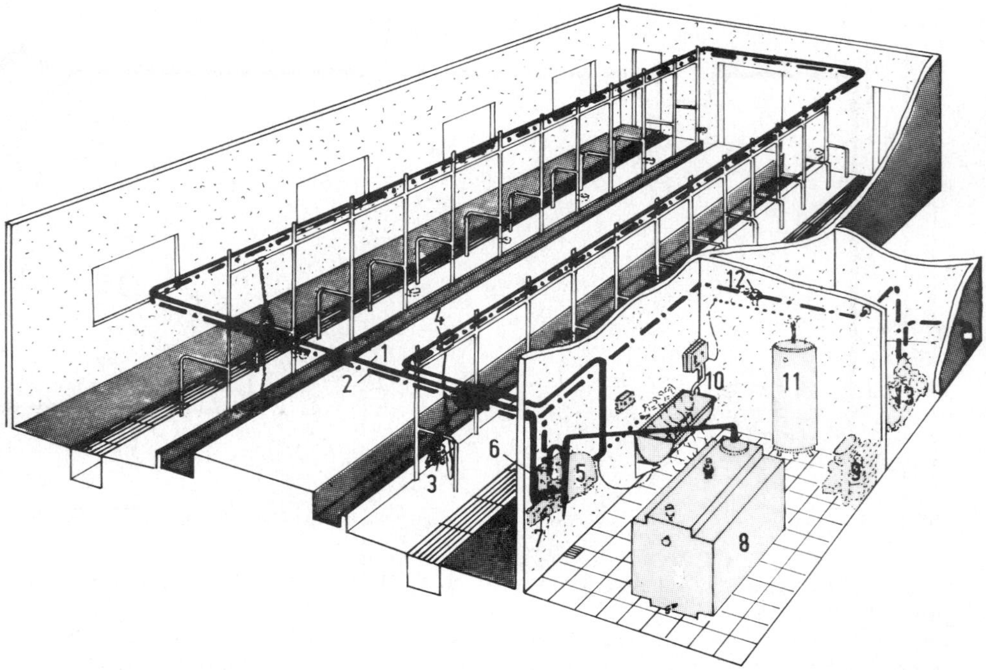

Fig. 4.11 A milking installation with milking pipeline.
Legend : 1 = milking pipeline ; 2 = vacuum pipeline ; 3 = milking unit ; 4 = pulsation unit ; 5 = releaser ; 6 = overflow protector ; 7 = milk pump with electric motor ; 8 = milk cooling tank ; 9 = cooling unit of the milk tank ; 10 = cleaning installation ; 11 = boiler ; 12 = vacuum regulator ; 13 = vacuum pump.

condensing unit only starts to function at the beginning of the milking and further only works now and then for maintaining the milk at a temperature below 5°C during storage. Since the available period for ice building extends far beyond the milk cooling period, a smaller refrigeration unit is required with the ice building chilled water bulk tank than with direct expansion. The electricity consumption however is higher than with a direct expansion cooling system viz. an average of 22 Wh per litre versus 15 Wh per litre (Petit and Van Der Biest, 1980). An indirect cooling tank, with chilled water, can however operate during the night (ca. 75 % of its working time) and can therefore take advantage of the low-cost night tariff electricity. On the contrary a tank for direct cooling functions 90 % of its working time during the day at the higher cost electricity tariff. Gradually more use is being made of heat recuperation with the cooling of milk (see later).

The automatic, programmed cleaning of the milking equipment by means of hot water (95°C) and detergents (e.g. sulphamic acid) is more and more applied : it allows a perfect and thorough disinfection

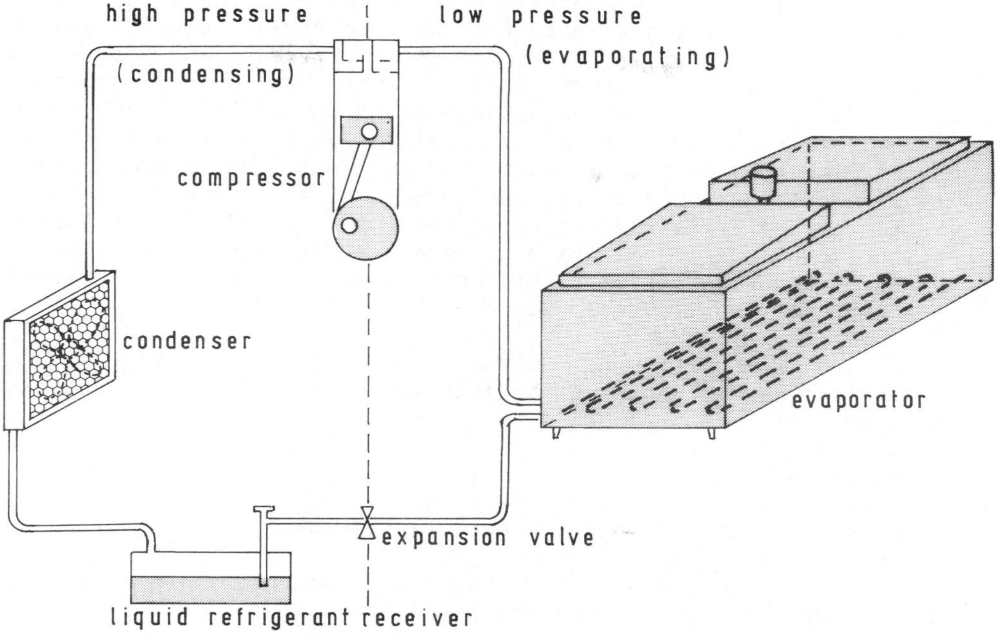

Fig. 4.12 The working principle of a refrigerated farm milk tank.

of milking machine and pipeline installations in a short time (ca. 5 min). Farm bulk milk tanks are also cleaned often and are preferably provided with an automatic cleaning installation.

The *natural lighting* of stanchion barns is provided by windows which are often equipped with a ventilation threshold. A window area at the rate of 1/15th of the floor area should furnish ample light.

The *artificial lighting* is provided by means of incandescent lamps (13 lumens per watt) or preferably by fluorescent lamps (40 lumens per watt) which are installed above the feeding passage (feeding) and the service passage (milking). The lighting of the service passage or milking stalls is rated at 120 lux per m^2 (i.e. ca. 3 W/m^2 with fluorescent lamps and ca. 9 W/m^2 with incandescent lamps) and for the feeding passage at 30 lux per m^2 (i.e. ca. 0.8 W/m^2 with fluorescent lamps and ca. 2.3 W/m^2 for incandescent lamps) (Debruyckere and Neukermans, 1973).

The *ventilation* (see also chapter 3) is mainly carried out by means of stacks (air outlets) and ventilation thresholds at the bottom of the windows (air inlets). Ventilation can also be done by an open ridge or - seldom - mechanically. The electric installation in the barn must be waterproof.

In the *two-row stall barn* two arrangements are possible : either the cows are facing-in or facing-out. In the facing-in arrangement one single wide feeding passage is provided while in the other arrangement one single service passage is found. The facing-out arrangement - which has the advantage that the walls are less befouled and that the animals can more easily be let into the barn - is sometimes used with the gradually less applied (noise, exhaust gases) mechanical mucking-out with tractor and front loader and with summer milking. Preference however is given to the facing-in arrangement with a wide central feeding passage along which the feeding can quickly and easily be carried out by a tractor with feed trailer (fig. 4.1). Since littering and mucking-out are laborious and rather unpleasant tasks and because mechanical mucking-out of a littered stanchion barn is not completely satisfactory (a delicate mechanical installation susceptible to wear and which is often rather expensive) this type of barn is nowadays less built and strawless housing is preferred : one type is the strawless stanchion barn with grids.

4.2.1.2 The stanchion barn with grids for dairy cattle

In this type of housing the following main parts can be distinguished :

- the feed passage ;
- the manger ;
- the stall ;
- the grids with the underlying slurry channel ;
- the service passage.

For the *feed passage* and the *manger* we refer to the description given for the littered stanchion barn.

The dimensions and the design of *the stall* together with the grids are of great importance for the achievement of a good barn : negative results have been encountered (lesions of claw and teat) in our country and elsewhere mainly due to a wrong design or construction.

According to our research (Maton and De Moor, 1973) and guidelines issued by the Belgian Ministry of Agriculture (Anon., 1974a) the length of the stall for black-and-white and red-and-white cattle has to amount to 1.45 m and the width of the stall to 1.10 to 1.20 m with a partition made of tubes between every two animals. The tying system has to enable an easy rising and lying of the animal. Mainly the vertical tying, mounted in such a way that it can be adjusted at the top (fig. 4.13) and the American articulated yoke are taken into consideration. The tying system is to be anchored in the floor 10 cm behind the rounded kerb (which has a height of 20 cm). The stall floor shall form a comfortable lying area for the animal and consists of a rubber mat (20 mm thick) laid on an underlayer of concrete of circa 10 cm. Behind the stall and at the same level are the *metal grids* which cover the slurry channel.

The grids are usually made of galvanized iron ; cast iron grids are also used but they are more fragile. The grids consist preferably of three flat bars in the front with a T-profile, having a width of 2 cm (to support the hoofs) and which are 3.5 to 4 cm apart, followed by a number of round bars (Ø 1 cm) with a spacing of 3.5 to 4 cm (allowing the dung to fall through). The bars are supported by a number of cross bars. Grids composed of rubber covered bars are also satisfactory (fig. 4.14).

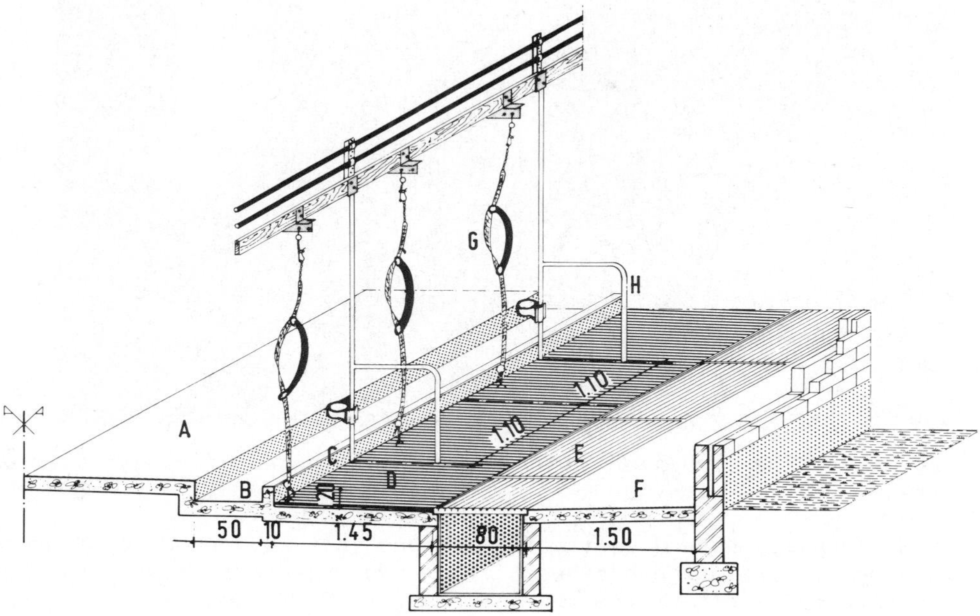

Fig. 4.13 The stanchion barn with grids.
Legend : A = feed passage ; B = manger ; C = kerb ; D = stall ; E = grids with underlying slurry channel ; F = service passage ; G = tying system ; H = separation, made of tubes.

The *slurry channel* is made of rendered brickwork and has a rectangular or quadrangular section. It is designed in such a way that no flushing or cleaning work is required. The fall of the bottom of the slurry channel is zero or 1 % in the opposite direction (away from the discharge opening). At its end the bottom of the channel is provided with a weir or lip, with a height of 20 cm, over which slurry continuously overflows towards the slurry reception pit. Before putting the barn into use, the slurry channel is filled with water up to the height of the weir or the lip in order to prevent crusting or adhering of slurry to the bottom and to permit a free flow of slurry at the beginning. The upper surface of the slurry has a slope of 1 - 2 % whereby the slurry reaches the lowest level at the discharge opening. Above the weir or lip a rubber flap is suspended which acts as an odour barrier between the slurry cellar and the barn. Sometimes a slide is additionally provided over the entire section of the discharge opening which can then be closed off during mixing and emptying of the slurry cellar in order to prevent poisoning of the animals due to evolving noxious gases. Taking into account that commercially available

Fig. 4.14 The stanchion barn with grids composed of rubber covered bars.

grids normally have a width of 80 cm and that they require an overlapping of 3 cm at both sides of the cement rendered slurry channel, the internal width of the channel ought to be 74 cm. Since the upper surface of the slurry shows a slope, the depth of the slurry channel will depend on the length of the channel ; the depth of the channel shall be at least 50 cm for a 10 m channel and 80 cm for a 20 m channel ; its length however should not exceed 25 m.

It can sometimes be necessary to clean the grids : this can best be done with a hard rubber broom specially designed for grids and without flushing. The slurry channel leads into the *slurry cellar*. A production of 1.5 m^3 slurry per cow and per month storage is taken as an average. The emptying of the slurry cellar can be accomplished in three ways. A vacuum tank spreader equipped with an air pump can be used which creates a vacuum in the tank whereby the slurry is lifted from the cellar. The same air pump pressurizes the tank enabling the spreading of its contents onto the fields. The pump will also provide the necessary agitation. This method is widely propagated, especially where no straw and/or hay is present in the slurry.

In case hay or straw, or both, are present in the slurry the use of a submerged shredding and mixing pump is indicated. This pump is mounted on a trolley or on the three-point lift of a tractor. The pump is submerged in the slurry and can, by switching a valve, be brought into the agitating or pumping position. Since the propeller acts as a chopper in the wide suction mouth, the presence of hay and/or straw forms no obstacle for the pump. This pump is not installed on the slurry tanker. The tank trailer is, in this case, equipped with a horizontal axle driven by the power take-off of the tractor. This axle is, at its extremity, provided with a number of propeller blades along which the slurry is forced outwards through a slit which can be closed. Finally the use of a displacement pump can be mentioned although this system is no longer used and has disappeared almost completely. Today it is also possible to store the slurry in metal or concrete silos above the ground instead of in under-ground slurry cellars. Fig. 4.15 represents such an installation, whereby the slurry is collected in a reception pit containing a pump which agitates and pumps the slurry to a silo. Recycling the slurry is possible since the silo is connected by means of a return-channel (with slide-valve) to the reception pit. The diameter of the slurry silo shall not exceed 15 m in order to obtain a sufficient mixing of its contents.

The *drinking water supply*, the *milking*, the *ventilation*, the *natural and artificial lighting* of a strawless stanchion barn with grids are accomplished in a similar way as described for the strawed stanchion barn.

4.2.1.3 The littered loose house

This type of housing was imported from the U.S.A. in the beginning of the Fifties. We distinguish the open and closed loose house according to whether the animals are allowed outside or remain indoors. This type of house consists of the following units :

- the *bedded court* : is a littered area situated 75 to 100 cm under the level of the feeding passage ; in this area the manure pack is allowed to build up and is removed once or twice per year, which explains the necessary difference in level between the resting area and the feeding passage ;
- the *feeding area* : is either a part of the resting area where the animals stand and take their roughage from the crib through a feed fence or, in a later stage of development, a paved area for the same purpose connected to the resting area. Sometimes a slatted floor with slurry channel was provided but this led to difficulties when removing the straw-rich slurry from the channel (fig. 4.16 lower drawing) ;
- the *feeding passage* : is situated in front of the crib and connects directly to the straw and hay storage areas. Self-feeding of silage and bunker-feeding of hay are also possible (see later) ;
- the *collecting or holding area before milking* : (1.1 to 1.3 m^2/cow) is mostly a concrete area to where the cows proceed and wait to be milked. It can be separated from the rest of the house by means of a chain or a gate ; the already milked cows are similarly separated from the cows waiting to be milked ;
- the *milking parlour* : where the cows are milked and receive concentrates, is inseparably connected to the loose house and will be discussed in detail with the cubicle house ;
- the *dairy* : situated beside the milking parlour and containing the milk tank ;
- in some cases the *exercise area or paved lot* : which is a confined site outside the house with a concrete floor where the cows can get fresh air and exercise but which requires a lot of cleaning ;

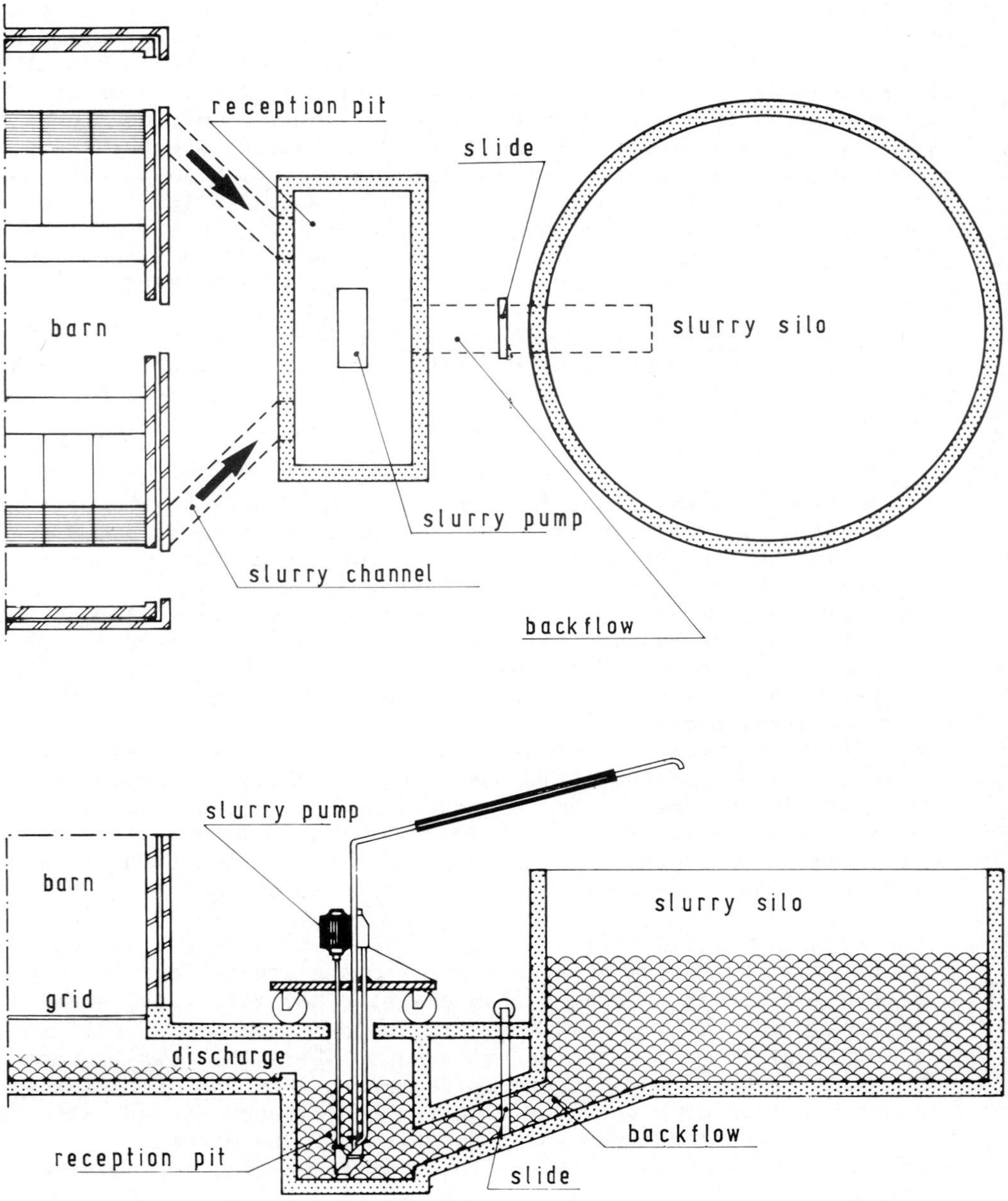

Fig. 4.15 The storage of slurry in a silo.

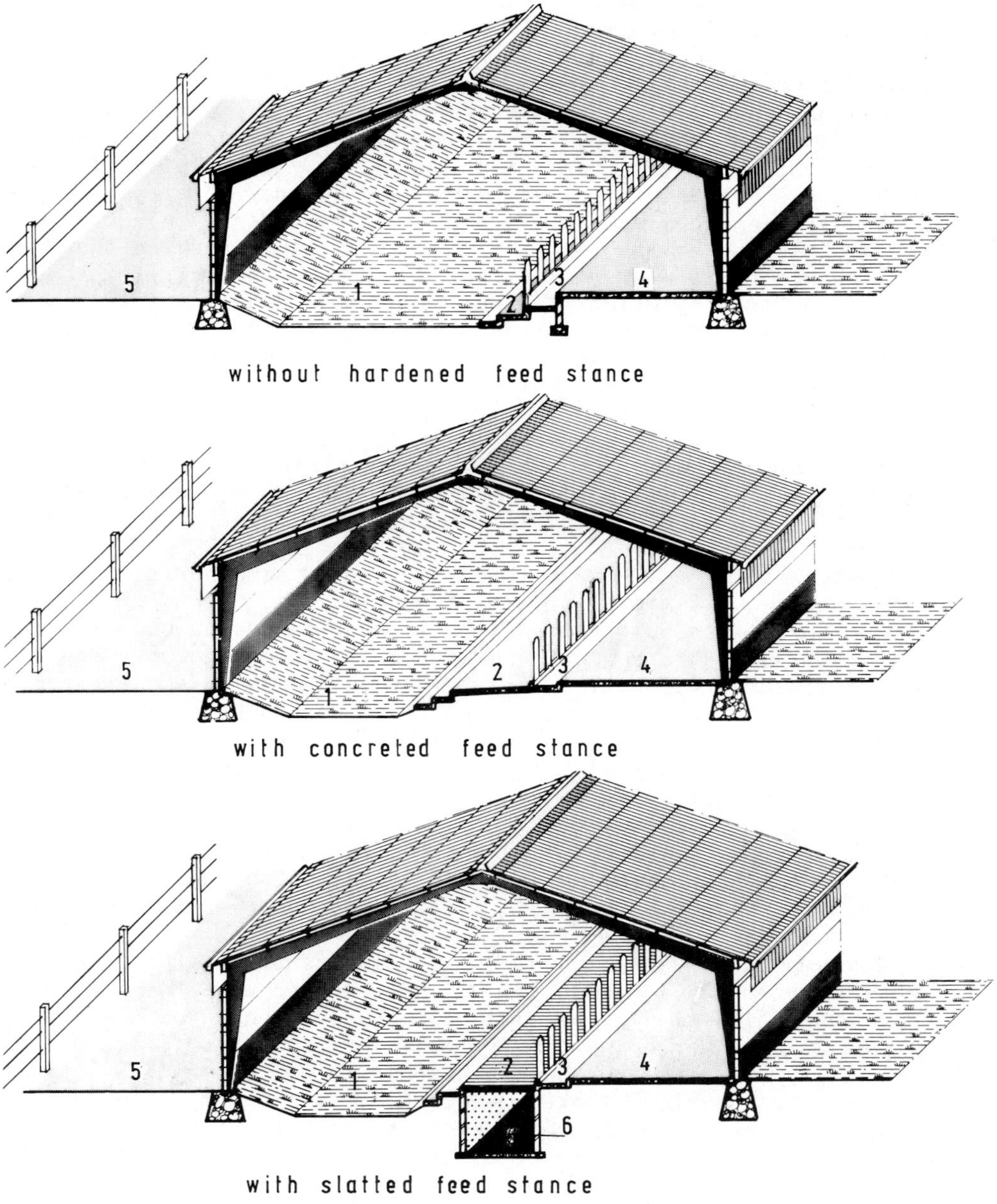

Fig. 4.16 The littered loose house.
Legend : 1 = bedded court ; 2 = feeding area ; 3 = manger ; 4 = feeding passage ; 5 = exercise area ; 6 = slatted floor with slurry channel.

- the *isolation pen* : for calving and sick cows (mainly a stanchion barn).

In such a loose house a floor space was allowed at the rate of ca. 10 m² per animal (without crib or feed pass) in the absence of an exercise area and 5 to 6 m² per animal if this area was present and in turn occupied 5 to 6 m² per animal.

The main obstacle to this type of loose house, which has not been widely adopted in Europe, is the considerable bedding requirement at the rate of 9 to 10 kg per animal and per day in a fully littered loose house and 5 to 6 kg per animal and per day in a loose house with a feeding area and an exercise area provided with a concrete floor. In areas of mainly pasture-land but also elsewhere the specialized dairy farms were confronted with the necessary purchase, storage and distribution of large amounts of high-cost straw which had a negative influence on the profitability of the loose house. This and other factors have led to a further evolution of the strawed loose house : the cubicle house.

4.2.1.4 The cubicle house for dairy cattle

The cubicle houses are, for new constructions, increasingly replacing the stanchion barns. In the Netherlands for example about 20,000 cubicle houses were counted in 1983 of which ca. 7 % were feed cubicle houses.

The *cubicle house* has more or less the same parts as the strawed loose house but differs in form. The lying area is composed of individual boxes, the number of which is equal to the number of animals (fig. 4.17). Single row cubicles are 2.3 m long. A double row of cubicles measures 4.3 m. They are 1.15 to 1.20 m wide and are separated from each other by means of a metal framework. These partitions exist in a large variety of forms. The partitions represented in figs 4.18 and 4.19 are the "classic" separations between cubicles and are made of a framework of metal tubes. These types were largely used in earlier days.

Nowadays the so-called "comfort-cubicles" have become popular. Those cubicles of which a number of forms are shown in figs 4.20 and 4.21 provide somewhat more comfort to the animals when they are lying down.

Between two rows of cubicles a bar is installed at a height of 50 cm (see figs 4.17 and 4.19) which, together with the headrail prevents the animals from leaving the cubicles. The headrail is a tube placed at shoulder height in front of the row of cubicles and which can be adjusted forwards and backwards (on average ca. 1.70 m from the rear of the cubicle). Thereby the animal is obliged, when rising, to step backwards so far that it defecates on the service passage and not in the cubicle. The level of the cubicle is therefore raised 20 cm above the service passage.

A special type of cubicle is the *feed cubicles* (fig. 4.22) : the animals take their feed, standing in the cubicle, from a manger. The headrail then acts as a feed fence. This type of cubicle is sometimes applied when a narrow stanchion barn is converted into a cubicle house.

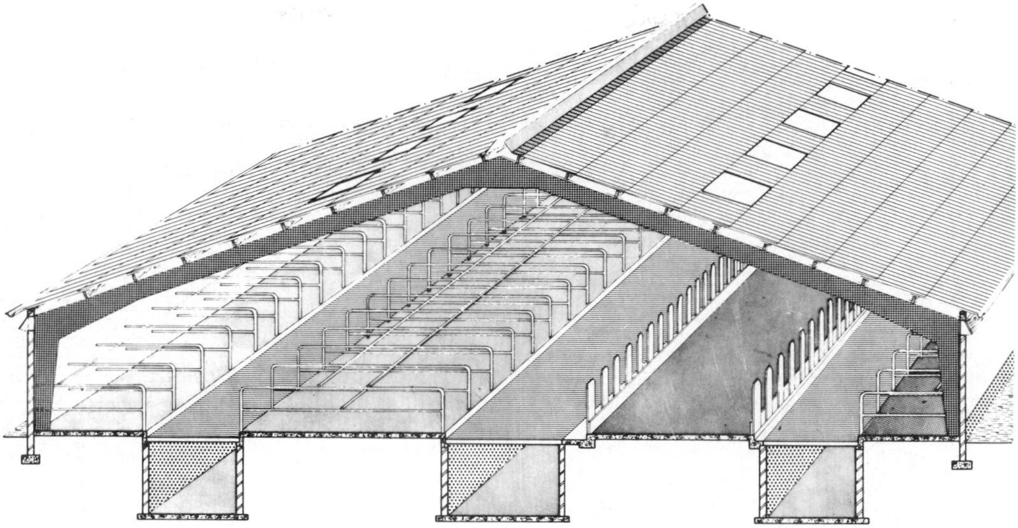

Fig. 4.17 Cubicle house with slatted passages and Swedish feeding fence for dairy cattle and accompanying young stock.

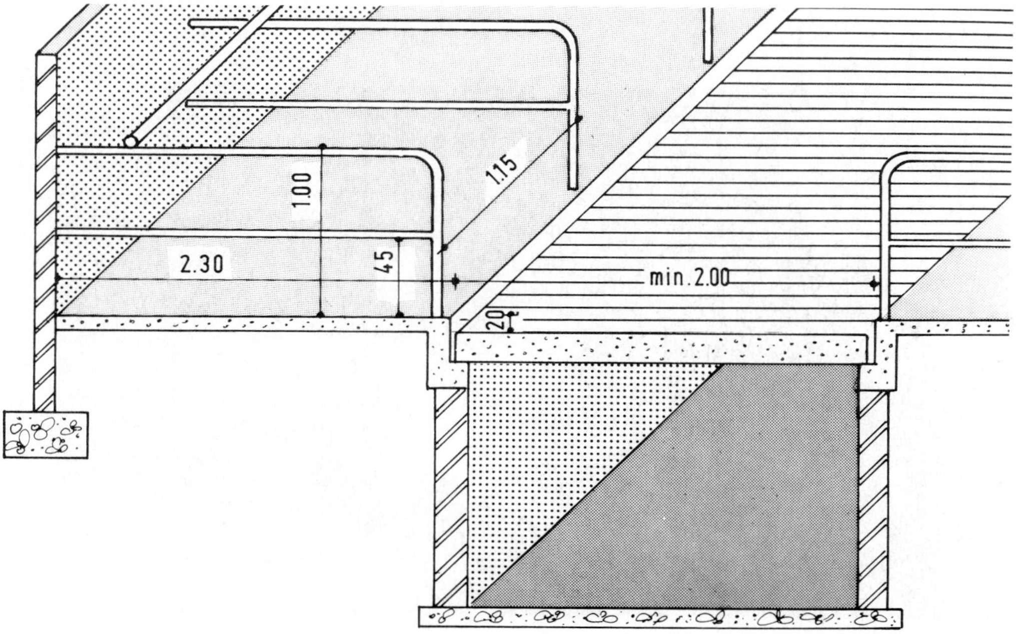

Fig. 4.18 A single row of "classic"-cubicles.

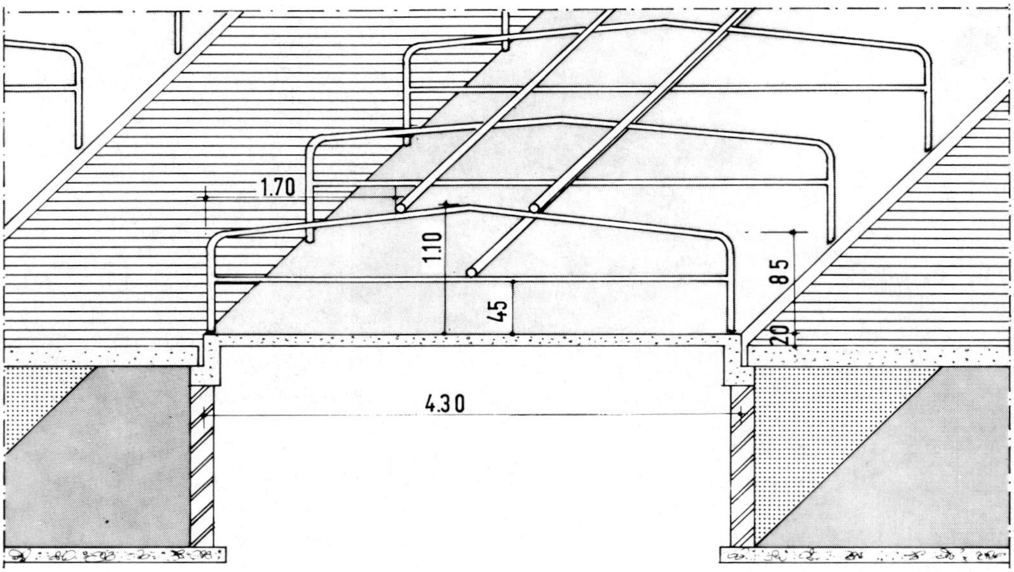

Fig. 4.19 A double row of "classic"-cubicles.

Fig. 4.20 "Comfort-cubicle" with a U-shaped partition made of tubes.

Fig. 4.21 "Comfort-cubicle" of the so-called R-type.

Fig. 4.22 Feed-cubicles.

The *flooring of the cubicles* can sometimes pose problems. If the floor of the cubicle is made of compacted earth, pits will inevitably be formed after some time owing to extensive usage. The bumpy soil forms an uncomfortable lying area for the animal, it rests against the heelstone of the cubicle and often gets swellings on the ischium or otherwise refuses the cubicle and rests in the passage where the udder can become foul. For this reason and in spite of the higher investment cost, the cubicle floor is often made of a 10 cm concrete layer, littered with sawdust or sand. The bedding with sawdust creates on one hand a comfortable lying area, preventing animals of lying elsewhere than in the cubicles and on the other hand absorbs the faeces which adheres to the hoofs, thereby contributing to a drier and cleaner bed. It is also possible to apply different insulating or non-insulating top layers on the concrete floor (cement-sand screed, Bernit, Stallit etc.) or 2-cm thick mats can be fitted (rubber mats, Enkamat-K etc.).

From our research (Maton et al., 1981) it seems that the cows show a distinct preference for a certain type of lying area. Soft floors are significantly more occupied than hard floors. The cows preferred significantly an Enkamat-K carpeted floor above all others. The cows make no systematic difference between hard floors whether they are insulated or not. The earlier opinion that cubicles should be provided at their rear with a vertical wooden board of ca. 15 cm high preventing animals from entering the cubicles backwards and soiling the beds was also investigated by us (Maton et al., 1981). The use of such a wooden board or lip clearly troubled the cows and exercises a negative influence on the degree of occupation of the cubicles and on the behaviour of the animals and should therefore be discouraged. A similar conclusion was reached by Westendorp and Hakvoort (1977). If possible, the animals generally return to the same cubicles, whereby the hierarchical order is of great importance. The most dominant animals clearly appropriate the most comfortable cubicles.

As a general rule the number of cubicles equals the potential of cows to be housed. In loose houses, designed for large herds, more animals are sometimes kept than if there were cubicles. From our unpublished investigations it became clear that a temporary overcrowding of 5 to 10 % does not impede the normal functioning. Wieringa (1982) found that an overcrowding of 25 and 55 % led to a reduction of the lying-in time of 4.7 and 11.2 % respectively. A substantial overcrowding leads to an increased number of animals, especially the lower ranking ones, lying down in passageways. Wieringa (1982) therefore advocates care when overcrowding a cubicle house.

At the rear of a row of cubicles or between two rows of cubicles (distance 2 to 2.5 m) a concrete passage or a slatted floor is provided. In the first case the dung has to be scraped from the floor once or twice a day (fig. 4.23). This can be accomplished by means of a dung scraper mounted in front of or behind a tractor. The dung is scraped either towards the dung pit or the slatted floor area. This work can be and is preferably replaced by the installation of a scraper driven by an electric motor, or by the use of a folding scraper (fig. 4.24). Since the folding scraper moves slowly (± 3 m/min), the animals are allowed to step over it and its silent operation will not disturb them.

Fig. 4.23 Tractor with scraper mounted in the rear.

Fig. 4.24 A cubicle house equipped with a folding scraper.

The installation of *slatted floors* will however exclude all labour and maintenance and is the best solution (fig. 4.17). The slats are made of reinforced concrete. A profile of a slatted floor is represented in fig. 4.25. The upper surface of the slats has a width of ca.16 cm and is roughly finished (non-slip). The gaps are 4 cm wide. Slats up to a length of 3 m are commercially available and can be installed without intermediate support (± 10 cm overlapping). Under the slats a slurry channel is constructed which at the same time acts as a slurry cellar (1.5 m^3 slurry per cow and per month of storage).

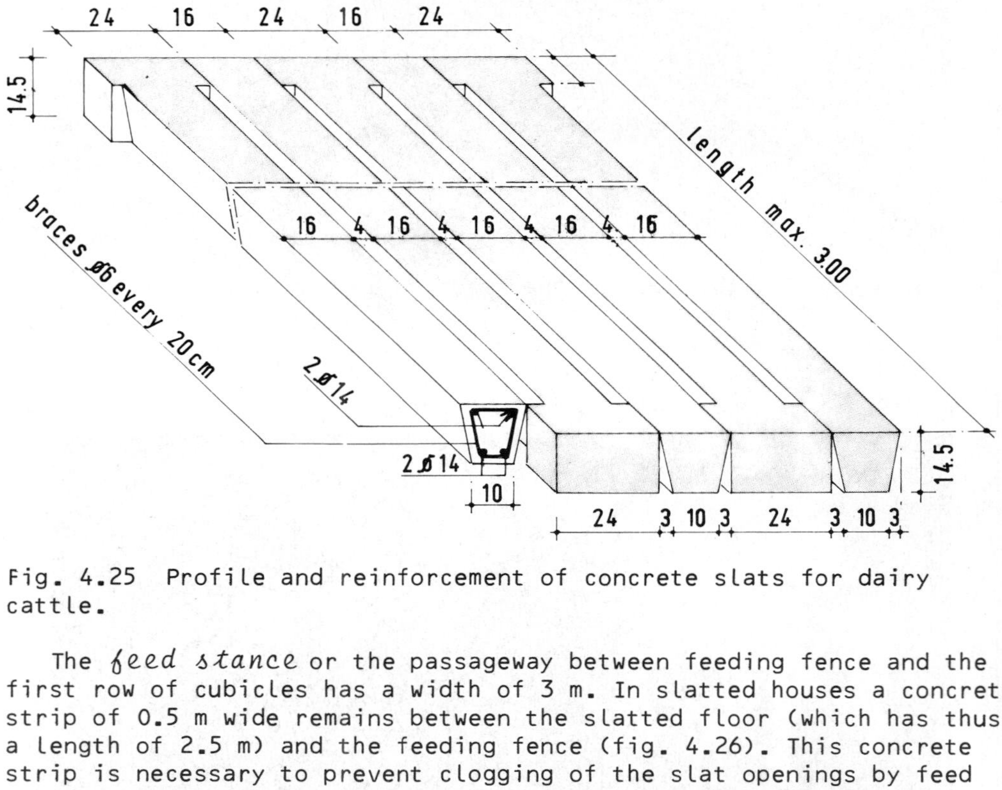

Fig. 4.25 Profile and reinforcement of concrete slats for dairy cattle.

The *feed stance* or the passageway between feeding fence and the first row of cubicles has a width of 3 m. In slatted houses a concrete strip of 0.5 m wide remains between the slatted floor (which has thus a length of 2.5 m) and the feeding fence (fig. 4.26). This concrete strip is necessary to prevent clogging of the slat openings by feed residues and at the same time it provides a better support to the front feet of the animals during feeding.

The *feeding fence* forms the separation between the feed passage with the manger and the area occupied by the animals. The Swedish feeding fence (figs 4.26 and 4.27) was largely used in the past. It was made of timber or metal tubing and in the last case a closing system was sometimes provided above the feeding fence ; in the downward position the animals were refused access to the crib while in the upward position the animals were allowed to take their feed from the crib. Nowadays a self-closing feeding fence is mostly used whereby the animals fence themselves in and are freed by manipulating a

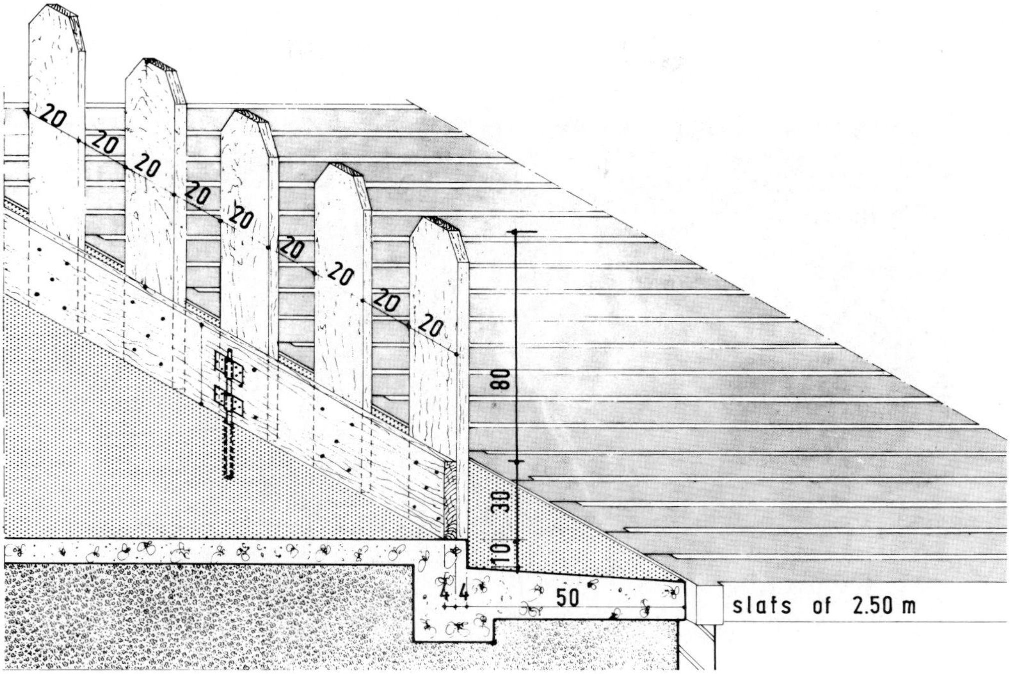

Fig. 4.26 The Swedish feeding fence made of timber.

Fig. 4.27 The Swedish feed fence made of steel.

lever (fig. 4.28).

Fig. 4.28 A self-closing feed fence.

In front of the feeding fence a shallow manger is provided adjacent to a wide, concreted feeding passage. In some cases the manger is omitted and the wide feed pass connects directly to the feeding fence. A trough frontage of 65 cm per animal is provided at the feed fence.

A number of possibilities exist for *feeding roughage* to the animals. The unloading of silage and the transport to the feed passage can be carried out by means of a front loader. Further distribution is then done by hand. A disadvantage of the use of a front loader is the increased risk of an excessive deterioration of the silage (sometimes referred to as secondary fermentation) owing to excessive disruption of the face. Nowadays preference is given to the use of a silage cutter whereby large rectangular blocks of silage are cut (ca. 400 kg with wilted grass silage and 600 kg per block of maize silage) and transported by means of the fork lift to the feed passage. The cut blocks can be kept for several days in the feed pass limiting the unloading of silage to once or twice a week and is therefore beneficial to the labour organization on the farm. Another possibility is the side-delivery feed wagon, which is filled by means of a front loader. This wagon distributes the feed along the feed barrier. This method excludes all manual labour but requires a greater investment.

Interesting also by its marked labour saving is self-feeding. In this case the animals have to fetch their own feed at the silage clamp faces. Self-feeding silage is possible at rectangular, flat silos (fig. 4.29) in front of which a movable feed barrier is installed (sometimes covered with a roof made from corrugated sheets). The feed barrier is gradually moved as the animals take feed from the silo.

Fig. 4.29 Self-feeding of cows at the silage clamp.

The feed fence is placed ca. 0.5 m in front of the silo which should not be higher than 2 m to ensure that all feed is within reach of the animals. A frontage of 15 cm per animal is normally allowed (Overvest, 1978). From our photographic observations we could conclude that with a frontage of 10 cm congestion occurred around the self-feeding silo at certain moments (especially after milking) and that mostly the lower ranking animals were prejudiced. A frontage of 20 cm gives ample room at the feed fence but could result in too slow an intake of the feed from the silo whereby the feed could deteriorate. According to Overvest (1978) maize silage and wilted silage should be removed at a rate of ca. 1 m and 1.0 to 1.5 m per week respectively. The results concerning labour saving obtained with this type of self-service are excellent and spillage can be reduced to a minimum.

With self-feeding the animals consume ca. 45 kg of silage each per day. Self-feeding has nevertheless found no large acceptance. Self-feeding requires the provision of a yard, where some animals will lie down and become dirty while the yard itself needs regular cleaning. Another disadvantage is the lack of information about the individual intake of feed.

Mechanical and automatic feeding can be carried out in different ways. The feed can be supplied either by a reciprocating scraper system or a chain and slat continuous feeder (fig. 4.30). A conveyor belt, made of synthetic material and which is pulled through the manger by means of an electric motor can also be used as feeder.

Fig. 4.30 Mechanical feeding by a continuous feeder with chain and slats.

Another type of feeder is the overhead cascade feeder which consists of a channel in which a chain with slats is installed. The channel, of which the length is half the length of the trough, moves on wheels forwards and backwards above it and distributes the feed uniformly (fig. 4.31). The overhead cascade feeder can also be constructed with a conveyor belt (fig. 4.32).

A feeding system combined with a tower silo (fig. 4.33), equipped with a top or bottom unloader allows completely automated feeding. The investment is however very high and its working is limited by the frequent occurrence of breakdowns and interruptions.

A cubicle house can or cannot be provided with a yard and is therefore an open or a closed cubicle house with respectively 9 m² per animal (of which ca. 2 m² of yard per cow) and 6 m² per animal. The cubicle houses built nowadays are almost all of the closed type (labour saving).

The *watering facilities* are provided by automatic drinking bowls installed inside the house and connected to the main or to a private water distribution system. Per 20 animals one bowl is provided or one central drinking trough with constant level (with float) filled from a water tank (fig. 4.34).

The *natural lighting* is achieved by translucent plastic sheets installed in the roof and of which the area is rated at 1/15th of the floor area and/or by a translucent covering of the open ridge at the top of the cow house.

The *ventilation* is accomplished in a natural way. The air inlet either takes place through an opening which can be closed more or less by a wooden trap or through a movable curtain, woven with nylon thread, which has a height of 50 cm and is installed under the gutter over the entire length of both side walls of the house or through thresholds under the windows. The air outlet takes place through an adjustable open ridge or through stacks. In areas with a mild climate it is also possible to leave one of the side walls open and preferably the one along the feed passage. A so-called open fronted cow house is thus obtained as shown in fig. 4.35. The open front is directed opposite the most prevailing wind direction. Besides the savings in construction costs a well-ventilated house is thus obtained.

The *cow-kennel* is probably the simplest type of cubicle house and is already common, in the United Kingdom for instance. It is constructed from a series of stanchions which support the roof made from corrugated metal sheets and which also form the partitions between the cubicles.

Against the outer stanchions wooden laths are nailed which will form the side and front walls of the cow-kennel. The stanchions are connected with horizontal and slanting girders in such a way that cubicles are created whereby the animals face the wooden front wall (fig. 4.36). The floor consists of compacted earth covered with sawdust. Between two rows of cubicles a concrete passageway is provided. Ventilation and lighting take place respectively through an open-ridge and translucent plastic sheets.

In other designs corrugated metal sheets were employed as walls and roof and even a hall covered with a half-round roof with plastic foil fitted to round curved tubing has been used. Prefab versions are also available. Feeding takes place in the kennel or outside at the feeding fence of a manger or at the self-feeding silage clamp.

According to Dutch investigations (Anon., 1972) the intake of dry matter was 3 % higher in cow-kennels with outside feeding, the milk

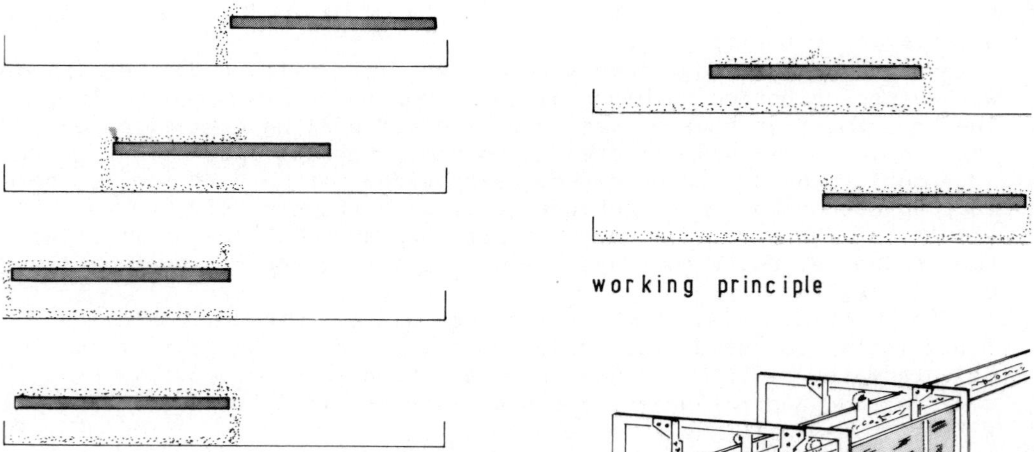

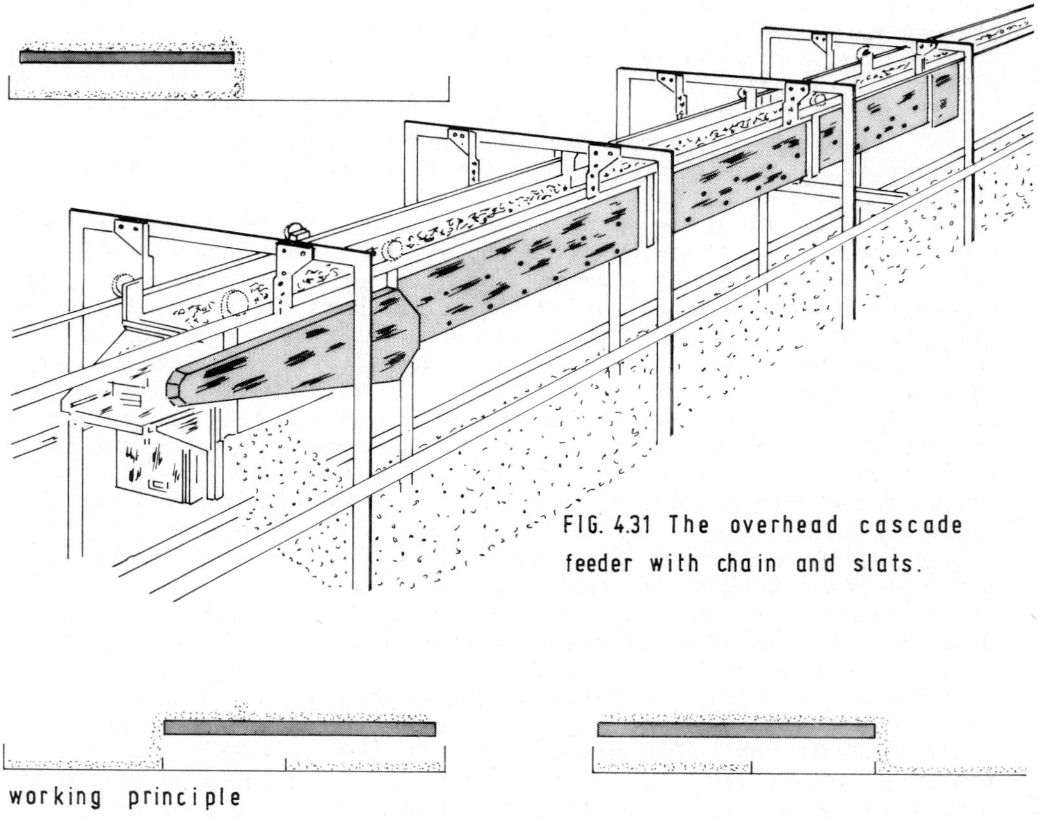

FIG. 4.31 The overhead cascade feeder with chain and slats.

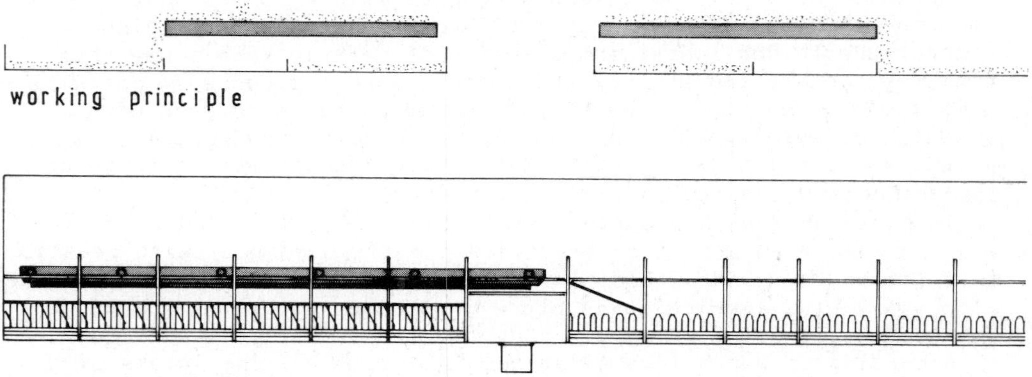

Fig. 4.32 The overhead cascade feeder with conveyer belt.

Fig. 4.33 A tower-silo for the storage of silage.

Fig. 4.34 A drinking trough in an open fronted cubicle house.

yield was 2.4 % lower and the fat content of the milk was 5 % lower than in an insulated stanchion barn.

Although the cow-kennel allows a considerable reduction of the

Fig. 4.35 An open fronted cubicle house.

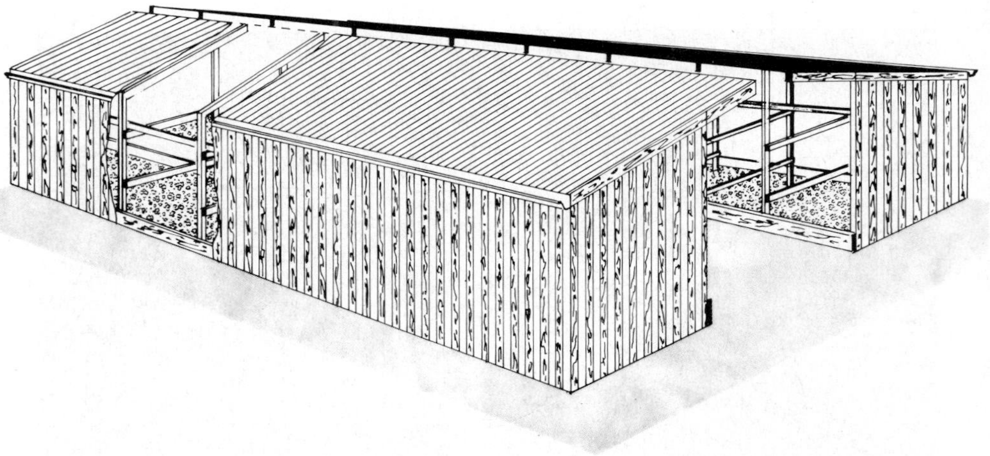

Fig. 4.36 The cow-kennel.

building costs it has also some disadvantages. The working conditions of the farmer in this type of housing are rather unpleasant and some Town-Planning Authorities are opposed to this type of housing which may spoil the landscape.

The *milking parlour* is an inseparable part of the loose house, the main difference being that the milking is carried out in a separate room compared to the stanchion barn where milking is done in the barn itself. This allows a hygienic milk production as well as a reduction of the labour demand and an improvement of the work conditions. Since the floor of the stalls is raised some 0.8 to 1.0 m above the floor area of the milker it is possible to carry out milking more efficiently, faster and with a minimum of fatigue. The cows enter the milking parlour from the collecting yard which often consists for the larger part of passages between the cubicles. After milking they leave the milking parlour and go straight to the cubicle house or perhaps to a yard outside the house.

The collecting yard and the area where the already milked cows are held are separated from each other by a chain or fence across the passage.

The milking parlour has to be sufficiently insulated ($k \leq 1.1$ W/(m^2.K)), well illuminated (10 to 15 % of the floor area) and ventilated (stack) while slight heating is recommended (e.g. infra-red heater). Supplies of cold and warm running water have to be on hand allowing the cleaning of the milking parlour and its ancillary material and udder washing. Different types of milking parlour equipment can be distinguished.

Loose houses for cows built in the Fifties and the early Sixties seldom holding more than 30 cows were often provided with a tandem type of milking parlour (fig. 4.37) and later the double chute type of milking parlour, as shown in fig. 4.38, was employed and is nowadays sometimes applied in transformed buildings.

The cubicle cow houses built in the last decade are however developed for a herd of 50 to 100 or more cows and are equipped with a herringbone milking parlour. In the 2 x 4 herringbone layout shown in fig. 4.39, a batch of 4 cows is allowed in each time in one of the two rows of four concreted stalls along both sides of the pit in which the milker works. The depth of the pit is 0.8 to 1.0 m. The cows form up in herringbone fashion with their heads facing outwards while a fence in front and behind the batch of cows keep them in the stalls. Fig. 4.39 represents the so-called saw-tooth version whereby the edge of the stalls has a saw tooth-shape which allows excellent access to the udder and facilitates the milker's job.

Fig. 4.40 shows the straight version of a 2 x 6 herringbone parlour whereby the edge of the stalls is straight thus simplifying its construction compared to the saw-tooth but making the accessibility of the udder less satisfactory. Also 2 x 5, 2 x 8, 2 x 10 and even the 2 x 12 herringbone layouts are found.

In a herringbone milking parlour the cows are milked in batches of 4, 5, 6, 8, 10 or more. Each cow is subsequently subjected to : a preliminary treatment (udder washing), putting-on the teat cup cluster and removing it, in as far as the first and last mentioned operations are still carried out manually (see later).

The dimensions of the herringbone parlour depend on the number of stalls and the way in which the cows enter and leave the parlour. In table 4.1 some dimensions are given (Maatje and Swierstra, 1978). A disadvantage of a 2 x 8 or more herringbone parlour are the long changing times. One becomes also much too group-bound : cows requiring long

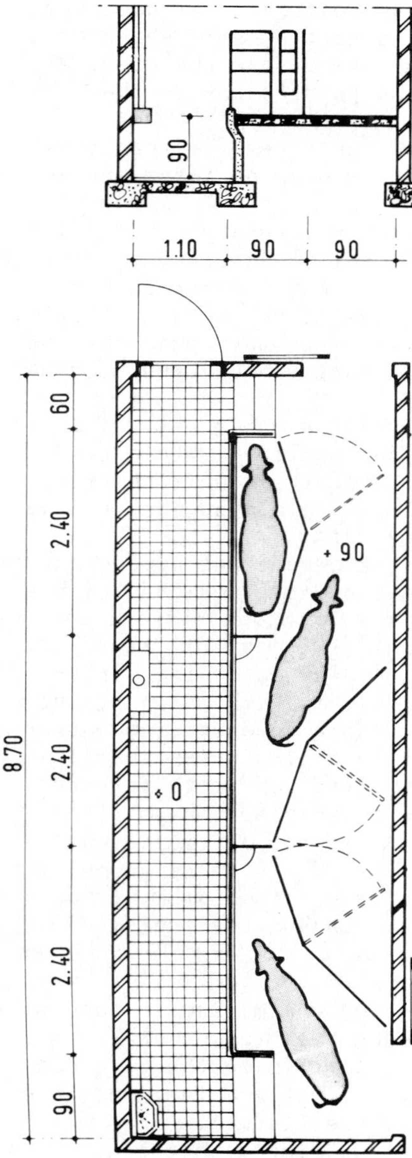

Fig. 4.37 The tandem milking parlour.

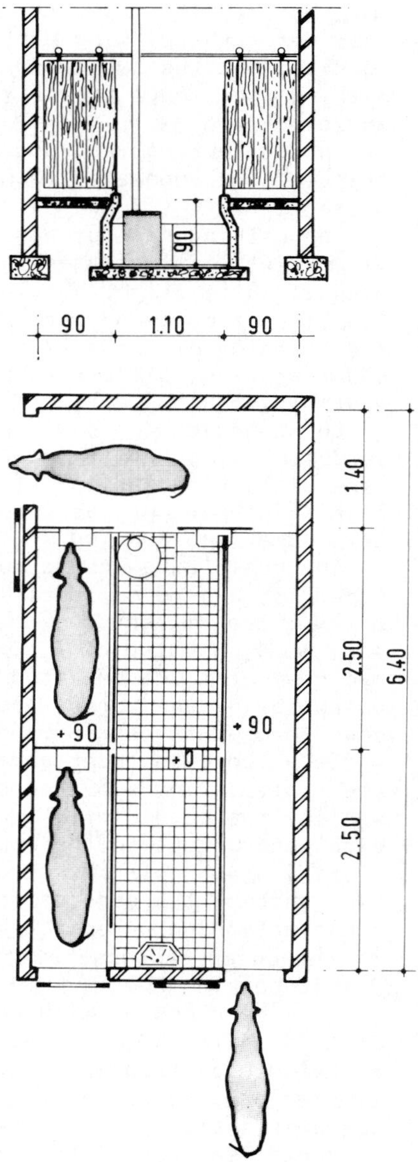

Fig. 4.38 The (double-) chute milking parlour.

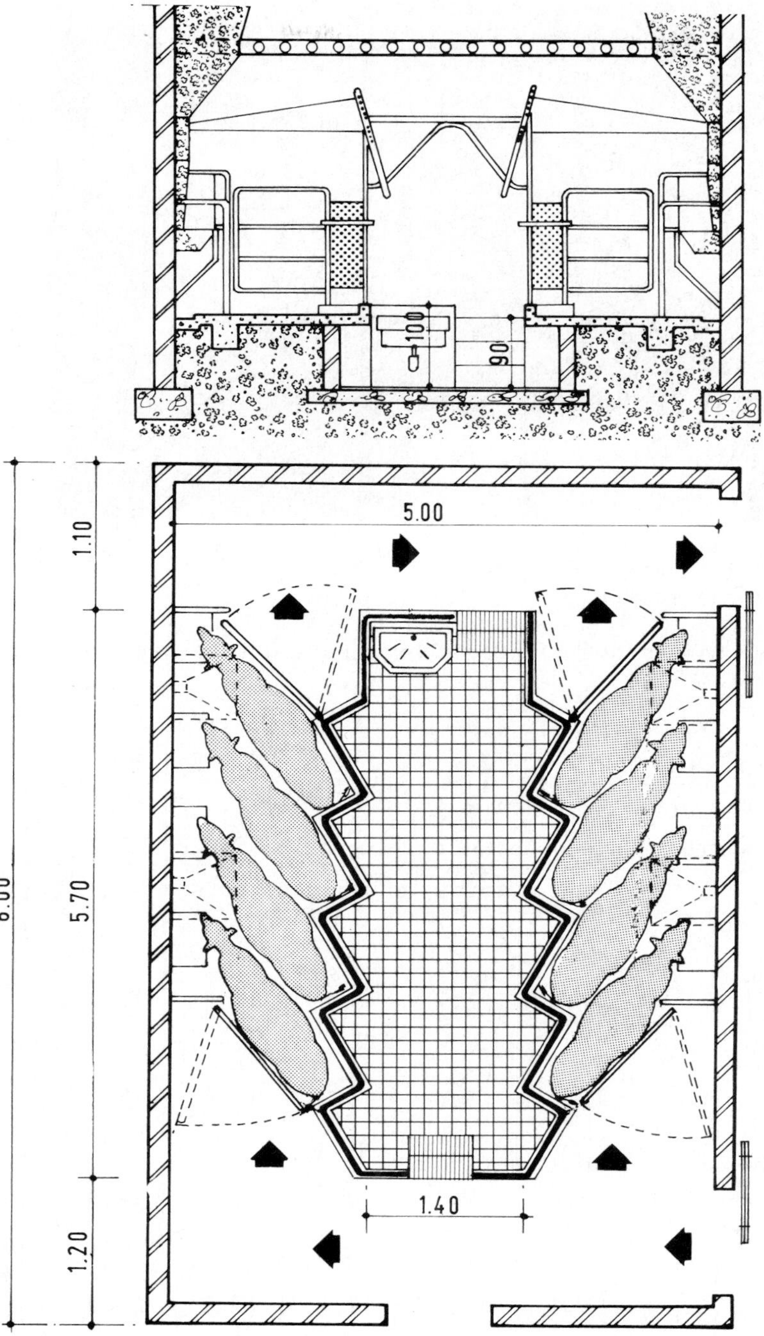

Fig. 4.39 A 2 x 4 herringbone milking parlour.

Fig. 4.40 A 2 x 6 herringbone milking parlour with a straight-line arrangement of the stalls.

TABLE 4.1 Some dimensions of herringbone parlours (expressed in metres).

Type of herringbone	2 x 4	2 x 5	2 x 6	2 x 8	2 x 10
min. width of the pit	2.0	2.1	2.2	2.4	2.6
minimum stall width	4.6	4.7	4.8	5.0	5.2
ground-plan	total length of stalls				
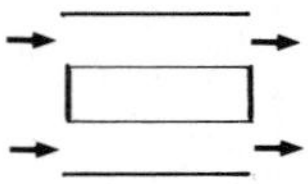	7.0	8.2	9.3	11.6	13.9
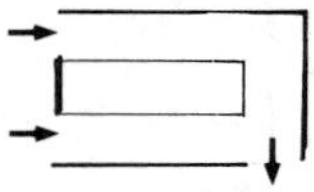	8.2	9.4	10.5	12.8	15.1
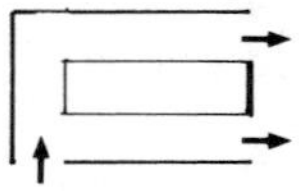	7.9	9.1	10.2	12.5	14.8
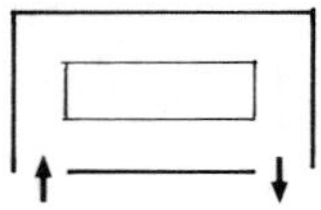	9.1	10.3	11.4	13.7	15.0

machine-on times (heavy yielders or slow milkers) hold up the whole batch which will inevitably lead to a reduction of the labour productivity. In order to solve those problems, trigon (fig. 4.41) and polygon milking parlours (fig. 4.42) have been designed for large dairy farms. A 4-6-6 trigon parlour has the same number of stalls as a 2 x 8 herringbone parlour but has a higher efficiency. Likewise a 4 x 6 polygon parlour has a higher throughput than a 2 x 12 herringbone parlour.

A number of years ago rotary or carousel parlours have known a limited market but this has come to a halt.

A rototandem milking parlour (figs 4.43 and 4.44) is composed of a rotary metal platform upon which a number of stalls are installed and in which the cows are held for milking. This plate is supported by a number of wheels and is driven by an electric motor. Each cow enters a stall through the same entry after which the milker carries out his normal routine-activities on the animal. Each cow goes through a complete revolution of the carousel whilst being milked and finally ends up in front of the exit. The carousel is stopped and the cluster is either manually or automatically removed. The exit door is opened and the cow returns to the cow house. The carousel moves somewhat further so that an empty stall faces the entry and another cow is allowed in. In larger units, a cow enters a stall simultaneously with another leaving a stall at each stop of the carousel.

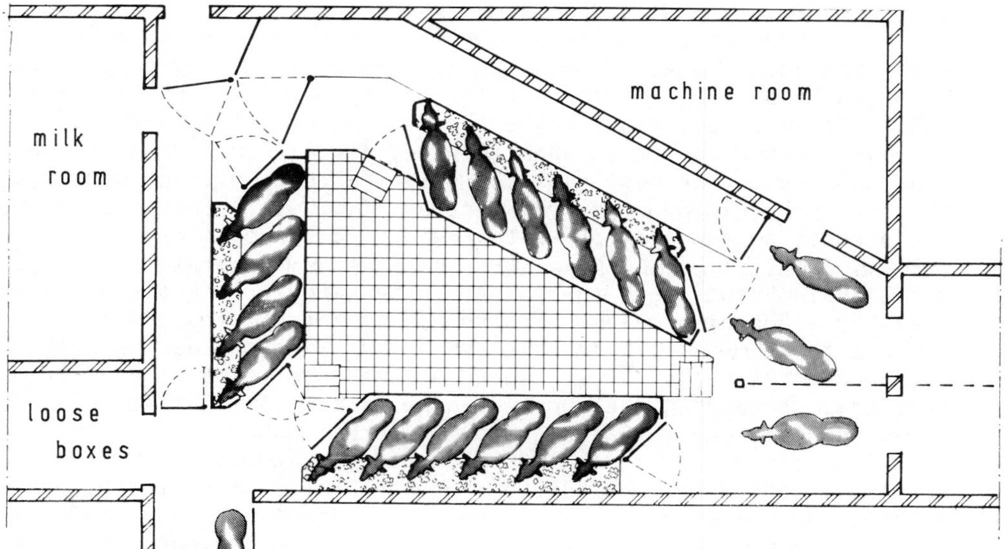

Fig. 4.41 A 4-6-6 trigon milking parlour.

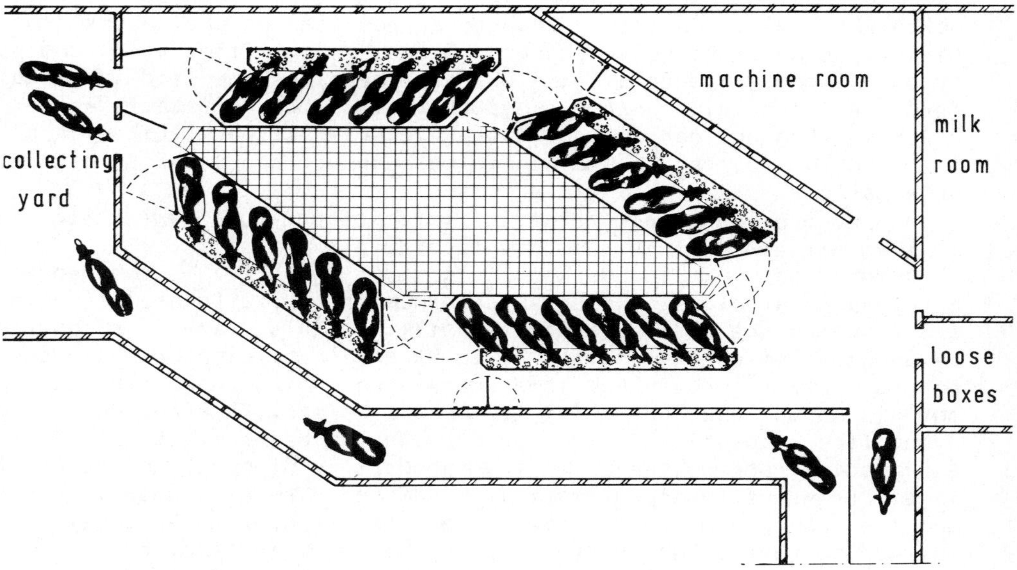

Fig. 4.42 A 4 x 6 polygon milking parlour.

Besides the above-mentioned rototandem milking parlour a number of other less spread types of milking parlours are in use, according to the positioning of the cows on the rotary floor.

In a rotary abreast milking parlour the cows face the centre. They enter the rotary plate frontwards and leave it backwards (fig. 4.45).

At large dairy farms the rotolactor is sometimes applied (fig. 4.46). The cows enter and leave the turning platform frontwards and pass through a tunnel under it when leaving. The milkers stand on the outer side as is also the case with the rotary abreast milking parlour. In the rotary herringbone milking parlour (fig. 4.47) the cows are positioned at an angle of 30° with respect to the platform while they face outwards. The cows step onto the platform frontwards and turn until their heads reach the outward side. They are retained in the stall by means of a gate until the end of the milking and then leave the platform frontwards. In this type of milking parlour as with the rototandem the milker is stationed at the inner side of the rotary platform. Concerning the choice of the milking parlour we can refer on one hand to our research and on the other hand to comparative research carried out in the Netherlands by, amongst others, van der Gaast (1981) and Hop (1979). Both the costs of buildings and milking equipment as well as labour requirement are taken into account. Table 4.2 gives the most suited type of milking parlour for different sizes of farms.

The normal types of herringbone milking parlours are suited for farms with up to ca. 130 dairy cows. The automatic removal of the cluster can be taken into consideration starting from a 2 x 5 herring-

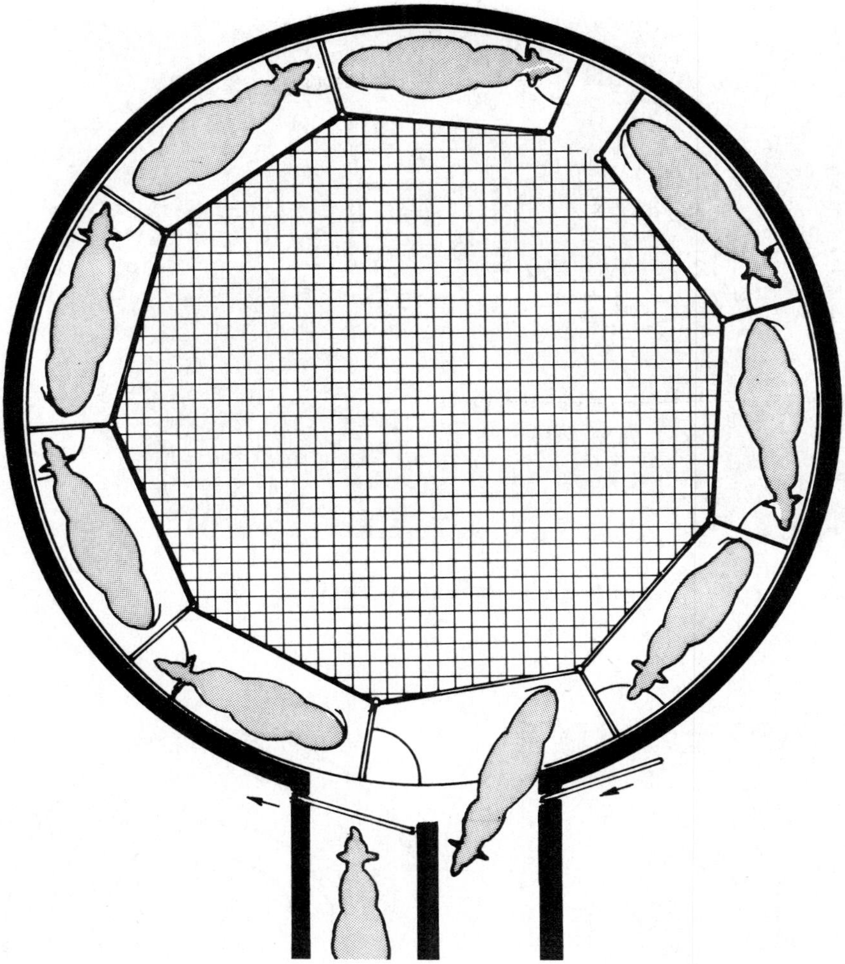

Fig. 4.43 A rotary nine-stall milking parlour.

bone parlour. For dairy farms with more than 130 cows the trigon milking parlour is to be preferred instead of the herringbone parlour. A rotary milking parlour is only to be considered for very large farms and challenges comparison with the polygon parlour. The labour time requirement involved with milking is treated elsewhere in this chapter.

It is common practice to provide a manger with each stall to allow each cow to be fed concentrate food whilst in the parlour. A dosed quantity of concentrates is allowed to the manger in front of the cow from a silo situated above the milking parlour by means of a number of vertical pipes which at the lower end are provided with

Fig. 4.44 A rotary eight-stall milking parlour.

Fig. 4.45 Rotary abreast parlour in which cows back off the platform.

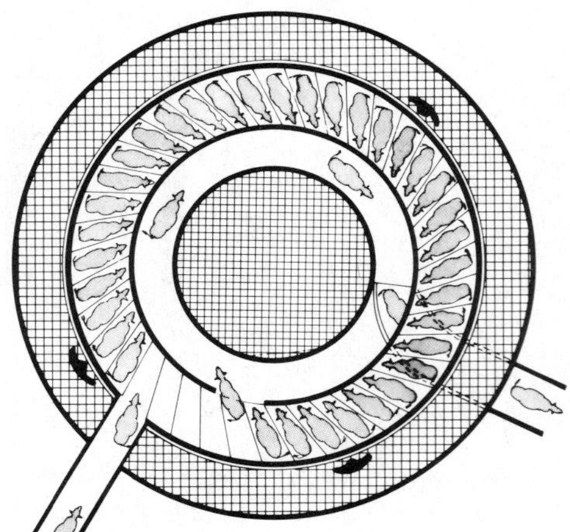

Fig. 4.46 Rotolactor milking parlour.

a dispenser. The distribution of concentrates is activated by the milker from the milking pit by means of a lever.

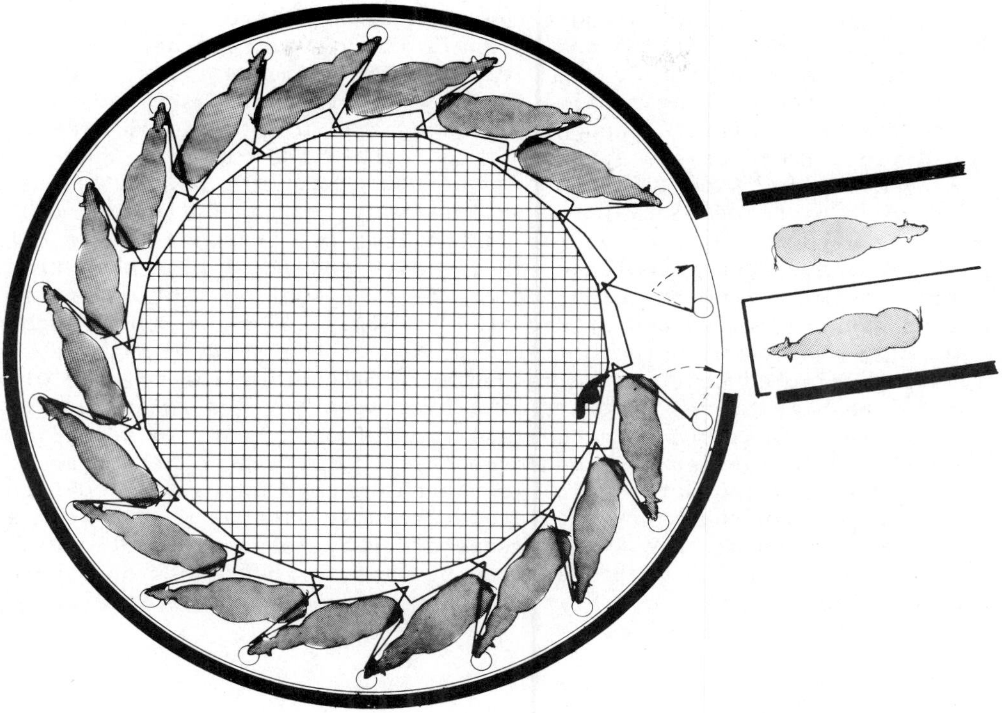

Fig. 4.47 The rotary herringbone milking parlour.

TABLE 4.2 The most-suited milking parlour in relation to the size of the farm.

Type of milking parlour	Number of clusters	Number of cows
herringbone 2 x 4	4	30 - 50
herringbone 2 x 4	8	45 - 80
herringbone 2 x 5	10	50 - 95
herringbone 2 x 6	12	60 - 110
herringbone 2 x 8	16	70 - 130
trigon 5-5-6	16	80 - 150
trigon 6-6-7	19	100 - 160
polygon 4 x 6	24	120 - 180
polygon 4 x 7	28	140 - 200
rotary 12	12	100 - 160
rotary 16	16	120 - 180
rotary 20	20	140 - 200

Full automatic feeding systems are also found and the concentrates are sometimes administered outside the milking parlour (see later).

In modern milking parlours the milk is collected in a glass or stainless steel pipeline and this via individual recorder jars which allow the measurement of the yield of each individual cow. The operation of the milking machine is identical to that described for the stanchion barn. Although there are still plenty of highline milk pipelines there exists a tendency to install the pipeline under the platform (fig. 4.48). The main advantage of lowline pipelines, as compared to highline pipelines, is the less accentuated fluctuation of the vacuum appearing at the udder. This results in a faster milking-out and less turbulence in the milk stream. A milk pump transfers the milk from the pipeline to the milk cooling tank, where it is subsequently cooled and stored at 4°C. Fig. 4.49 shows a herringbone milking parlour with the different parts of the milking machine.

The rejected heat in the condenser from cooling the milk is blown into the atmosphere. By means of a heat recovery system (a heat pump) it is possible to recuperate the released heat. In a specially built heat exchanger the condenser heat is fully or partly transferred to cooling water which is thereby warmed up to 50 - 60°C and is stored in a thermally insulated reservoir. The warm water can eventually be brought to a higher temperature by means of an electric boiler or immersion heater and can be used for household or farm purposes. Approximately 0.7 l of water at a temperature of 60°C is obtained per litre cooled milk.

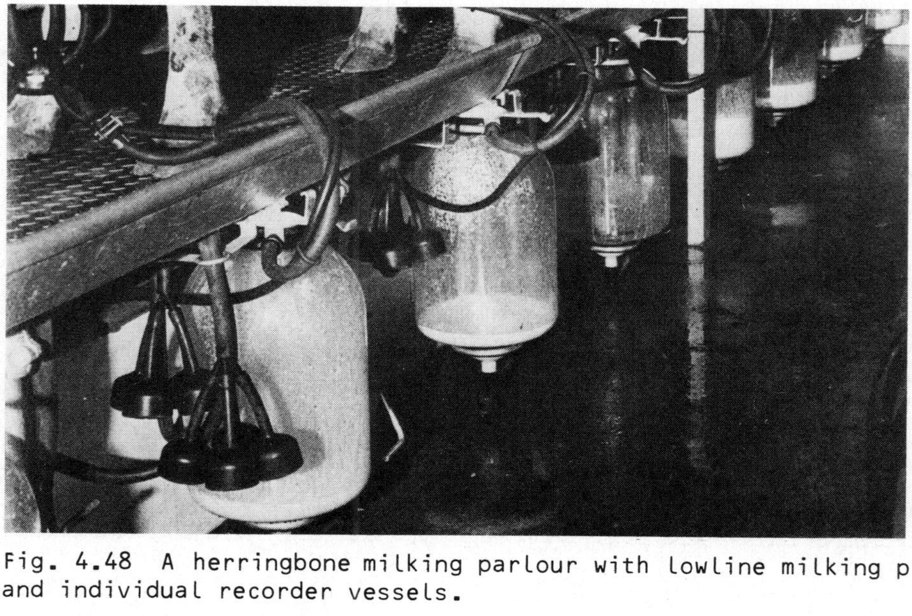

Fig. 4.48 A herringbone milking parlour with lowline milking pipeline and individual recorder vessels.

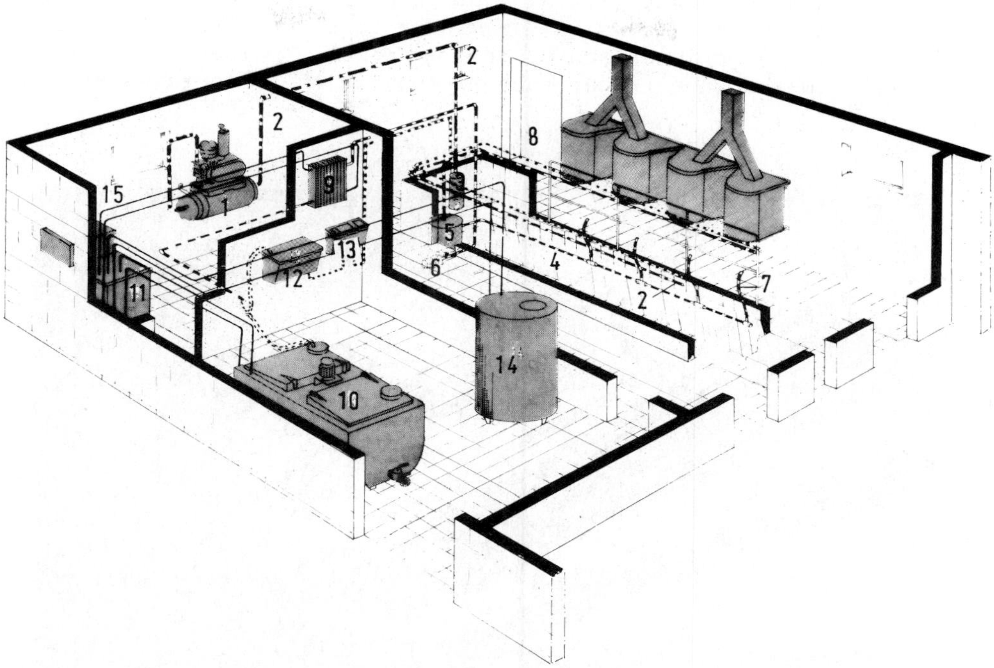

Fig. 4.49 Perspective of a herringbone milking parlour with milking machine and ancillary equipment.
Legend : 1 = vacuum pump ; 2 = vacuum line ; 3 = interceptor of condensed water ; 4 = milking pipeline ; 5 = releaser ; 6 = milk pump ; 7 = long milking tube ; 8 = cleaning pipeline ; 9 = plate heat exchanger ; 10 = refrigerated farm milk tank ; 11 = cooling unit (ice building unit) ; 12 = wash trough ; 13 = circulation cleaning automat ; 14 = heat recovery unit ; 15 = chilled water pipeline.

The pre-cooling of milk is nowadays increasingly applied, for it has been observed that an addition of warm milk (35°C) to the cooled milk (4°C) in the cooling tank causes the temperature of the mixture to rise to above 10°C and even to more than 15°C. From the bacteriological point of view this is not really harmful since the duration of this temperature rise is rather short. The mixture of warm milk with cold milk however brings about a certain churning of the warm milk and increases the lipolysis which has an adverse effect on the quality of the milk : lipolysis causes the transformation of lipids into glycerol and volatile fatty acids which are detrimental to the taste of the milk. By cooling the milk fully or partly before it enters the cooling tank, not only an improvement of the quality of the milk but also an important saving of energy is obtained.

This pre-cooling can be carried out by means of cold spring-water, well-water, tap water or even by ice water. Use can be made of a plate heat exchanger or a tubular pre-cooler.

A tubular pre-cooler (fig. 4.50) consists either of three coaxial tubes whereby milk flows in the middle tube and cold water counter-currently in the inner and outertubes or of a number of small milk-tubes situated in a larger tube through which the cold water flows.

Fig. 4.50 The tubular pre-cooler (two upper tubes on the photograph).

The length of the tubular pre-cooler depends on the quantity of milk to be cooled and is ca. 5 to 10 metres. Ubbels (1979) mentions an energy-saving of 4.5 to 5 Wh/l of milk for a tubular pre-cooler with a length of 7 m and 6 Wh/l of milk for one having a length of 10 metres. A plate pre-cooler (fig. 4.51) consists of a number of thin corrugated stainless steel plates which are fitted into a frame. The plates are sealed from each other. The milk flows between two plates whilst cold water flows counter-currently between two adjacent plates. With the use of cold water a milk temperature of ca. 3°C higher than that of the cold water is obtained. A saving of electricity of ca. 6.4 Wh/l of milk is thus possible (Ubbels, 1979). The water consumption amounts to 2 to 3 litres per litre of cooled milk. This water, heated by the milk, can be used as drinking water for cattle. This requires a reservoir. Instead of pre-cooling with water it is also possible to use ice water. The so-called instant cooling is then obtained. This ice water is produced by means of an indirect cooling tank (with ice-bank arrangement) or a separate refrigerating unit. A plate pre-cooler with a minimum of 51 plates and an ice water quantity of 3 to 3.5 litres per litre of milk allows a cooling of the milk

Fig. 4.51 The plate pre-cooler (at left on the photograph).

to a temperature of 4°C before entering the milk cooling tank. The refrigerated farm bulk tank must then only keep the milk at this temperature which will enable an energy saving. Fig. 4.52 represents a complete schematical design with instant cooling and heat recovery.

The pre-coolers are simultaneously cleaned with the milk installation. Both tubular as well as plate pre-coolers can easily be cleaned according to a Dutch investigation (Ubbels, 1979) although this is somewhat more difficult with the plate pre-cooler because of its more complex construction.

Both the total bacterial count of the milk as well as the acidity of the milk fat showed no differences with and without the use of a pre-cooler. French investigators (Mahieu, 1981) found a slight difference in the bacterial count in favour of the pre-coolers on the condition that they were thoroughly cleaned. In the opposite case the bacterial count can be quite high. The increased resistance, due to the use of a pre-cooler, necessitates a heavier milk pump but on the other hand the cooling capacity of the milk tank can be less.

Automation of the milking has developed enormously. In the most-advanced designs the udder is automatically washed in a preparation box with warm water directed upwardly by means of nozzles fitted in the floor of the passageway to the milking parlour (fig. 4.53). The passing cow interrupts a light beam which is directed to a photo-electric cell. The pump is actuated by means of a relay and sprays water via the nozzles onto the udder allowing the omission of hand washing the udder or at least the simplification of it.

In the milking parlour the milker has to operate several milking machines and he cannot always remove the clusters at the precise time

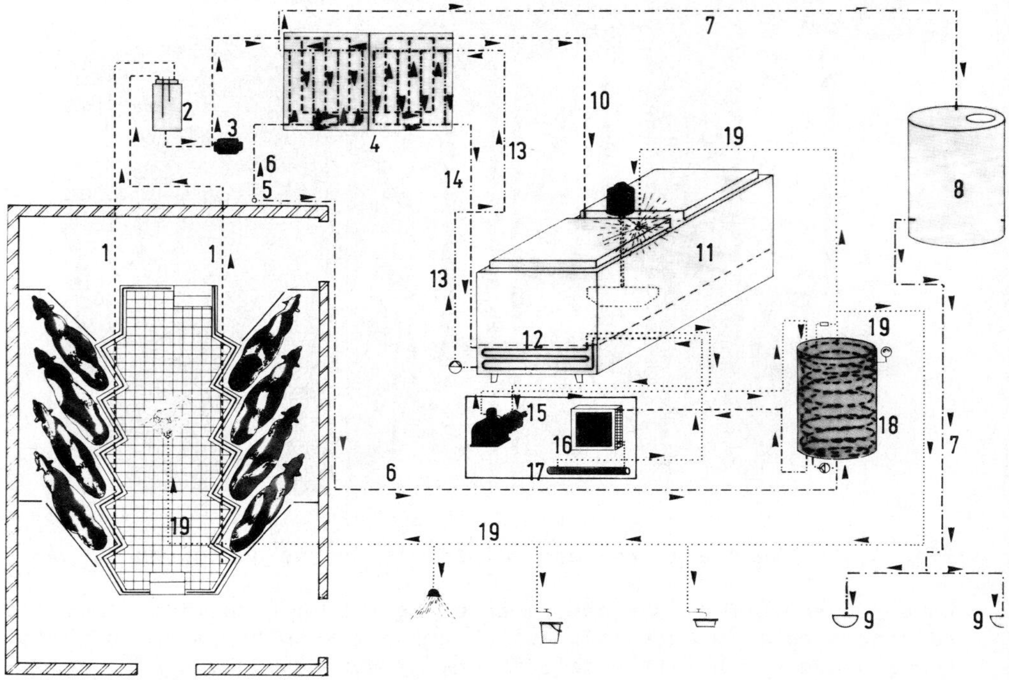

Fig. 4.52 Some possibilities for energy-saving by means of instant cooling and heat recovery at the cooling tank.
Legend : 1 = warm milk ; 2 = interceptor of condensed water ; 3 = milk pump ; 4 = double acting plate pre-cooler ; 5 = well water ; 6 = cold water ; 7 = warmed-up water ; 8 = reservoir for warmed-up water ; 9 = drinking bowls ; 10 = cooled milk ; 11 = refrigerated farm milk tank ; 12 = evaporator ; 13 = ice water ; 14 = warmed-up ice water ; 15 = compressor ; 16 = air-cooled condenser ; 17 = refrigerant receiver ; 18 = heat recuperation unit ; 19 = warm water.

corresponding with the milking-out.

In a rotary milking parlour the milk flow might have ceased before a complete revolution of the platform. This will cause overmilking i.e. the milking machine will continue its suction on the udder. Overmilking causes damage to the udder tissues. Besides the risk of damage the manual removal of the cluster imposes additional work on the stockman. These disadvantages can be avoided by providing the milking parlour with an installation for the automatic removal of the cluster. As soon as the milk flow becomes less than 0.2 kg/min the milk flow detector (float, electrode or photo-electric cell) triggers the electronic switching unit. In the installation as represented in fig. 4.54 the float indicator (4) will actuate the switching unit (6) as soon as the milk flow drops below the predetermined end point

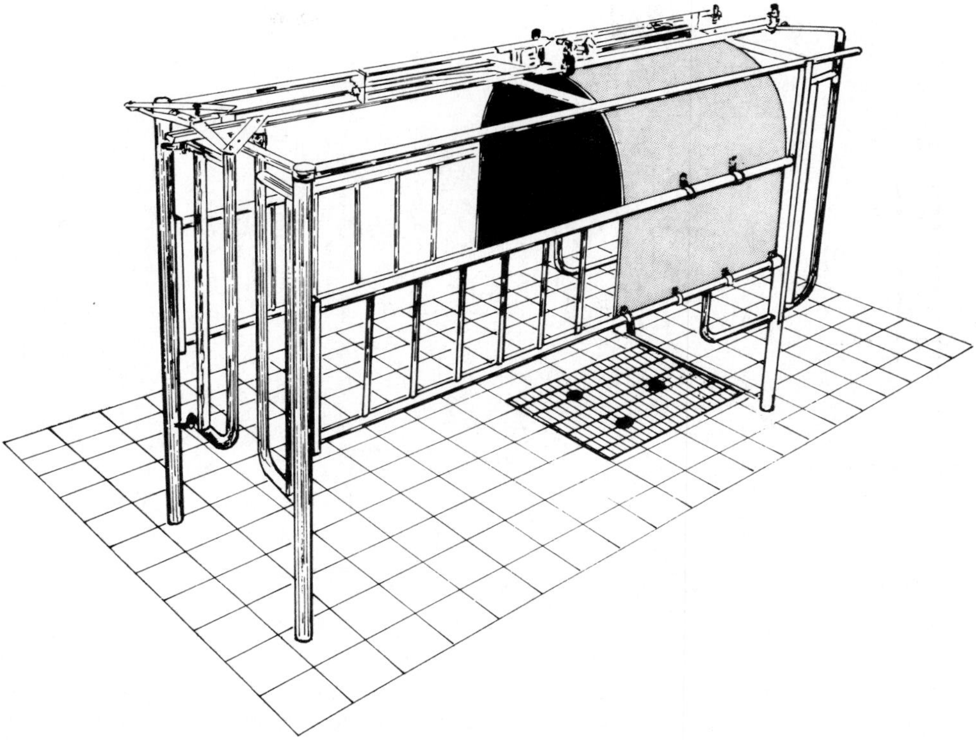

Fig. 4.53 The preparation box or "prep-stall" for automatic washing of the udder.

of 0.2 kg milk per minute. An electro-magnetic valve (7) closes off the vacuum-line (8) and allows air in the indicator and thus in the lower part of the cluster. By removing the vacuum the cluster will automatically drop off. In rotary milking parlours the cluster is then mechanically turned away from the udder and brought in reach of the milker. In a herringbone milking parlour however there is no such rotary movement of the stall. Therefore a cylinder (9) with piston is installed with each cluster. The piston is attached to the claw (2) by means of a rope (10). With the closing-off of the vacuum in the claw the vacuum is directed to the cylinder and the piston will pull away the cluster from the cow.

The milker activates the start button at the moment of putting on the cluster. A delay-relay in the switching unit (6) is triggered which will, during 0.5 to 2 min prevent the automatic removal of the cluster since in the beginning of the milking the flow is likely to be less than 0.2 kg milk per minute and this might otherwise establish a premature cluster removal at the start of the milking. A second time

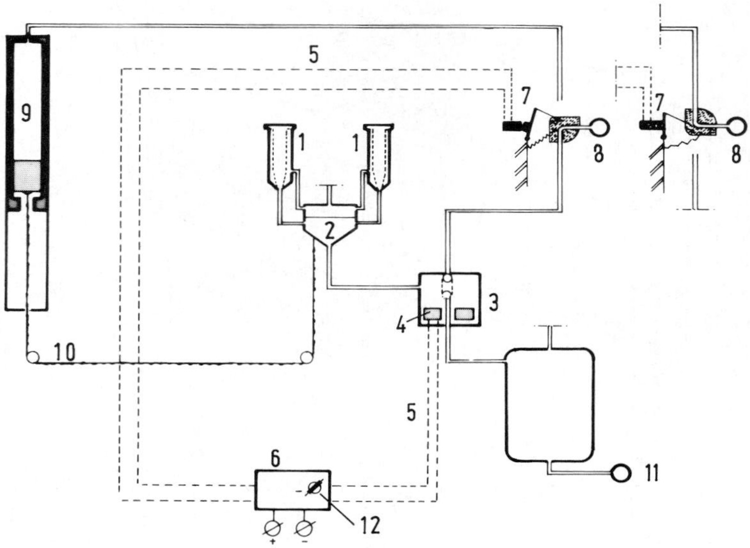

Fig. 4.54 The principle of operation of an automatic cluster removal system.
Legend : 1 = teat cups ; 2 = claw ; 3 = detector ; 4 = floating contact ; 5 = electric wiring ; 6 = switching unit ; 7 = electro-magnetic valve ; 8 = vacuum line ; 9 = cylinder ; 10 = rope ; 11 = milk pipeline ; 12 = start knob.

relay with a delay of 0.5 to 1 min prevents the automatic removal of the cluster in case of a short power failure. The installation of the automatic cluster removal gives complete satisfaction provided that the milking machine is in a good condition and is not liable to sudden important fluctuations in the vacuum level. The possible fouling of the float installation forms a weak point in the set-up.

Recent cluster removal systems are therefore carried out electronically. The long milking tube is provided with a sensor composed of two tubular electrodes connected to each other by means of a non-conducting synthetic component. They are thus electrically isolated. One of the tubular electrodes is fed with a constant voltage. The milk surface in the sensor conducts the electrical current to the second electrode. The larger the milk flow the greater the conductivity will be and the higher the voltage at the other electrode.

The pulsed character of the milk let-down, as a result of the pulsations, is electrically corrected in order to get an exact image of the milk flow.

Another form of automation which is based on the same principle as the preparation box is the automatic installation for the disinfection of the udder and this is installed in the return-passageway which leads to the cow house. At the moment that the cow interrupts the light-beam of a photo-electric cell a pump is actuated for a short time. By means of nozzles a disinfection agent is sprayed onto the udder. Commonly applied disinfection agents are a 0.5 % iodine solution or a mixture of chlorhexidine, alcohol and glycerol. The disinfecting can also be mechanized in the milking parlour. The disinfectant is carried through a gummy tube to a nozzle and the milker washes the udder after milking by means of this spraying tube.

The constant rise of the number of cows on the farm increases the interest in the automation of the whole milking process by means of a microcomputer.

Already in 1975 Rossing and Ploegaert (1975) developed an automated milking system allowing the recognition and identification of the cows in the milking parlour and the individual administering of concentrates according to their milk yield. Heavy yielders however have too little time to consume the required amount of concentrates during the milking. Additionally the consumption of large amounts of concentrates twice a day is not so beneficial for the digestive tract. From those facts the programmed out-of-parlour feed dispensers originated.

Each cow carries a collar around the neck to which a transmitter is attached with a preset (electronic) cow number (fig. 4.28). The transmitter is passive i.e. is activated when it approaches the receiver which is built-in under the manger for concentrates. As soon as the cow moves her head into the manger the transmitted electronic cow number is picked-up by the receiver and stored in the process computer. The computer transmits the amount of concentrates to be administered to the programmable installation for the dispensing of concentrates. The cow receives this rationed amount of concentrates automatically. Fig. 4.55 shows such a feed box.

Most of the computer systems work with adjustable concentrate portions e.g. 100 to 250 g each time. The day can be divided in several periods and the cattle keeper can store the maximum number of portions allowed to each cow per period. As long as the cow remains in the feed box she receives a portion at regular intervals e.g. every 30 sec. and this until she has consumed the complete portion of the current period. Left-over portions from a previous period are added to the following period. In that way a spreading of the consumption of concentrates is obtained over a whole day. At any moment the cattle keeper can ask the computer the most important information such as a list of the cows which have not eaten their ration of concentrates. This can be an indication of illness or heat. If a cow leaves the feed box earlier e.g. when she is pushed aside, as happens frequently, the bottom of the manger tips over after a few seconds and the contents fall into the reservoir beneath it.

Fig. 4.55 Box for the programmed individual administration of concentrates outside the milking parlour in a cubicle house.

The dominant cow, who chased out the lower ranking cow now faces an empty manger since she wears a different cow number than the pushed away cow. One feed box is provided for each group of ca. 25 cows.

At least once a month the cattle keeper updates the amounts of concentrates for the corresponding cow numbers in the computer and this largely on the basis of the results of the milk yield control.

According to Meijer (1980) heavy yielders can also be adequately fed concentrates without the use of a programmable installation for dispensing concentrates. It is hereby essential to split up the animals in production groups. The heavy yielders are fed a basic amount of concentrates at the feed fence while the remainder is given in the milking parlour. The division into production groups has however a number of drawbacks. Each time an animal changes group, social unrest is produced in the group. The use of such production groups is limited to large scale farms only and the construction of the cow house may not always be suitable for this purpose.

This programmed feeding of concentrates forms only part of a much larger computerized management system which is schematically represented in fig. 4.56 and which also controls both the productivity as well as the health condition of the whole dairy herd.

A central computer provided with a display, a printer and floppy discs (fig. 4.57) is connected to the necessary equipment in the different buildings (both in the milking parlour as well as in the loose house). In the milking parlour a keyboard and a display are available

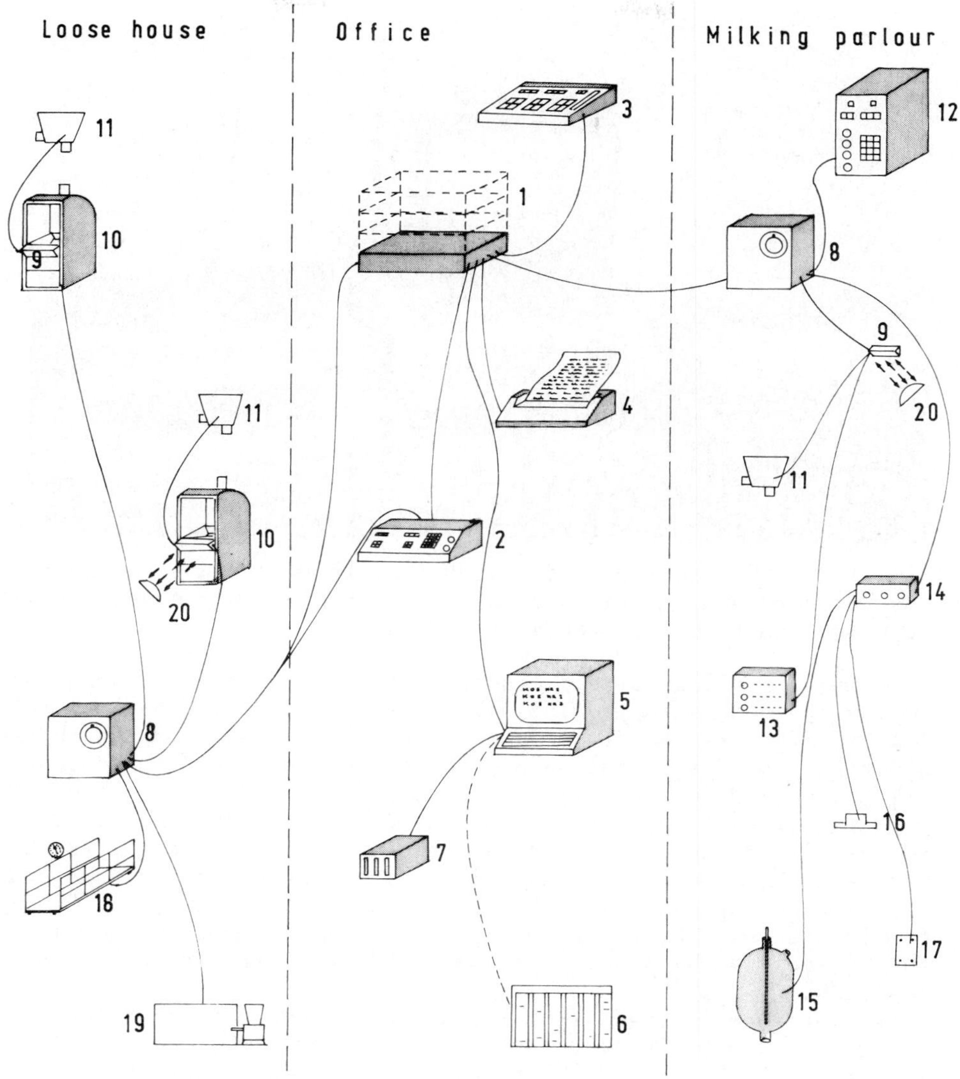

Fig. 4.56 Design of a computer controlled management system in dairy cattle husbandry.
Legend : 1 = central management unit ; 2 = central feeding unit ; 3 = keyboard ; 4 = printer ; 5 = personal computer ; 6 = cow calendar ; 7 = floppy disc unit ; 8 = power supply 220 V/24V ; 9 = transmitter-receiver for identification ; 10 = feed box ; 11 = dispenser for concentrates ; 12 = central unit in dairy ; 13 = display ; 14 = interface ; 15 = milk yield recording unit ; 16 = body-temperature sensor ; 17 = mastitis-detector ; 18 = weigher ; 19 = feeding automat for calves ; 20 = cow collar.

Fig. 4.57 Farm home computer with from left to right the printer, the keyboard and the video display, the concentrate processor with a keyboard and a double floppy disc unit.

allowing the milker direct access to the computer for storage or demand of information during milking (fig. 4.58). Each cow entering the stall is identified and the computer then commands the administration of a certain amount of concentrates (if concentrates are still supplied in the milking parlour). Each milking stall possesses a display giving instructions or information on that particular cow to the milker. In a recent experimental development the milk temperature and the electrical conductivity are measured and at the end of the milking the milk yield is registered. This information is directly processed by the computer and an extra large deviation warns the milker via the display allowing him to take the necessary measures.

An increase of the milk temperature can indicate an illness or heat. In a study, Rossing (1980) was able to trace nearly all cows in heat on the basis of an increased milk temperature. Mastitis or inflammation of the udder is a serious illness which affects many cows. The occurrence of mastitis causes an increase in the sodium and chlorine levels in the milk and a decrease in the potassium and lactose levels. This change in the composition of the milk causes a rise of the electrical conductivity. In this way it is possible to trace infected udder quarters when the electrical conductivity of one quarter is at least 16 % higher than that of any other quarter (Rossing, 1980). A deviating milk production can also be the indication that something is wrong. If this is the case with one or several cows it might indicate an illness or a casual event. A supervision of the concentrate

Fig. 4.58 Herringbone milking parlour with underneath the electronic individual milk yield recorder and above it the power unit for the automatic cluster removal with a display at the side showing the cow number and the yield.

intake is strongly recommended. If many cows show a deviating milk yield it might reflect an external cause such as bad weather, poor roughage quality etc. A systematic deviation at a particular milking stall is possibly due to a defective milking unit or a fault with the recording of the milk yield. Outside the milking parlour a number of concentrate feed boxes are provided. The bottom of each feed box can incorporate a weigher for the automatic registration of the weight of the animals. The cow calendar is also stored in the computer in order to complete the zootechnical supervision.

At the appropriate time the computer prints an attention list (action list) giving a number of instructions for the immediate future such as a prospective heat (ca. 80 days after calving), a check for gestation (three and six weeks after mating), drying off (after seven months of pregnancy), supplementary feeding (after eight months of gestation), calving (after nine months of gestation). Besides all these instructions the computer can also be used to store additional data about the animals such as : registration number, date of birth, descent, monthly and daily production, protein and fat content of the milk, the consumption of concentrates and diseases.

In future it will be possible to store and process a lot more useful information with the aid of the computer, such as the grassland calendar, calculation of the ration, young stock management and financial bookkeeping. The computer has thus made its entrance in agriculture and could cause a complete revolution. The computer can easily sense, process and store large amounts of information but in the end it is the farmer who takes the final decisions. Or, as Biegman (1981) writes it : "A bad farmer cannot be turned into a good one by purchasing a computer".

4.2.1.5 Some layouts of cow houses

A *stanchion barn with grids, for 40 cows* (fig. 4.59).
It is a two-row stanchion barn, with dimensions 12.3 m x 29.7 m. A wide central feeding passage of 3.60 m will facilitate the feeding. Opposite the dairy a number of individual calf boxes are provided.

A *cubicle house for 40 cows* (fig. 4.60).
Fig. 4.60 shows the layout of a two-row cubicle house for 40 cows with slatted cubicles for young stock and/or beef cattle. A frontage of 65 cm per head is provided. Cows are milked in a 2 x 4 herringbone milking parlour. Nine individual calf boxes and three isolation tie-stalls for cows are provided. By enlarging the house and extending the number of cubicles the layout of a cow house for more animals is easily obtained.

A *feed cubicle house for 56 cows* (fig. 4.61).
The feed cubicle cow house represented in fig. 4.61 has an identical internal breadth as the stanchion barn shown in fig. 4.59 namely 12.30 m. This simplifies the transformation of a large number of existing stanchion barns into feed cubicle cow houses. The central feeding passage and the manger can often remain while the lying and exercise areas have to be adapted.

A *cubicle house for 62 cows* (fig. 4.62).
It is a three-row cow house with continuous feeding passage. The young stock are in slatted boxes ; it may also be possible to provide cubicles there. The cows are milked in a 2 x 4 herringbone milking parlour. An isolation area (three stalls with grids) and eight individual calf boxes are provided.

A *cubicle house for 78 cows* (fig. 4.63).
This narrow cubicle house consists of two single rows of cubicles along the two longitudinal walls. There is a wide central, non-continuous feeding passage. All animals are able to eat simultaneously. Centrally and in the continuation of the feeding passage there is a 2 x 6 herringbone milking parlour. An advantage of this layout is the high density that is obtained, but the house is very long.

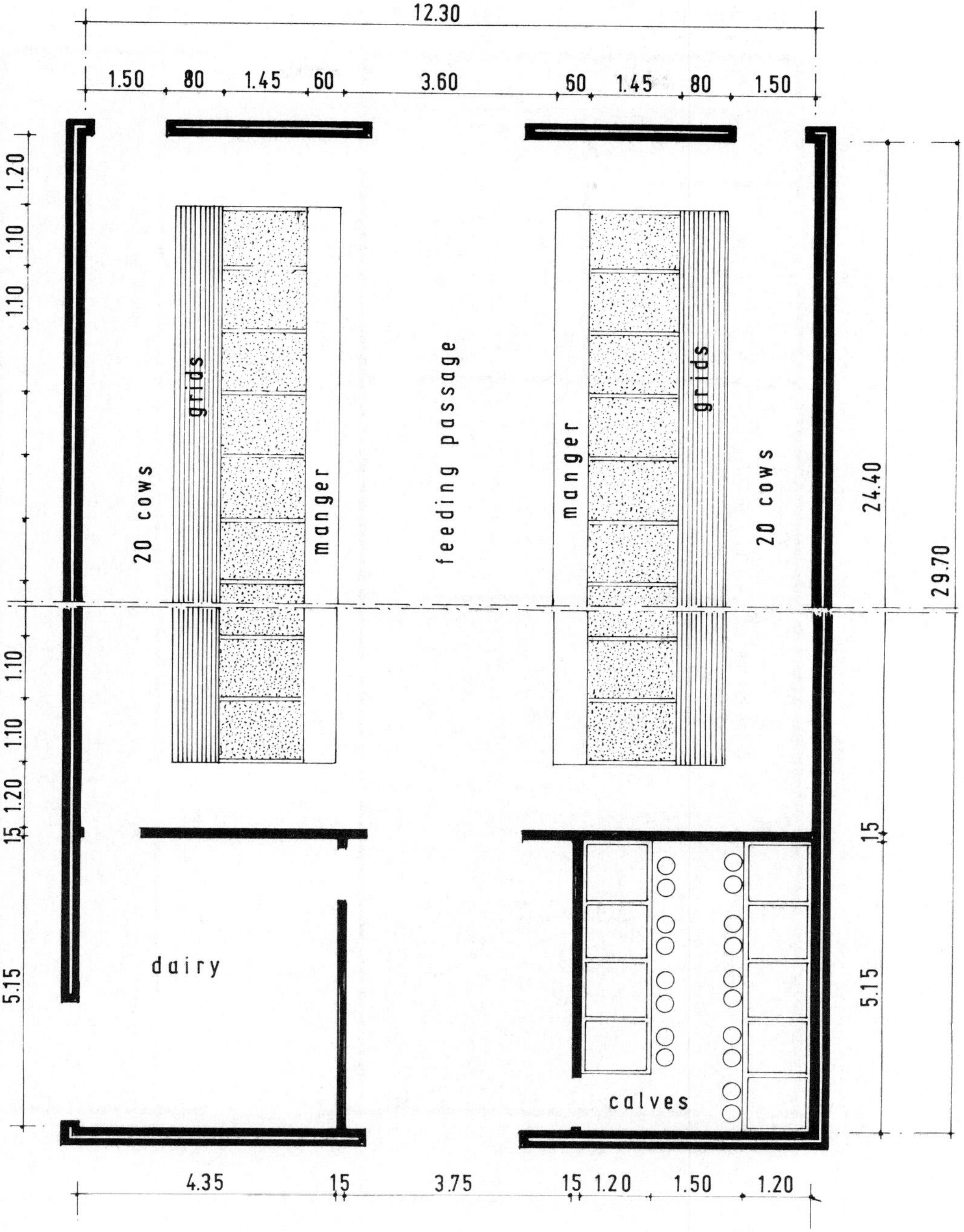

Fig. 4.59 A stanchion barn with grids for 40 cows.

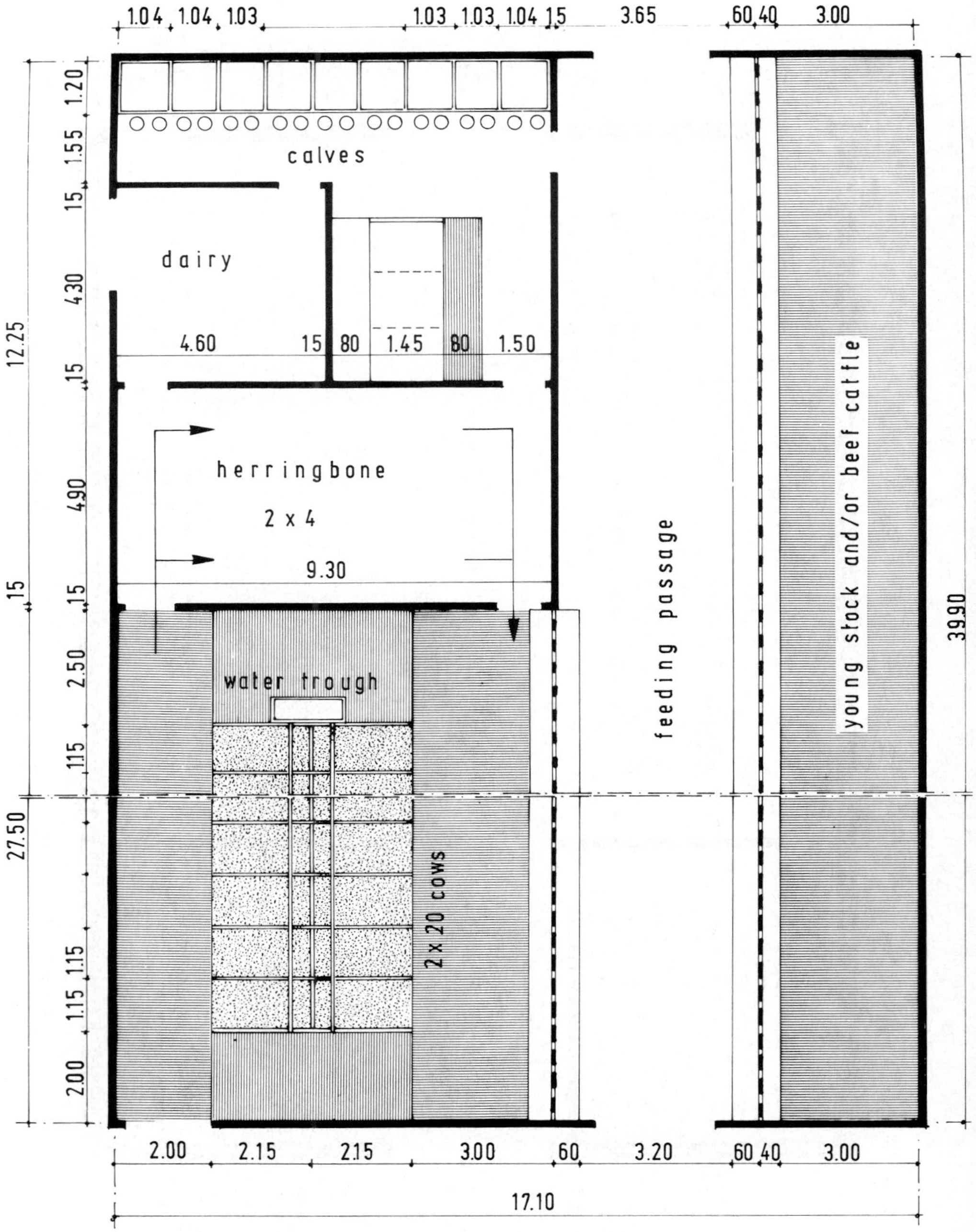

Fig. 4.60 A cubicle house for 40 cows.

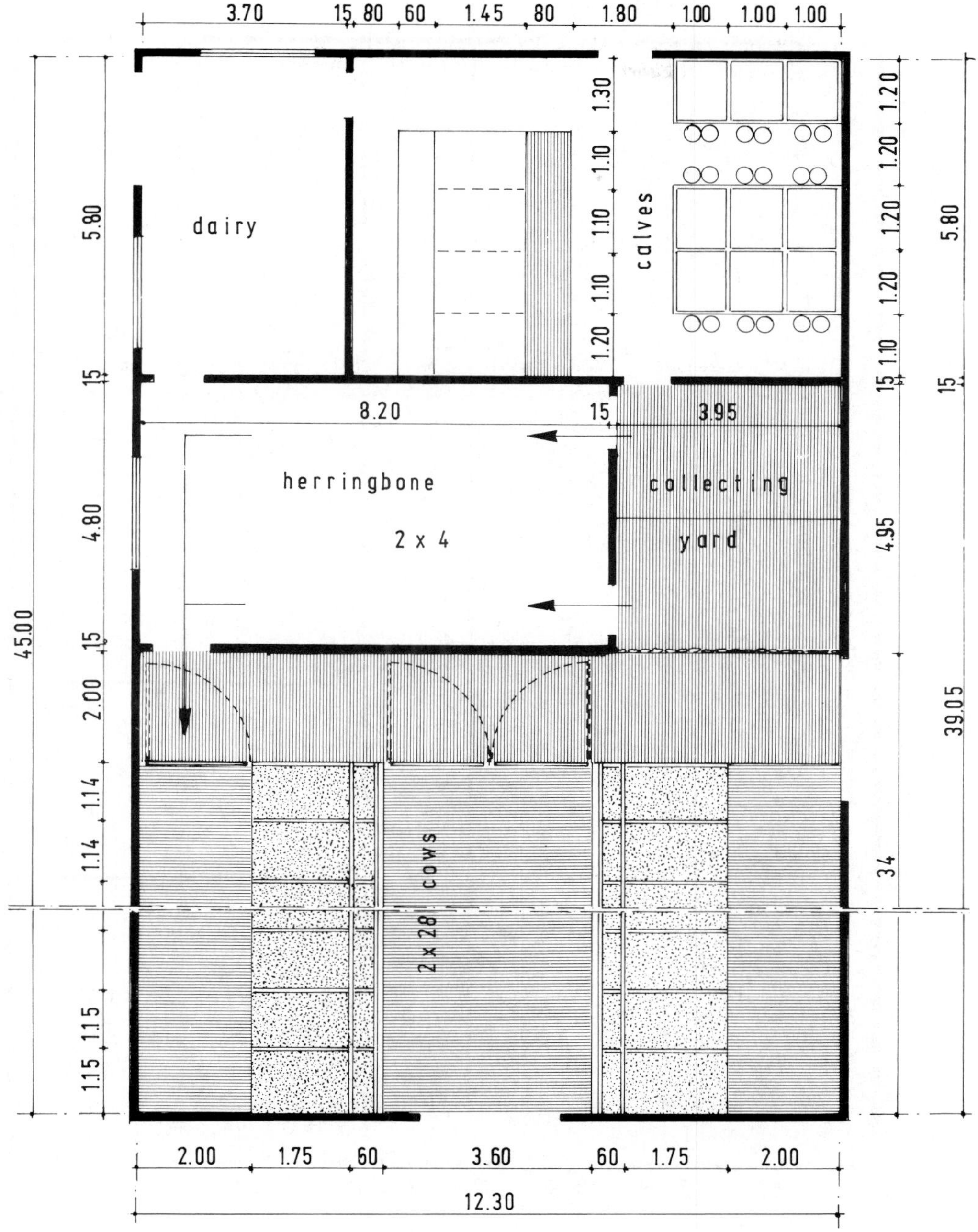

Fig. 4.61 A feed cubicle house for 56 cows.

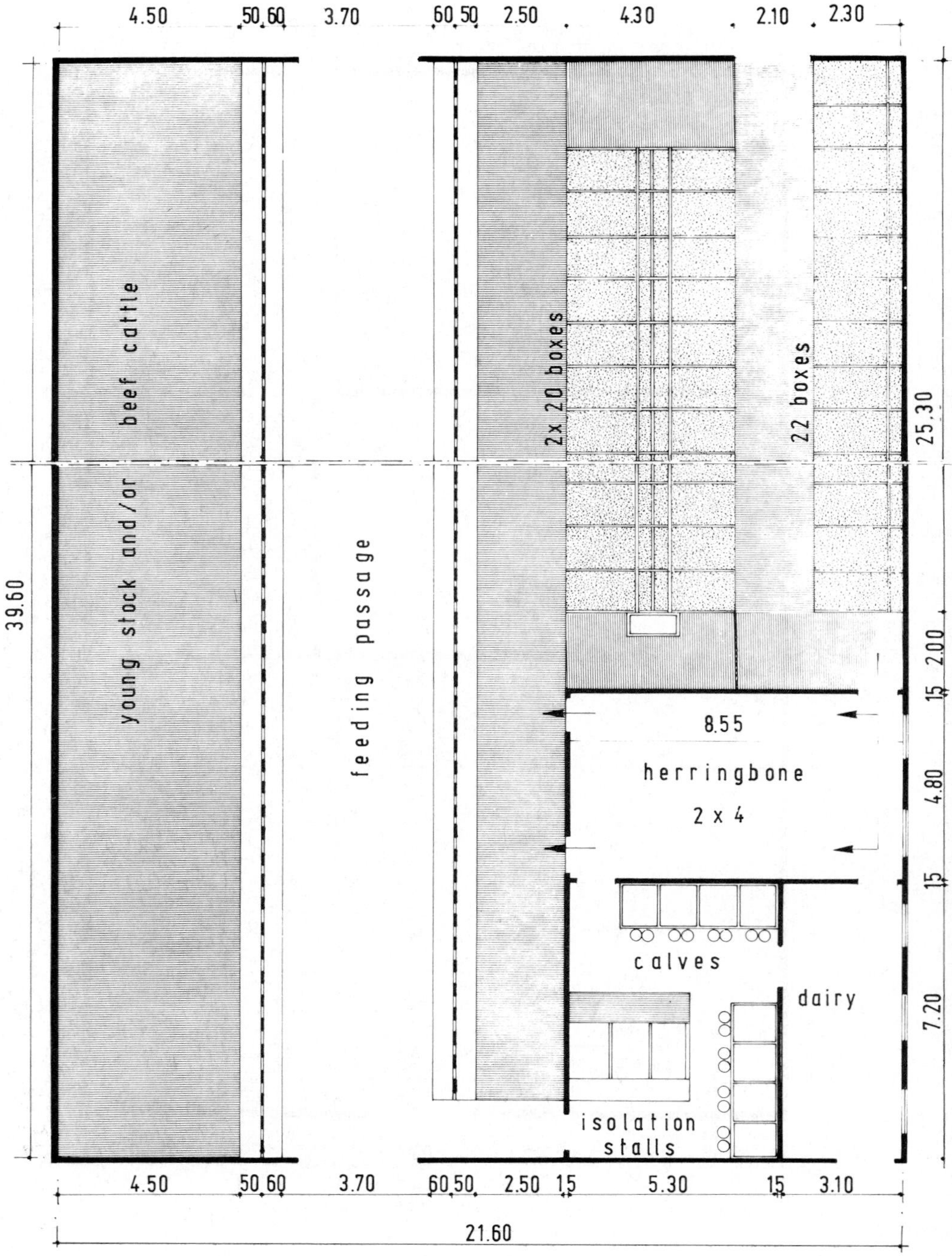

Fig. 4.62 A cubicle house for 62 cows.

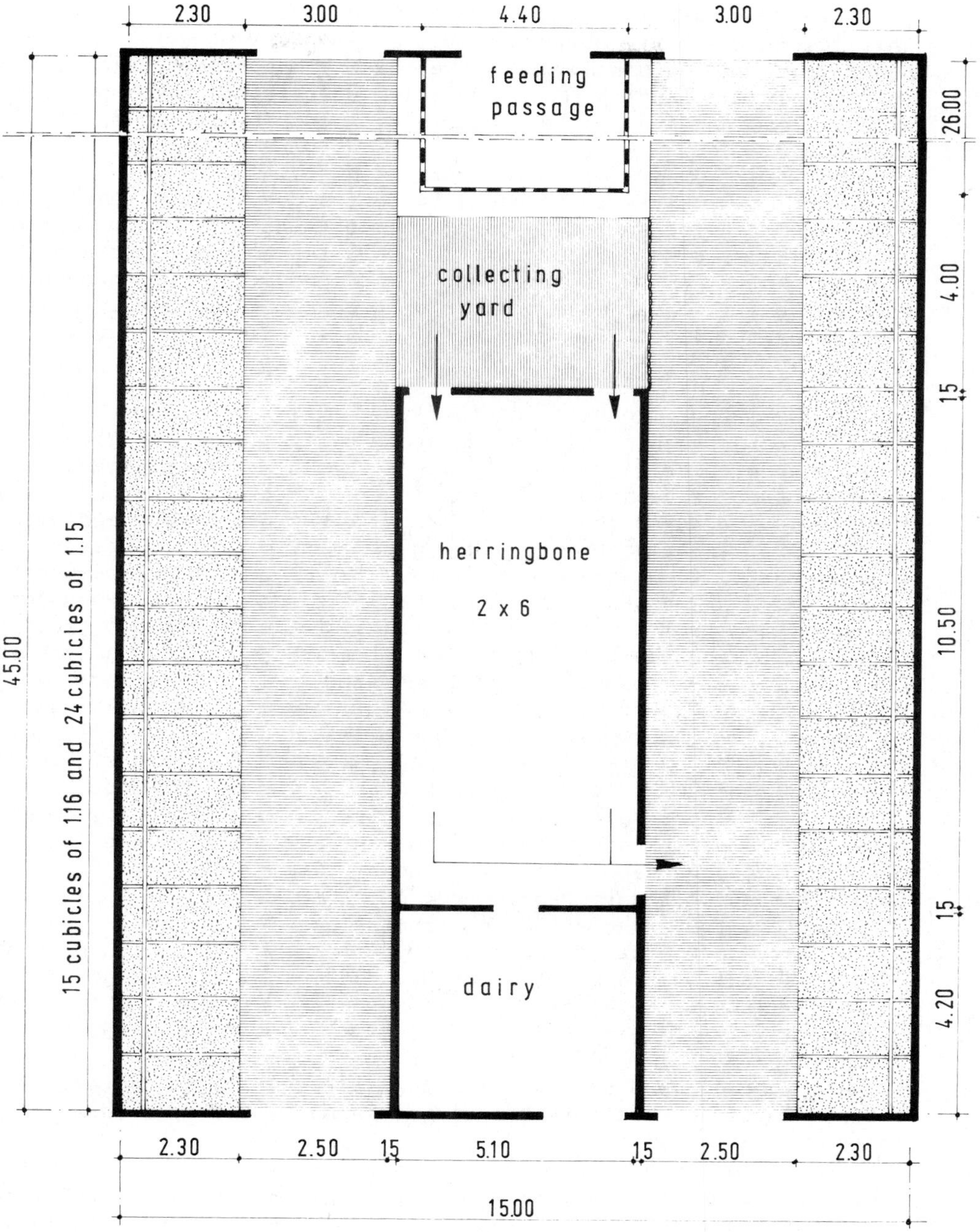

Fig. 4.63 A cubicle house for 78 cows.

A cubicle house for 117 cows (fig. 4.64).
This four-row cubicle cow house has, as the previous one, a non-continuous feeding passage. All the cows cannot line-up simultaneously in front of the feeding fence and this necessitates the application of bulk feeding. A 2 x 6 herringbone milking parlour was chosen for this house.

A cubicle house for 156 cows (fig. 4.65).
This four-row cubicle cow house with central continuous feeding passage is suited for splitting up of the dairy herd into production groups. The milking parlour with the appurtenances are attached to the main house. A 2 x 10 herringbone milking parlour was chosen to enable the milking of each production group in two cycles. An isolation area with 12 stalls together with 24 individual calf boxes are provided.

4.2.2 The labour organization in dairy houses

It is not possible, within the scope of this work, to give an outline, however brief it might be, about the way the activities can be organized and mechanized in the different types of dairy houses. In the previous description concerning the construction and equipment of dairy houses we have however highlighted most of these aspects. It appears useful to give a global survey, based on our own research, of the labour time requirement in the different housing types for dairy cattle, from which it will be possible to select the types of housing allowing the greatest labour time saving. Since dairy husbandry greatly ties the dairy keeper down and since he can less and less call in outside labourers who are quite expensive, he will seek a reduction of the labour in the dairy house, especially during the weekend. It is therefore of utmost importance to know which types of housing require a minimum of labour for the care of the dairy herd.

For a number of daily activities, carried out in the above-mentioned types of cow houses and according to certain labour methods already described we have been able, by means of time studies, to fix average labour times, which are represented in table 4.3.

From this table the following conclusions are possible :

- the daily care of a cow during the winter in a stanchion barn amounts to a minimum of 8.4 minutes per day, against a maximum of 6.9 minutes per day in a loose house. From labour-technical point of view the loose house for dairy cattle is certainly to be preferred above the stanchion barn ;
- the stanchion barn with grids allows a labour saving of ca. 17 % or 1.7 minutes per head and per day compared to the strawed stanchion barn ;
- the milking, both in the stanchion barn as well as in the loose house, amounts to about half of the total labour time requirement; various other activities take about 1/4th to 1/5th of that time ;
- the milking in a stanchion barn equipped with a pipeline milking installation allows a labour saving of ca. 8 % or 1 min per head and per day compared to a bucket milking installation ;

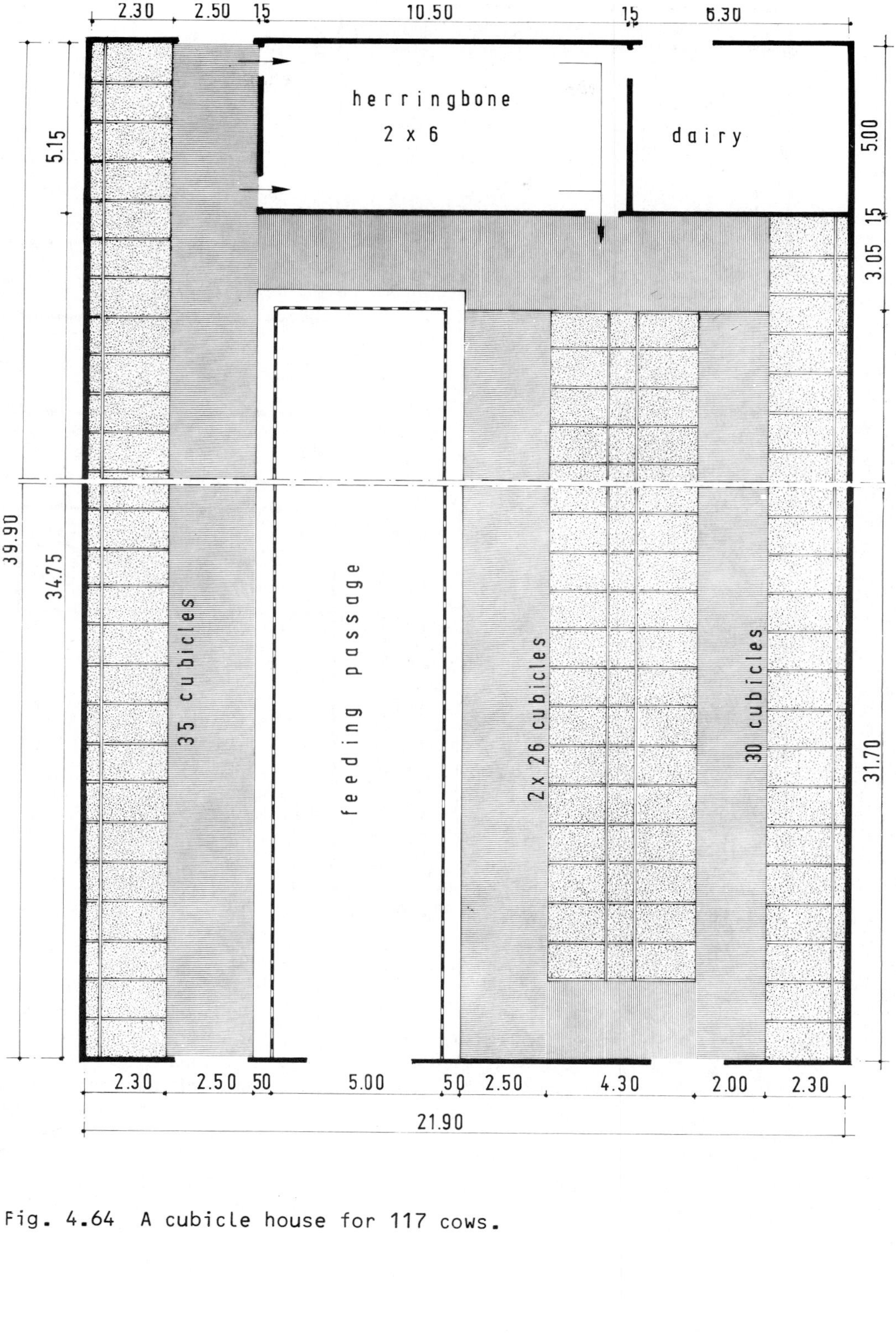

Fig. 4.64 A cubicle house for 117 cows.

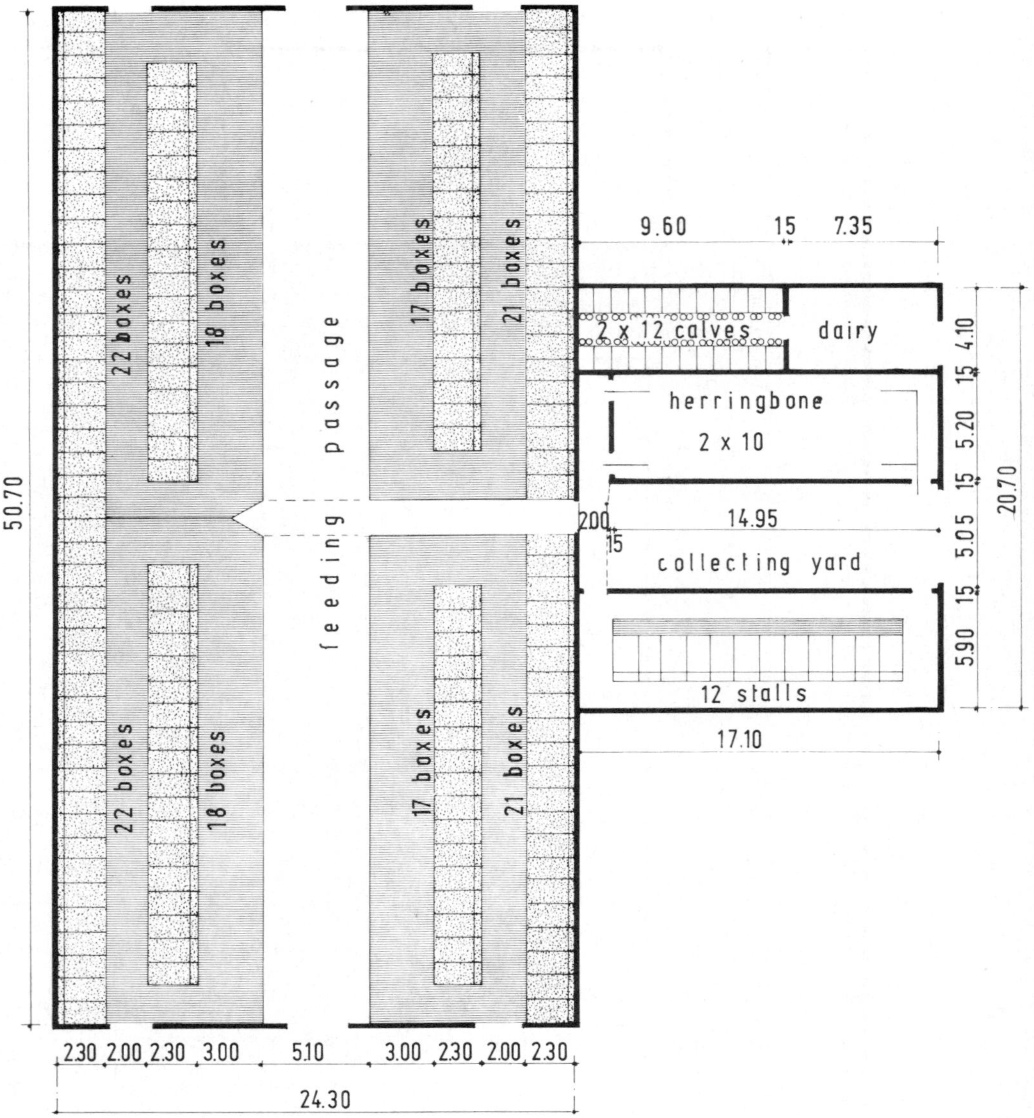

Fig. 4.65 A cubicle house for 156 cows.

- the automatic removal of the cluster in a herringbone milking parlour allows a labour time reduction of 1 minute per head and per day or 14 %.

Fig. 4.66 shows the annual labour time requirement in the different types of dairy houses according to the size of the herd. From this figure it seems that with an increase of the herd size the labour time

TABLE 4.3 The labour requirement in dairy houses in minutes per cow and per day (dairy farm with 60 cows).

Type of cow house		Stanchion barn			Cubicle house, slats, silage cutter, pipeline milking	
Work method		littered, manual mucking-out		with grids	herringbone milking parlour	
		HF BM	MF PM	MF PM	Manual removal	Automatic removal
Over a year (365 days)	Milking	5.0	4.0	4.0	3.0	2.0
	Add. work with milking	0.9	0.9	0.9	1.2	1.2
	Various	1.5	1.5	1.5	1.5	1.5
	Subtotal (1)	7.4	6.4	6.4	5.7	4.7
Winter work (180 days)	Sweeping crib & passage	0.3	0.3	0.3	0.1	0.1
	Feeding concentrates	0.4	0.4	0.4	–	–
	Feeding silage	2.0	1.0	1.0	1.0	1.0
	Littering & mucking-out	2.0	2.0	0.3	0.1	0.1
	Subtotal (2)	4.7	3.7	2.0	1.2	1.2
Summer work	Moving animals	1.5	1.5	1.5	0.9	0.9
Winter work (1 + 2)		12.1	10.1	8.4	6.9	5.9
Summer work (1 + 3)		8.9	7.9	7.9	6.6	5.6
Man hours/cow/year without roughage production		63.7	54.7	49.8	41.1	35.0

where HF = hand feeding ; BM = bucket milking installation ; MF = mechanical (silage cutter) feeding ; PM = pipeline milking.

requirement decreases faster in cubicle houses than in stanchion barns. The labour requirement for a littered stanchion barn intended for 40 to 60 cows and with hand feeding, bucket milking and manual mucking-out amounts to more than twice that of a modern equipped cubicle house.

Table 4.4 represents the labour productivity of milking according to different methods. If one knows that handmilking, which was quite usual up to about 25 years ago, allowed the milking of only 7 to 8 cows per hour compared to 100 nowadays in a rotary fifteen-stall milking parlour, it is obvious that enormous progress has been made in dairy husbandry.

In conclusion we can say that *the care of a cow (daily and periodic activities) in a medium sized cubicle house for dairy cattle* with a far advanced rationalization of the labour

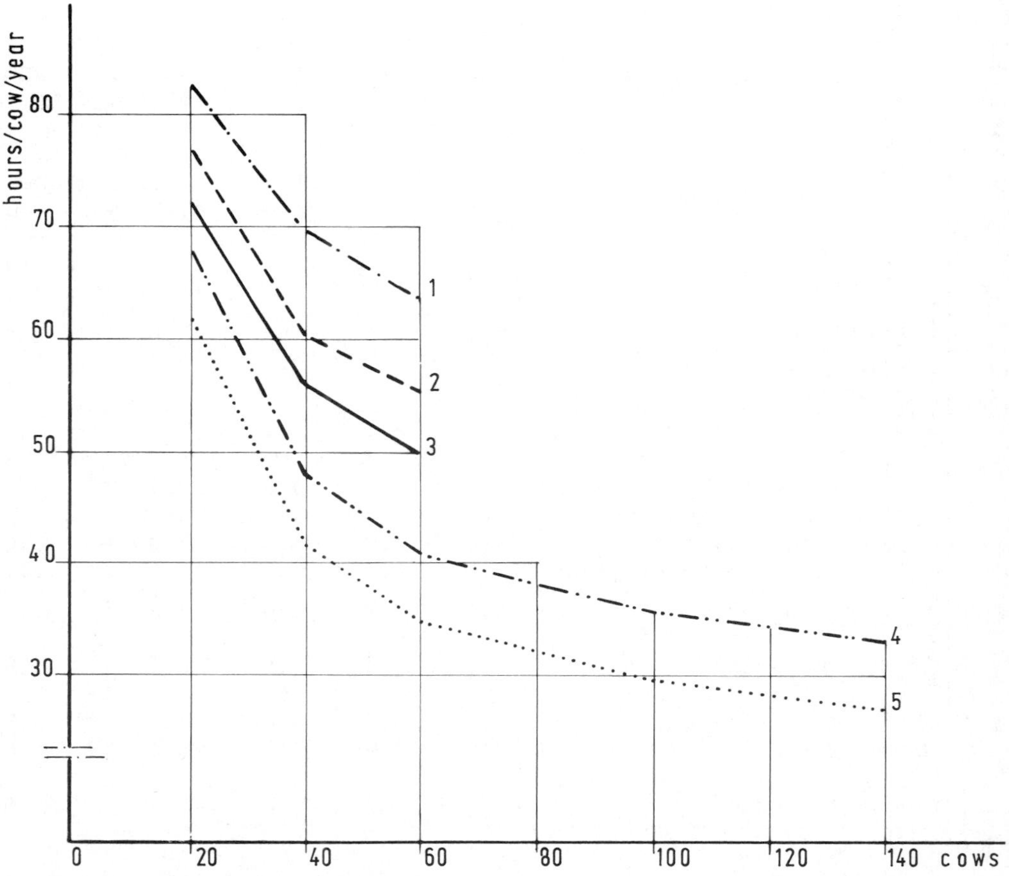

Fig. 4.66 The labour requirement (expressed in hours/cow/year) in different types of dairy houses and according to their size. Legend : 1 = littered stanchion barn, hand feeding, milking machine ; 2 = littered stanchion barn, mechanical feeding, pipeline milking installation ; 3 = stanchion barn with grids, hand feeding, pipeline milking installation ; 4 = cubicle house, mechanical unloading of silage, herringbone milking parlour ; 5 = cubicle house, mechanical unloading of silage, herringbone milking parlour and automatic cluster removal.

(cubicles, slats, herringbone milking parlour with pipeline milking into the refrigerated farm milk tank with automatic cluster removal, silage unloading machines) *requires less than 40 hours per annum* compared to 60 hours in the strawed stanchion barn.

TABLE 4.4 The labour time requirement for milking according to different methods.

House type	Milking method	Number of cows per man and per hour
Stanchion barn	Milking by hand	7.5
	Machine milking, 1 person, 2 units, buckets	24
	Machine milking, 1 person, 3 units, pipeline	30
Loose house	Tandem milking parlour (3 x 1), 1 person, 3 units, pipeline	30
	Herringbone milking parlour (2 x 4), 1 person, 4 units, pipeline	40
	Rotary milking parlour, 6 stalls, 1 person, 6 units, pipeline	48
	Herringbone milking parlour (2 x 6), 1 person, 12 units, automatic cluster removal	66
	Herringbone milking parlour (2 x 8), 1 person, 16 units, automatic cluster removal	80
	Rotary milking parlour, 12 stalls, 1 person, 12 units, automatic cluster removal	85
	Rotary milking parlour, 15 stalls, 1 person, 15 units, automatic cluster removal	100

In a specialized family dairy farm, the cattle keeper is able to look after 70 cows (excluding the care of the young stock) during the stabling period, in a fully equipped cubicle house, assuming an 8 hours working day. During the grazing period the labour time requirement for the care of the dairy herd decreases considerably at farms with a stanchion barn while it only diminishes slightly in a well equipped cubicle house. During summer the cattle keeper is also obliged to take care of the pastures and the production of roughage often with the assistance of a contractor.

4.2.3 The construction costs of dairy cow houses

The *construction and equipment costs* of different types of new cow houses were calculated on the basis of Belgian materials and wages in 1982. They are given in table 4.5. It has to be stressed that these figures have only an approximate value. Indeed construction costs vary largely according to the chosen layout, the used building materials and equipment, the inclusion of own labour, the selected ventilation system etc. We compared all the costs of building and equipping various types of houses for herds of 20 up to 140 cows and taking into account the costs required for the proper house, the straw storage barn, the manure storages, the milking parlour and the dairy, inclusive the milking machine, the refrigerated farm bulk tank, the watering facilities and the electricity supply, but exclusive the storage of roughage (Martens et al., 1980). Indoor feeding was taken as a general rule for all types of housing.

TABLE 4.5 The investment, expressed in St £/cow (a) for different types of two-row dairy houses in relation to the size of the house.

Type of cow house	Number of cows							Increment per cow (b)
	20	40	60	80	100	120	140	
Cubicle house								
- without slats	1,560	1,030	853	765	712	677	651	500
- with slats	1,640	1,085	900	808	752	715	689	530
Feed cubicles								
- without slats	1,410	943	787	709	662	631	609	475
- with slats	1,515	1,013	845	761	711	678	654	510
Stanchion barn								
- littered	1,080	780	680	630	600	580	566	480
- with grids	1,265	963	862	811	781	761	746	660

where a = inclusive a 3-month storage of manure and exclusive roughage storage, enlargement was carried out only by lengthening the house ; b = from 20 to 140 cows a constant increment is added per additional cow which covers the costs of enlargement of the house (lengthening) and adaptation of the various additional rooms.

From table 4.5 we can deduce the following facts :

- the littered stanchion barn requires the smallest investment irrespective the size of the cow house; this conclusion was acknowledged in the Netherlands by Swierstra and Van Ooyen (1979) ;
- the feed cubicles require, taking into account the saving of house space, a smaller investment than cubicle houses with indoor feeding ;
- stanchion barns with grids require a significantly higher investment than littered stanchion barns and this is mainly due to the more expensive below-ground liquid manure cellars ;
- cubicle cow houses with slats require a higher investment than cubicle cow houses without slats and this because the liquid manure storage in a long narrow channel costs more than an outdoor below-ground liquid manure cellar; the calculations were based on mucking-out by means of a tractor scraper, the purchase price of which is much lower than of an electric dung scraper ; the same conclusion is also valid for the feed cubicles ;
- from ca. 80 cows a cubicle house with slats requires a lower investment than a stanchion barn with grids.

In the past years there has been a considerable rise in building costs : indeed, in November 1970 the construction and equipment of a cubicle cow house amounted to £ 485 Sterling per cow, in November 1975 to £ 625 per head and in 1982 to £ 1,000 per head thus prices have more than doubled in 12 years. This and other facts mean that the construction and equipment of a new cubicle house for 50 cows requires an investment of ca. £ 50,000 (1982).

The annual costs arising from the construction and equipment of dairy houses are approximated below.

	Buildings	Machines
- depreciation	6.7 %	14.0 %
- interests	5.0 %	5.0 %
- maintenance, repairs, insurances	0.8 %	3.5 %
Total	12.5 %	22.5 %

of the original value

Based on tables 4.3 and 4.5 the annual costs of cow houses have been calculated including the cost of labour and the expenses made for buying straw (5 pence/kg, inclus. the cost of storage) and expressed per head and per year, in relation to the type of housing. Wages were calculated at £ 3.40 per hour. These costs are summarized in fig. 4.67.

The following conclusions can be drawn :

- a littered stanchion barn incurs the highest costs from a herd of over 30 cows ;
- the stanchion barn with grids is the cheapest house for a herd of ca. 30 cows ;
- a loose house with slats whether it is the cubicle type or the feed cubicle type requires lower annual costs than a loose house without slats ;
- from ca. 30 cows, the feed cubicles with slats in the alleys are the least expensive in annual costs, the difference with a cubicle house with slats in the alleys is however very small and decreases as the size of the house increases ;
- the decrease in annual costs per head in relation to the size of the house is more important for loose houses than for stanchion barns ;
- the yearly costs per cow of a cubicle house for 140 cows are less than 2/3 of those for a similar house for 40 cows.

Remarkable is the high level of annual costs of new-built dairy houses. They amount to 4.5 p to 9 p/litre of milk according to the type of house and the size of the herd and are based at a rated milk yield of 4,500 litres per cow and per year. Since the major budget item, the cost of feed, is not included in the above-mentioned price it becomes clear that only a small percentage of dairy stockmen with new-built houses can earn the aforesaid £ 3.40 per hour at the actual milk prices.

In any case, our calculations clearly demonstrate that the cubicle cow house is from 30 cows onwards, the most labour saving and economical way of housing dairy cattle.

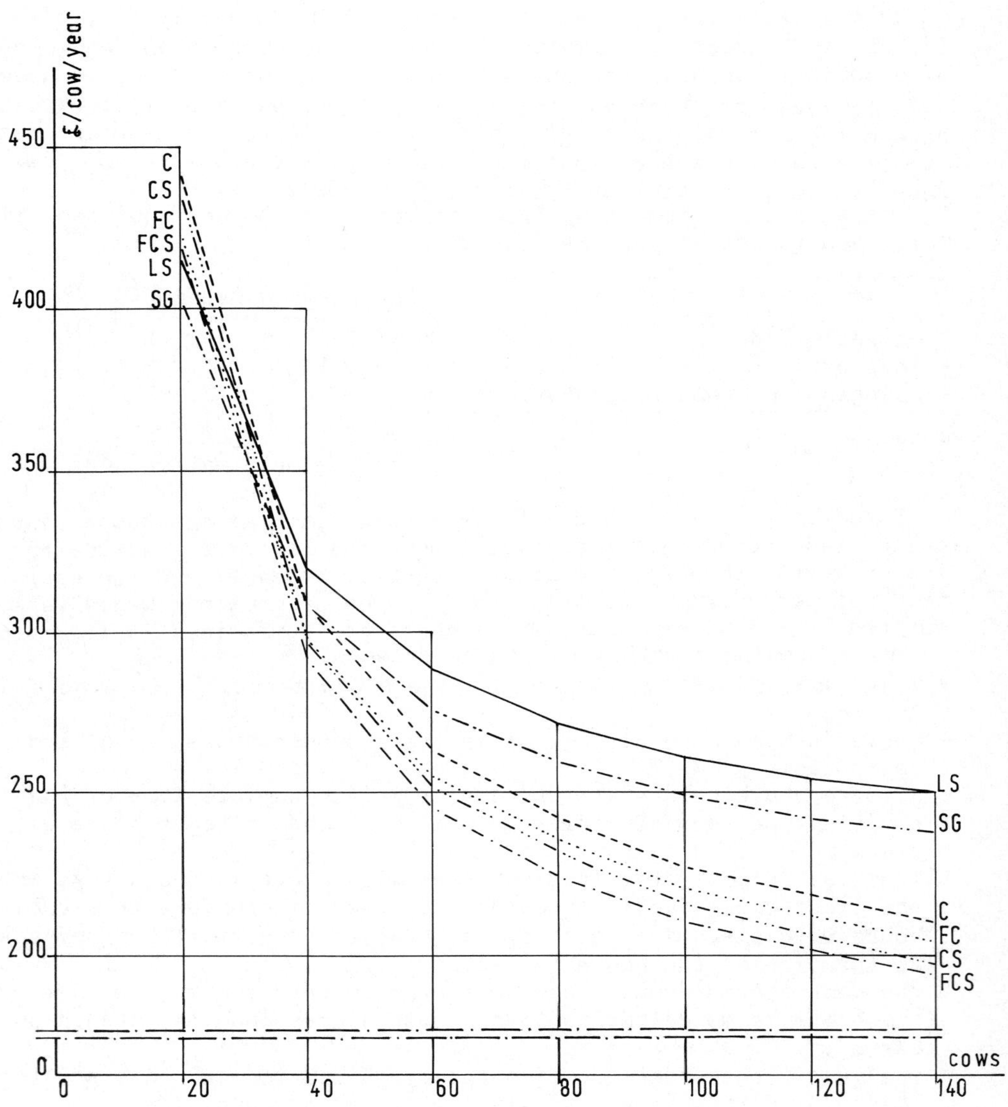

Fig. 4.67 The annual costs (expressed in £/cow) of a dairy house (buildings, equipment, labour, straw) in relation to its size. Legend : CS = cubicle house with slatted floors in the passages ; C = cubicle house without slats in the passages ; FCS = feed cubicles with slats in the passages ; FC = feed cubicles without slats in the passages ; SG = stanchion barn with grids ; LS = littered stanchion barn.

4.2.4 Zootechnical and veterinary aspects of the housing of dairy cattle

Although the relationship between housing on one hand and health and performance on the other is still not completely known in all its aspects, we are nevertheless able to give a brief insight in this based on both our own research (Maton and De Moor, 1973 ; 1975) and on information from the literature on this subject. The housing of dairy cattle influences both the behaviour of the cows as well as the incidence of traumata and claudications, of reproduction and birth disorders and of respiratory disorders (Maton and De Moor, 1973 ; 1975).

Up to a few years ago we used the method of time lapse photography. A special camera takes a picture at regular intervals (e.g. every five minutes) of the animals in the house and this during 50 hours. During the night the pictures are taken with the aid of a flash-light. The film is analyzed, after development, by means of a projector. This method is based on discontinuous observations whilst the time passes continuously. Since 1979 we have adopted a second method : the closed circuit television. An infra-red sensitive television camera is installed inside the animal house. The behaviour of the animals, filmed by the T.V.- camera, is recorded on magnetic tape by means of a video recorder. By slowing down the tape speed from normally ca. 50 to ca. 0.7 images per second it is possible to record the animal behaviour during ca. 89 hours on a normal videotape (the so-called time lapse video recorder). The illumination is by infra-red flood light. With the inclusion of a time generator (clock) it is possible to include the time (hours, minutes, seconds) and the date (day, month, year) on the recording. In such a manner continuous observations are possible. The video tapes can later be watched on a television monitor and the behaviour of the animals can then be analyzed. The first method was applied by us in a research covering four stanchion barns viz. two strawed ones and two with grids and four loose houses of which two were littered, the other two being cubicle houses (Maton and De Moor, 1975). All cow houses were, except for a few details, built secundum artem. The results of these observations are summarized in tables 4.6 and 4.7.

We can observe that dairy cows take 6 to 7 % more rest by lying down or ca. 1 1/2 hour more per day in loose houses than in stalls of stanchion barns with a breadth of 1.0 m. There is no difference in the total rest time per day between strawless and littered housing. Eichhorn (1965) also came to the conclusion that dairy cows spent identical times lying down both in cubicle houses and in littered loose houses. We found however that the average duration of each lying period was 15 to 30 minutes longer when straw was applied than in strawless houses, which means that in the latter type of housing the cows will get up and lie down more frequently.

This makes us believe that loose houses offer more comfort to dairy cattle than stanchion barns with a lying bed having a width of 1.0 m and that the use of straw contributes in achieving an even higher com-

TABLE 4.6 The activities of cows in different types of houses (expressed in hours and % of a day).

Cow house type	Lying down		Standing up, incl.milking without (b) (a)		Eating and drinking (b)		Not lying down (a) + (b)		Number of cows
	% of 24 h	h	% of 24 h	h	% of 24 h	h	% of 24 h	h	
Stanchion barn									
- with straw	46.4	11.1 ± 1.8	35.3	8.5	18.3	4.4	53.6	12.9	11
- with grids	46.1	11.1 ± 1.7	39.5	9.5	14.4	3.5	53.9	12.9	12
Loose house									
- with straw	53.0	12.7 ± 2.1	29.2	7.0	17.8	4.3	47.0	11.3	27
- with cubicles and slats	53.1	12.7 ± 1.0	29.5	7.1	17.4	4.2	46.9	11.3	39

TABLE 4.7 The total and average lying down time of cows in different types of housing (expressed in hours per day).

Cow house type	Total time of lying down	Number of lying down periods	Average time per lying down period
Stanchion barn			
- with straw	11.13	7.1	1.57 ± 1.03
- with grids	11.07	9.6	1.15 ± 1.00
Loose house			
- with straw	12.72	8.1	1.57 ± 1.40
- with cubicles and slats	12.75	10.1	1.27 ± 0.95

fort of the cows. Our research (Maton et al., 1978) has however demonstrated that the difference in lying time between a stanchion barn and a loose house is partly due to the fact that the width of the stall is only 1.0 m in the examined stanchion barn.

A stall width of 1.1 m in a stanchion barn was found to give a higher lying down time than a width of 1.0 m. It also appeared that lying beds of 1.0 m made it rather difficult for cows to lie down simultaneously. Wider stalls result in a significant increase of the lying down time. The analysis of the films clearly showed that in stanchion barns with lying places having a width of 1 m a cow was often obliged to get up when her neighbour wanted to lie down. A lying place of 1 m thus seems unsatisfactory. In a cubicle house it sometimes happens that a lying cow has to interrupt her rest prematurely because a higher ranking cow comes into her vicinity (Maton and De Moor, 1975).

Table 4.8, based on research carried out by Eichhorn (1965), shows that the cows are rather restless during the first three weeks of their stay in the cubicle house (too few lying down and too many standing up in feed places and slatted passages). Lying down on the slatted floors disappears after one week and from the fourth week onwards the cows make ample use of the cubicles, mainly for lying down.

TABLE 4.8 Utilization of the different spaces in a cubicle house by the cows (5 cows, 6 weeks, observation time : 21 hours per day).

Duration in h	Cubicles		Slatted passages		Feed stance	
Week nr	lying down	standing up	lying down	standing up	lying down	standing up
I	7	32	3	38	0	20
II	16	24	0	20	0	40
III	7	67	0	0	0	26
IV	52	32	0	12	0	4
V	53	24	3	10	0	10
VI	48	28	0	15	0	9

Furthermore, Wander (1974) found that with increasing noise levels in the cubicle house, the lying time of the cows decreased whilst the standing up time increased. He found no difference in the degree of occupation (= % of cubicles occupied) between a cow-kennel and an insulated cubicle house. Research carried out by Maton et al. (1981) using an infra-red television camera concerned the lying behaviour of animals in a cubicle cow house. Six different types of lying floors were scattered over an entire house with 80 dairy cows. Identical floors were never installed next to each other. The results of this research are summarized in table 4.9.

TABLE 4.9 The degree of occupation (in % of a day) of cubicles furnished with different floor top layers (Maton et al., 1981).

Floor top layer	Degree of occupation in %
A. Cement-sand screed	29.4
B. Bernit	19.6
C. Enkamat-K	59.8
D. Eterspan	29.2
E. Rubber mats	40.8
F. Stallit	34.8

Statistical (in-) equalities C > E > F > A = D > B with $p < 0.05$.

From table 4.9, concerning the degree of occupation of the different lying floors the following conclusions can be drawn :

- the cows show an apparent preference for a certain type of lying floor if they can make a choice from several floors ;
- soft floors are significantly more used by the cows than hard floors, whereby Enkamat-K obviously has the highest preference ;
- no systematic preference is made by the cows between hard floors, whether they are insulated or not.

Similar conclusions were reached by a number of other researchers. According to Hedren (1966) cubicles with a soft lying floor are more often used. Andreae and Papendieck (1971) report that cows do not prefer hard floors. In a comparative research over three whole days, seven cubicles covered with rubber mats were compared with 5 cubicles provided with a hard insulating top layer. The cubicles however were not littered. The rubber mats attained a degree of occupation of 41.5 % compared to only 5.4 % for the different hard insulated top layers. Gjestang and Gravås (1976) obtained similar results. In a feed cubicle house they compared concrete lying floors with two different types of rubber mats. For both rubber mats they found a degree of occupation of 42.9 % and 46.1 % compared to only 25.9 % for concrete floors. Wander (1971a ; 1971b ; 1974), who carried out a large number of behavioural studies, came also to the conclusion that the softness of a lying floor is more important than its thermal properties. This explains the preference of the cows towards the Enkamat-K, a soft synthetic mat.

Westendorp and Hakvoort (1977) found that cows preferred to lie down on a place with a covering in the form of a mat rather than on a hard concrete floor with only a little littering ; this appeared also from our investigations. Bedding can however play an important role in the degree of occupation of the cubicles. Especially on hard floors, a thick layer of littering can create a higher degree of occupation.

From our observations (Maton et al., 1981) it also appeared that cubicles in a double row in the middle of the house are significantly more occupied by the cows than cubicles aligned in one row along a wall. Also the cubicles, situated at an extremity of a row are significantly less occupied by the cows than those situated in the middle of a row, which is confirmed by several investigators (Dregus et al., 1979 and Westendorp and Enneman, 1975). Cubicles situated at places where animals pass frequently are also less used such as those in the neighbourhood of the milking parlour, near the water troughs and crossings of passageways, etc., which has also been confirmed by Porzig (1969) and Dregus et al. (1979). A board or raised edge at the rear of the cubicle also exercises a negative influence on the degree of occupancy and the behavioural pattern and should therefore be excluded ; this is confirmed by Westendorp and Hakvoort (1977). The animals will usually lie down in the same cubicle, if that is possible and the hierarchical ranking order is here also of importance. The higher ranking animals obviously appropriate the best cubicles, which was also shown by Porzig (1969) and Thines et al. (1975).

From the traumata, partly or fully originating from the conditions of housing, teat lesions are certainly the most important besides the claudications which are also often of traumatic origin and bound to the housing. Teat lesions are often caused by teat tramping (fig. 4.68), mainly by the cow itself or sometimes by her neighbour. They are well-known predisposing factors in the development of infections and especially of mastitis. Claudications are in 90 % of the cases caused by claw lesions, mainly of the hind legs (Carotte, 1974; Nygaard and Birkeland, 1975). These lesions can be seen as very important and between the different claw lesions there exist clear interdependencies.

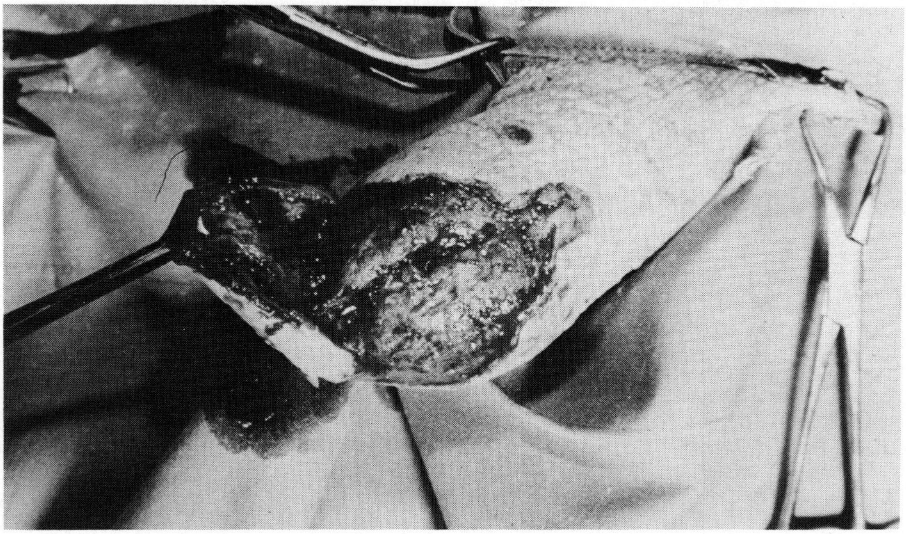

Fig. 4.68 Teat trauma caused by teat tramping.

Double soles are the result of an extensive blister of the corium and a loose horn wall is sometimes caused by a serous or purulent blister at the height of the white line ; moreover a blister can develop under a traumatic loosened horn wall. Ulcers are often a result of an infectious interdigital dermatitis (fig. 4.69) while an infectious arthritis of the claw joint can be caused by an ulcer or by panaritium (Maton and De Moor, 1973). Purulent bursitis of the bursa podotrochlearis is generally a result of a secondary inflamed ulcer. A comparative survey done by Maton and De Moor, (1975) in a large number of cattle houses regarding the incidence of teat lesions, claw lesions resulting in claudications and lesions of the limbs has led to the results summarized in table 4.10.

This research was carried out on a large number of cows and on Belgian farms with an average of 25 to 50 cows. The farms can be described as modern and well-equipped which can only enhance the value of the obtained results.

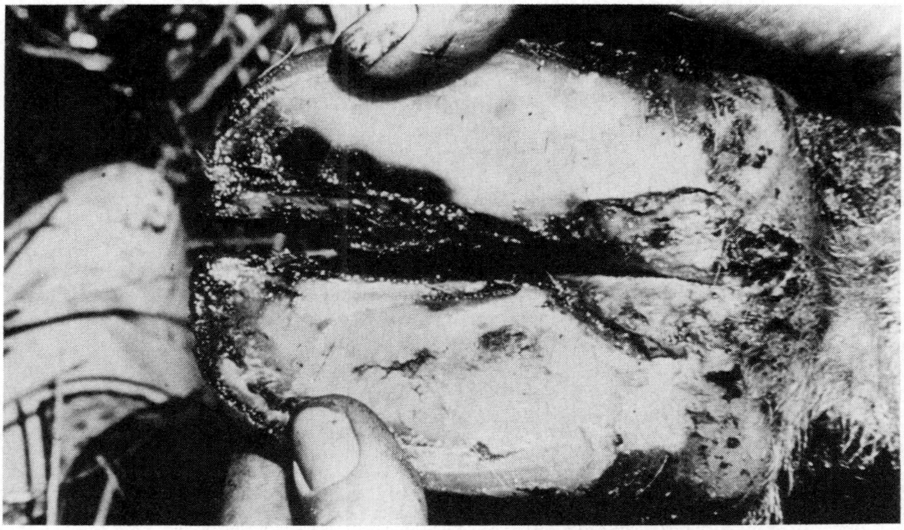

Fig. 4.69 V-shaped fissures and undermined sole with infectious interdigital dermatitis.

TABLE 4.10 Traumata and claudications with dairy cattle as related to the type of housing.

Cow house type	Number of farms	Number of cows	% cows with teat trauma	% cows with claudicat. caused by claw lesions	% cows with limb lesions
Littered stanchion barn	1,149	28,048	3.40	3.03	1.09
Stanchion barn with grids	706	20,995	5.35	1.86	1.01
Strawless cubicle house	279	13,236	1.85	5.61	0.58
Littered loose house	88	3,454	1.09	2.90	0.62
Total or average	2,222	65,733	3.59	3.17	0.94

First of all, it appears that under Belgian conditions and in well-equipped animal houses the incidence of both teat traumata as well as limb injuries are relatively low : according to the type of housing mostly less than 2 %, sometimes up to 5 %, of the cows are affected. Minor disorders were however not taken into account. G. Carotte (1974) mentions the results of an inquiry carried out in France. It seems that 6.8 % of the cows are affected by bursitis praecarpalis and 9 % by traumata caused by tramping ; some houses were well-equipped while others were less.

Moreover, it seems that the incidence of teat tramps is 2 to 5 times higher in stanchion barns than in loose houses. This partly explains the higher incidence of mastitis in stanchion barns compared to loose houses (Eichhorn, 1965).

Amongst the stanchion barns, those provided with grids showed a greater incidence of teat tramping than the strawed type : this confirms the results of a previous study (Maton and De Moor, 1973).

Limb injuries occur twice as much in stanchion barns than in loose houses whereby no differences were found between littered and unlittered houses. The lack of movement thus plays an important role. Grommers (1968) found that carpal-swellings and stanchion barn deformed limbs occurred fifty times more in littered stanchion barns than in littered loose houses.

From our research (Schepens et al., 1980) carried out in collaboration with the Faculty of Veterinary Science of the State University of Ghent (Belgium) it seems that claw lesions are much more frequent in cubicle houses than in all other stanchion barns. The lowest incidence of claw lesions is found in good stanchion barns with grids. This can probably be explained by the fact that in this type of housing the claws are the least in touch with humid manure. Standing in humid manure causes the softening of the claws and especially of the interdigital interstice of the claw whereby certain infections (Spherophorus necrophorus, Corynebacterium pyogenes, Fusiformis nodosus), penetrating via small injuries, can more easily occur. We have to mention however that the soles become especially hard and dry in stanchion barns with grids and this will aggravate the care of the claws considerably. Further details of our study (Maton and De Moor, 1975) show that from a total of 13,158 cows, housed in cubicle houses equipped with slatted alleys 628 or 4.77 % were affected with claw lesions, while from 15,096 cows in cubicle houses with full concrete alleys 956 or 6.33 % manifested such injuries. This difference - which could have been more pronounced if not so many badly constructed concrete slats had produced injuries and inflammation of the claws - can be explained by the observations of Spindler (1973) who clearly demonstrated that the horn of the claw wears out twice as fast on humid concrete than on dry concrete. A fully concreted alley, mucked-out by means of an automatic scraper (fig. 4.24) remains wet. An alley made from concrete slats is clearly less humid since the urine is immediately drained and the danger of claw lesions is less pronounced. This is acknowledged by Coenen (1980) who found 6.4 % cows infected with interdigital dermatitis in a cubicle house provided with slatted alleys compared to 17.2 % in a cubicle house with fully concreted alleys. Roughly finished floors in alleys and exercise areas or jagged edges of slats inevitably lead to claw injuries. From our research (Schepens et al., 1980) we were however able to conclude that the management level and the ability of the farmer were contributory in obtaining large differences in the incidence of claw lesions and this irrespective of the type of house.

Johnson et al., (1969) also stressed the importance of management and the environment in the occurrence of claw lesions. Prange (1968) attaches great importance to the quality of the run : a surplus of remaining liquid, damaged floors in the collecting yard, the stalls and the passages can give rise to interdigital dermatitis. Koller et al., (1979) explicitly stresses the fact that the influence of the cattle keeper on the health and production of the animals is much greater than that of the type of cow house.

From an analysis and an extensive study of the literature done by Brandsma (1982) it appears that cows on the whole enjoy a healthier way of life in a loose house than in a stanchion barn : Jörgensen (1972) found that cattle keepers with loose houses had to call in a veterinary surgeon 131 times/100 cows/year, while those with cows in a stanchion barn with grids 188 times and those with a littered tying barn 175 times. Of all the calls 42.2 % were in relation to mastitis, which however occurred less frequently in loose houses. Thamling (1980) found that the annual expulsion in stanchion barns in Schleswig-Holstein amounted to 33.7 % compared to 30.2 % in loose houses. Farms with a stanchion barn showed according to Vecht et al. (1980), on average a higher incidence rate of subclinical inflamed udder quarters and increased cell counts compared to cubicle houses. Since Dijkhuizen and Renkema (1977) demonstrated that these abnormal udder quarters produce on average 20 % less milk than healthy udders, Brandsma (1982) attributed a loss in milk yield of 50 to 100 kg/cow/year to differences in the health condition of the udders of cows in a stanchion barn compared to those in a loose house. Interesting is also the fact that acetonaemia - a disease of the metabolism which leads to severe disturbances of the digestion and characterized by a strongly increased level of ketones in the blood - can be avoided much easier in loose houses than in stanchion barns. The fact that the animals have more exercise in the loose houses indeed leads to an increased muscular cetolysis (Cottereau, 1974).

A number of parts of the littered stanchion barn were especially examined by us for their influence on teat traumata and claudications. The results of our investigation concerning the type of tie, the dimensions of the stall (length and width) and the use of stall partitions are summarized in table 4.11. In the stanchion barn, the tying consisting of U-shaped tubes and whereby the cows wear a chain around the neck, fitted to a bar which in turn glides over two inverted U'-s (U-tubes with transversal bar) turned out to cause more teat traumata and claw disorders. The ordinary "vertical" chain (fig. 4.5) in this context is one of the best tying types. If this tie is provided with a nylon strap, eczema in the neck area can sometimes occur with certain cows.

Further analysis of our observations concerning the American yoke (fig. 4.3) gave the following results and are represented in table 4.12. The application of the American articulated yoke clearly reduces teat tramping compared to the American yoke with stiff bars. It appears thus that tying types which hamper the smooth rise of a lying cow will cause more teat tramping and claw disorders.

The incidence of teat traumata and claw injuries diminishes as the stall becomes longer, further research is however necessary to establish the appropriate length of the stall in relation to the breed. With too short a stall platform the cow often stands and lies with her feet respectively teat on the rear edge of the stall platform which can be the cause of irritation or trauma. Too long a platform will cause excessive soiling of the cow. The incidence of teat traumata

TABLE 4.11 The incidence of traumata and claudications in relation to the different parts of a littered stanchion barn.

Tying type	Vertical chain	Vertic. tie with nylon strap	American yoke	American yoke with group fastening	U-tubes with chain	U-tubes with transversal bar
n =	18,332	10,134	3,236	11,878	7,478	4,889
% cows with teat trauma	2.93 %	3.37 %	3.28 %	3.83 %	3.01 %	4.15 %
Claw lesions with claudications	2.42 %	3.36 %	2.29 %	2.51 %	3.88 %	5.15 %
Injuries of the limbs	1.03 %	1.05 %	1.39 %	1.77 %	0.63 %	0.72 %
Stall length (cm)	145-150	150-155	155-160	160-165	> 165	
n =	2,972	9,365	14,067	19,529	13,695	
% cows with teat trauma	4.85 %	4.21 %	3.67 %	3.19 %	2.55 %	
Claw lesions with claudications	3.77 %	4.39 %	2.77 %	2.68 %	2.61 %	
Injuries of the limbs	1.38 %	0.97 %	0.73 %	1.43 %	1.09 %	
Stall width (cm)	95-100	100-105	105-110	110-115	≥ 115	
n =	22,534	33,052	5,051	1,032	421	
% cows with teat trauma	3.21 %	3.62 %	2.99 %	1.84 %	1.90 %	
Claw lesions with claudications	2.43 %	3.39 %	3.68 %	1.84 %	2.61 %	
Injuries of the limbs	1.28 %	1.02 %	0.83 %	0.67 %	0.71 %	
Stall partitions	present	absent				
n =	46,015	15,965				
% cows with teat trauma	3.30 %	3.57 %				
Claw lesions with claudications	2.85 %	3.54 %				
Injuries of the limbs	0.92 %	1.16 %				

TABLE 4.12 Incidence of teat traumata with the American yoke.

Tying type	Individual tying type		Group tying type	
	with stiff bars	with articulated bars	with stiff bars	with articulated bars
Number of cows	938	873	2,744	5,917
% cows with teat traumata	5.22 %	1.49 %	4.74 %	3.50 %

and also of injuries to the limbs decreases as the width of the stall becomes larger and the stall should be 110 cm wide. With too narrow stalls there is mutual hindering between the animals and this increases the risk of injuries while it decreases the comfort of each individual animal (see earlier). The absence of stall partitions also seems to influence, although to a lesser extent, the occurrence of teat traumata and injuries of claws and limbs. The provision of such a partition between every two animals will decrease this incidence.

A thorough investigation carried out by us in situ (Maton and De Moor, 1973) in twenty strawless stanchion barns with grids showed that the incidence of teat tramping and/or disorders of claws and limbs was increased after the omission of the rubber carpeting of the concreted platform of the stall and of the stall partitions, and by the use of either a tying system made from U-shaped tubes with a transversal bar or the non-articulated American yoke.

In the same study (Maton and De Moor, 1973) we were able to prove the necessity for a stall in a strawless stanchion barn with grids to be 140 to 150 cm long and 110 cm wide whilst the grids must consist of three flat galvanized bars with a width of 2 cm (bars under the hoofs) behind which round galvanized rods of 1 cm diameter are installed (free passage of manure) ; the spacing between the rods must be 3.5 to 4 cm. The presence of grids is however troublesome for the veterinary surgeon when carrying out difficult deliveries.

Besides claw injuries, which occur quite often as explained before, a number of other traumata can occur in the cubicle cow house. The dimensions and the form of the stall partitions, which are made of metal tubes, play an important role in this connection. The animals must indeed enjoy a maximum of comfort. Too short a stall, a neck bar which is put too far to the rear in the cubicle, or too hard cubicle flooring, will force the animals to lie at the rear upon or against the concrete lip of the box and this will frequently result in bruises or swellings to the ischial area (fig. 4.70). If the cubicle is too narrow and if the lower horizontal tube of the cubicle partition is mounted too low the cow will easily bruise the ilium flank (fig. 4.71). If the lower tube however is installed too high the cow will try to pass under it, becomes stuck and gets hurt when she tries to get free. Westendorp (1973) also pointed to these dangers. We know, from experience, that the cubicles as shown in fig. 4.17 give complete satisfaction. Dehorning the cows is strongly recommended in a loose house especially to prevent traumata due to butting with the

Fig. 4.70 Swelling at the ischial area due to a badly designed cubicle.

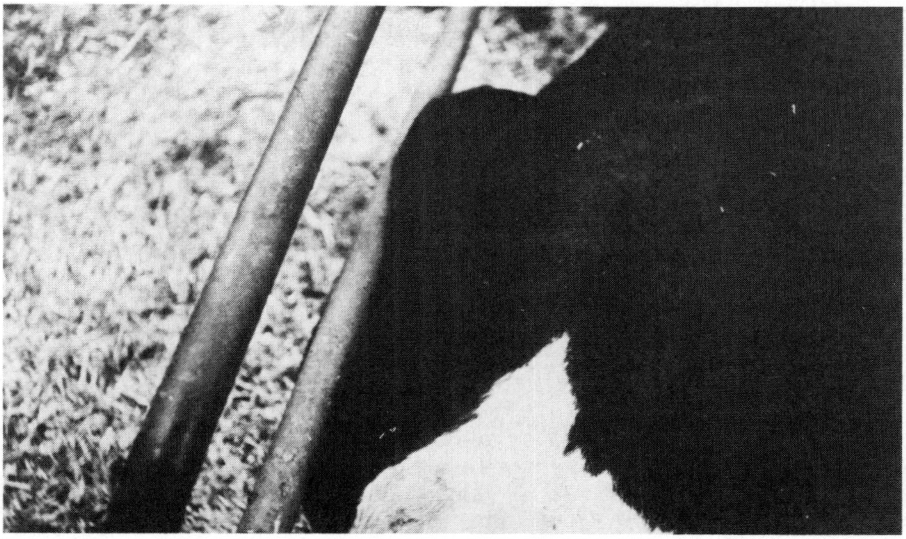

Fig. 4.71 Contusion of the ilium flank due to a badly designed cubicle.

horns : although in most of the cases this will not result in injuries it can sometimes lead to superficial or deep contusions and even to penetrating wounds of the abdomen and to abortion.

In relation to reproduction and birth we have to mention first of all that the oestrus is much easier to detect in a loose house than in a stanchion barn, where a "silent heat" and subsequently late servicing can sometimes occur. Mounting can in a loose house lead to traumatic lesions. According to De Kruif (1975) cows come earlier on heat after calving in a loose house than in a stanchion barn. The insemination results should be better in cubicle houses than in stanchion barns : Willems (1971) observed an average insemination result, after a first mating, of 66.7 % in an inquiry carried out at 49 farms with cubicle houses compared to an average of only 57.6 % in 306 stanchion barns. In the same inquiry, Willems (1971) compared the insemination results, obtained in 1969 - 1970, in cubicle houses with those obtained in 1966 - 1967, at the same farms which then housed their cows in stanchion barns : the insemination results rose from 56.3 % to 66.7 %.

In a comparative research of the course of the births in littered stanchion barns compared to littered loose houses and covering well over 1,200 calvings, Grommers (1968) found that with primiparae the partus progressed normally (i.e. without aid) in 21.5 % of the cases in loose houses compared to only 15.9 % of the parturitions in stanchion barns ; with pluriparae the results were respectively 57.7 % and 33.7 %. In a research with 43 uniovular twins he was not able to confirm these differences. The same author however proved that the expulsion of the calf progressed faster in a loose house than in a stanchion barn (Grommers, 1968) as illustrated in table 4.13.

TABLE 4.13 Average duration of the parturition in relation to the type of housing (Grommers, 1968).

Parity	Average duration of the expulsion from the moment of observation of the allantois until the end of the spontaneous parturition (in min)	
	Loose house	Stanchion barn
Primiparae (10)	154	161
Pluriparae (25)	98	129

Grommers (1968) also observed, in littered loose houses 5 cases of prolapsus uteri on a total of 1,281 calvings, compared to 20 cases on a total of 1,233 calvings in strawed stanchion barns. In the latter type of house it is sometimes observed that cows stand with their hind feet on the lower situated dung plate instead of lying or standing on the stall and this can encourage a prolapsus. Metritis however is more easily spread in loose houses than in stanchion barns.

Concerning the dangers for the occurrence of respiratory disorders the following can be said. In earlier days some thought that low temperatures, might be harmful to dairy cattle and there was even a tendency towards the insulation of loose houses. In our opinion this is not necessary in a temperate climate : even the open loose houses offer no danger to cattle health if the exercise yard is directed away from the direction of the prevailing cold winds. Souty (1961) mentions an investigation of Bianca which has in fact shown that the "comfort-zone" for the European cattle is situated between 0° and 15°C and that the milk yield reaches its maximum in this temperature interval although the optimum is 10°C. According to Yeck et al., as mentioned by Hoogerkamp (1971), who investigated the relationship between ambient temperature and milk yield, with the Holstein breed, it appears that the milk yield at -16°C is equally good as at higher temperatures. Very high temperatures (25°C and higher) can however cause an important drop in the milk yield. Zeeb and Schmidt (1973) found that temperatures below 12°C were more "cow-friendly" than temperatures above 12°C. It is a proven fact that high yielders tolerate low temperatures better than high. The anatomo-physiological explanation herefore includes the fact that cattle possess anastomoses i.e. short-circuiting connections between arteries and veins which are mainly situated in the extremities. With severe cold the arterial blood can thus return via the veins to the trunk before reaching the extremities. The peripheral blood circulation is slowed down whereby too strong a cooling of the body of the cattle is avoided (Comberg and Hinrichsen, 1974).

It is however necessary to avoid draughts which are often the cause of respiratory disorders.

In a temperate climate one doesn't have to be afraid of too low temperatures in loose houses, but it might be necessary to administer a larger feed supply in cold periods to keep the milk yield at its normal level.

The ventilation always needs attention (open ridge, stacks) both in the insulated stanchion barn as well as in the uninsulated loose house. The air inlets have to be designed in such a way that the cold entering air causes no down draughts on the animals.

4.3 THE HOUSING OF BREEDING CALVES AND YOUNG STOCK

4.3.1 The construction and equipment of a house for breeding calves and young stock

Suitable housing for breeding calves and young stock is important in obtaining a lower death-rate and a better growth of the animals while it also enables a rational organization of the care of the animals. The housing of calves and young stock can, depending on the age and the care, take place in different ways as illustrated in table 4.14.

The breeding of young stock is mostly carried out at the dairy farm, whereby a separate house is reserved for the young animals. Exceptionally the breeding of young stock to heifers and the dairy cattle husbandry are considered as two separate specializations run at two different farms. In this way a few specialized farms originated, which against a daily allowance breed female calves to heifers.

TABLE 4.14 The possible types of housing for young stock of different age categories.

Age	Types of housing	
0 - 1/2 month	individual	- calf pen
1/2 - 3 months	individual	- calf pen - tying stall
	group housing	- pens with straw on a grating of laths - pens with littered lying bed and slatted-floor alley - cubicles - feed cubicles
3 - 22 months	individual	- tying stalls
	group housing	- fully slatted floor - cubicles - feed cubicles
22 - 24 months	adapted to the housing system of the dairy cattle	

4.3.1.1 The calving pen

In a stanchion barn the cows preferably remain in the same stall as they calve. In a loose house however it is essential that the birth of a calf takes place in a suitable environment. A copious strawed calving pen is therefore indispensable in a loose house. The cow is allowed in shortly before calving and is free to walk around in it. The most suitable place to build the calving box with respect to the loose house remains a point of discussion. An open, isolated box can be preferred which is situated inside the dairy house and which allows swift calving. But since the construction of a cubicle house implies a rather spacious area per cow and uninsulated walls one might also prefer a more insulated room, outside the loose house where the temperature can be higher.

The calving box (fig. 4.72) has to be sufficiently spacious to prevent the cows of getting blocked during calving and to allow efficient assistance during the delivery. The size of the box has to be at least 3 m x 3.5 m. The calving box needs an easy access and has to be draught-free, well-lit and easy to reach and clean.

In practice it is sometimes found that the calf remains with the dairy cow in the calving box for three more days. A number of disadvantages are attached to this usage :

- extra space is required in the calving pen ;
- there is no supervision whatsoever on the intake of colostrum by the calf ;
- the risk of infection is greater ;
- there is unrest both with the calf as well as with the cow when

Fig. 4.72 A strawed calving pen.

they are separated from each other after three days ;
- the cow needs to be milked additionally.

Therefore it is advisable to remove the calf from the cow immediately after birth. The cow can be kept in the calving pen for a few more days after which she is returned to the loose house. A calving box is spacious and therefore also expensive. As a consequence it is sometimes replaced by tie stalls, mostly with grids to avoid littering. Its construction is similar to the above-mentioned stanchion barn with grids. A somewhat loosened tying is provided together with a stall width of 1.20 metre and a removable stall partition to avoid any possible hindrance to the veterinary surgeon whilst performing a caesarian section.

4.3.1.2 The housing of new-born calves in individual pens

New-born calves have to be individually housed during the first weeks. Since a calf is very susceptible to all kinds of infections, the death rate is the highest during the first weeks. Special attention has to be paid to hygiene and the climatic conditions. Individual housing is therefore essential and offers the following advantages :

- licking, ear sucking, navel sucking or tramping is avoided ;
- there is less risk of a spread of infection ;
- the stockman can exercise a closer supervision on the health and the appetite of the animals.

The milk feeding system has a dominant influence on the behaviour of calves during their first weeks. If the suckling need of the calves is satisfied at an early age, problems with cross-suckling between calves are lessened when they get older ; the use of a teat-feeding device is therefore advisable and it also increases the daily growth and the daily lying time (Gjestang, 1983 ; Mees and Metz, 1984). Although group-pens, provided with confinement to the feed-front and teat-pails, lead to less skin lesions and improved cleanliness with the calves than individual boxes, these advantages do not match those mentioned above for individual pens.

The dimensions of the individual calf pens depend on the time that the calves will spend in them. When they stay only 2 weeks a width of 60 to 80 cm and a length of 1 m to 1.25 m are sufficient (fig. 4.73).

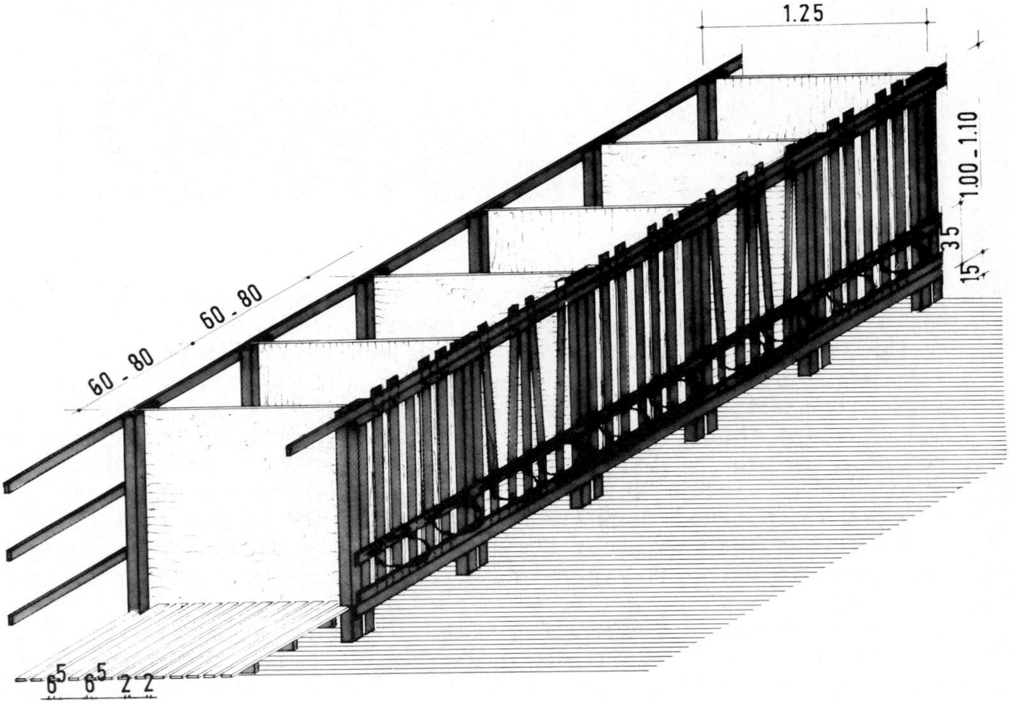

Fig. 4.73 Individual boxes for calves up to two weeks old.

If calves are kept in pens for 4 to 8 weeks the dimensions ought to be respectively 75 cm x 150 cm and 100 cm x 180 cm (Mitchell, 1976). The calf needs a sufficient degree of freedom of movement to allow a good development of the osseous system of the muscles. For this reason a stanchion barn is not suitable at this early age. In Sweden, the use of a stanchion barn is even legally forbidden for calves under the age of 12 months (Anon., 1974b).

Individual calf pens are generally built along a wall. The width of the feeding passage is ca. 130 cm for a single row of boxes and ca. 160 cm for a double row. They have to be constructed in such a way that they can easily be cleaned and disinfected. The partitions are often made of boards such as asbestos cement boards which can be removed and thoroughly cleaned.

If the calf remains in the pen up to an age of 8 weeks a bucket with concentrates has to be supplied besides a bucket filled with water and a hay rack (fig. 4.74). The required number of boxes is then much higher and one has to provide at least 1 individual pen per 4 dairy cows, depending on the calving pattern (Toren, 1978).

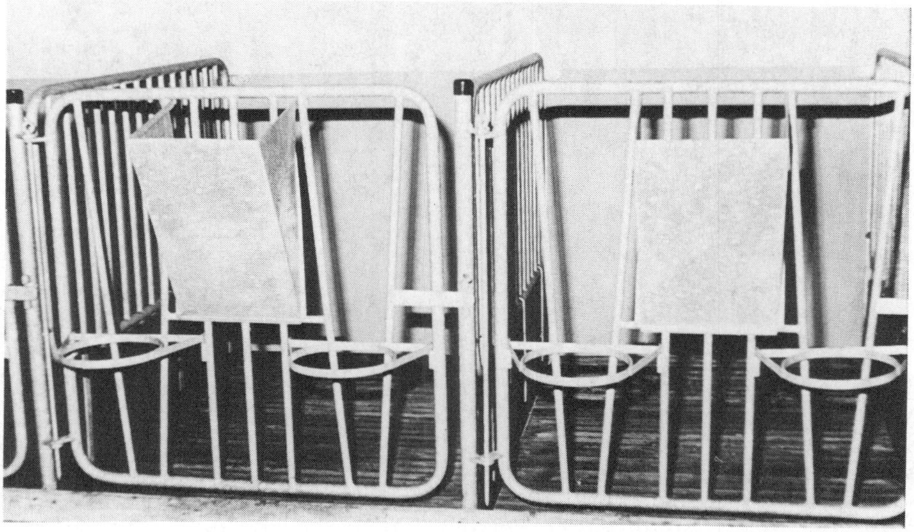

Fig. 4.74 Individual calf pens provided with a wooden slatted floor and a front equipped with 2 bucket holders and a hay rack.

For supplying milk, the milk bucket replaces the water bucket for 0.5 to 1 hour and then the water bucket is returned. When the width of the box is 1.20 m three buckets viz. for milk, water and concentrates can be placed in the front. The hay rack can then be mounted on one of the side walls of the box (fig. 4.75).

The floor of the calf boxes can either be fully concreted or slatted. When a full concrete floor is chosen (fall : 5 cm/m) a covered drainage channel has to be included. The concrete floor is then heavily bedded and in order to keep the calves clean new bedding material must be added regularly. As soon as the calf is transferred to another house the manure has to be removed from the calf box.

A better but more expensive solution is the application of slatted floors in the calf boxes. Bedding material is also distributed on the slatted floor. The use of slats will keep the calves drier whilst the consumption of straw is less. Wooden and concrete slats can be distinguished : for wooden slats the tread is normally ca. 6.4 cm wide

Fig. 4.75 A breeding calf nursery with a row of wide calf pens built along a wall.

whilst the gap is 2 cm. If concrete slats are preferred, the slats as used in pig houses are suitable, they have a gap of 2 to 2.3 cm. The slatted floor is raised 30 to 35 cm above the floor of the house allowing easy access to the area under the slats for the removal of the dung which is trodden through the slats. The floor under the slats has a fall of 5 cm per metre towards an uncovered gully in front of the row of pens allowing easy drainage of the liquid manure. The concrete slats are installed in the transversal direction whereby the supporting walls, made of blockwork of concrete are at right angles to the passageway and form no obstacle to the removal of the manure.

4.3.1.3 The tying stall barn for calves and young stock

Two types of stanchion barns can be distinguished : the littered stanchion barn and the strawless stanchion barn with grids. As said before, the accommodation of young calves in a stanchion barn is to be discouraged. Moreover, it is difficult to keep the calves clean in a tying stall, even if provided with grids ; the building costs and the labour time requirement are high in comparison with other types of housing.

In the *littered tying stall barn* the stall length varies between 1 m to 1.5 m and the stall width between 0.65 m to 1 m, according to the age of the animals. If one wants to keep the calves clean, the consumption of straw will be rather high. For the further equipment of this house we refer to the dairy cow house (see earlier).

The *strawless tying stall barn with grids* is a somewhat better solution. The stall length is 1.10 m for calves up to 1.5 years and 1.35 m for animals between 1.5 and 2.5 years, whilst the stall width is respectively 0.7 m and 1 m. Behind the stall three wooden beams can be installed, each 10 cm wide, together with a grating : by placing 3, 2, 1 or 0 wooden beams between the stall and the grating it is possible to adjust the length of the stall to the length (age) of the animals ; the beams which are not installed between stall and grating are placed behind the grating. Here also we refer to the corresponding house for dairy cattle for details of other equipment of the calf house.

4.3.1.4 The group housing for young stock

The ideal type of housing, both from the labour organizational point of view as well as technically is certainly the *loose house for young stock*. Young growing cattle require sufficient movement for good physical development. The animals are divided into groups according to their age, certainly when a large herd is involved. After a period of individual housing in calf pens the calves are moved to a follow-on calf house with group penning.

Group pens with straw and a grating of laths, for calves between 0.5 to 3 months old, viz. up to the end of the period in which they receive milk are interesting as straw and labour requirements are then relatively low. The dimensions are chosen in such a way as to allow the accommodation of five to six calves of the same age per pen. A pen area of 0.75 to 1.00 m² per calf is provided (Anon., 1979). An example is shown in fig. 4.76.

The front of such a group pen consists of a feed fence and a small door. Each pen needs one hay rack, one bucket for concentrates, one bucket for milk and one automatic drinking water bowl. The milk can also be supplied in conserved form (mixture of milk powder and water) by means of a canister, a tube and a teat. This form of milk supply is already widely used.

The width of the feed fence needs to be 35 to 40 cm per calf while the depth of the pen is normally 2.50 to 3.00 m (Anon., 1979). There is a considerable saving on straw since the urine is evacuated through the laths and a drainage channel under the pens.

Instead of a wooden grating, concrete piggery slats can be used. They have a gap of 2 to 2.3 cm.

In *group pens with a slatted passage and a littered lying bed* (fig. 4.77) the calf requires an area of 1.0 to 1.5 m². This type of calf house is normally used for calves up to the age of three months. A slatted passage having gaps of 3 cm is provided behind the feed barrier. Connected to the slatted passage and separated from it by an upstanding timber is the lying bed which consists of concrete covered with bedding material. The width of the passageway must be 1.75 m. The length of the bed is 1.25 m to 1.50 m (Anon., 1979).

The total pen length, bed plus passage, is thus 3.0 m to 3.25 m. The frontage at the feed barrier is 35 to 40 cm per calf. It is recommended to put 6 to 10 weaned calves in one pen. The use of straw as bedding material will increase labour and costs, while some straw can also get through the slats, but from zootechnical point of view this type of follow-on calf house is excellent.

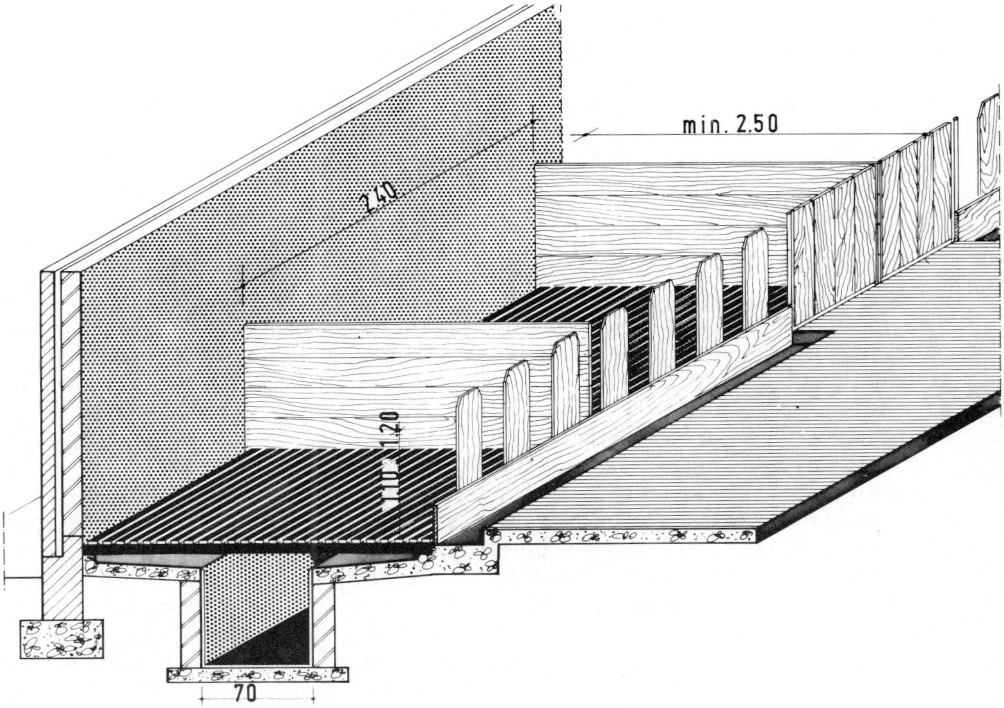

Fig. 4.76 Group pen, with straw and grating, for young stock of 0.5 to 3 months old.

Fully littered loose houses are seldom used at dairy farms for young stock beyond the age of 3 months, because of the high costs involved with the bedding material (5 kg of straw/animal/day).

The *cubicle house* is a suitable type of intermediate house for young stock from the age of 2 months. Certainly when the dairy cattle is also accommodated in a cubicle house it is desirable to give the young stock a similar accommodation so that they are already accustomed to the cubicles before their transfer. Indeed, it sometimes happens that young stock which were previously housed in a fully slatted house refuse cubicles in a dairy cow house and lie down in the passageways.

Young stock can spend the winter in a cubicle house without any problems but fixed boxes can hardly be adapted to the ever-changing sizes of the animals. As a consequence, several sizes of cubicles are required and the animals have to be transferred periodically (2 - 3 times) to a larger sized cubicle. Cubicles for young stock are therefore only suited for large farms possessing a large number of young stock and whereby each age-group is sufficiently large. Table 4.15 gives some dimensions for cubicles intended for young stock.

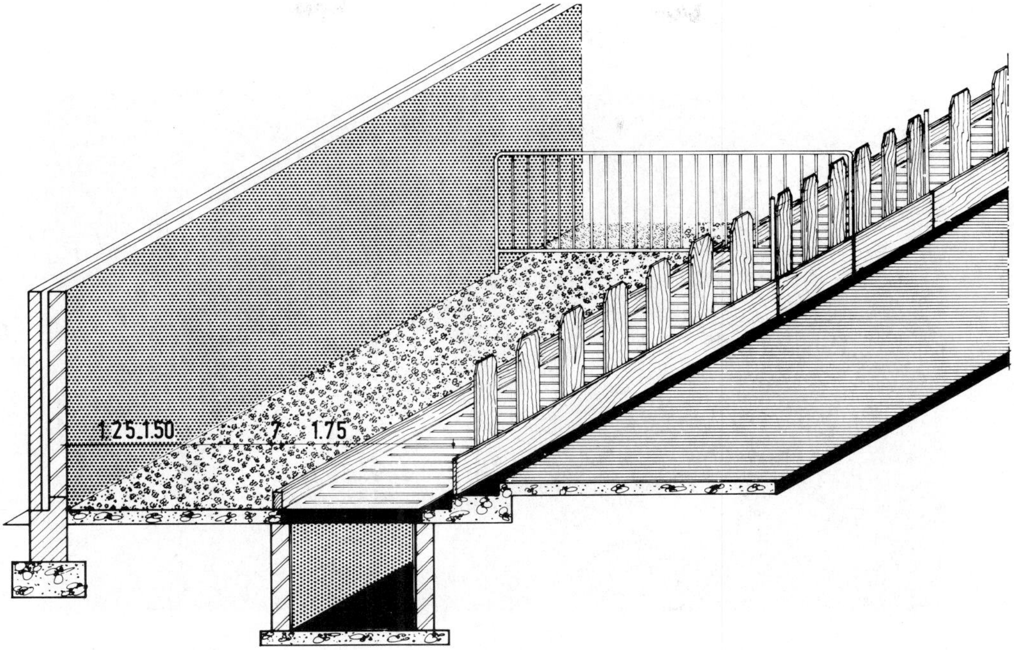

Fig. 4.77 Group housing with fully slatted passage and littered lying area, for young stock of 0.5 to 3 months old.

TABLE 4.15 The breadth and length of cubicles, the frontage and the dimensions of passages (expressed in cm) for the different age-groups of young stock when housed in a cubicle house (Smits and Swierstra, 1979a).

Age in months	Cubicle		Passage		Frontage per animal
	length	width	alley	alley behind feed fence	
0.5 - 2	130	60	150	175	35
2 - 5	150	70	150	175	40
5 - 12	170	80	200	225	50
12 - 18	190	90	200	225	55
18 - 22	210	100	200	250	60
22 - 24	210	110	220	300	65

In practice, it is nearly impossible to provide six different sizes of pens as mentioned in table 4.15 and therefore one is often satisfied with about three different sizes. Commonly used sizes for pens are :

- 150 cm x 70 cm for young stock, up to 6 months (fig. 4.78) ;

- 190 cm x 90 cm for young stock, up to 18 months ;
- 200 cm x 100 cm for young stock older than 18 months.

Fig. 4.78 Cubicles for calves up to an age of 6 months.

It is also possible to construct cubicles which are adjustable in length and breadth and which are therefore suited to the different age groups. In practice however they are seldom applied since their cost-price is too high and they require additional labour for the adjustment.

A disadvantage of the cubicle house in comparison with the fully slatted loose house is the higher space requirement per animal so that a cubicle house involves a higher investment than a fully slatted loose house.

The floor of the cubicle is preferably carpeted, although hard floors, e.g. concrete, are also satisfactory if they are sufficiently littered with sawdust. The passageways are preferably slatted. From practice it appears that young stock are rather quickly accustomed to the cubicles and that they keep them clean. Young stock, accommodated in a cubicle house are usually very clean.

A special form of cubicles are the *feed cubicles*, which are however less suitable for young stock because of their changing size.

The *fully slatted loose house* (fig. 4.79) is, considering its building costs and the labour time requirement per animal, very often applied for the accommodation of young stock. It is however not suited for calves under the age of 3 months. This type of housing is indicated when the young stock are later to be transferred to a stanchion

Fig. 4.79 The slatted loose house for young stock.

barn for dairy cattle, but is less appropriate if the animals are to be accommodated in a cubicle house. The great asset of this type of housing is the efficient utilization of the available space. The animals walk, stand and lie on slats with a width of 16 cm and a gap of 3 to 4 cm according to their age-group. The animals tread the manure through the gaps. Under the slatted floor a liquid manure cellar is provided.

Contrary to littered loose houses the cleanliness of slatted houses is only slightly influenced by the occupancy of the house.

The animals should be able to lie unhindered. The space allocated to each animal depends on its age or weight as is the required frontage at the feed barrier. Table 4.16 gives some information on the fully slatted house.

4.3.1.5 An example of a layout for a young stock house

Before considering some layouts of houses intended for young stock it is necessary to obtain an insight into the required stocking capacity. A guideline therefore is given in table 4.17.

From this it appears that the number of young stock of different ages, between 0 and 24 months which have to be housed is approximately equal to the number of dairy cows present at the dairy farm.

With the building layouts given for dairy cattle, some attention was already given to the housing of young stock.

Large dairy farms, however, keep their young stock in a separate house because of the extent of the herd. There is of course a great variety of possible types of housing for young stock. Fig. 4.80 represents a design of a cubicle house for young stock suitable for a

TABLE 4.16 Dimensions of slatted loose houses for young stock.

Age in months	Space requirement m^2/animal	Trough frontage m/animal	Depth of the pen (m)	Gap (cm)*
3 - 6	1.0 - 1.5	0.4	3.00	3 to 4
6 - 12	1.5 - 2.0	0.5	3.20	3.5 to 4
12 - 24	2.0 - 3.0	0.6	3.80	4
24 - 30	3.0 - 3.2	0.6	4.75	4

*slat width in all categories : 16 cm.

TABLE 4.17 The number of places required for young stock of different ages, expressed in percent of the total number of dairy cows.

Age in months	Number of places per 100 dairy cows
0 - 0.5	15
0.5 - 3	20
3 - 12	30
12 - 22	30
22 - 24	10
Total	105

farm with 100 dairy cows.

The internal dimensions of the house are 45.0 m x 12.9 m and a wide central feeding passage is provided. There are 6 tie-stalls with a breadth of 1.20 m each, intended for animals about to calve. These stalls can also be used for sick cows or cows with claw disorders. Beside it are two strawed calving boxes. One of the calving boxes can sometimes serve as a pen for the stock bull. The new-born calves are immediately transferred to the individual calf crates where they remain up to the age of two weeks after which they are brought to group pens provided with a bedded lying place and a slatted passage along the feed fence. At the age of 3 months the animals are brought to a loose house with cubicles and slatted passages. Three groups of cubicles,each with different dimensions are provided e.g. 20 cubicles of 1.50 m x 0.70 m, 25 cubicles of 1.70 m x 0.80 m and 25 cubicles of 2.00 m x 1. 00 m.

4.3.2 Zootechnical and veterinary aspects of the housing of rearing calves and young stock

Healthy calves lie down 60 to 80 % of the time, which is considerably more than full-grown cattle (Mitchell, 1976). A comfortable lying bed for calves is therefore essential. Dry straw is the most appropriate bedding for young calves, although its use demands considerable labour. Of paramount importance to avoid calf mortality is the strin-

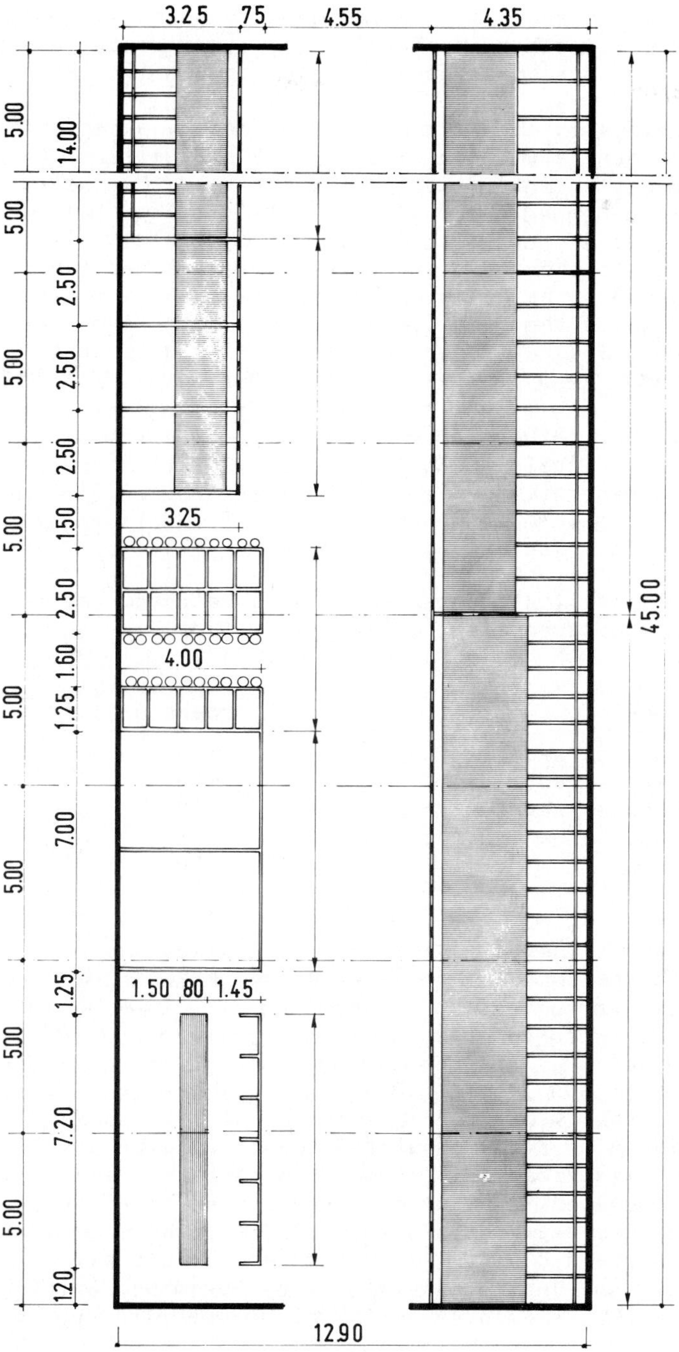

Fig. 4.80 Design of a calf and young stock house suitable for a farm with 100 dairy cows.

gent hygiene which has to start before the calf is born. A clean, adequately littered calving box is therefore no superfluous luxury. Between each batch the calf boxes have to be scrupulously cleaned and disinfected.

Calves are gregarious in nature and it is generally accepted that communication with other calves contributes to their welfare. Group penning also encourages a competitive spirit for food (Mitchell, 1976). Where calves are group-penned a social hierarchy develops during the first few weeks. It is however advisable to keep the calves individually at least to the age of two weeks as explained earlier.

According to table 3.4 the optimum temperature in a calf rearing unit is 20° to 15°C while the relative humidity should be maintained between 60 % and 80 %. The calf however tolerates much lower temperatures, even below freezing point if it is adequately fed, protected from wind and rain and given a dry bed. Mitchell (1976) mentions the critical temperature for calves for different types of houses ; these are represented in table 4.18. The lower critical temperature of an animal is generally defined as that environmental temperature below which the animal has to increase its heat production in order to maintain its body temperature.

TABLE 4.18 The lower critical temperature for a well-fed calf of 50 kg in a draught-free house, according to Mitchell (1976).

Type of housing	Critical temperature, in °C
calf lying down on dry straw	-13
calf lying down on wet straw	-10
calf lying down on wooden laths	-10
calf lying down on uncovered concrete	+2
standing-up calf	-10

According to Mitchell (1976) "climate-houses", i.e. houses where the internal temperature approaches the external temperature such as open fronted houses, are suited for the housing of calves during whole the year in Scotland. He compared the health condition and the growth of calves accommodated in an insulated and uninsulated closed house (Mitchell and Broadbent, 1973). No significant differences were found in the growth and mortality but the consumption of milk powder and concentrates was higher in the uninsulated house. Jongebreur (1974) found in Holland no significant differences in mortality and growth between calves accommodated in an uninsulated closed house having a temperature which varied between -6°C and +23°C and those in an insulated closed house where a constant temperature of +12°C was maintained. From research carried out in the Netherlands by Boxem and Smits (1977 ; 1979) and Smits and Swierstra (1979b) in open-fronted calf houses, it appeared that the mortality was low and that the animals were all in good health. Kommerij (1979) has, also in Holland, placed simple covered individual calf crates in the open air.

The calves have even been housed in those pens immediately after birth without any problem. Those littered "outside" pens can be used when there is a shortage of individual boxes inside the house or, according to the author, when there are calves inside the buildings with diarrhoea. They then come in a "clean and healthy" environment and can without problems be transferred to the closed calf house after their sickness period. Whatever type of calf-house is used, either an open or closed house, it should always be draught-free.

Not everyone, however, believes in the suitability of an open calf house. According to Toren (1978) a minimal temperature of +10°C is required for newborn calves, with which we completely agree. This temperature can be obtained by means of e.g. an infra-red heater installed above the calf during the first day of life. It will allow the animal to become dry faster. After a few days the calf is sufficiently hardened and low temperatures are then no longer detrimental to its health.

4.4. THE HOUSING OF GROWING BULLS AND STOCK BULLS

4.4.1 The houses for rearing stock-bulls

The rearing of bull calves to adult stock bulls does not require special provisions at the dairy cow farms. Tying stalls are not suited except when the lying bed is partly made of concrete slats or has a well placed and ample sized concrete or metal grating allowing a fast drainage of the urine. In loose houses on the contrary, bulls are reared in groups according to their ages. On dairy farms fully slatted pens are used for beef breeds, but for the double muscled breed littered pens are preferred. The layout resembles that for the housing of beef and we therefore refer to the description given for the housing of beef cattle. Future artificial insemination bulls are actually produced through a special breeding programme implying rational matings (sire dams). The male calves are then reared in specialized centres. The rearing of male calves from 3 to 12 months can be carried out in groups of 5 calves in littered pens (7 m^2/bull) provided with a concreted outside exercise area (7 m^2/bull). They can also be housed during that time in individual fully slatted pens which measure 2.5 m x 2.5 m and are provided with slats of 14 cm and with a gap of 4 cm. The partitions between the pens consist of four galvanized tubes with a diameter of 6 cm which are mounted one above the other at respectively 0.4 m, 0.8 m, 1.2 m and 1.5 m above the floor. This form of housing allows a severe selection on the strength of the limbs and minimizes the labour involved with the daily care. The pen is only cleaned after the departure of the bull, the daily mucking-out is completely omitted. Slatted floors cause regular wear of the hoofs whereby hoof-care becomes superfluous. We refer to Chapter 3 for the equipment (ventilation, lighting, materials).

4.4.2 The housing of stock-bulls

Three basic requirements are to be fulfilled in relation to the housing of stock-bulls :

- to contribute to a good condition of the stock bull and furnish him with a healthy and quiet accommodation ;
- to limit the labour requirement to a minimum and keep the building costs within reasonable limits ;
- to provide and guarantee safety to the stockman.

If the dairy cattle are housed in a cubicle house, the stock bull is accommodated in a littered or fully slatted pen. In a stanchion barn for cows the stock bull is tethered in a stall. In specialized stock bull husbandry (artificial insemination centres) the littered stanchion barn excites interest : the space requirements are reduced to a minimum while the bulls can easily be caught for service.

4.4.3 The littered tie stall

The littered stanchion barn for stock bulls consists of pairs of stalls of which the breadth and length are adjusted to suit the age and breed. The width varies from 0.9 m for one-year bulls to 1.20 m for adult bulls, while the length of the stall increases from 1.8 m to 2.4 metres. Between every two bulls a full-length partition is provided while in the front an escape-gate is installed having a width of 0.4 m. The latter allows a safe tying and loosening of the bull. Behind the middle of the escape-gate at the rear of the stall there is an inverted U-shaped tube having a length of 70 cm and a height of 90 cm which prevents the bulls of standing across whilst it keeps them clean and guarantees a free passage at the return of each bull to his stall. Fig. 4.81 shows such a type of stall and one can remark that the stockman is well protected during feeding by the tubular cage in front of the stall. The lying bed is made of a 15 cm layer of concrete with a fall of 5 cm/m towards the drainage channel. The tying system consists of a neck-collar attached to two short chains which can glide over two braces installed along both sides of the kerb in front of the bulls. Mucking-out can be carried out mechanically or by means of a tractor (see dairy cattle). Every 10 to 20 bulls, a gate is installed which fences off the service passage so that should a bull get loose it is enclosed in a limited run. Those gates are installed 40 cm away from the outer walls thereby forming an escape way.

4.4.4 The littered individual pen

The individual littered pen for stock bulls is at least 3.5 m wide along the feed passage and 5 m long. The floor is made of a 15 cm concrete layer which slants towards the centrally situated drainage channel. The side walls are constructed from framework with an escape gate having a width of 40 cm in each corner. For details we refer to the fully slatted pen. For stock bulls of beef breeds this housing is indicated besides the tying stall. For bulls of dairy breeds the fully slatted pen is to be preferred in order to stimulate the inheritance of strong limbs and claws via the male line of breeding and thus allowing descendants to be housed in cubicles with slatted passages.

4.4.5 The fully slatted individual pen

Each bull possesses a pen (fig. 4.82) measuring 3.5 m x 5.0 m.

Fig. 4.81 The littered tying stall for bulls.

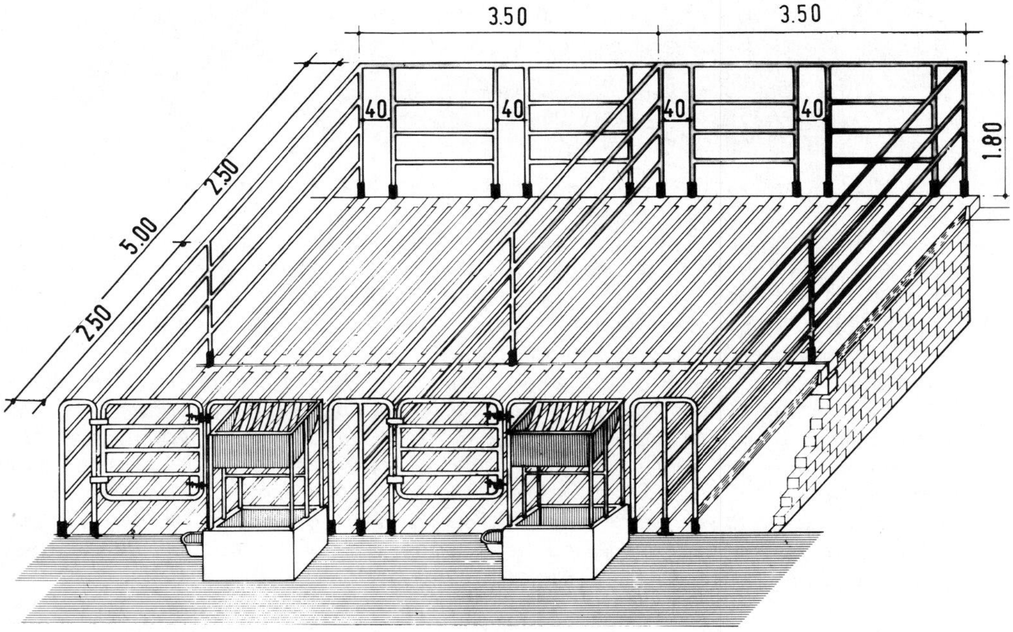

Fig. 4.82 The pens for stock bulls.

The whole floor is provided with concrete slats (fig. 4.83) having a length of 2.5 m, a breadth of 16 cm and a gap of 4 cm. The slats have to be of durable construction, show an equable non-slip surface while rounding of the edges will reduce the risk of injuries. A liquid manure cellar, of ca. 1 m deep is found under the slats and can store the faeces and urine produced over several months. The cellars of adjacent pens are separated from each other by means of walls of 100 cm high which allow a thorough mixing of the contents and a complete emptying of each of them.

Fig. 4.83 Individual fully slatted pen for a stock bull.

Three sides of the pen are fenced by means of a framework of galvanized tubes with a diameter of 6 cm (wall thickness 3 mm) and a height of 1.80 m. The vertical tubes are, up to a height of ca. 15 cm above the floor, protected against rust by providing them with a waterproof plastic tube or a concrete upstand. The fence at the rear passage (= escape passage) is provided with two or three escape gates of 40 cm width. The fence along the feed passage is composed, from left to right of an escape gate of 40 cm, a door, a trough and a second escape gate. The door is built from the same galvanized tubes and has a width of 1.20 m whilst it is 1.80 m high. The door, when opened, turns over an angle of 180° and is provided with a safety lock. The automatic drinking bowl is situated on one of the sides of the feeding trough and as close as possible to the front to facilitate inspection. The hay rack, made from galvanized tubes, contains a supply of straw or hay to which the animal has free access.
Under the hay rack, outside the pen (to prevent the animal soiling

the trough and also to simplify the construction of the pen) a concrete trough is installed having dimensions of 100 cm x 60 cm x 60 cm which is filled twice a day with a ration of roughage and concentrates. The trough is completely surrounded by galvanized tubes, so that there is no danger to the passers-by.

4.5 THE HOUSING OF SLAUGHTER CATTLE

The production of beef cattle is an important branch of cattle husbandry. We can distinguish : veal calves, suckling calves and slaughter cattle, comprising bullocks or bulls and female animals not intended for milk production. Meat production with reform-cows (finished dairy cows) is of less importance.

4.5.1 The construction and equipment of veal calf houses

Veal calves are preferably kept in individual pens (fig. 4.84 to fig. 4.88) until they are slaughtered i.e. after 16 to 20 weeks when they have come from an initial birth weight of ca. 50 kg to a final weight of 180 to 200 kg. Those individual pens are 1.60 m long and 0.65 m wide. Jongebreur and Smits (1978) compared the performances of veal calves housed in individual pens of 50, 55, 60 and 65 cm wide and 150, 160 cm long. They demonstrated that the growth was better if the calves were housed in boxes of 160 cm x 65 cm than in boxes of 150 cm x 55 cm and also that lying down and rising was easier. The pen divisions consist of wooden planks while the front is made of wooden planks provided with an opening through which the calf has access to a plastic bucket with milk substitute.

This opening can often be closed so that the stockman is not hindered when filling the bucket which is supported by a bucket holder. The rear wall of the pen is composed of demountable laths fitted into iron braces. This facilitates the inspection of the animals (diarrhoea). The calf stands on a wooden grating. The rear half consists of laths of 40 mm x 20 mm with gaps of 20 mm while the front half consists of wooden laths of 80 mm x 20 mm with gaps of 20 mm : the front laths are wider to prevent draught and the risk of pneumonia. This latter deck is ca. 40 cm above the floor.

All iron is scrupulously kept out of reach of the calf in order to obtain white veal. Veal calves are therefore exclusively fed with milk substitute. Automatic artificial nipple feeders do not give satisfactory results. Bucket feeding is still the best feeding method. The milk substitute is prepared in an agitating tank by mixing milk powder with warm water. The buckets are filled with a hose-pipe, fed by a pump which delivers an individually adjustable volume of warm milk substitute to each bucket. The feed passage ought to have a width of ca. 2 m since part of it is occupied by the buckets.

The faeces on and under the grating is removed by a hose pipe. The service passage behind the row of pens is 1 m wide.

A type of housing for veal calves which is applied in Germany is represented in fig. 4.89. The calves are not kept individually in pens but are housed as in a stanchion barn. The partitions are 70 cm long and made - against common practice - of galvanized iron. The front is

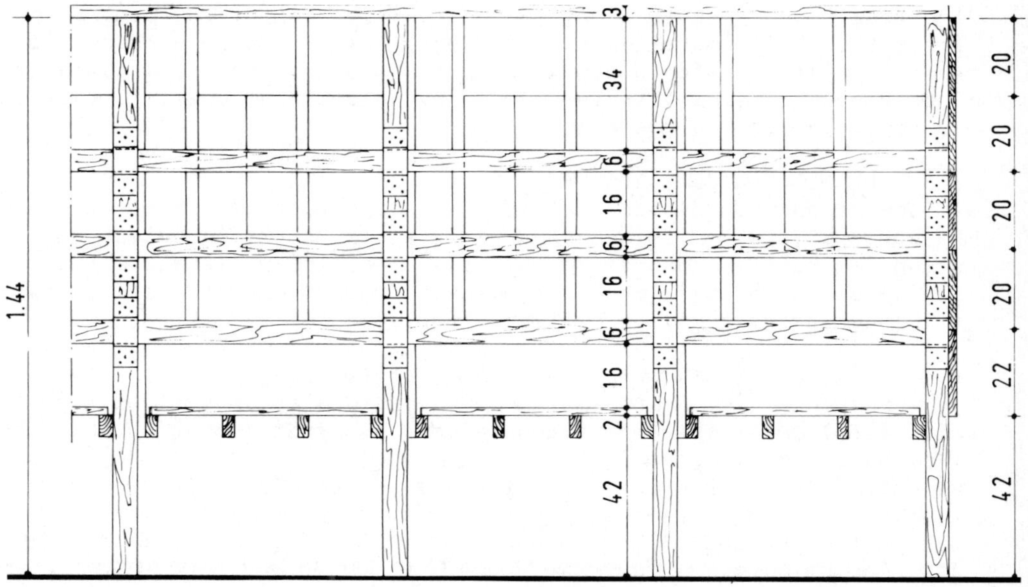

rear view

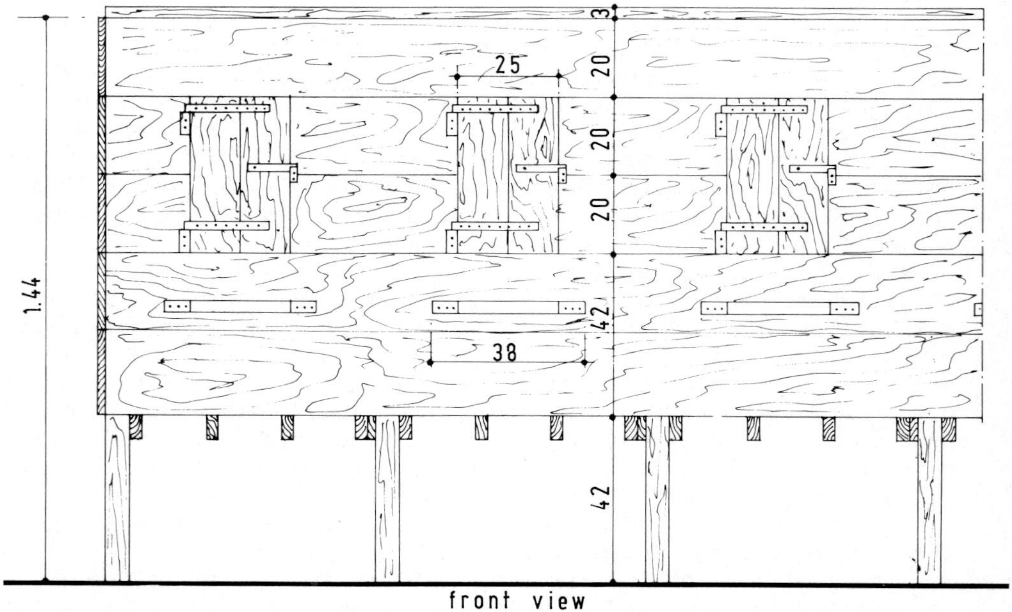

front view

Fig. 4.84 The veal calf pen.

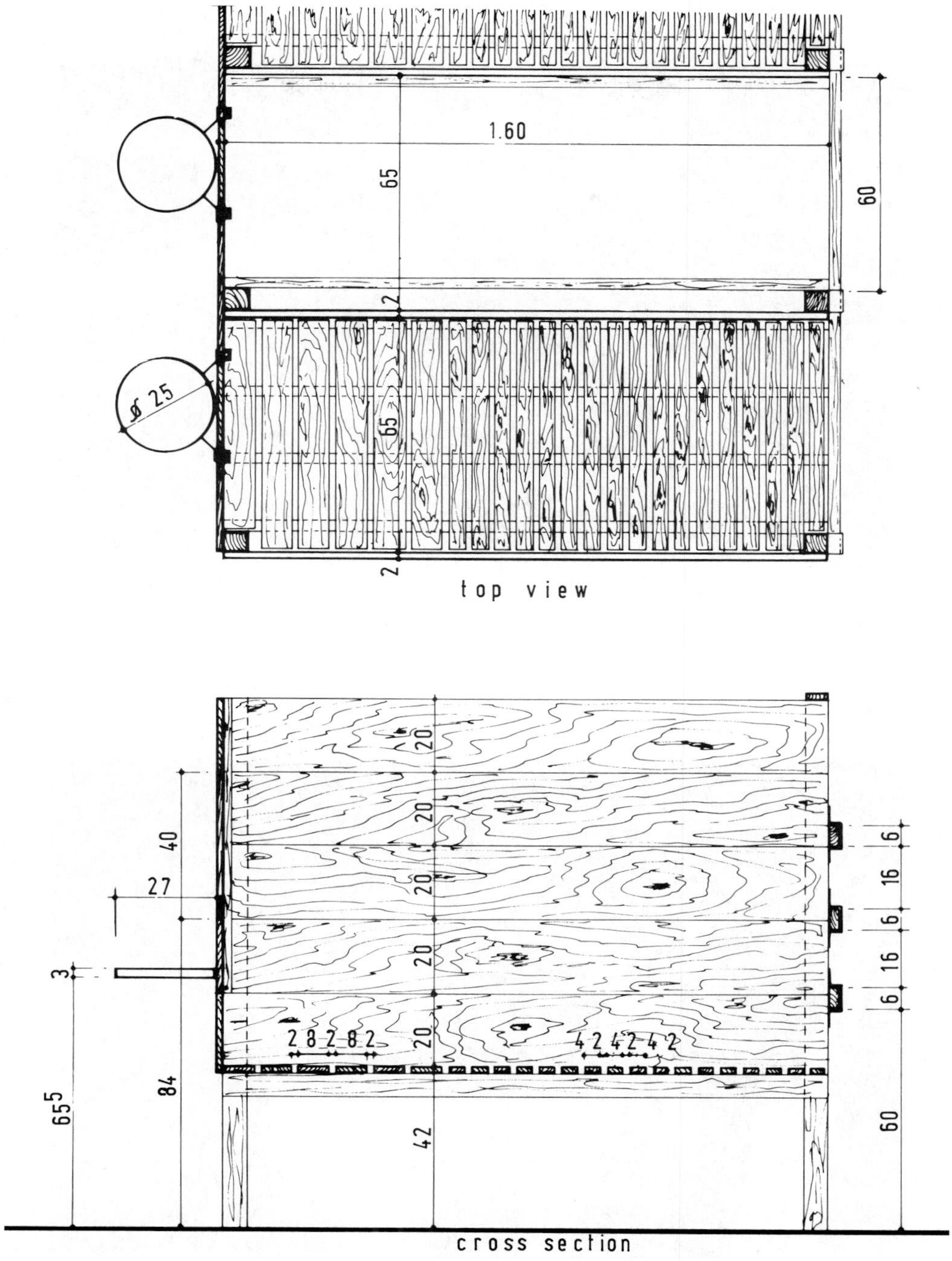

Fig. 4.85 The veal calf pen.

Fig. 4.86 A two row centre pass veal calf house with individual pens.

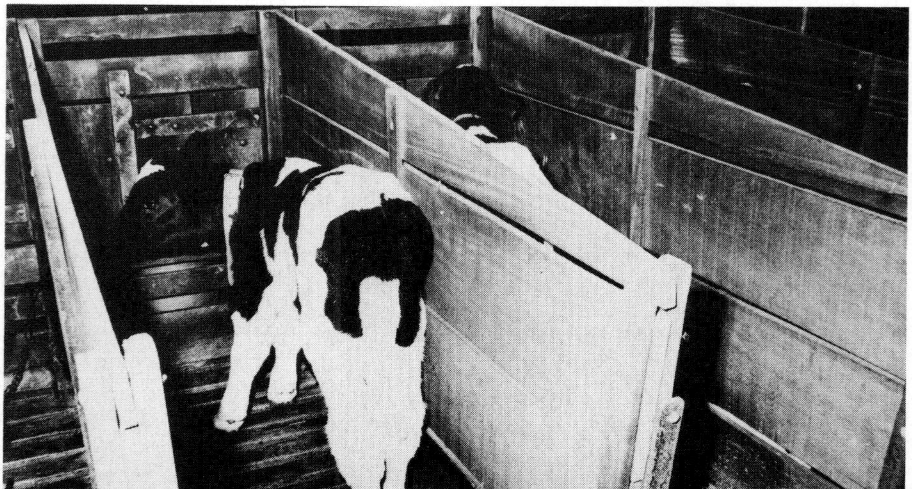

Fig. 4.87 Individual tethered box with slats.

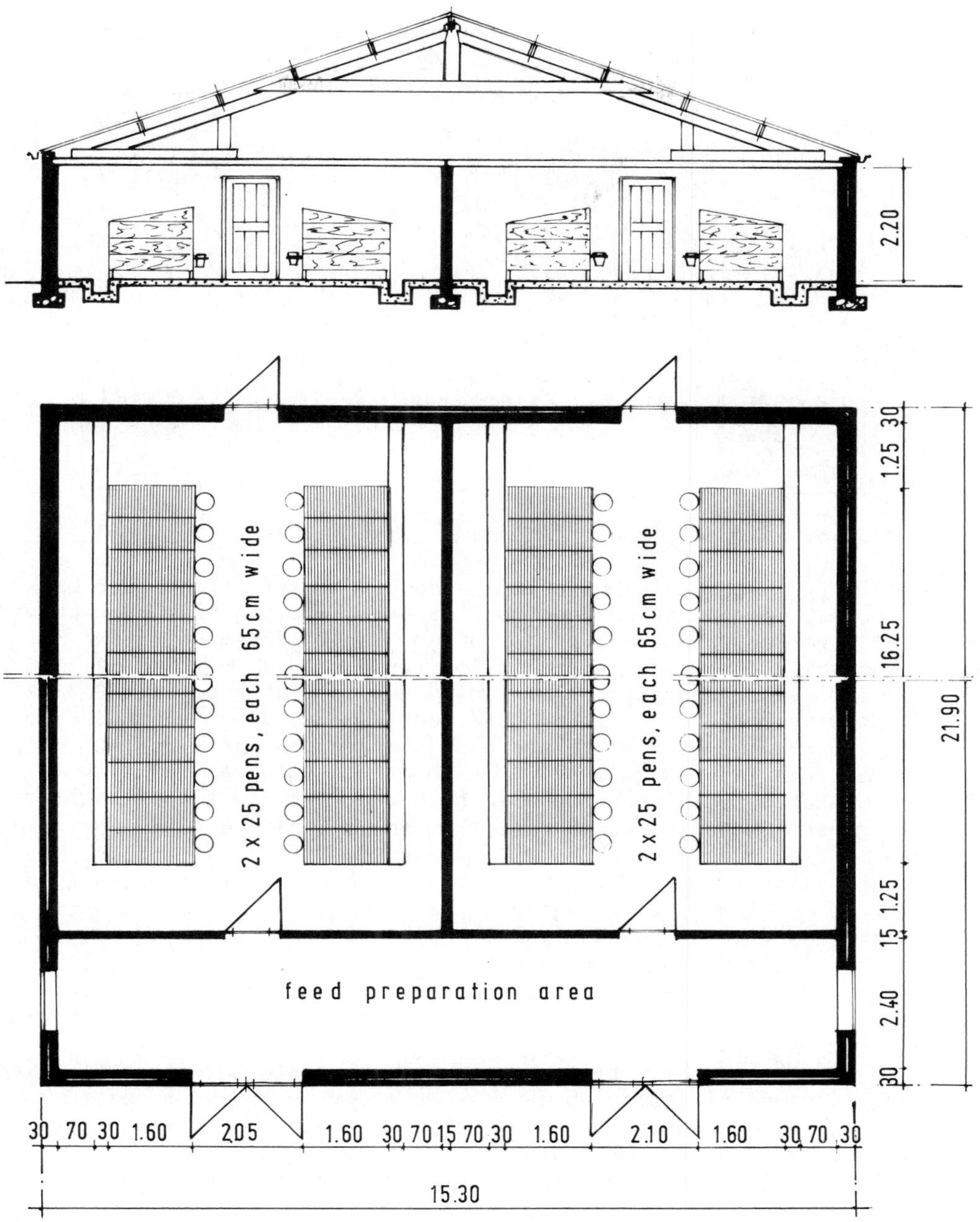

Fig. 4.88 Ground-plan and cross section of a house for 100 veal calves.

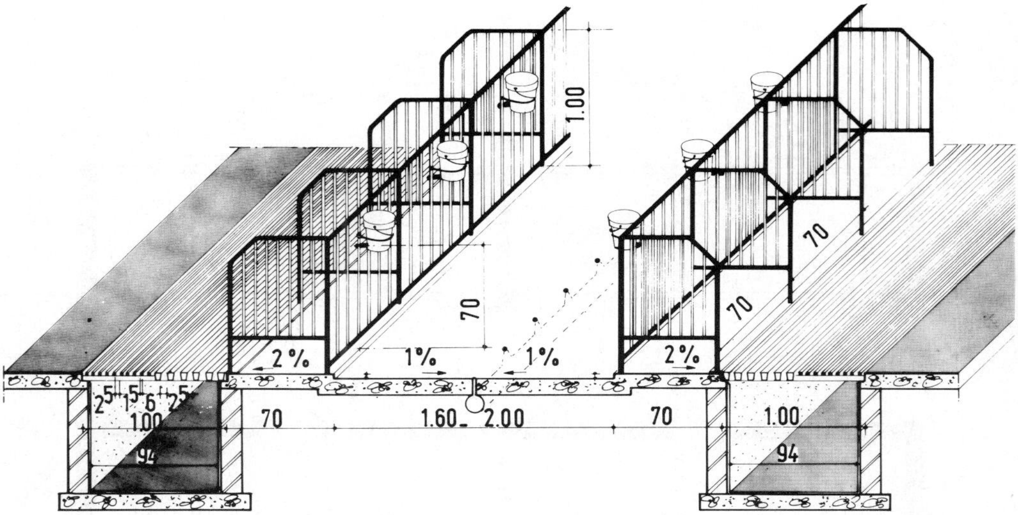

Fig. 4.89 A novel type of house for veal calves.

similarly fenced since no crib is needed. A calf self-feeder bucket with teat is mounted on the front panel. The lying bed is 68 cm wide per animal up to a final weight of 180 kg and 72 cm up to a final weight of 220 kg. The length of the lying bed is respectively 160 cm and 170 cm. It consists of a grating made of laths of ca. 60 mm x 20 mm with a gap of 25 mm. In the rear a metal grid is provided with ca. 1.5 cm wide bars which are 2.5 cm apart. In the first 70 cm of the bed the laths are installed with a spacing of only 1.5 cm. An insulated full floor can also be used there. On a fully slatted lying bed however milk residues from the dripping of calf self-feeders can not accumulate which is quite possible with a full floor. A liquid manure channel is situated below the lying bed. The animals can either face in or face out. The whole is easy to clean, is conveniently arranged and does not hinder the air circulation (Koller et al., 1979). In our opinion the partitions and the front can better be made of wood.

According to Fiems (1982) the walls of individual pens should not be closed completely, so that social contact between the calves is still possible.

Group penning of veal calves is also possible but is rarely used. The advantage of saving space is completely nullified by disadvantages such as poorer supervision of the animals, an uneven growth and a worse feed conversion (Burgstaller et al., 1981).

According to table 3.4 the calf house can best be heated (by means of a central heating installation) since at the start of the veal production a temperature of 20°C is desirable which may drop gradually to 16°C towards the end of the production period. This was confirmed by Burgstaller et al. (1981). Adjustable fans take care of the ventilation and extract up to 150 m^3/h/head. Since veal calves transpire heavily, draught and temperature fluctuations are to be avoided. A thorough insulation of the veal calf house (k $\leq$ 0.7 W/(m^2.K)) is therefore necessary. Some (Koller et al., 1979) even mention k $\leq$ 0.5 W/(m^2.K). Cellular concrete is, in this case, a common wall material.

Calf fattening is mainly carried out on large specialized farms and often under contract. The capital investment for the house amounts to £ 200 to £ 250 per head according to the size of the farm. The labour time requirement is 4 to 5 hours per finished calf.

4.5.2 The houses for suckler cows and suckling calves

A special way of breeding calves, intended later on for meat production is to allow them to suckle the udder of a suckler cow also called beef cow. Suckler cows, together with the bull, can be kept outside the whole year in those regions with a temperate climate and where the animals possess natural shelters such as hedges, shrubs, woods (Peyraud, 1974) and provided that they are given supplementary feed. In this case a littered shelter (5 m² /animal) is provided, which is open on one side and of which the walls are made of wooden or cement plates and the roof of corrugated asbestos cement sheets. Calving takes place outside or in this shelter and mortality is very low. This method is however not applicable in regions with a severe climate where the cattle have to be housed during the winter.
A suitable type of accommodation is then the cubicle house for suckler cows and attached to it, a large littered creep for calves (1.5 m^2 /calf). This creep is not accessible for cows. The calves can leave the creep for suckling or can be locked-in (fig. 4.90). It is also possible to house the suckler cows in a stanchion barn (fig. 4.91) where they are tethered and where the calves are kept in littered group pens (i.e. creeps) behind the cows. The calves can leave their pens and reach the suckler cows by crossing the service passage situated between the cow stalls and the calf creeps. Above the trough a horizontal bar is provided to keep the calves out of the trough and the feeding passage. The service passage is closed, at every 4 to 6 cows, by means of gates in order to keep the calves in the vicinity of their mother.

Keeping suckler cows is a form of extensive cattle husbandry which is mainly practised in deserted and infertile regions. Table 4.19 illustrates the importance of beef cow keeping and its important differences throughout various European countries in 1981. West-Germany and the Netherlands have practically no beef cows, in Belgium the number of beef cows amounts to 12 % of the total number of cows, while in the United Kingdom and France it rises to 30 % of the total.

4.5.3 The construction and equipment of slaughter cattle houses

The production of beef cattle can be carried out extensively, semi-intensively or intensively. Extensive beef cattle husbandry has declined although this does not necessarily mean that the (fattening) pastures have disappeared. The semi-intensive beef cattle husbandry is aimed at the production of slaughter-ready animals of ca. 560 kg in about 18 months whereby partly roughage (silage, hay) and partly concentrates are fed. The intensive production of beef cattle is aimed at raising slaughter-ready animals weighing 450 - 480 kg in one year

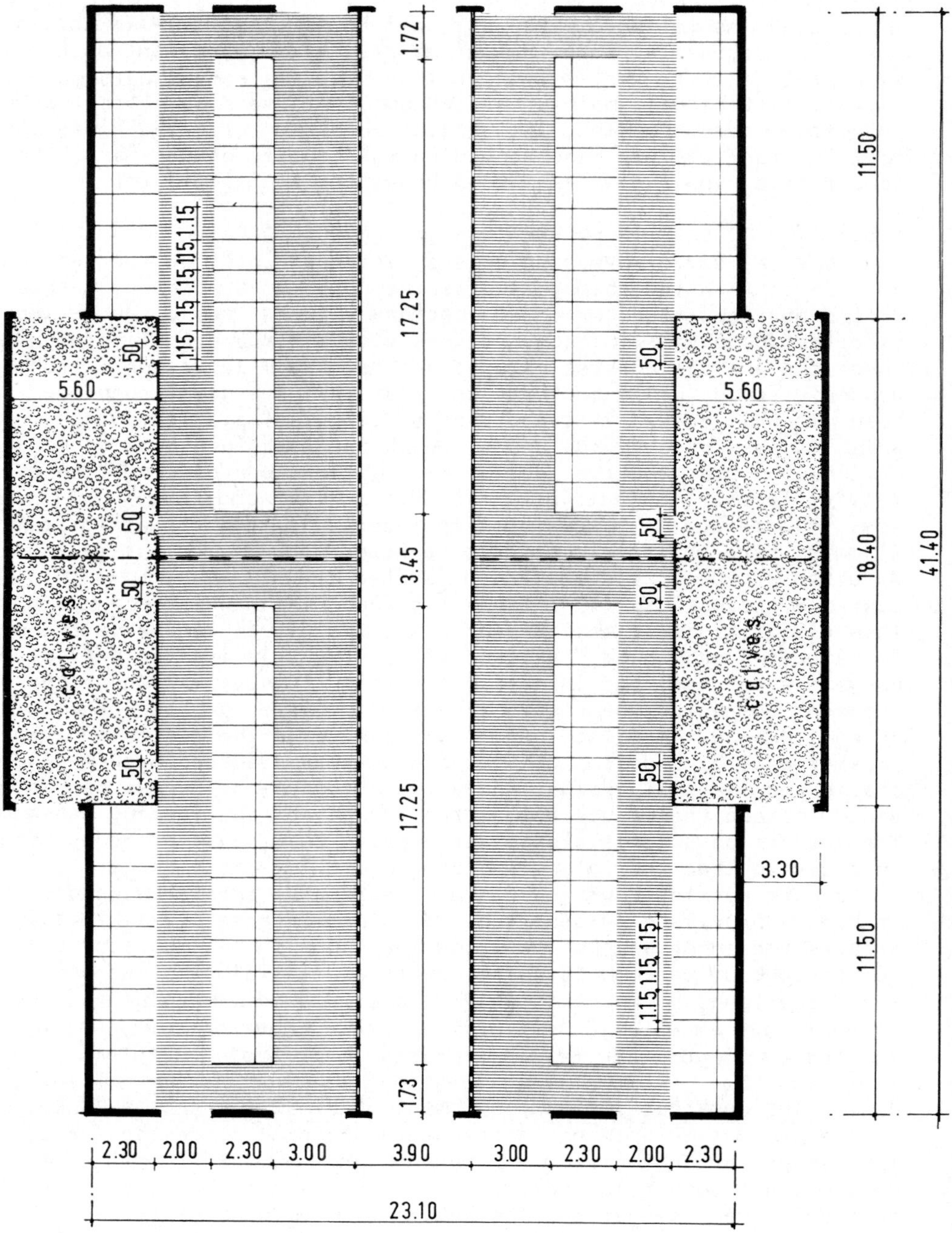

Fig. 4.90 Cubicle house for suckler cows.

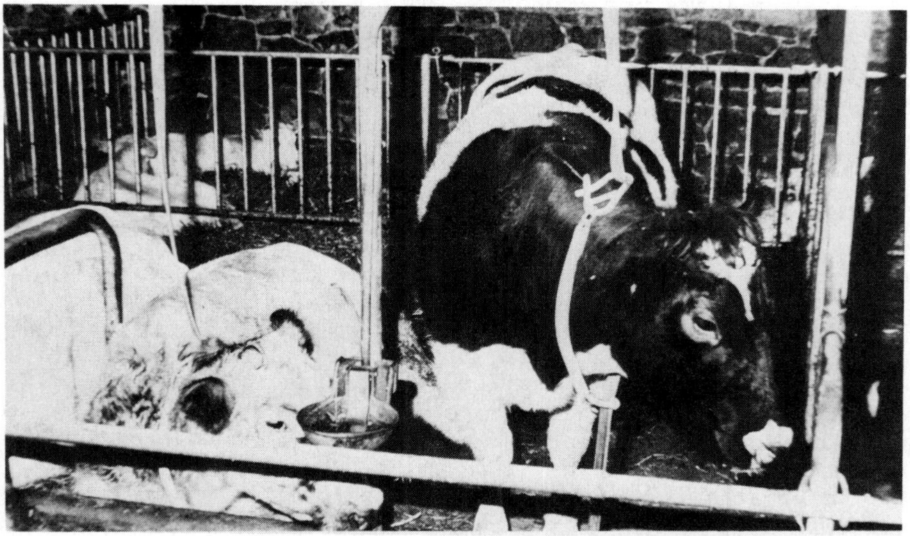

Fig. 4.91 A stanchion barn for suckler cows.

TABLE 4.19 The stock of cows in some West-European countries (x 1,000) in 1981 (Anon., 1983).

Country	Total number of cows	Dairy cows	Beef cows	% of beef cows (total number of cows = 100 %)
W.-Germany	5,593	5,438	155	3
France	10,026	7,054	2,972	30
Netherlands	2,407	2,407	-	-
Belgium	1,096	965	131	12
United Kingdom	4,714	3,302	1,412	30

(baby-beef) whereby solely dry feeds, mainly concentrates, dry pulp and some straw are used. There are also other methods of intensive beef production such as for example the fattening in a 10-month period of thin pasture bulls (200 kg) to slaughter-ready animals, 16 - 17 months of age and weighing finally ca. 620 kg. The housing types for beef cattle are similar to those for dairy cattle and we will therefore discuss only the major differences. A division of the animals into groups depending on their weight is certainly advisable for large herds.

4.5.3.1 The tying stall barn

The littered stanchion barn for beef cattle

The central feeding passage in a two-row house has to be at least 2.70 m wide when one wants to distribute feed by means of tractor and trailer. A raised feeding passage is used i.e. the front of the trough is the same height as the feeding passage. The trough is 0.5 m wide while the trough-bottom is 2 to 5 cm above the lying bed. The kerb has a height of 20 cm and a width of 10 cm. A solid tying system is used (see later). The littered stall has a fall of 1.5 cm/m towards the dung plate. For bulls, a urine discharge grid is provided in the middle of the stall. The length of the stall varies from 1.10 m to 1.65 m while the width varies from 0.65 m to 1.00 m per animal, whereby the animals are split up according to their age. The dung plate is 50 cm wide and is 20 cm below the stall. Mechanical mucking-out is possible as is described for dairy cattle houses. The dung plate has a slope of 3 % towards the rear, where a covered or uncovered foul drain collects the urine and drains it into the catchpit.

The litter alley is 1.20 m wide if it is built along a side wall and 1.50 m when a central litter alley is preferred. The litter alley is 15 to 20 cm higher than the dung plate and shows a slope of 2.5 % to the latter. When the muck is removed by means of a tractor they are at the same level.

Because of the high building costs and the high labour requirement involved, littered stanchion barns are less applied for the housing of beef cattle. An exception however is the housing of double muscled beef cattle. They are valuable and are therefore sometimes kept in littered stanchion barns (fig. 4.92).

Natural lighting is provided through a window area of 1/15th of the floor area. Artificial lighting is rated at 1 fluorescent lamp of 40 watts per 6 livestock units. For the ventilation we refer to chapter 3.

The stanchion barn with grids for beef cattle

This type of house represented in fig. 4.93 is also used for slaughter-cattle since it allows the omission of straw and the mucking-out labour. A special construction is required when bulls or oxen are to be housed. A "full" stall made of gummi or stone is indeed unsuitable since it would be wetted and fouled by the urine of the bulls. Therefore a 45 cm wide strip of bricks or concrete and resting on a light concrete layer is provided behind the trough, followed by a number of concrete slats, each 12 cm wide and with a gap of 4 cm and finally a metal grid of 80 cm wide. The concrete slats start at 45 cm behind the trough to prevent clogging of the gaps by feed residues. The exact number of slats depends on the length of the stall which in turn is related to the weight of the animals. This length is 1.00 m for animals up to 180 kg, 1.35 m for animals up to 450 kg and 1.45 m for animals up to 560 kg (metal grids not included). A better solution is the installation of six concrete slats each having a width of 12 cm and a slot width of 4 cm and of which the length corresponds to the

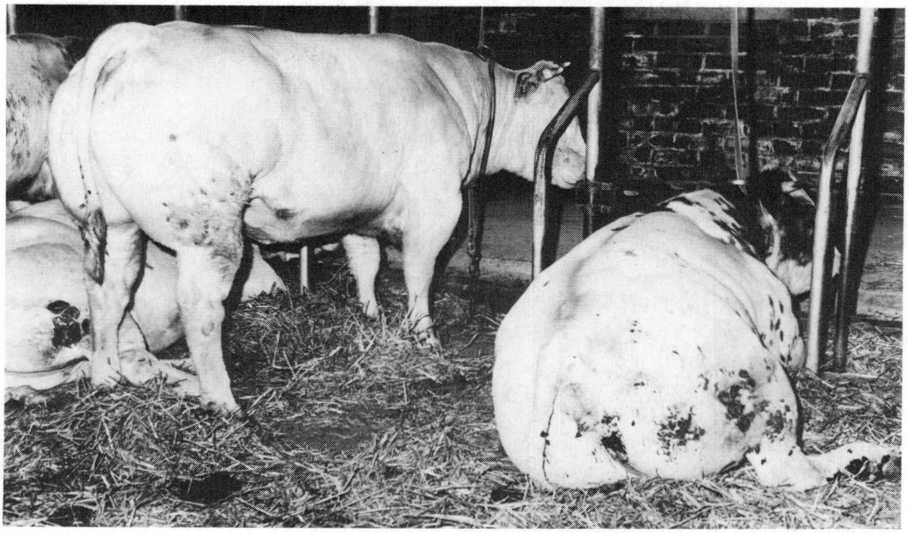

Fig. 4.92 Double muscled beef cattle in a littered tie stall barn.

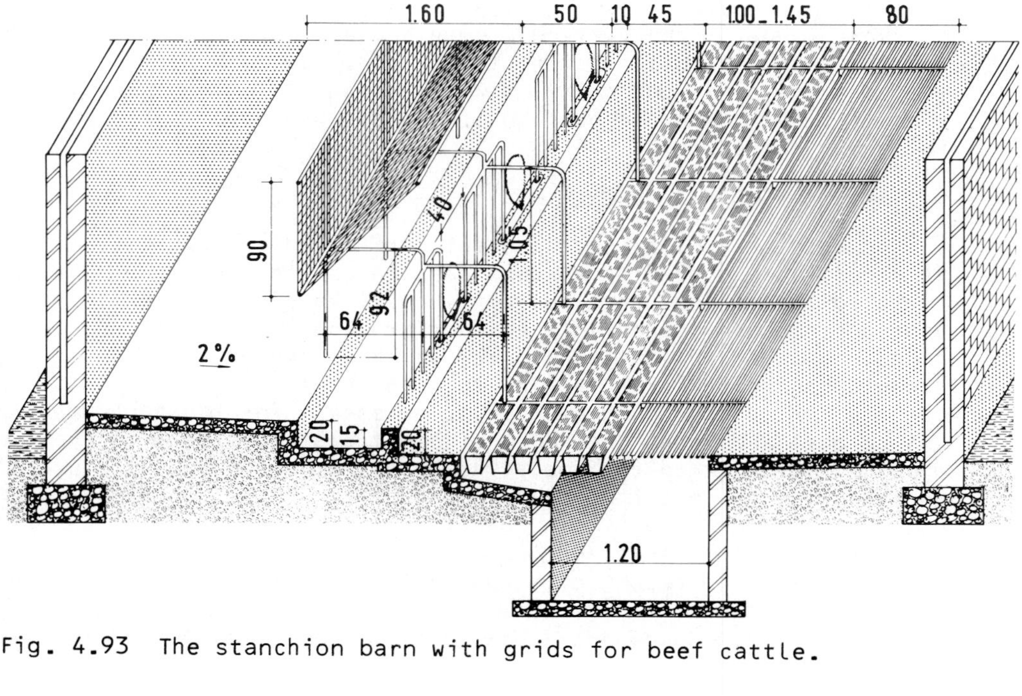

Fig. 4.93 The stanchion barn with grids for beef cattle.

width of the stall (100 cm). They are supported at their ends by a concrete beam which is mounted across the dung channel. These six slats are placed in front or behind the grid according to the age of the animals.

In this way the stall is tailored to each individual animal. If such slats are not obtainable they can be substituted by those applied in the cubicle house which are 16 cm wide but the above-mentioned way of construction must be respected. Under the slats and grids a liquid manure channel is provided (see higher : dairy cow houses) and a production of 20 l/animal/day is reckoned when dry feeding is applied. The tying system for fattening bulls must be more solid than the one for dairy cattle. As shown in fig. 4.93 it consists of a framework fixed to the kerb and forming a complete set with the partition : an iron bar with two rings at its extremities and carrying in its centre a ring to which a leather strap with safety-closure is attached moves along two vertical tubes which are 40 cm apart. It is also possible to make use of a vertical tying which allows a much smoother rising and lying down of the animals than the previous tying system. Each animal is separated from its neighbours by a partition wall which extends at least 65 cm into the stall. The latter is 0.65 m to 1.00 m wide. For ventilation and lighting we refer to the stanchion barn with feed passage and without grids.

The constructions described require however high investments and this type of house is therefore, certainly for bulls, economically less interesting and a loose house is more appropriate.

4.5.3.2 <u>The loose house</u>

The loose houses can be divided into three different types according to their layout : the littered loose house, the slatted loose house and the cubicle house. Each of these types can be built as an open or a closed loose house but in the last few years mainly the closed type of loose house has been built (labour saving). It is advisable to dehorn the animals.

The littered loose house

The original loose house was fully littered and is still in use. It comprises a number of strawed pens in one or more rows. The fully littered loose house requires daily ca. 6 kg of straw per animal which explains the diminished interest for it. The partly littered loose house has a concreted feed stance, which allows a reduction of the straw consumption for littering to 4 kg per animal and per day. An electric or tractor dung scraper is necessary for the daily mucking-out of the feed area. The application of slatted floors in the feed area is difficult to combine with the littering because the use of an expensive propeller pump is then required. The *strawed loose house with sharp rising concrete floor* (7 to 10 %) was recently introduced in France and Germany.
Every day a small amount of straw (1 to 2 kg per animal and per day) is distributed at the high side. The animals gradually tread the manure and straw downwards (= Tretstall). The manure has to be removed

daily at the lowest point. The animals are generally more soiled in the loose house with sharp rising floor than in other types of loose houses.

The pen area depends on the number of animals per pen, the weight of the animals and, if fed individually, on the required frontage as shown in table 4.20.

TABLE 4.20 The dimensions of littered loose house pens.

Animal weight (kg)	200	300	400	500	600	700
Pen area/animal (m^2)	2	3	4	5	6	7
Frontage/animal* (m)	0.4	0.5	0.6	0.65	0.65	0.7
Pen depth** (m)	5	6	6.7	7.7	9.2	10

*If every animal needs a place at the trough.

**Minimum dimensions. They are increased in order to give all pens in one row an equivalent "depth".

The lying area is mostly mucked-out only once viz. at the end of the fattening period. The accumulated straw dung may rise up to 1 m above floor level. The entire equipment (trough, partitions) has therefore to be removable whereas one has also to take into account that front loader mucking-out must be possible (height of doors and roof, width of passages).

The cubicle house

This type of housing is also used for slaughter cattle. As for young stock, the measurements of the cubicles have to be adapted to the age of the animals. They must therefore be transferred to cubicles of different sizes. According to Koller et al. (1979) those are : 1.60 m x 0.75 m up to 300 kg, 1.75 m x 0.85 m from 300 to 400 kg, 1.90 m x 0.95 m from 400 to 500 kg and 2.00 m x 1.05 m for 500 kg and more. The floor of the cubicle consists, when bulls are present, of wooden or concrete beams with a slot of 4 cm wide through which the urine can be drained (fig. 4.94). The passageways behind the cubicles or between the rows of cubicles shall be 1.5 m wide for animals up to 200 kg, 2.0 m for animals from 200 kg and more and even 3.0 m wide when they are also used by the animals to take their feed from the feed trough which is situated parallel to the row of cubicles. These passages are preferably made of concrete slats with an underlying liquid manure channel in which the urine from the cubicles can also be drained. Cubicles with a concreted floor and impermeable top layer are also common.

The passages may be fully concreted but this involves regular cleaning (cfr. dito dairy cow houses). They have to be at least 2 m wide to allow tractor scraping. The urine is drained in a gutter on the passage. Mucking-out here can also be carried out by a folding scraper.

The construction as shown in fig. 4.94 is however too expensive and therefore little used in practice. It is better to accommodate beef bulls on fully slatted floors (Boucqué et al., 1971 ; Daelemans et al., 1972). Cubicles can be of use for female slaughter cattle only but even then age range spread and transferring remain problematic.

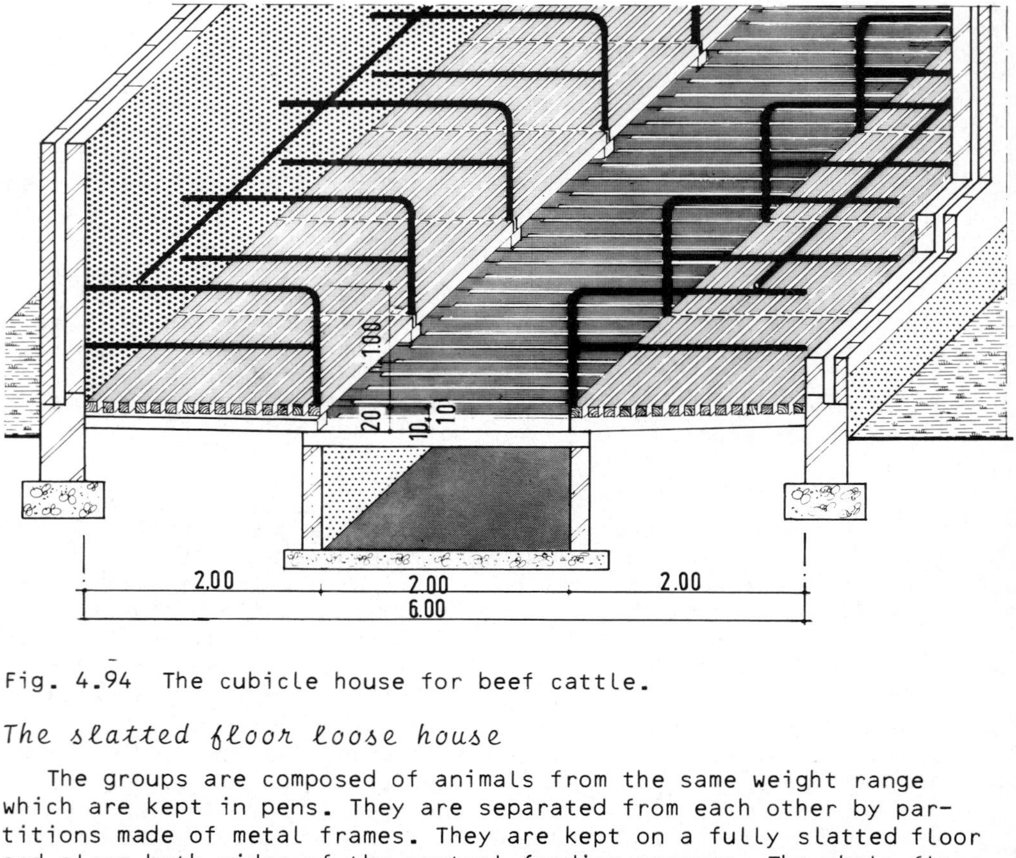

Fig. 4.94 The cubicle house for beef cattle.

The slatted floor loose house

The groups are composed of animals from the same weight range which are kept in pens. They are separated from each other by partitions made of metal frames. They are kept on a fully slatted floor and along both sides of the central feeding passage. The whole floor consists of slats of 16 cm wide and which are 3.5 to 4 cm from each other. The slats have to be made of concrete since wooden slats have the disadvantage of rapid wear and slipperiness and are therefore unsuitable. Concrete slats can be used in a free span of 2.5 m for animals exceeding 500 kg and of 3 m for animals weighing less than 500 kg. The area allocations of the pens are 1.1 m² /animal for beef cattle between 8 and 13 weeks of age, 2.0 m² /animal for the age of 14 to 30 weeks, and 3.5 m² /animal for beef cattle from 31 weeks to 52 weeks if the animals are kept in groups of at least 10 units. In the past few years only closed, uninsulated houses of this type were built (fig. 4.95).

Fig. 4.95 The slatted loose house for beef cattle.

In the semi-intensive beef husbandry, silage can be administered in the trough or via self-feeding. With trough-feeding the animals take their supply of roughage, pulp etc. through the self-closing feed barrier of which a few elements can be removed to allow the animals to enter and leave the pen. Self-feeding takes place on a concrete apron (cleaning !) at trench silos (frontage : 10 cm/animal) which have already been described for dairy cow houses. Also other roughage installations with trough, as described for dairy cattle, can be applied for beef cattle if a trough-frontage of 50 to 70 cm per animal is provided. Dry feeding can be carried out ad lib in all loose house types in large feed hoppers of which the content corresponds to the estimated weekly consumption by a batch of 10 animals. The concentrate box is provided with a trough and a 15 cm trough frontage per animal is suggested. Hay feeding by means of hay racks is also possible.

In fig. 4.96 a possible layout is given for the housing and care of 84 beef cattle intended for the production of baby beef. The animals are from the age of 3 months housed in fully slatted floor pens with an adjustable area. By moving the common partition between two pens it is possible to allot each batch of 7 animals a space of 2.3 m^2 per animal up to the age of 30 weeks and 3.5 m^2 per animal from this age onwards. In this way the time consuming and stress-creating transfer

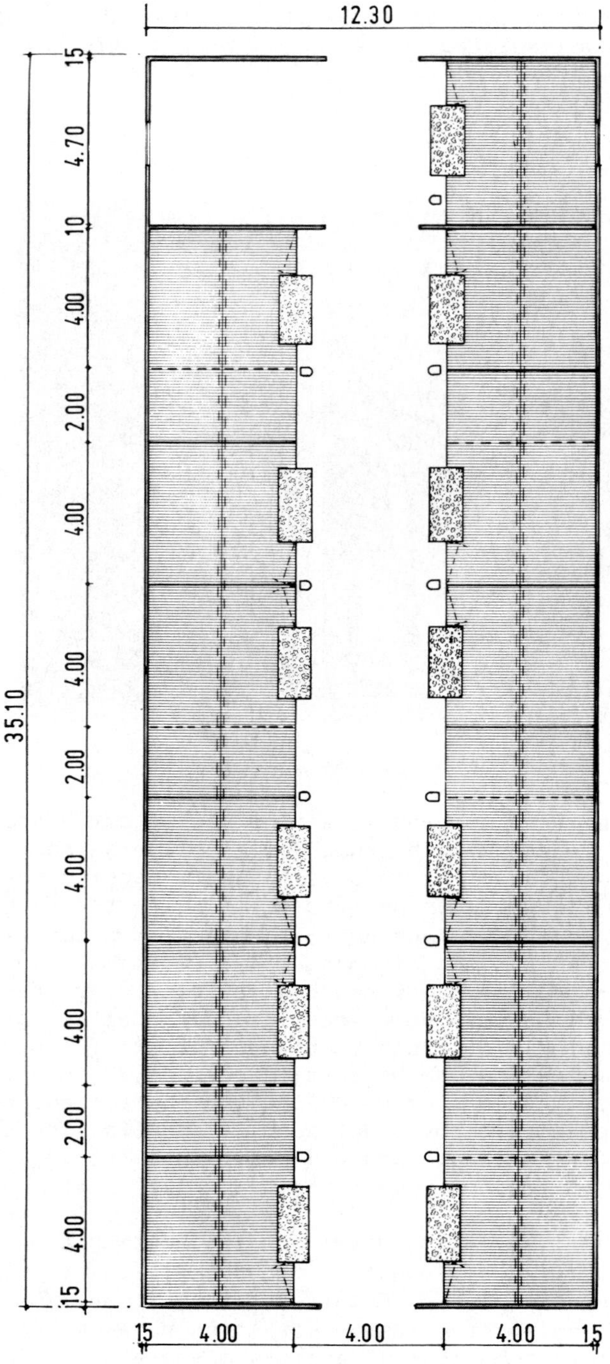

Fig. 4.96 A slatted loose house for 84 beef cattle, split per two pens, in two different age groups.

of animals can be avoided. If on the other hand the farmer chooses a house with fixed pen partitions i.e. fixed pen dimensions (e.g. a pen area of 24 m² per 7 animals) the building costs increase considerably.

Self-feeding is done in concentrate or dry pulp feed hoppers and at a rack for straw or hay. One automatic water bowl is installed per 7 animals. The ventilation is carried out through an open ridge and inlets in both longitudinal walls.

4.5.4 Labour organizational and economic aspects of the housing of beef cattle

According to our time studies (Daelemans and Boucqué, 1972) we can state that the daily care of beef cattle, exclusively fed with dry feed requires 0.8 min per animal and per day in a fully slatted (3.5 m² per head) and 0.9 min per animal and per day in a cubicle house with slatted passages (4.9 m² per head). Periodical activities such as dehorning calves, weighing and selling slaughter-ready animals, have to be added to these times.

The fattening of a young calf to a slaughter-ready animal of 500 - 550 kg in a fully slatted house requires a mere 5 hours of labour of which ca. 30 % consists of supervisional activities. In less equipped beef cattle houses and with semi-intensive beef husbandry (dry feeding + roughage) the labour time requirement can amount in loose houses to 10 - 12 hours and in stanchion barns to 14 - 16 hours per beef animal.

The construction costs of a fully slatted house for beef cattle. amount to ca. £ 250 per animal. The littered loose house is the cheapest construction but this does not imply that this type of house allows the most economical exploitation. The labour requirement for littering together with the costs of the straw increase the cost price of the final product to such an extent that this type of house finally becomes uninteresting.

We were able to demonstrate (Daelemans and Boucqué, 1972) that the annual costs of beef cattle husbandry in fully slatted loose houses amount to only about half of those in littered loose houses and that they are ca. 40 % lower than those of cubicle houses.

4.5.5 Zootechnical and veterinary aspects of the housing of beef cattle

Boucqué et al. (1979) compared the meat production results obtained with bulls in a littered stanchion barn and in a littered loose house whereby eight different feed rations were applied. The animals were housed in the same, uninsulated building. The following conclusions could be reached :

1. the daily growth rate was significantly higher in the littered loose house than in the stanchion barn. The animals in the stanchion barn showed an average daily growth rate of 1,092 g per animal compared to 1,160 g per animal in the loose house ; which makes a difference of 68 g per day ;
2. this higher growth rate leads to a significantly shorter fattening period of the animals housed in a loose house ($p < 0.01$) ;
3. the daily feed intake was always higher with the animals in the loose house, than with those in the stanchion barn ;

4. the average feed conversion amounted to a starch equivalent of 4.51 for tied bulls and 4.79 for bulls housed in a loose house ;
5. no relation could be established between the type of housing and the slaughter quality.

From literature ad hoc it appears that there is no uniformity between the results obtained by different researchers concerning the influence of the housing type on the growth rate and feed conversion of beef bulls.

In another comparative research (Boucqué et al., 1971 ; Daelemans and Boucqué, 1972) the performances of bulls fed with dry feeds and kept in a littered loose house, in a cubicle house and in a fully slatted house were evaluated.

In the first and last mentioned type of house, the density varied while in the cubicle house a number of cubicles were provided with wooden slats, at the same level as the slatted passages while others were provided with concrete slats, raised about 30 cm above the slatted passages. Table 4.21 shows the obtained results. It appears that in all types of loose houses very good daily weight gains were obtained, which however in cubicle houses were no better and sometimes significantly worse than in other types of loose houses. Furthermore the daily growth was significantly better in littered loose houses than in densely stocked slatted houses (2.5 m^2/animal) but was no better in spacious slatted houses (4.3 m^2/animal). It therefore seems that the density influences the growth performance. The feed conversion was not significantly different for the various types of houses.

Hoogerkamp (1971) and Koller et al. (1979) also reached the same conclusions that housing beef bulls in cubicle houses is not interesting. Givens et al. (1968) also showed that the use of cubicles did not favourably influence the meat production results. Mahoney et al. (1967) obtained significantly worse results in a fully slatted house of 1.4 m^2/animal than in one of 2.3 m^2/animal. Carotte (1974) sometimes observed docking by tramping with a density of less than 2 m^2/animal.

In the same research (Daelemans and Boucqué, 1972) we observed, through time lapse photography (table 4.22), that in the same hangar, animals which were accommodated in a loose house and especially in a fully slatted loose house were lying down much longer during the day than those housed in tie stalls, which is in complete agreement with our findings concerning dairy cattle (Maton and De Moor, 1975). Furthermore it appeared that the animals utilized the raised cubicles with concrete slatted floor only to a slight extent : 97.3 % and 84 % of the lying time, respectively half-way and at the end of the meat production period, was spent outside the cubicles.

It appeared clearly during our research, that the animals in fully slatted houses were remarkably clean as those in littered loose houses, although the straw consumption in the latter amounted to 6.1 kg per animal and per day. This conclusion, which a fortiori, is valid in comparison with the littered stanchion barn has also been confirmed in practice.

TABLE 4.21 Results of the meat production of beef cattle in relation to their type of housing.

Type of housing	fully slatted floor		slatted floor + cubicles		littered	
			wood	concrete		
Group	I	II	III	IV	V	VI
area/bull (m^2)	4.3	2.5	4.9	4.9	7.7	4.5
number of bulls	14	24	13	14	13	23
initial weight (kg)	230.3	230.5	231.2	230.6	229.9	231.5
final weight (kg)	567.6	544.5	544.3	544.1	569.8	568.1
duration of experiments(d)	258.1	260.3	267.8	258.4	246.8	250.8
daily growth (g)*	1,307ac	1,206ab	1,169b	1,213ab	1,377c	1,342c
daily intake (kg)						
- feed	8.88	8.70	8.87	8.98	9.52	9.45
- straw	0.32	0.38	0.31	0.33	0.28	0.23
feed conversion						
- feed	6.80	7.22	7.59	7.40	6.91	7.04
- straw	0.24	0.32	0.27	0.27	0.21	0.17

*The average values on one line and indicated with a different letter are significantly different with $P \leq 0.05$ or 0.01 (Duncan, 1955).

Very pronounced differences in magnitude and character of the claw lesions in relation to the type of housing were found by Schmidt and Andreae (1974) as shown in table 4.23.

Table 4.23 shows that claw lesions appear very frequently in all types of beef cattle houses. In the littered stanchion barn, all animals showed claw disorders. In littered loose houses without exercise area a minimum of claw disorders were noted. In loose houses with an exercise area (permanent wet manure film on the concrete) and self-feeding at a trench silo a large number of claw lesions occurred, especially when they were provided with fully slatted floors inside, although it was less than in the littered stanchion barn. Typically, the nature of the claw lesions differed between the stanchion barn and the loose house. In the stanchion barn all animals showed medial claws with overgrowth of the claw horn - which was also utilized for standing and walking - and whereby supersession of the enlarged sole horn in the direction of the interdigital interstice of the claw and necrosis of the sole horn were apparent. In loose houses the so-called "wedge sole claw" (overgrowth of the claw horn) on the lateral claws of the hind limbs, with enlarged sole horn and bent-over interdigital wall penetrating into the interstices of the claw is frequently observed. The above-mentioned claw lesion with beef bulls in stanchion barns leads often to damage of the fetlock-joint and sometimes also of the pastern-joint. From a research by Carotte (1974) it appeared that in 40 % of the loose houses in which the gap between the slats was less than 3 cm, claw lesions occurred. These claw lesions were present in all loose houses (100 %)

TABLE 4.22 The behaviour of intensively fed bulls in various types of houses, expressed in percent of the time (100 % = 24 h = 1,440 min).

Type of housing	Group	Number of bulls	Area per bull (m^2)	Lying down	Standing up, excl. eating & drinking	Eating	Drinking
Loose houses							
fully slatted (1)	I	7	4.3	69.0	18.9	11.0	1.1
fully slatted (2)	I	7	4.3	63.4	25.6	9.3	1.7
fully slatted (1)	II	12	2.5	65.7	23.1	10.0	1.2
fully slatted (2)	II	10	2.5	64.1	27.6	7.4	0.9
cubicles (1)	III	6	4.9	64.4	26.8	7.9	0.9
cubicles (2)	III	6	4.9	63.2	30.0	6.4	0.4
raised cubicles (1)	IV	7	4.9	61.9	26.8	9.4	1.9
raised cubicles (2)	IV	7	4.9	61.1	30.1	7.6	1.2
littered (1)	V	7	7.7	61.7	26.8	10.5	1.0
littered (2)	V	7	7.7	59.0	32.1	7.3	1.6
littered (1)	VI	12	4.5	60.2	29.3	9.0	1.5
littered (2)	VI	12	4.5	61.0	29.9	8.4	0.7
Stanchion barn							
littered (3)		7	-	54.5	39.3	5.2	0.9

where (1) = half-way through the experiment ; (2) = at the end of the experiment ; (3) = in the same building as the loose house.

where the gap was greater than 4 cm. Interesting is also the conclusion reached in two studies of De Boer (De Boer et al., 1971 ; De Boer and Ham, 1975) from which it appears that tied animals, with comparable feed rations, showed a worse carcass quality and in particular became fatter than animals which were kept in loose houses.

The problem of the area allowance per animal in loose houses was investigated by Andreae et al. (1980). They demonstrated that the total lying down time of yearling beef bulls (ca. 400 kg in weight) decreased from 14.3 h at 3 m^2/animal to 13.6 h at 2 m^2/animal. The obtained difference was however not significant. They also stated that the concentration of the hormone cortisol in the blood plasma was higher with 2 m^2/head than with 3 m^2/head, which indicates more stress. This also leads to a more intensive metabolism which results in a higher energy consumption and hence a slower growth. The reason why animals in densely populated houses live in a state of increased stress is probably due to an insufficient individual distance between the animals. Several researchers such as Hafez (1969) and Porzig et al. (1969)

TABLE 4.23 Claw lesions in different types of houses for beef cattle.

	Littered stanchion barn	Loose houses		Littered house with trough but without exercise area
		littered lying area, exercise area & self-feeding	slatted floor with exercise area & self-feeding	
Number of animals	25	14	20	15
healthy claws				
- number	0	4	3	11
- percentage	0	29	15	73
"wedge sole claws"				
- number	0	10	15	3
- percentage	0	71	75	20
claws with sole horn necrosis				
- number	25	0	0	1
- percentage	100	0	0	7
partial peeling of the sole wall				
- number	0	0	2	0
- percentage	0	0	10	0

have already demonstrated that one of the most important requirements for a living together in harmony within a batch of animals is the respecting of the "individual distance" i.e. a certain distance between individuals within a group. When the distance between animals decreases and becomes less than this individual distance the risk of aggression increases. When on the contrary a maximum distance is exceeded, such as is the case on pastures, the animals will approach each other to form a more closed community. In densely stocked houses, the animals are permanently within the individual distance from each other which results in a permanent state of stress.

In general we can conclude that the housing type for beef cattle must be chosen as a function of the beef production method. The littered tying stall can only be taken into consideration for adult beef cattle which are only kept for a short period and whose body dimensions will change only slightly. For such animals the dimensions of the stall can be adapted to the "final product". The fully slatted loose house is indicated for baby-beef production, provided that a sufficiently large area is allowed for each animal. In such a house claw lesions can be kept within reasonable limits, the meat production results are good, while the labour requirement and annual costs are at a minimum. Despite its rather large straw consumption the littered loose house is more appropriate for accommodating valuable animals (double muscled cattle) which are normally brought to a higher final weight. Both the cubicle loose house and the stanchion barn with grids are less suited for housing beef cattle.

REFERENCES

Andreae W. and Papendieck T., 1971. Verhalten von Milchkühen bei der Wahl ihrer Liegeboxen im Laufstall, Der Tierzüchter, 23 : 432-435.

Andreae U., Unshelm J. and Smidt D., 1980. Verhalten und anpassungsphysiologische Reaktionen von Mastbullen bei unterschiedlicher Belegungsdichte von Spaltenbodenbuchten, Der Tierzüchter, 32 : 467-468, 473.

Anon., 1972. Verslag over 1971 van het Instituut voor Landbouwbedrijfsgebouwen te Wageningen, the Netherlands, nr. 56, 40 pp.

Anon., 1974a. Richtlijnen voor bindstallen voor melkkoeien, Direktie voor Landbouwtechniek, Ministerie van Landbouw, Brussels, 24 pp.

Anon., 1974b. Handboek voor de Rundveehouderij, I.L.R.-I.L.B., V.O., the Netherlands, 257 pp.

Anon., 1979. Huisvesting en verzorging van jongvee, IMAG publikatie, nr. 112, Wageningen, the Netherlands, 60 pp.

Anon., 1983. Yearbook of Agricultural Statistics 1978-1981, Statistical Office of the European Communities - Eurostat, Brussels, 286 pp.

Biegman A., 1981. Het geprogrammeerd krachtvoersysteem, Bedrijfsontwikkeling, 12 : 31-38.

Boucqué C.V., Daelemans J., Casteels N.R., Cottyn B.G., 1971. De invloed van ingestrooide- of roostervloerloopstallen op de rundvleesproduktieresultaten, Landbouwtijdschrift, 24 : 1043-1058.

Boucqué C.V., Fiems L.O., Cottyn B.G., Buysse F.X., 1979. The effect of straw-bedded loose houses or tie stalls on the performances of finishing bulls, Livestock Production Science, 6 : 369-378.

Boxem T. and Smits A., 1977. Eerste ervaringen met de opfok van jongvee in open frontstal, Bedrijfsontwikkeling, 8 : 1027-1033.

Boxem T. and Smits A., 1979. Opfok van kalveren in open stal is een goed bruikbaar alternatief, Landbouwmechanisatie, 30 : 977-980.

Brandsma S., 1982. Een analyse van de verschillen in melkproduktie tussen bedrijven met een ligboxenstal en die met een grupstal, Bedrijfsontwikkeling, 13 : 623-627.

Burgstaller G., Hüffmeier H., Kalich J., Schlichting M., Nack M., Van den Weghe H., 1981. Empfehlungen zu technischen Einrichtungen für die Kälberhaltung, Der Tierzüchter, 33 : 76-78.

Carotte G., 1974. Les taurillons mal logés et les traumatismes, L'élevage bovin, nr. 11, 31-35.

Coenen J., 1980. Mehr Augenmerk den Klauen schenken, DLG-Mitteilungen, 95 : 1270-1272.

Comberg G. and Hinrichsen K., 1974. Tierhaltungslehre, E. Ulmer Verlag, Stuttgart, W. Germany, 464 pp.

Cottereau F., 1974. Acétose et tétane d'herbage : Animaux Sains, Animaux Productifs, L'élevage bovin, numéro spécial, pp. 83-87.

Daelemans J. and Boucqué C., 1972. Enkele bouwkundige en arbeidstechnische aspekten van loopstallen voor mestvee, Landbouwtijdschrift, 25 : 403-417.

De Boer F., Smits B. and Dykstra K., 1971. Voederhoeveelheid, groei- en slachtkwaliteit bij jonge vleesstieren, Landbouwkundig tijdschrift, 83 : 354-359.

De Boer F. and Ham G., 1975. Biureet en gedroogde batterijmest als stikstofbron voor vleesstieren, Bedrijfsontwikkeling, 6 : 603-609.

Debruyckere M. and Neukermans G., 1973. Algemene richtlijnen in verband met de klimaatregeling in gesloten stallen, Landbouwtijdschrift, 26 : 251-282.

De Kruif A., 1975. De vruchtbaarheid op rundveebedrijven, Bedrijfsontwikkeling, 6 : 615-620.

Dijkhuizen A.A. and Renkema J.A., 1977. Economische aspecten van ziekten en ziektebestrijding, in het bijzonder Mastitis, in de Nederlandse Melkveehouderij. Tijdschrift Diergeneeskunde, 102 : 1239-1248.

Dregus J., Szucs E., Szollosi I., 1979. The preference for and occupancy of cubicles with different types of floors by dairy cows under loose housing conditions, 30th Annual Meeting of the European Association for Animal Production, Harrogate, England, 4 pp.

Duncan D.B., 1955. Multiple range and multiple F tests, Biometrics 11 : 1.

Eichhorn H., 1965. Arbeitswirtschaft, Technik und Gebäude bei der Planung neuer Stallformen für Milchvieh, A.L.B. Schriftenreihe, Heft 26, 122 pp.

Fiems L., 1982. Het welzijn van kalveren, De Belgische Veefokkerij, nr. 5, pp. 4-5.

Givens R.L., Garrett W.N., Bond T.E., Morrison S.R., 1968. Activity of beef cattle with stalls, Transactions A.S.A.E., 11 : 374-375, 383.

Gjestang K. and Gravås L., 1976. Golvbelegg i bås for mjølkeku, Institutt for bygningstecknikk, As. Norway, nr. 87, 43 pp.

Gjestang K., 1983. Sammenligning av innredningssystemen for kalver (0 - 6 mndr), Meldinger Norges Landbrukshøgskole, Norway, nr. 20, 22 pp.

Grommers F., 1968. Dairy cattle health in loose housing and tying stalls in the Netherlands, World Review of Animal Production. Special issue 1968, 18 : 88-90.

Hafez E.S.E., 1969. The Behaviour of Domestic Animals, Baillière, Tindall and Cassell Ltd., London, G. Britain, 647 pp.

Hedren A., 1966. Einige Erfahrungen über das Verhalten von Milchkühen in Ställen mit Liegeboxen, Kolloquium CIGR, Lund, Sweden, 10 pp.

Hoogerkamp D., 1971. Stallen voor mestvee, Landbouwmechanisatie, 22 : 797-802.

Hop J., 1979. Ontwikkelingen beïnvloeden keuze van de melkstal, Boerderij, Veehouderij, 63 : 30-33.

Johnson D.W., Dommert A.R. and Kiger D.G., 1969. Clinical investigations of infectious foot rot of cattle, Journal of the American Veterinary Medical Association 155 : 1886-1891 (Coll. Vet. Med. State Univ. St.-Paul, Minn 55101).

Jongebreur A., 1974. Die Unterkunft von Zuchtkälbern während ihrer Aufzucht, Proceedings Intern. Congress Agric. Engineering, Flevohof vol. VIII, the Netherlands, 4 pp.

Jongebreur A. and Smits A., 1978. Boxafmetingen voor vleeskalveren, IMAG, Wageningen, the Netherlands, nr. 110, 16 pp.

Jörgensen M., 1972. Influence of environment on udder health of dairy cows, Report : Commission Internationale du Génie Rural, Sect. 2. Working Conference Gent 9 - 12 November 1970 : "The Influence of the Environment in Animal Housing". Meddelelser fra Statens Veterinaere Serumlaboratorium, Copenhagen, Denmark, 513 : 1-10.

Koller G., Hammer K., Mittrach B., Süss M., 1979. Rindviehställe, D.L.G.-Verlag, Frankfurt, W. Germany, 174 pp.

Kommerij S., 1979. Buitengeplaatste hokjes : goede aanvullende kalverhuisvesting, Waiboerhoeve, the Netherlands, 14 : 42-45.

Maatje K. and Swierstra D., 1978. Standaardisatie van standmaten in visgraatmelkstallen, Landbouwmechanisatie, 29 : 55-58.

Mahieu H., 1981. Préréfrigération et réfrigération instantanée, Le technicien du lait, nr. 5, pp. 3-12.

Mahoney W., Nelson G., Ewing S., 1967. Performance of experimental close confinement cattle feeding systems, A.S.A.E., Annual Meeting, Saskatoon, Oklahoma, U.S.A., 10 pp.

Martens L., Daelemans J., Maton A., 1980. De investering in de melkveestal en zijn uitrusting, Landbouwtijdschrift, 33 : 429-455.

Maton A. and De Moor A., 1973. Diergeneeskundige aspekten van de huisvesting van melkvee in een aanbindstal met roosters, Vlaams Diergeneeskundig Tijdschrift, 42 : 417-441.

Maton A. and De Moor A., 1975. Een onderzoek naar de samenhang tussen de huisvestingsvoorwaarden en gedragingen van en letsels bij melkvee, Vlaams Diergeneeskundig Tijdschrift, 44 : 1-18.

Maton A., Daelemans J., Lambrecht J., 1978. De gedragingen van melkvee in een ingestrooide bindstal in funktie van de standplaatsbreedte, Landbouwtijdschrift, 31 : 827-835.

Maton A., Daelemans J., Lambrecht J., 1981. Onderzoek naar de invloed van de ligboxbevloering op het liggedrag van melkvee in een ligboxenloopstal, Landbouwtijdschrift, 34 : 963-983.

McKnight D.R., 1978. Performance of Newborn Dairy calves in Hutch Housing, Canadian Journal of Animal Science, 58 : 517-520.

Mees A.M.F. and Metz J.H.M., 1984. Onderzoek omtrent het zuiggedrag van kalveren, Bedrijfsontwikkeling, 15 : 131-134.

Meijer A., 1980. Veevoedings- en produktieaspekten van automatische krachtvoerverstrekking, Boer-Koe-Computer samen op weg in de jaren 80, IMAG, Wageningen, the Netherlands, nr. 146, pp. 34-36.

Mitchell D., 1976. Calf housing handbook, Scottish Farm Buildings Investigation Unit, Aberdeen, Scotland, 73 pp.

Mitchell C. and Broadbent P., 1973. The effect of level and method of feeding milk substitute and housing environment on the performance of calves, Animal Production, 17 : 245-256.

Nygaard A. and Birkeland R., 1975. Bein- og Klanvtilstanden hos kyr, Report nr. 79, Agric. Univ. of Norway, 30 pp.

Overvest J., 1978. Vreetbreedte bij zelfvoedering van voordroogkuil en snijmaïs, Bedrijfsontwikkeling, 9 : 863-868.

Petit K.L. and Van Der Biest W., 1980. Melkkoelen, Vereniging der Elektriciteitsbedrijven in België, Brussels, Belgium, 43 pp.

Peyraud J., 1974. La vache allaitante, L'élevage, Paris, France, nr. 28, pp. 81-89.

Porzig E., Tembroek G., Engelmann C., Signoret J.P., Czakó J., 1969. Das Verhalten landwirtschaftlicher Nutztiere, VEB Deutscher Landwirtschaftsverlag, Berlin, W. Germany, 430 pp.

Prange H., 1968. Incidence of diseases of foot in Black Red Lowland cattle in a vet. practitioners area, Monatschrift für Veterinärmedizin, 24 : 281-287.

Rossing W., 1980. Koeherkenning en apparatuur voor de automatisering in de melkveehouderij, Boer-Koe-Computer samen op weg in de jaren 80, IMAG, Wageningen, the Netherlands, nr. 146, pp. 7-33.

Rossing W. and Ploegaert P., 1975. Automatic cow identification, recording milk yield and feeding concentrate, Research report IMAG, Wageningen, the Netherlands, nr. 4, 10 pp.

Schepens M., Maton A., Lambrecht J., Declerck D., 1980. Een oriënterende studie van het verband tussen klauwgebreken en huisvesting van melkvee, Landbouwtijdschrift, 33 : 1225-1230.

Schmidt V. and Andreae U., 1974. Über den Einfluss von haltungsbedingten Klauenveränderungen auf die Kron- und Fesselgelenke bei Mastbullen, Berl. München, W. Germany, Tierärzl. Wochenschr., nr. 87 : 1-5.

Smits A. and Swierstra D.,1979a.Huisvesting van jongvee, Landbouwmechanisatie, 30 : 1103-1105.

Smits A. and Swierstra D.,1979b. Onderzoek open kalverstallen, IMAG, Wageningen, the Netherlands, nr. 138, pp. 44-46.

Souty J., 1961. La stabulation libre à logettes, Ministère de l'Agriculture, Paris, 70 pp.

Spindler F., 1973. Le béton use les onglons, les résultats d'expériences allemandes, L'élevage, nr. 19, pp. 73-75.

Swierstra O. and Van Ooyen J., 1979. Gebouwenkosten van rundveestallen, Landbouwmechanisatie, 30 : 1171-1173.

Thamling, C.H., 1980. Anbinde- oder Laufstallhaltung, Der Tierzüchter, 32 : 403-411.

Thines G., Soffie M., De Marneffe G., 1975. Aires de résidence préférentielles d'un groupe de vaches laitières en stabulation libre, Annales de Zootechnie, Paris, France, 24 : 177-187.

Toren G., 1978. Huisvesting van jongvee, Bedrijfsontwikkeling, 9 : 146-150.

Ubbels J., 1979. Buizenvóórkoelers, Bedrijfsontwikkeling, 10 : 1017-1021.

van der Gaast G., 1981. De melkstalkeuze op het bedrijf, Bedrijfsontwikkeling, 12 : 1014-1020.

Vecht U. et al., 1980. Diagnostiek, oorzaak en bestrijding van Mastitis bij het rund. Stichting Centraal Diergeneeskundig Instituut, Lelystad, Nederland, Annual Report, 1981-1982, pp. 84-88.

Wander J., 1971a. Tierverhalten als Beurteilungsmasstab für Stallbauten Der Tierzüchter, 23 : 243-245.

Wander J., 1971b. Der "rindergerechte" Stallfussboden, Mitteilungen der DLG, 86 : 1072-1074.

Wander J., 1974. Ergebnisse von Wahl- und Leistungsversuchen mit Kühen in Leichtbauten, Der Tierzüchter, 26 : 497-499.

Westendorp T., 1973. Stallen met ligboxen of voerligboxen, I.L.B., Wageningen, the Netherlands, nr. 62, 74 pp.

Westendorp T. and Enneman G., 1975. De juiste standbedekking in voerligboxen, Bedrijfsontwikkeling, 6 : 924-925.

Westendorp T. and Hakvoort B., 1977. Strooisel en standbedekking in Ligboxenstallen, Landbouwmechanisatie, 28 : 1101-1104.

Wieringa H., 1982. De invloed van overbezetting in ligboxenstallen op het gedrag van melkkoeien, Bedrijfsontwikkeling, 13 : 627-631.

Willems C., 1971. Een vergelijking van de bevruchtingsresultaten met K.I. op ligboxenstallen en andere stallen, I en II, Tijdschrift voor Diergeneeskunde, 96 : 215-217, 1457-1459.

Williams P.E.V., Day D., Raven A.M., 1981. The effect of climatic housing and level of nutrition on the performance of calves, Animal Production, 32; 133-141.

Zeeb K., 1973. Die Anbindung von Rindern im Kurzstand, Bauen auf dem Lande, 24 : 322-323.

Zeeb K. and Schmidt B., 1973. Systematik der Milchviehhaltung aus der Sicht des Rindergesundheitsdienstes, Tierärzliche Umschau, 12 : 654-659.

Chapter 5

THE HOUSING OF PIGS

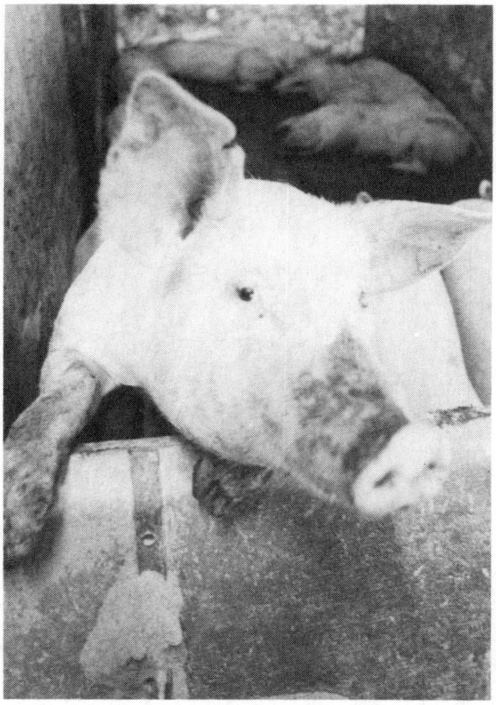

Chapter 5

THE HOUSING OF PIGS

5.1 GENERALITIES

The relationship between construction and equipment of pig houses and the life cycle of the pig is even closer than that of cattle. It is therefore necessary to know the most important biological characteristics of the pig since the housing of the pig in all its aspects is based upon them. The given characteristics are of course only guidelines since variability is inherent to all living species.

The young and female breeding pig is naturally or artificially inseminated at the age of ca. 8 months and when she weighs at least 100 kg. After a pregnancy of 3 months, 3 weeks and 3 days (115 days) she normally delivers a litter of 7 to 12 piglets. The newly born piglets weigh ca. 1.5 kg each. The sow remains with her litter for a period of 4 to 5 weeks and the piglets then weigh 7 to 8 kg at weaning. The weaned piglet is mainly kept for the production of meat whereas only a small number of them (ca. 5 %) are raised as breeding pigs. To prevent the pork having a bad odour and taste male piglets destined for fattening are castrated at the age of 4 weeks. The weaned piglet, intended for meat production is nowadays intensively fed to a final weight of ca. 95 kg which from an initial weight of 20 kg requires about 120 days. The fattener is slaughtered at about 6 months of age. In this way ca. 2.5 fattening cycles per year are possible in the fattening house.

In about 80 % of cases the sow comes on heat 4 to 8 days after weaning. The heat lasts 2 to 3 days and the normal interval between two subsequent heats is 21 days although divergences are frequently found. Annually she produces two litters and after usually four litters, at the age of about three years, she is slaughtered. Productive sows however are kept for up to 8 to 10 litters. A sow can become 130 to 150 cm long and 40 to 50 cm wide.

A boar is first used for mating at the age of ca. 8 months and weighs then 120 to 140 kg ; he is kept for 2 to 2.5 years and is thus destined for the abattoir, at about 3 years old.

5.2 THE CONSTRUCTION OF PIG HOUSES

Nowadays the low profile hangar type of building is generally used for pig houses i.e. a hangar without storey. The hangar can either consist of a composite wall, which supports the roof construction and has then both an insulating and supporting function or of columns supporting the trusses whereby the composite walls are merely an insulating filling between the supporting columns. A good insulating roof and an insulating flooring at the lying places (6 cm hardcore concrete, 15 cm expanded clay beads, 2 cm screed of cement and sand) complete the construction. We refer to Chapter 3 for the choice of building materials.

With extensive pig production the animals are mainly kept in the open air whereby rudimentary shelters are provided and often rather primitively equipped houses are available during the winter. This way of raising pigs has largely been abandoned and replaced by intensive pig production whereby the animals are accommodated in well-equipped, climatized houses adapted to their life cycle and whereby a rational and labour-saving care of the pigs is possible. Young, dry and pregnant sows may however have access to the open air e.g. paddocks or pastures. The layout of the house must be adapted to the life cycle of the pig and suitable for the particular production rhythm. Fig. 5.1 represents schematically this adaptation of the construction of a pig house to the life cycle of the pig. This leads to a number of specific houses which are mentioned below.

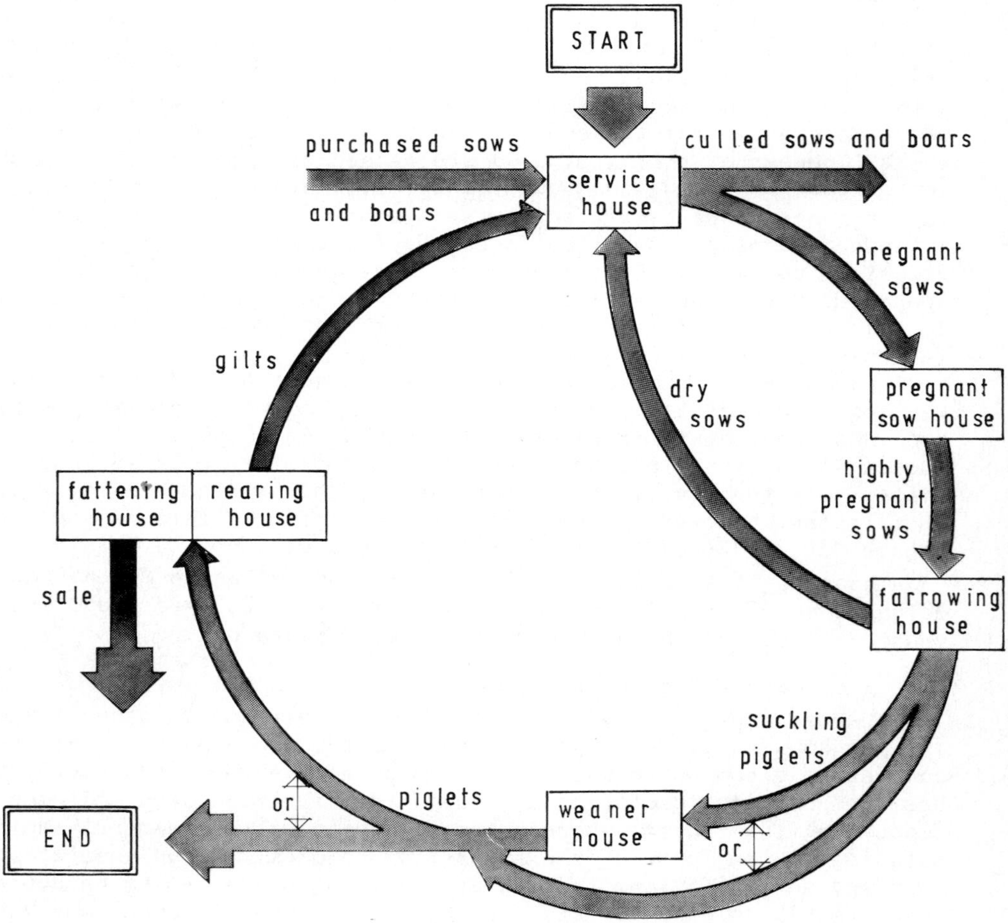

Fig. 5.1 The pig house adapted to the reproduction cycle.

The *breeding house*, where the breeding takes place, can further be divided into the following units :

- the service house : for dry and just-served sows and boars ;
- the house for pregnant sows ;
- the farrowing house : for sows and piglets of 0 to 4 - 5 weeks of age; the sows are brought in no later than the 110th day of pregnancy ;
- the weaner house : for weaned piglets viz. until their transfer to the fattening house or the rearing house or until selling ;
- the rearing house : for young breeding pigs (from ca. 10 weeks of age) which are bred to sows or boars ;

The *fattening house* : where the meat production takes place.

5.3 THE LAYOUT OF THE BREEDING PIG HOUSE

The different types of breeding pig houses will now be described in detail.

5.3.1 The service house

The service house accommodates dry sows, boars and just-served sows (figs 5.2 and 5.3). After weaning all sows are transferred from the farrowing house to the service house where they remain until pregnancy is confirmed i.e. some four weeks after serving. Optimum social environment for gilts entails group housing and daily boar contact. The boars are therefore kept in the proximity of the sows to stimulate in a natural way the heat through sound, odour and if possible also sight. Dry sows are preferably housed in group pens or sometimes in individual pen stalls or tie stalls whilst served sows are housed in individual pen stalls or sometimes in individual tie stalls.

Group pens

A number of sows are simultaneously removed from the piglets and accommodated in a partly slatted pen situated in the service house. The pen area amounts to 1.5 m^2 per sow of which the rear half is provided with a concrete slatted floor (10 cm slat width at the top, 2.5 cm slot width) above which a drinking bowl or nipple drinker is installed. The first 24 hours after weaning, the sows are not fed. After this period a feed hopper is installed in the pen and the sows are fed ad lib until mating (flushing). The number of sows in a group pen is limited to five. With 5 sows an area of 7.5 m^2 is required and this can for instance be obtained with a 2.5 m wide pen, which is 3 m long and of which the rear 1.5 m is provided with a concrete slatted floor (see also fig. 5.2). The partitions are 1.1 m high. The sows are removed from the pen for service and are served in the boar's pen, in the passage or in a special pen (mating pen) after which they are moved to individual pen stalls.

The individual pen stalls and individual tie stalls

The individual pen stalls and individual tie stalls are dealt with in detail in the description of the housing of pregnant sows.

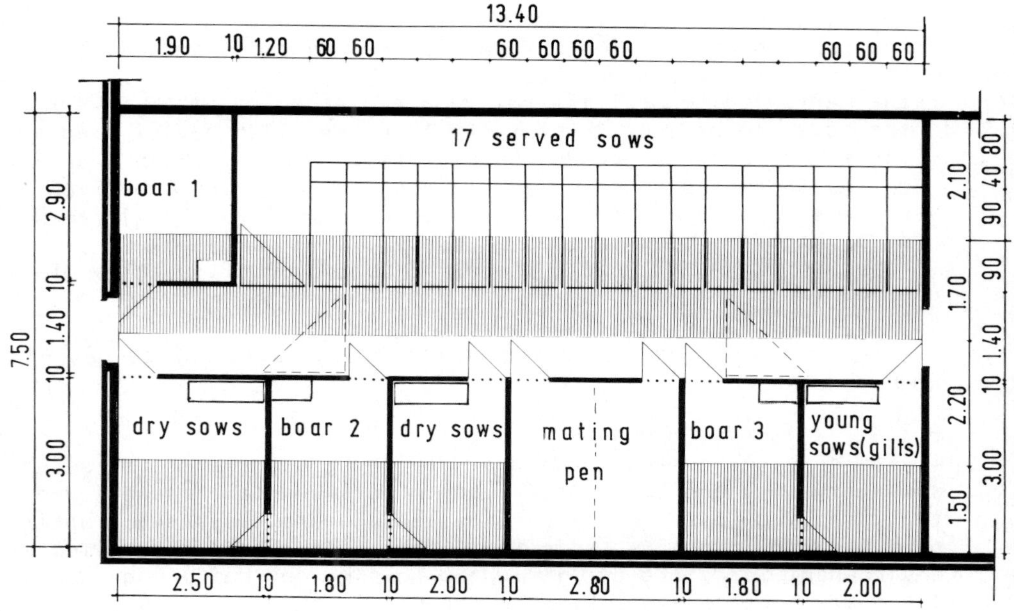

Fig. 5.2 The ground-plan of a service house.

Fig. 5.3 A service house.

The boar pens

The boar pens are situated in the service house. Each boar is individually housed in a pen measuring ca. 3 m x 1.8 m (fig. 5.2). The different pens are separated from each other by a brick wall or concrete plates 1.2 m high. In the front a concrete trough and a gate, made of galvanized tubes, (Ø 1"), are provided. Half of the floor is an insulated lying area whilst the remaining half is made of slats which are at the top 10 cm wide and which are 2.5 cm apart. The pen may be connected to a paddock, i.e. a hardened yard of 3 m x 2 m, which the animals can reach through an opening of 70 cm wide and 90 cm high. This opening is situated in the rear wall (brickwork) of the pen and is provided with a pig-operated rubber flap which prevents draught. The full floors have a fall of 3 % towards the slatted floor if it concerns a lying place and away from the house if it concerns an outside run.

In the past boars were housed outside the sows'house, in individual huts with walls made of brickwork and with an insulated concrete floor which is connected to a hardened yard fenced with concrete plates of 1.2 m height. This way of housing can still be of use for the temporary housing of purchased boars (quarantine).

The service area

The sows are mainly served in the passageway behind the individual pens or in the boar pens. In some cases a mating pen has been provided to which sows and boars are led for copulation. The minimum dimensions of this pen are 3 m x 2.5 m and it is furnished with a full non-slip concrete floor which has a fall of 2 % towards the drainage pit or dung channel.

The floor is covered with a copious layer of sawdust or wood shavings . The walls are 1.1 m high and are made of concrete plates or brickwork (10 cm). Fig. 5.2 shows a service house with one mating pen. The simultaneous serving of several sows in mating pens or boar pens means a labour saving since the supervision during copulation can be carried out over a number of sows.

5.3.2 The house for pregnant sows

The served sows are, after confirmation of pregnancy (at least 4 weeks after copulating), transferred to the house for pregnant sows. In earlier days this house consisted of group pens with 1.5 to 2 m^2 of lying and exercise area per sow, inclusive the individual 0.5 m wide feed stalls of 1.85 m long, sometimes with about 70 % concrete slats and including the trough (fig. 5.4). Nowadays two possibilities exist : individual pen stalls or tie stalls.

The individual pen stalls

The individual pen stalls may be installed lengthwise or across the house. Each individual pen stall has a length of 1.80 m (excl. the trough) and a width of 0.60 m to 0.65 m.
The stall dividers have a height of 0.90 m and are constructed of a welded frame of vertically placed galvanized tubes (Ø 1"), flat iron

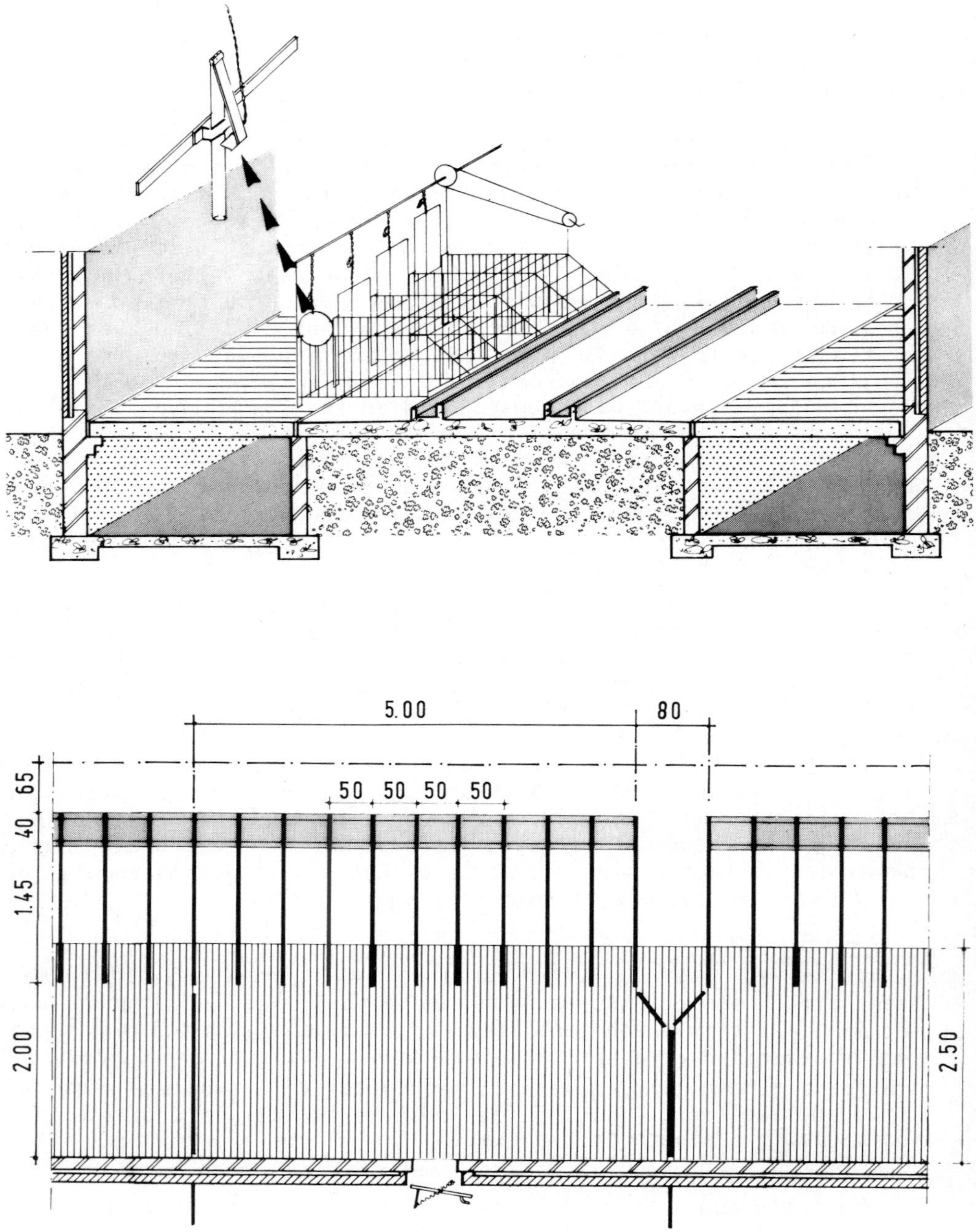

Fig. 5.4 Ground-plan and perspective of a group pen for pregnant sows, equipped with individual feeding stalls.

bars or horizontal tubes (fig. 5.5).

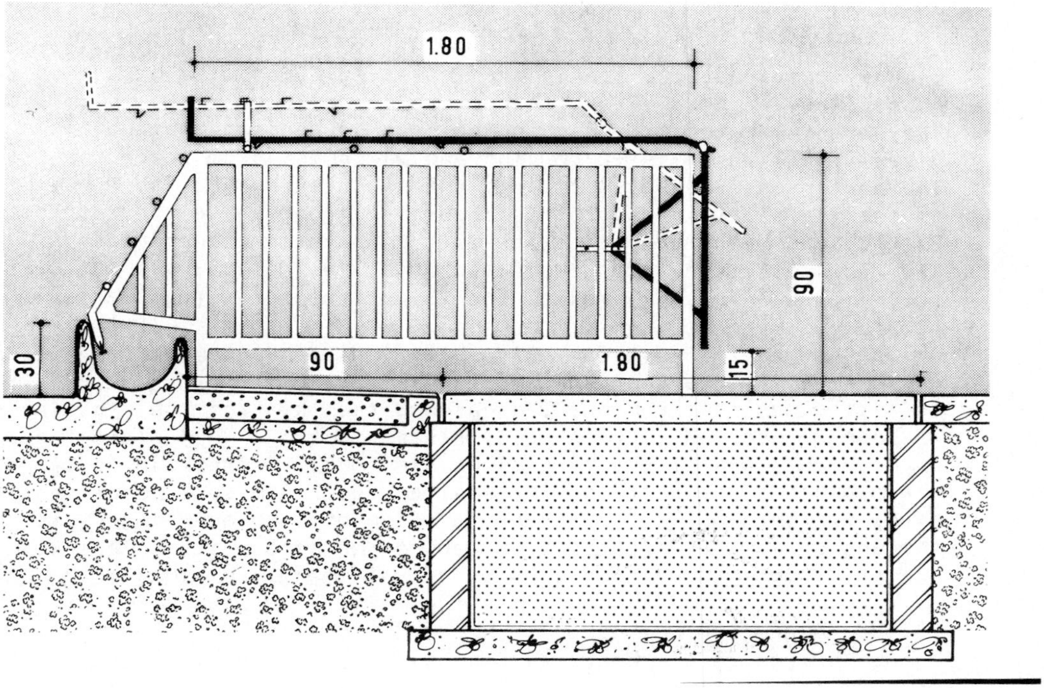

Fig. 5.5 Section of an individual pen stall for pregnant sows.

The front is provided with a concrete or glazed trough, 30 to 40 cm wide and of which the front and the rear are 30 cm and 20 cm high, respectively. A slanting framework of galvanized pipes (Ø 1") is installed above the trough which on one hand prevents the escape of the sows and on the other hand allows the stockman to distribute the feed without being hindered by the animals. The feed distribution can be carried out by means of a low feed trolley i.e. a metal platform supported by 2 swivelling and 2 fixed wheels. Three horizontal tubes resting on the partitions of the pens and parallel to the trough at distances of respectively 30, 60 and 90 cm from the kerb prevent pigs from climbing the framework which separates their stalls. Each stall is provided at the rear with a door, made of a framework of galvanized tubes, which can be closed from the feeding passage by means of a lever. The floor of the individual pen stall is insulated (15 cm expanded clay beads + 2 cm sand and cement screed) over a length of 90 cm and over its entire width. The remaining part of the floor is made of concrete slats i.e. conical slats with an upper slat width of 10 cm, a lower slat width of 6 cm, a height of 10 cm and a slot width of 2.5 cm positioned transversely to the trough. The liquid manure channel is situated under the slatted area ; its dimensions are based on the duration of the storage and on a liquid manure production of 8 l per sow and per day. A partly or fully slatted service passage with a width of 1.5 m is provided behind the row of individual pen stalls.

The individual tie stalls

The individual tie stalls are installed lengthwise or across the house. They are 1.8 m long (excl. the trough) and 60 to 65 cm wide. They are separated from each other by a galvanized framework of pipes (Ø 1") which is 90 cm high and which is 15 cm above the floor. The partitions extend 90 cm into the tie stalls. The front half of the tie stall is made of an insulated floor while the rear half is provided with concrete slats of the above-mentioned design. The tie stalls are arranged in two rows, facing out, without passage (figs 5.6 and 5.7) making the layout cheaper through a saving of space.

The sow is tied by means of a plastic strap connected through a chain to a floor anchor located 30 cm behind the kerb. The strap is fitted behind the front legs around the chest of the pig. The pig is soon accustomed to this belly-strap tie and enjoys ample freedom of movement, whilst the partitions limit this freedom in so far that the pig has to defecate on the slatted floor. In some countries (Denmark, Switzerland) a neck frame is often applied as tying system ; the anchor is then located 5 cm behind the kerb ; it is less satisfactory from the point of view of animal welfare (Aumaître, 1983).

5.3.3 The farrowing house

At the end of pregnancy (110th day) the sows are treated with an anthelmintic, washed and transferred to the farrowing house for farrowing and suckling. The farrowing house is divided into a number of compartments. Their size is chosen in such a way that a group of animals can be put into them at the same time and removed at weaning also at the same time (all-in all-out). In this way the compartments can be thoroughly and regularly cleaned and disinfected and a labour saving can be obtained by transferring sows and piglets in groups. The temperature of the house can now also be adapted to the requirements of piglets of the same age. Against these benefits, however, compartmentalizing requires a higher investment.

Each compartment consists of one or mainly two rows of farrowing pens. Where it was usual to provide three passageways viz. one feeding passage and two dunging passages it is now customary to include only one central passageway to save space and thus heating and building costs.

The farrowing pen (fig. 5.8) is a limited area allotted to each sow for farrowing and suckling, viz. a crate or a tie stall. Both sides of the sow space is reserved for the piglets to suckle, lie or run . A part of the creep area is heated by means of floor heating (100 or 150 watts), an infra-red lamp or a gas heater and during the first

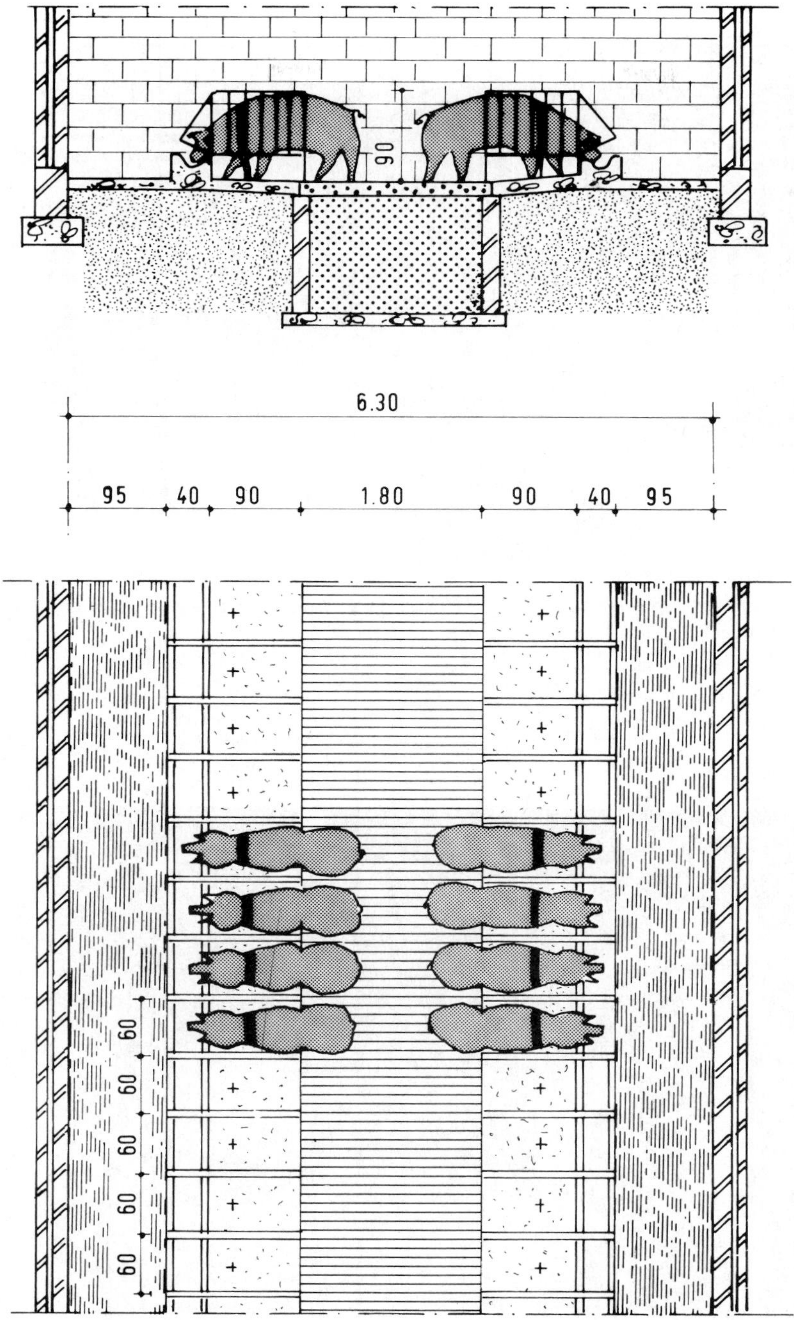

Fig. 5.6 Ground-plan and section of a double row tie stall house of the facing-out type, intended for pregnant sows.

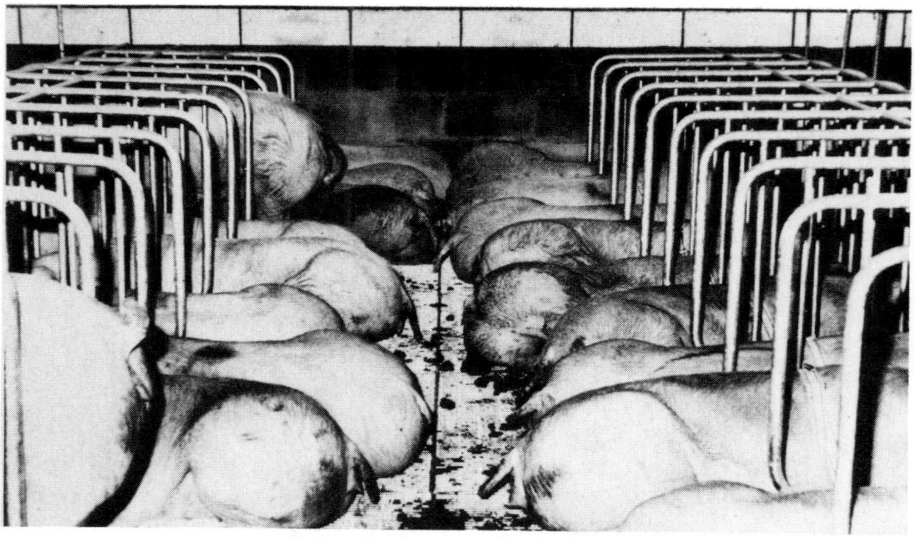

Fig. 5.7 Interior view of a tie house for pregnant sows, with facing-out arrangement.

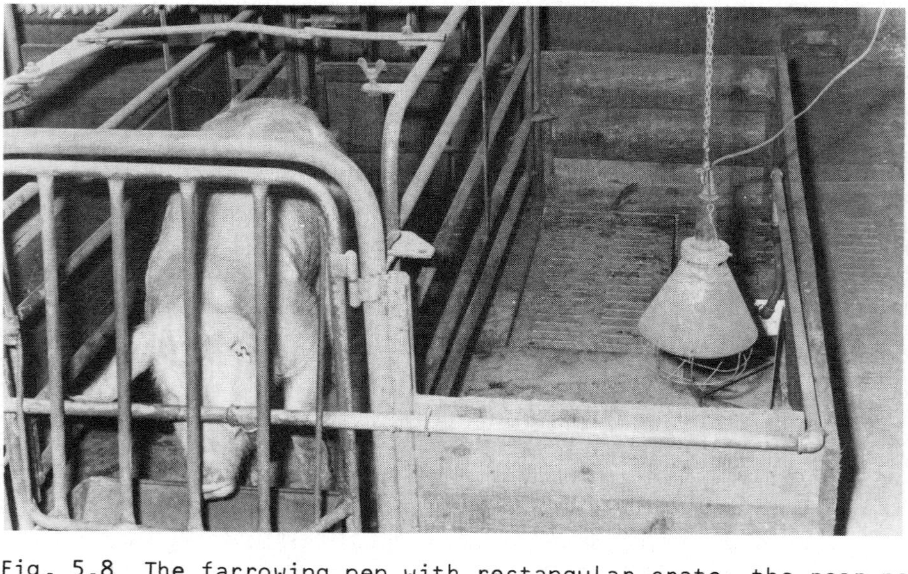

Fig. 5.8 The farrowing pen with rectangular crate, the rear part of which is slatted.

days even a combination of the first with the second or third heat source is used. This heated piglet creep must assure a suitable microclimate in the nestling area. Floor heating of the piglet creep by means of warm water, thermostatically kept at 40°C, is preferred. Seamless copper tubes (Ø 10 to 15 mm) or heat resistant plastic tubes (Ø 1/2") are laid on a compressed insulation layer which in turn rests on a plastic membrane on top of a hard underlayer or layer of concrete. The heating tubes are covered with a wire mesh on which a 5 to 6 cm thick layer of concrete is applied and which is then covered with a 2 cm thick top layer of ground volcanic stones (fig. 5.9).

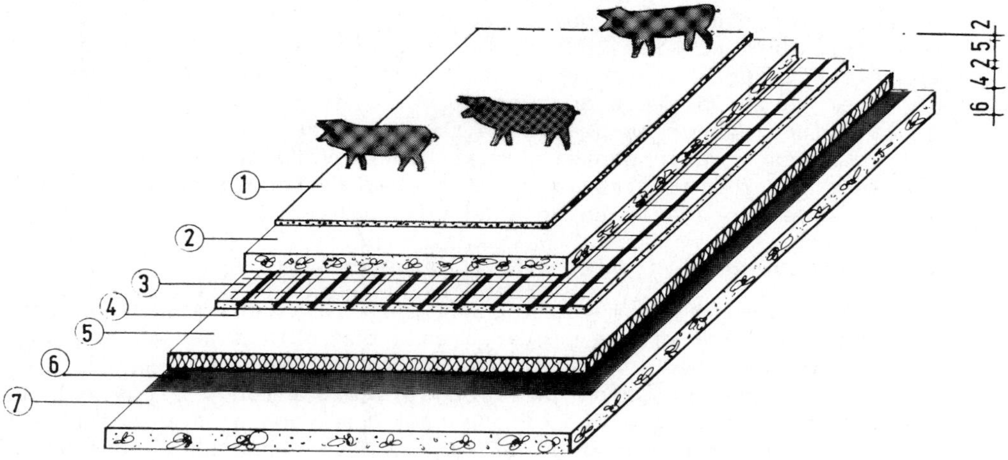

Fig. 5.9 The floor construction of a piglet creep using floor heating by means of warm water.
Legend : 1 = top layer ; 2 = concrete ; 3 = wiremesh ; 4 = heating pipes in a sand bed ; 5 = insulation plate ; 6 = plastic foil ; 7 = hardcore concrete (or compacted earth).

Moreover, the environmental temperature of the whole compartment has to be kept at 17 to 20°C (the macroclimate). A central heating system is therefore applied : it is composed of a boiler and, in the house, finned pipes or radiators installed against the long walls and controlled by a thermostat. The passage from which the air is sucked is pre-heated to about 8°C.

The farrowing pen is partly or fully slatted. Various slatted floor types are commercially available viz. : concrete, cast iron, galvanized welded mesh of triangular or round metal bars, galvanized punched metal sheets, punched aluminium or stainless steel sheets and plastic slats. Choosing the right slot width is difficult because of the large differences in claw dimensions of sows and new-born piglets. Sow claws are ca. 60 mm wide (both digits together) and also ca. 60 mm long, while the claws of new-born piglets are, according to Mitchell and Smith (1978), 16.5 mm wide (both digits together) and 11.2 mm long. At the end of the third week of life the claws of the piglets measure, according to the same authors, respectively 24 mm and 16.9 mm. From this it appears that a slot width of 10 mm or smaller is suitable for piglets and of course for sows. With concrete slats such a gap is unsuitable as it doesn't allow sufficient dung passage and therefore leads to dirty pigs. This is the reason why the other, more expensive slatted floors are preferred. With fully slatted floors concrete slats with a slot width of 10 mm can satisfactorily be used for an adequate drainage of the front half of the pen. The other half of the pen is then made of one of the above-mentioned floor materials. Since 1970, straw bedding has gradually lost its importance and it is no longer applied in new pig houses. A trough for the sow is installed in front of the pen. Either a nose or nozzle drinker situated in the trough or a bite drinker mounted above the trough is provided. After the first week supplementary feed is given to the piglets on a small metal dish and after three weeks a small feed hopper is placed in the pen. The piglets are fed with concentrates rich in proteins ; at high environmental temperatures the feed deteriorates rapidly and therefore should be administered in small portions and replenished regularly.

The farrowing pen with straight crate

This farrowing pen measuring 2.2 m x 1.6 m consists of a centrally situated sow cage with a piglet area along both sides of the cage (fig. 5.10).

The sow cage including the trough is 2.2 m long, 0.7 m wide and 0.9 m high. The long sides of the cage are made of four or five horizontal galvanized tubes of 1" diameter, placed one above the other. The lower horizontal tube must be 25 cm above the floor and may be adjustable in height (25 cm for young sows, 30 cm for multipara sows): in this way one prevents the sow from sliding under the rail, where she might get stuck and perhaps hurt when trying to get out. The second, third and fourth rail are respectively 45 cm, 65 cm and 90 cm above the floor. Three horizontal galvanized tubes are placed across the top side-rails, parallel with the trough at distances of respectively 30 cm, 60 cm and 90 cm. In this way one prevents the sow from climbing the cage. The front of the sow cage has either a metal trough with a width of 30 cm partially protruding into the feeding passage or a concrete trough with an inner width of 30 cm and an outer width of 40 cm. Above the trough a sloping framework is installed consisting of horizontally or vertically placed galvanized pipes of 1" diameter which prevents the sow from leaving the cage and which makes it possible for the stockman to distribute the feed into the trough without being hindered by the sow. The feed distribution can be carried out by means of a feed trolley with a content of ca. 200 kg and equipped with two swivel wheels and two fixed wheels fitted with rubber tyres. To the left and right of the cage are

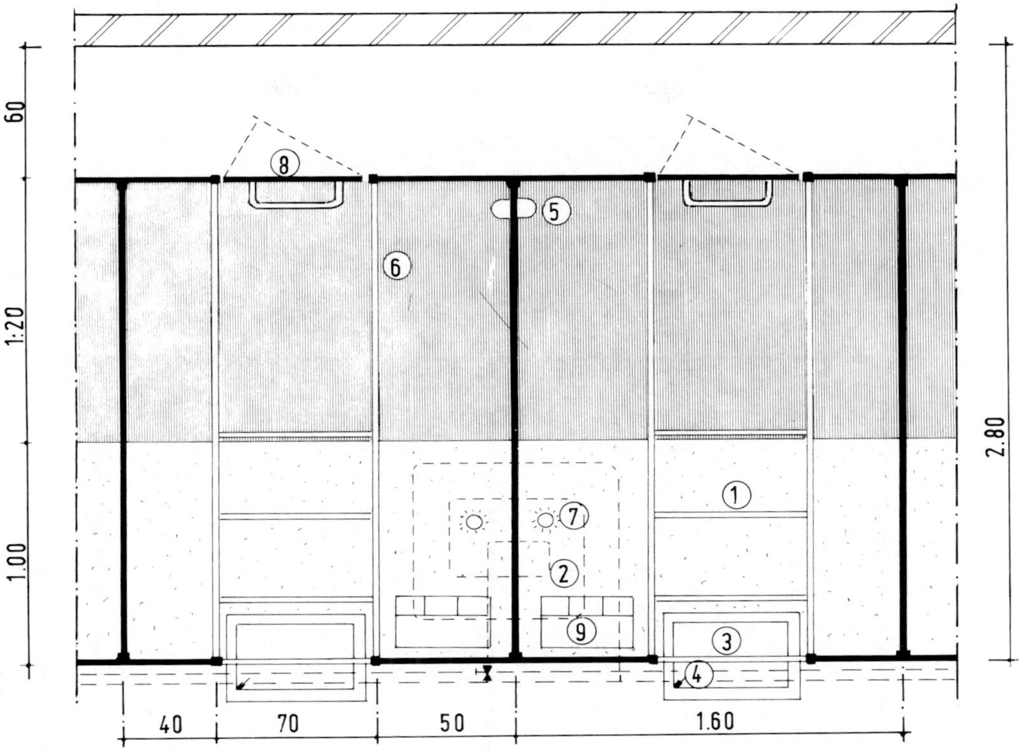

Fig. 5.10 Ground-plan of a part-slatted farrowing pen with straight crate.
Legend : 1 = sow cage ; 2 = floor heating ; 3 = trough ; 4 = nipple drinker ; 5 = creep drinker ; 6 = slatted floor ; 7 = infra-red heater 8 = anti-crushing guard ; 9 = feed hopper for the piglets.

the piglet areas, which are each 2.20 m long and 0.40 or 0.50 m wide according to the weaning age. A part of one of them is equipped with a supplementary heating (dull emitter, floor heating) and serves as a creep or nest. The partition walls between the different farrowing pens, together with the front and rear walls of the pen need only to be 60 cm high.This height is adequate to keep the piglets inside the pen. The partitions are preferably made of an easy-to-clean material such as asbestos cement plates or moisture-resistant hardwood planking mounted in steel U-channels.

The farrowing pen with angled crate

A special type of rectangular farrowing pen which has found some acceptance during the last few years is the angled-crate farrowing pen (fig. 5.11). This layout leads to a more spacious area for the piglets and adds to the comfort of the larger piglets as compared to

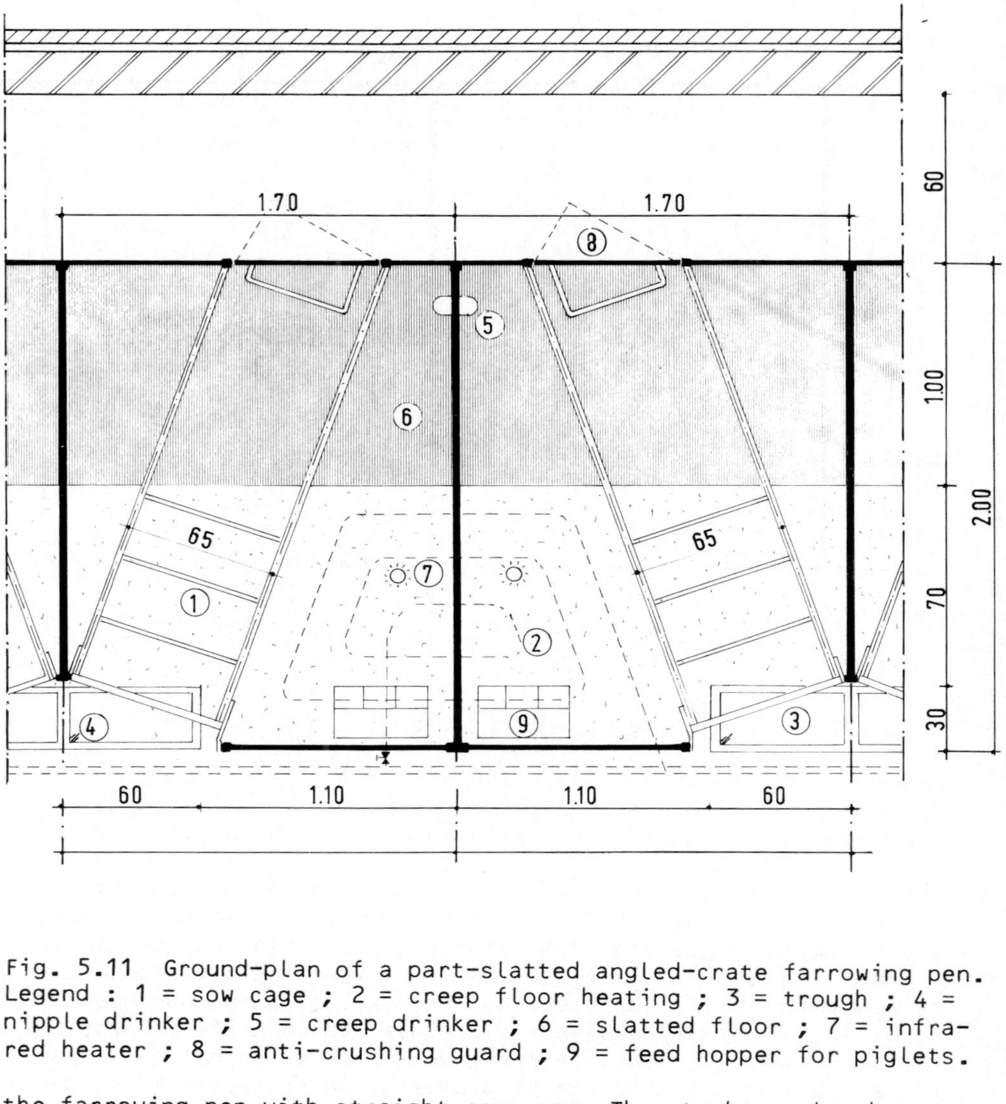

Fig. 5.11 Ground-plan of a part-slatted angled-crate farrowing pen. Legend : 1 = sow cage ; 2 = creep floor heating ; 3 = trough ; 4 = nipple drinker ; 5 = creep drinker ; 6 = slatted floor ; 7 = infra-red heater ; 8 = anti-crushing guard ; 9 = feed hopper for piglets.

the farrowing pen with straight sow cage. The stockman also has a larger freedom of movement when he has to enter the pen for instance to catch a piglet. The pen is 2 m long and 1.7 m wide. Construction and layout are otherwise similar to the farrowing pen with straight crate.

The farrowing tie stall with straight sow stall

This farrowing pen also includes an area for the sow and two areas for the piglets (fig. 5.12). The stall is 2.5 m long and 1.5 to 1.7 m wide, partially or totally slatted. Concrete or cast iron slats, galvanized punched metal sheets, galvanized welded mesh of triangular or round metal bars and plastic coated metal mesh with a similar supporting structure can all be applied with good result for the flooring of the pen. The front of the sow stall has a concrete or galvanized metal trough, a framework above the trough and a pair of partitions made of galvanized pipes of 1" diameter which are approximately 1.2 m long. The tying system is similar to the one described previously. The piglet areas have a width of 40 to 60 cm and are provided along both sides of the sow stall. The farrowing tie stall is bounded by four partitions of 60 cm high and consisting for instance of asbestos cement boards of 2 cm thick mounted in steel U-channels. The sow trough is installed in the front of the stall where the stall meets the feed passage which is 0.8 m to 1.0 m wide. The rear wall has a slide for letting the sow in and out of the stall and it abuts onto a narrow service passage of 0.6 m width. If the latter is omitted the partitions of the sow stall must be installed in such a manner that they can be turned up to allow the sow in through the front viz. by means of a removable board.

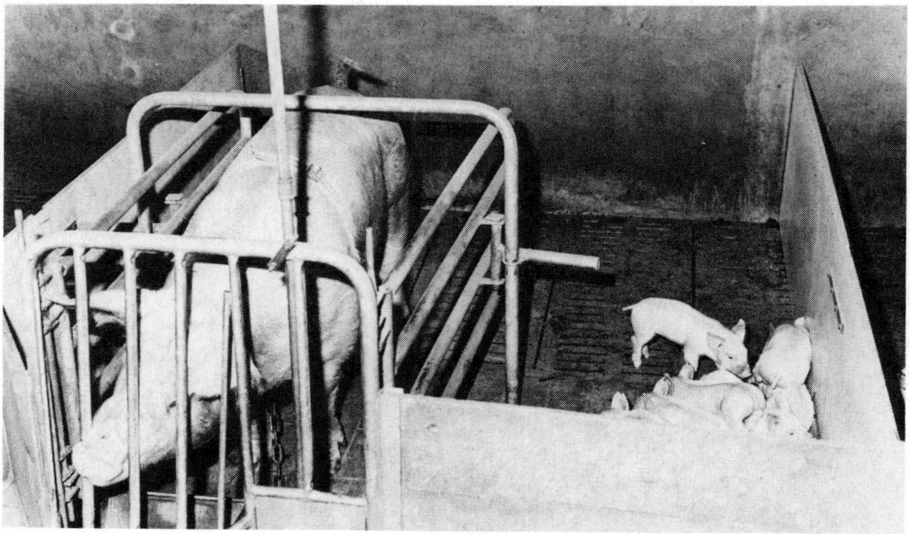

Fig. 5.12 A farrowing tie stall with straight sow stall.

The farrowing tie stall with angled sow stall

The farrowing tie stall with angled sow stall has the same dimensions as the farrowing pen with angled crate, the exception being the tying of the sow. Fig. 5.13 illustrates the layout of this type of farrowing pen.

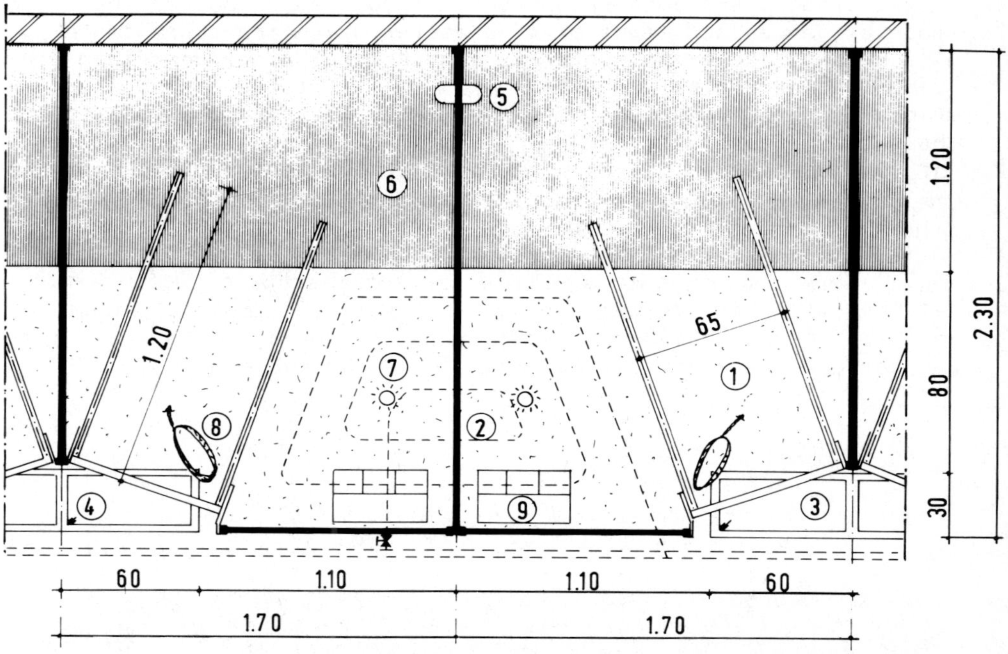

Fig. 5.13 Ground-plan of a farrowing tie stall with angled sow stall. Legend : 1 = sow stall ; 2 = creep floor heating ; 3 = sow trough ; 4 = nipple drinker ; 5 = creep drinker ; 6 = slatted floor ; 7 = infra-red heater ; 8 = chest tying ; 9 = feed hopper for piglets.

The sows are removed from the farrowing pens at weaning i.e. normally when the piglets are 4 to 5 weeks old. The piglets remain in the farrowing pens for one more week. At this stage two possibilities exist. The piglets can either remain in the farrowing pen until they are sold or moved to the fattening house or they can be accommodated in a specially designed house viz. the weaner house which consists of one or more compartments and which is described below. A lot more farrowing pens (twice as many) have to be provided if the weaner house is omitted. This of course requires substantial additional investment. In that case it must be possible to remove the lower cage rails or to turn up the cage partitions. Fig. 5.14 shows turned-up partitions of a farrowing tie stall in a house where piglets remain

Fig. 5.14 A farrowing tie stall, with angled sow stall in the turned-up position, where piglets remain after weaning.

in the farrowing house after weaning.

5.3.4 The weaner house

It is possible to leave the piglets in the farrowing house after weaning but usually they are transferred to a separate house with pens for weaners, the so-called weaner house, where they remain until the age of 10 weeks (for selling at ca. 20 kg) or 15 weeks (up to ca. 35 kg on the farrow-to-finish farms). The weaner house consists of one or two (exceptionally more) rows of pens. The row of pens abuts against a 0.8 m to 1 m wide feed passage. Each pen accommodates 10 to 12 weaners. The front includes a removable feed hopper which is 1.20 m long and 30 cm wide. The partitions between the pens can be made of 60 cm high asbestos cement boards above which, at 15 cm height, one horizontal galvanized tube (Ø 1") is installed. Another possibility is the use of 75 cm high concrete plates or wire fences. The floor of the pen consists fully or partially of slats, mainly concrete slats (slat width 10 cm, slot width 2 cm) which run perpendicular to the feed passage. The more expensive floor types such as cast iron slats, stainless steel slats, galvanized welded meshes of triangular or round metal bars or even plastic slats can be used successfully. A liquid manure cellar of ca. 1 m deep is provided below the slats (fig. 5.15). If a partially slatted floor is chosen a part of the lying place can be equipped with floor heating (warm water pipes) which is thermostatically kept around 40°C. This makes the lying area more attractive to the pigs and prevents or reduces the soiling of it. With transformed houses the pens are often raised with respect to the feeding passage whereby the passage

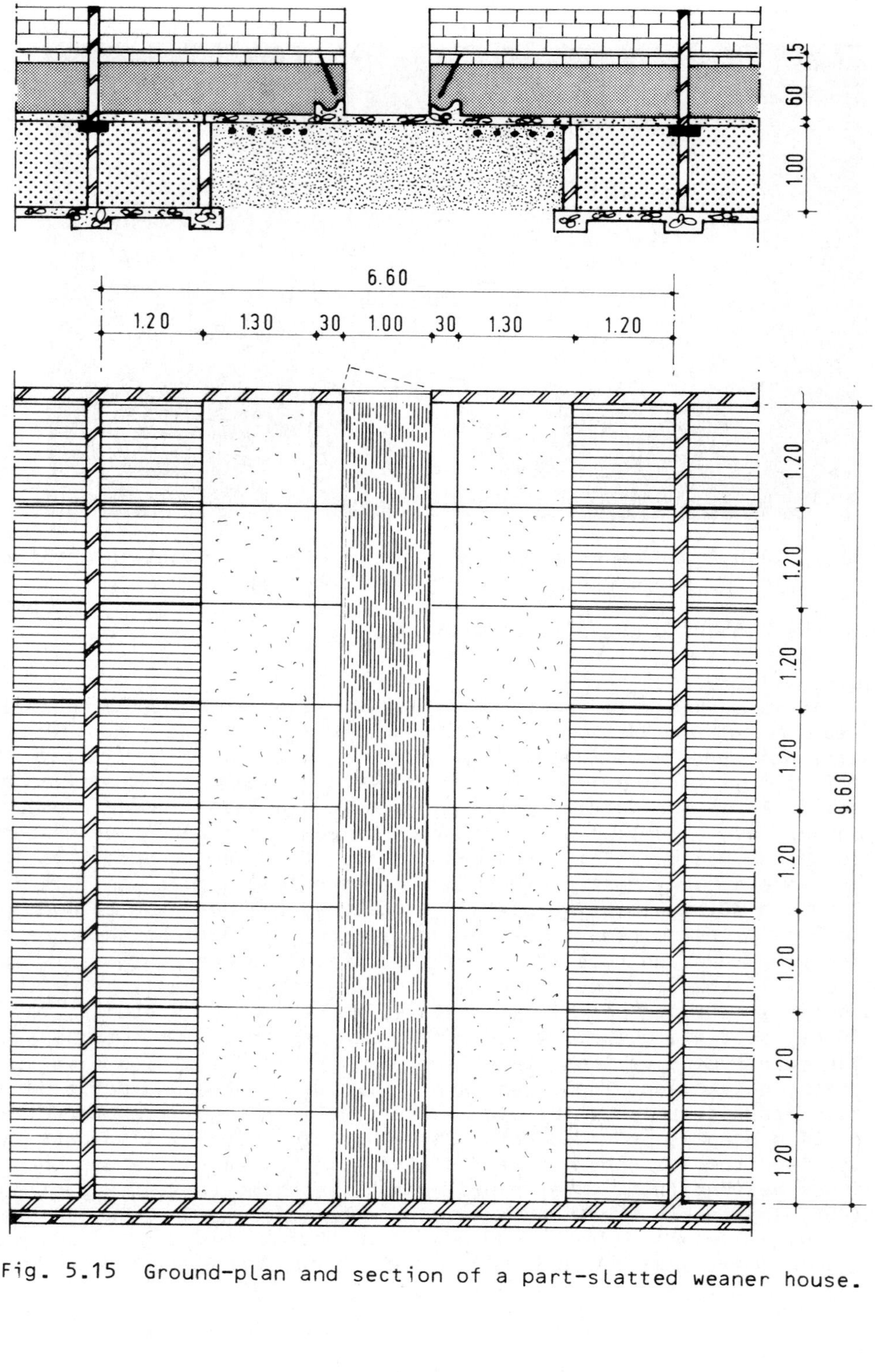

Fig. 5.15 Ground-plan and section of a part-slatted weaner house.

and the bottom of the liquid manure cellar are situated at the original level of the floor (fig. 5.16).

Fig. 5.16 The weaner house with central feed passage and raised slatted floor.

The weaner house, especially the compartmentalized one, has a high stocking rate (0.25 m² pen area per weaner up to 10 weeks or 20 kg and 0.33 m² pen area per pig up to 35 kg or 15 weeks) and a relatively low volume. This implies the need for good ventilation with a large number of air replacements per hour. All attention must be directed to the draught-free supply of fresh air and thermostatically controlled supplementary heating is unavoidable.

5.3.5 Special provisions for a breeding house

5.3.5.1 The ventilation, lighting and heating of breeding houses

The *service house and the house for pregnant sows* are both naturally ventilated : the air inlet takes place through a number of adjustable slots under the windows (which represent 5 % of the floor area) or through a number of adjustable openings in the walls. The outlet of foul air is through a number of stacks. A stack area of 0.5 m² to 1.0 m² per 100 m² floor area is recommended according to the stocking rate. The air inlets measure twice as much. The optimum temperature is between 12 and 16°C and the optimum relative humidity 60 to 80 % (table 3.4).

The *farrowing house* can be ventilated in a similar way as the house accommodating dry and pregnant sows. In compartmentalized houses with central feeding passage, the air is mostly drawn in from the central feed pass (during the winter pre-heated) and is extracted by means of fans at the other (lowest) end of the house. The maximum ventilation quantities are given on p. 79. It is necessary to heat this house, by methods described previously. The environmental temperature in the house is kept between 17 and 20°C (table 3.4).

It can be useful to provide the farrowing house with a *creep box* : it is 120 cm long and 50 cm wide, its total height being 50 cm (fig. 5.17). The creep box is made of wood or fibreboard, it has no floor and is equipped with an infra-red lamp heater of 100 W at the top, which, when the side-flap is turned up, can heat the inside of the creep box to 20°C above the environmental temperature. After ca. 14 days the heating can be switched off. It is also possible to use asbestos cement tubes with a rectangular section (30 cm x 50 cm) and a length of 80 cm. The tube is pushed against a wall and the remaining opening functions both as entry and exit (fig. 5.18). An infra-red lamp of 100 W, protected by a wire mesh, is hung in the tube. In farrowing pens with a straight sow cage there is no place available next to the sow for the inclusion of a creep box. The use of a creep box is an interesting solution for the large old-fashioned farrowing-and-weaning pens since it allows an important decrease in heating costs by permitting a lower house temperature.

Other possibilities for "local" heating of the piglets are, especially in totally slatted farrowing pens, the application of a movable electric heat pad (Baxter, 1974 ; 1984) or the use of electric (or with warm water) heated slatted floors in the creep.

The ventilation in the *weaner house* is preferably carried out by an exhaust fan in view of the high stocking density. An average of 1.2 m³ of fresh air per kg liveweight and per hour (p. 79) is considered sufficient. A continuous adjustment of the speed of the fans is carried out by a thermostat. It is not necessary to provide windows and this will facilitate the insulation of the house and at the same time will keep the weaners quieter. The house is heated to 26 to 20°C, according to the age of the weaners (see table 3.4).

5.3.5.2 The watering facilities

Drinking water is supplied by automatic drinking bowls (fig. 5.19) or automatic nipple drinkers (fig. 5.20) fed by the main or by a pressurized distribution system on the farm and operated by the animals. They are preferably installed above the slats or above a perforated concrete plate with drainage pit since some spillage of water always occurs. Per pen, normally one drinking bowl is installed. With nipple drinkers clean water is always available to the animals (Bekaert and Daelemans, 1970) : the bacterial count is only ca. 250/ml compared to several thousands and sometimes even millions per ml in automatic drinking bowls. Bite drinkers are becoming more and more common. Recently nose drinkers or nozzle drinkers have been introduced, especially for sows (fig. 5.21). They are installed in the trough and function as a valve stem activated nipple drinker.

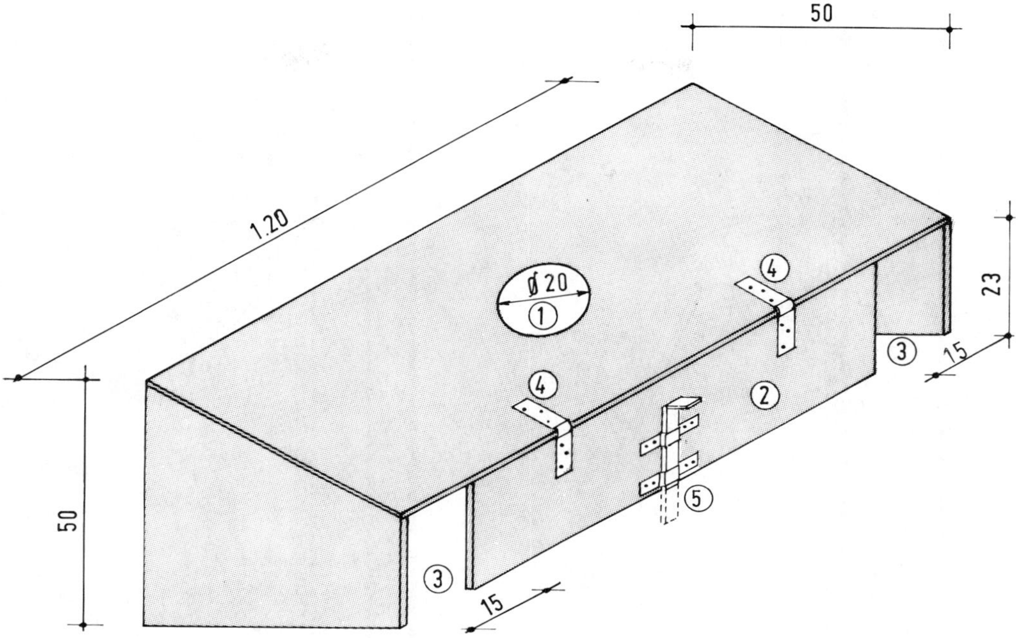

Fig. 5.17 The creep box.
Legend : 1 = opening for the dull emitter ; 2 = movable side flap ; 3 = entry for the piglets ; 4 = hinges.

Fig. 5.18 A tubular creep box.

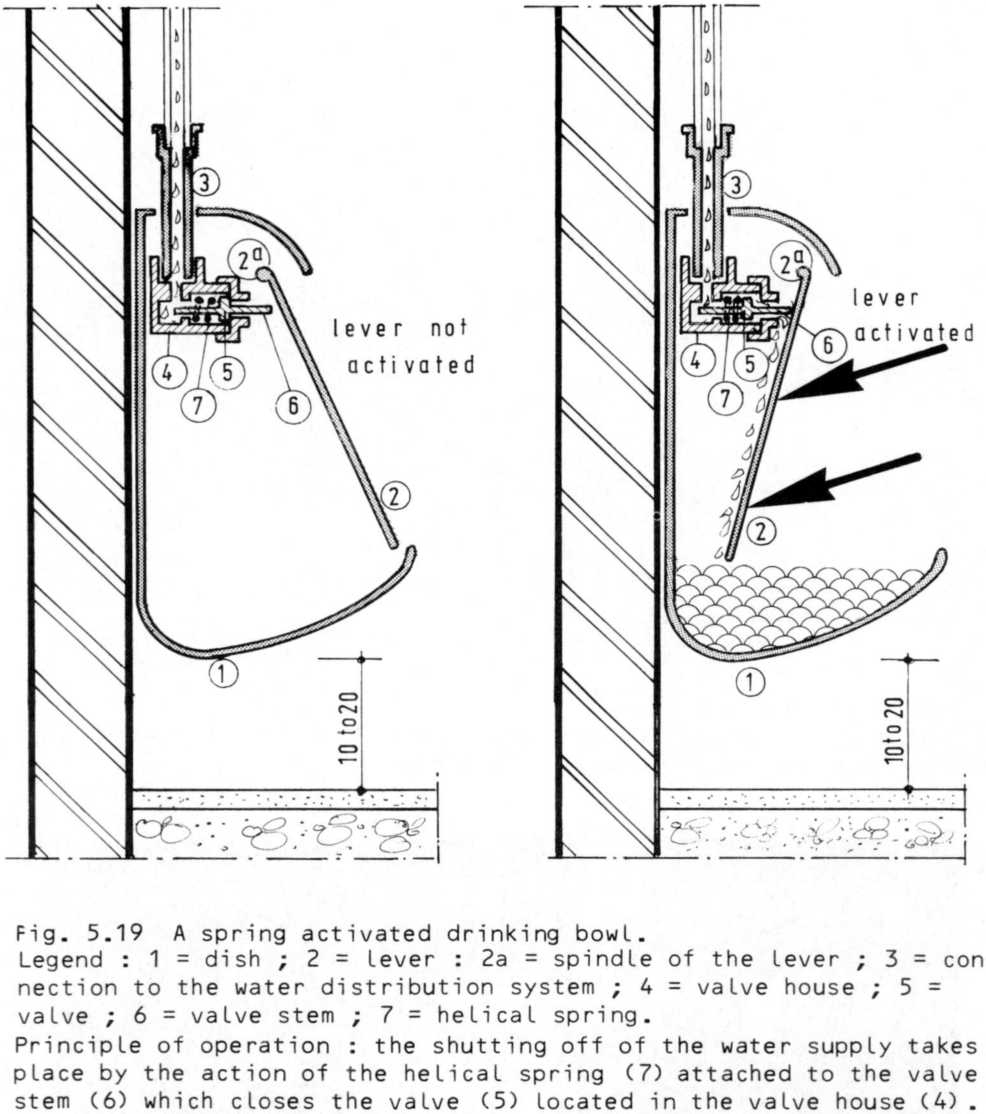

Fig. 5.19 A spring activated drinking bowl.
Legend : 1 = dish ; 2 = lever : 2a = spindle of the lever ; 3 = connection to the water distribution system ; 4 = valve house ; 5 = valve ; 6 = valve stem ; 7 = helical spring.
Principle of operation : the shutting off of the water supply takes place by the action of the helical spring (7) attached to the valve stem (6) which closes the valve (5) located in the valve house (4). The lever (2) revolves on a spindle (2a). When an animal activates the lever (2) the lever will push the valve from its seat (to the left) and water from the distribution system flows in the dish (1) through the connection piece (3) and the valve house (4). The animal can take in the water.
The drinking bowl without spring (cfr. housing of dairy cattle) and the one with spring can normally be used for dairy cattle as well as for pigs, although there is a tendency to develop different models for both species (lever-type and dish-size). The minimum water spillage for pigs is around 1 %.

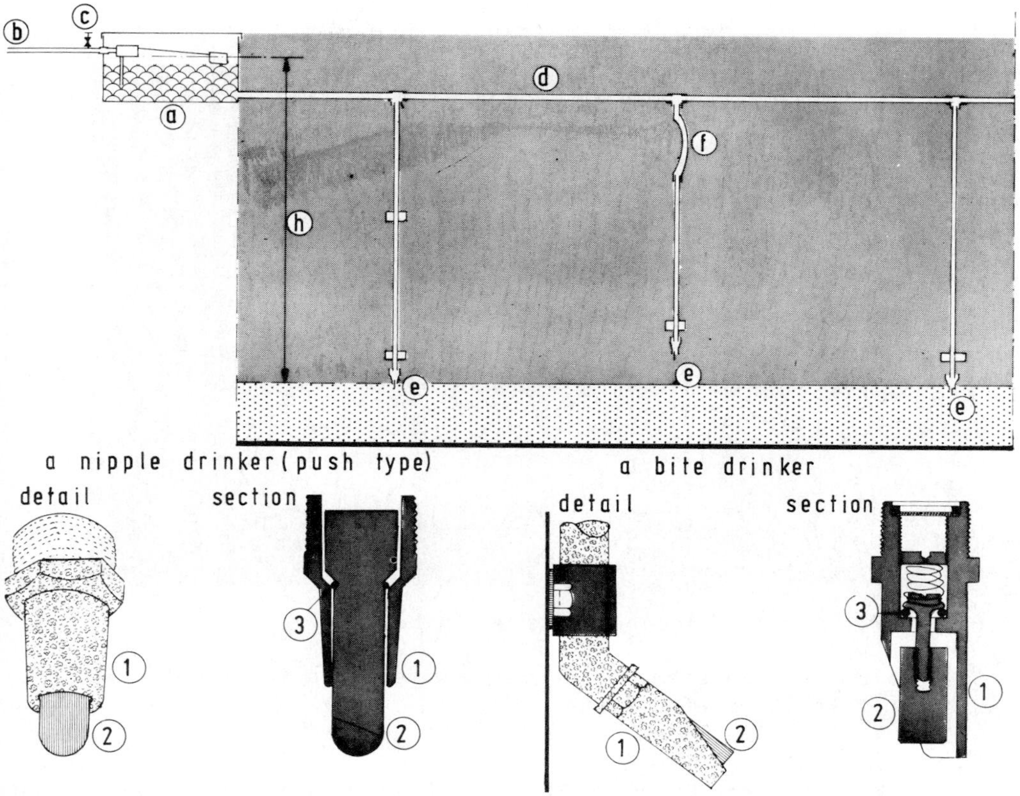

Fig. 5.20 A nipple drinker installation.
Legend : a = main reservoir with float valve ; b = high pressure water supply ; c = stop cock ; d = low pressure supply ; e = nipple drinker ; f = plastic piping ; h = height of the water column above the nipple drinker ; 1 = nipple body or valve house ; 2 = valve stem ; 3 = valve = rubber sealing ring.
Principle of operation : typically for the nipple drinker is the omission of the dish. The nipple is a cylindrical body which has to be taken into the pig's mouth during drinking. The system includes a shutting-off mechanism to interrupt the flow of water when no pigs are drinking. We distinguish nipple drinkers activated by the valve stem or the lever according to whether the pigs are pushing a valve stem or a lever. Tap water is brought to low pressure by means of a float valve in the main reservoir (a) which feeds the nipple drinkers (e) through the low pressure water pipes (d). The pig takes the nipple body (1) in its mouth and pushes the valve stem (2) upwards thereby releasing the valve (3), attached to the valve stem, from the valve seat and whereby water is allowed to flow in the mouth of the pig. As soon as the pig releases the nipple drinker the valve drops back on the seat by gravity and the water supply is shut off. The minimum spillage amounts to 12 %.

Fig. 5.21 A nose drinker in the sow's trough.

The sow pushes the stem up, allowing water into the trough, where she can take it up afterwards.

5.3.5.3 The sow shower

The sows are washed prior to their transfer to the farrowing house. A sow shower can give excellent service. It is built between the house for pregnant sows and the farrowing house. It is a corridor which is 1.2 to 1.3 m wide and of which the length varies according to the number of sows that have to be washed simultaneously viz. 3 m for 4 sows, 4 m for 6 sows, 5 m for 8 sows etc. (fig. 5.22).

The concrete floor has a sharp fall (5 cm per metre) towards the drainage pit which by means of a siphon is connected to the sewer. The walls are lined with tiles up to a height of 1.5 m or rendered with a 15 mm thick layer of cement-sand (volume ratio : 1 part cement, 5 parts sand and 2.5 parts water) which is applied in two stages. The last layer is smoothly finished. Nozzles are provided in the floor and above the sows with which the animals are showered for approximately one hour (ca. 10 l of warm water per sow). The sows will clean themselves largely through mutual friction. After showering the sows remain for a while in the sow wash stall for "drip-drying".

In order to dispose of the damp atmosphere and to prevent it passing to other buildings, water-resistant doors are provided while an adjustable exhaust stack with a minimum diameter of 60 cm or an exhaust fan extracts the humid air.

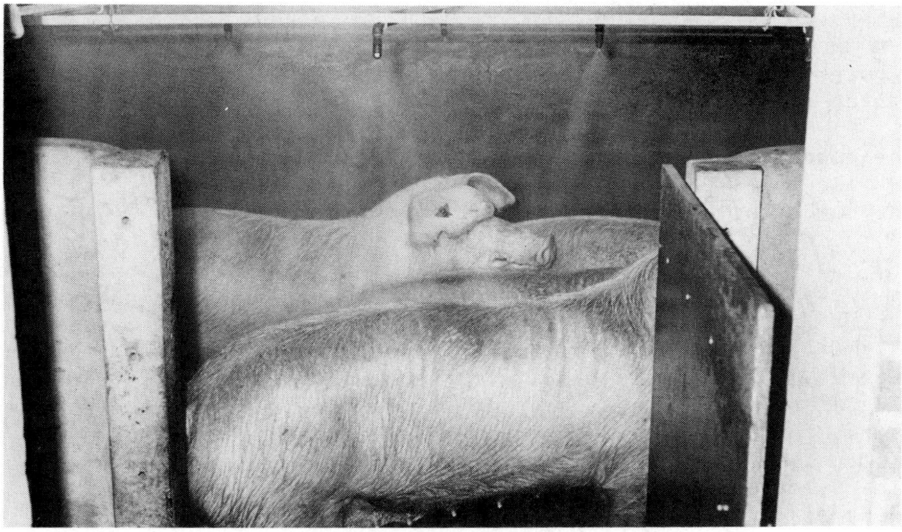

Fig. 5.22 The sow shower.

5.4 THE ORGANIZATION OF THE SOW FARM

Sows pass through a number of consecutive stages : emptiness, gestation, farrowing and suckling and thereby occupy the corresponding houses in the same sequence.

The number of pens required can differ according to whether the sow herd is split up in groups (batching) or not. Therefore a distinction has to be made in the calculation of the accommodation. In addition a number of practical layouts together with a description of the sow calendar and the herd recording system are given.

5.4.1 The required number of pens

5.4.1.1 Non-batching

The number of pens to be provided in each phase of the cycle depends on the following factors :

- the duration of occupancy, in weeks, per accommodation type, including cleaning and disinfecting (W) ;
- the total (the shortest possible) duration of a cycle, in weeks, (C) : C = 21 for weaning at 4 weeks, C = 22 for weaning at 5 weeks, C = 23 for weaning at 6 weeks ;
- the number of productive sows (S) : i.e. all sows of the herd from the very first mating onwards ;
- the coefficient for irregularity and lost days (Y) which is set at 1.05 for farrowing and weaner pens, at 1.10 for dry and pregnant sows and at 0.95 for fattening pigs.

The number of pens to be provided (N) for each phase of the cycle is given by the formula :

$$N = \frac{S \cdot W \cdot Y}{C} \qquad (1)$$

Besides this number an additional number of pens have to be provided for boars and gilts. Their number can be calculated by the following formulae :

- the number of boar pens = S : 20 or S : 25 (2)

- the number of gilt pens $= \frac{S \cdot W \cdot R}{52 \cdot G}$ (3)

where S = the number of sows ; W = the duration of occupancy in weeks (incl. cleaning) ; R = the replacement number, i.e. the number of maiden gilts divided by the number of productive sows, often set equal to 1 ; G = the number of maiden gilts per pen.

Let us, for example, calculate the number of farrowing and weaner pens, together with the number of stalls for dry and pregnant sows required for a herd of 100 sows, where the piglets are weaned at the age of 4 weeks.

The duration of occupancy of the farrowing pens (W) is, with weaning at 4 weeks, normally 6 weeks (4 weeks suckling + 1 week, because the piglets remain one more week in the farrowing pen + 0.7 weeks prenatal settling-in time + 0.3 weeks cleaning) ; the cycle duration (C) is 21 weeks and the irregularity coefficient (Y) is 1.05. The duration of occupancy of the weaner pens (piglets for sale at 9 to 10 weeks) is 7 weeks (6 weeks rearing + 1 week cleaning). For dry sows and just-served sows a residence time of 4 weeks and for pregnant sows of 13 weeks is taken into account. This gives the following results :

- the number of farrowing pens $= \frac{100 \times 6 \times 1.05}{21} = 30$;

- the number of weaner pens (up to 20 kg) $= \frac{100 \times 7 \times 1.05}{21} = 35$;

- the number of stalls for dry sows $= \frac{100 \times 4 \times 1.1}{21} = 21$;

- the number of stalls for pregnant sows $= \frac{100 \times 13 \times 1.1}{21} = 68$;

- the number of boar pens = 5 (for one boar per 20 sows) ;

- the number of gilt pens $= \frac{100 \times 21 \times 1}{52 \times 5} = 8.1$ or rounded up to 9.

Based on the above calculations a ground-plan of a farm for 100 sows is drawn and is represented in fig. 5.23. The calculation of the required number of pens is rather straightforward which can hardly be said of the calculations involved for the grouping of sows.

5.4.1.2 Batching

The batching or grouping of sows is intended as a simplification of the work scheme by synchronizing the heat (oestrus) of several sows. Instead of weaning each week, weaning will now be carried out every

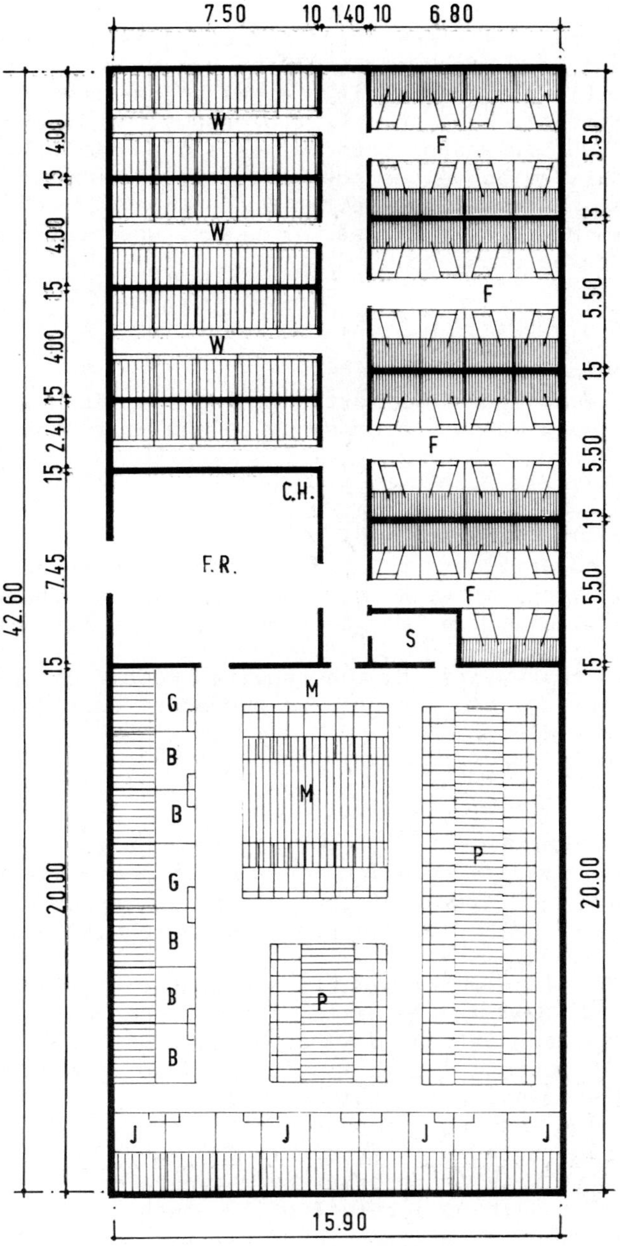

Fig. 5.23 Ground-plan of a house for 100 sows (6.8 m² /sow) : 30 farrowing pens, 4 sections (F); 35 weaner pens, 4 sections (W); 18 individual pen stalls for just-served sows (M) and 2 group pens (G) for dry sows; 68 individual tie stalls for pregnant sows (P); 5 boar pens (B) and 9 pens for breeding sows (J).
F.R. = feed room ; C.H. = central heating ; S = sow shower.

two or more weeks. Under these circumstances formula 1 is no longer valid since other factors will also determine the required number of places. An example will illustrate this. Let us assume that a breeding farm synchronizes the sows in only two groups. It is obvious that a farm with 100 sows needs then not less than 50 farrowing pens and not 30 as calculated above. We now have to take into account the relation between the batches (B), the cycle duration (C) and the interval (I) between two batches, expressed in weeks, when calculating the number of pens to be provided :

$$C = I \cdot B \tag{4}$$

where C, I and B have to be integer numbers.

The number of compartments (K) in a particular section is now given by the result of the formula below, rounded up to the next whole number :

$$K = \frac{B \cdot W}{C} \text{ or } K = W : I \tag{5}$$

where K = the number of compartments ; B = the number of batches ; W = the duration of occupancy, in weeks per batch ; C = the duration of the cycle, which is 20, 21, 22 or 23 weeks respectively for weaning at 3, 4, 5 or 6 weeks.

From this number of compartments (K) the required number of pens (N) can be determined, taking into account the number of sows (S) and the number of batches (B) :

$$N = \frac{S \cdot K}{B} \tag{6}$$

Let us now, by means of an example, illustrate the limitations of the principle of batching viz. in relation to the fact that the interval between the groups, the number of compartments, the number of pens, the number of groups and the duration of the cycle must all be whole numbers.

Let us assume that on a farm where piglets are weaned at the age of 5 weeks (cycle time = 22 weeks), one wants to group the sows. A cycle duration of 22 weeks can only be divided by 1, 2, 11 and 22 and therefore only the following four intervals (derived from formula 4) are theoretically possible for this farm :

a. 22 groups, corresponding with an interval of 1 week ;
b. 11 groups, corresponding with an interval of 2 weeks ;
c. 2 groups, corresponding with an interval of 11 weeks ;
d. 1 group, corresponding with an interval of 22 weeks.

From the practical point of view only the first two solutions are acceptable (a and b).

In case the stockman chooses the second solution (b) he has to make 11 groups (B = 11) with an interval of 2 weeks (I = 2) and the number of sows must be a multiple of the number of batches e.g. 165 (S = 165) which corresponds to 15 sows per batch. Fig. 5.24 shows the symogram (= synchronized motion graph) of that farm and illustrates the synchronization of compartments and batches. Formula 5 enables us to find

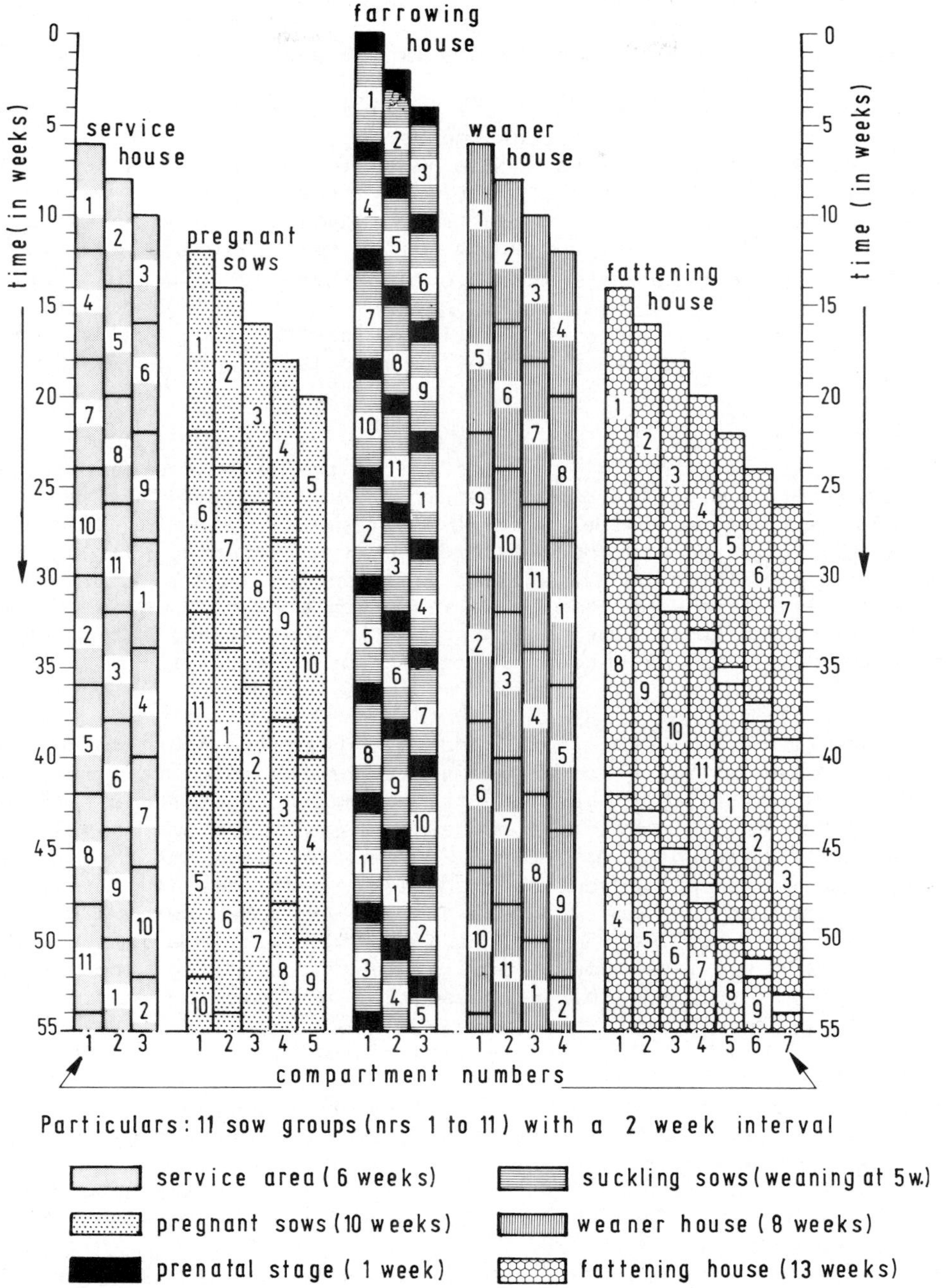

Fig. 5.24 Symogram (Synchronized Motion Graph) of a farrow-to-finishing farm using sow grouping.

the required number of compartments :

- the number of farrowing compartments (W = 6 weeks) equals (11 x 6)/22 = 3 (is an integer number) ;
- the number of weaner compartments (W = 8 weeks) equals (11 x 8)/22 = 4 (is an integer number) ;
- the number of sow batches in the service area (W = 6 weeks) equals (11 x 6)/22 = 3 (is an integer number) ;
- the number of sow batches in the house for pregnant sows (W = 10 weeks) equals (11 x 10)/22 = 5 (is an integer number) ;
- the number of compartments for fattening pigs (W = 13 weeks) equals (11 x 13)/22 = 6.5 or rounded up to the next whole number = 7.

The above whole numbers indicate that there is no waste of space in these particular compartments. However, with fattening pigs, the result is not a whole number, in so far that fig. 5.24 clearly shows that each compartment is left unused for 1 week per cycle.

5.4.1.3 Conclusions concerning the required number of pens

Batching leads to a rigid splitting of the sows into batches. This grouping is in practice often disturbed and this results in an irregular production scheme. Instead of proper grouping it seems to us more appropriate to apply a fixed weekly programme, whereby weaning takes place on Wednesday-afternoons or Thursdays. In this way services and farrowings can greatly be avoided during the weekends. The other activities, such as entering the sows, transferring the pigs, etc. can then also be spread over self-chosen weekdays.

5.4.1.4 Some layouts of houses

Based on the above-mentioned adaptations of the house to the different phases of the reproduction cycle, on the required number of accommodations in each phase and taking into account their rational disposition to one another, a number of layouts for sow houses accommodating respectively 50, 100 and 200 sows are given as an example. Batching is not applied but a fixed weekly programme is followed. Two layouts for housing 50 sows are given viz. one having a width of 9.9 m (fig. 5.25) and one of 18.6 m (fig. 5.26) and from these it appears that the wider house requires less "travelling" and has less outside walls than the smaller house. For 100 sows four different lay-outs are presented :

- a house for a breeding farm selling piglets which are weaned at 5 weeks (fig. 5.27) ;
- a house for a farrow-to-finishing farm where weaning is carried out at 5 weeks and where the weaners stay in weaner pens up to 14 weeks (fig. 5.28) ;
- a house for a breeding farm selling piglets which are weaned at 5 weeks but where the weaners remain in the farrowing pens (farrowing-weaner pens, fig. 5.29) ;
- a house for a farrow-to-finishing farm where weaning is carried out at 5 weeks and where the weaners remain in the farrowing pens up to 14 weeks (farrowing-weaner pens, fig. 5.30).

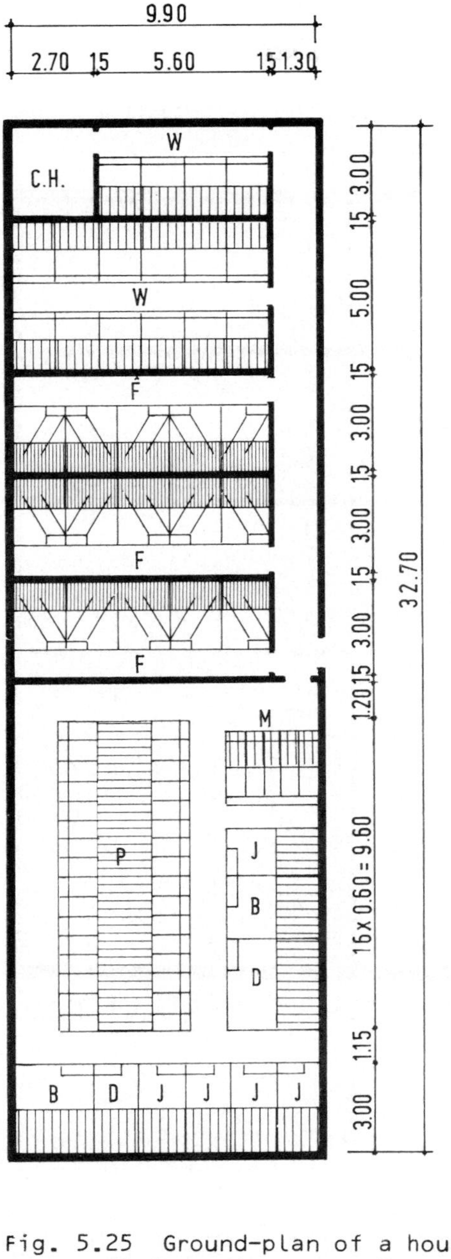

Fig. 5.25 Ground-plan of a house for 50 sows : narrow design (6.5 m² / sow) : 15 farrowing pens, 3 sections (F); 16 weaner pens, 2 sections (W); 5 individual pen stalls for just-served sows (M) and 2 group pens (D) for dry sows; 2 boar pens (B) and 5 pens for breeding sows (J); 32 individual tie stalls for pregnant sows (P).
C.H. = central heating.

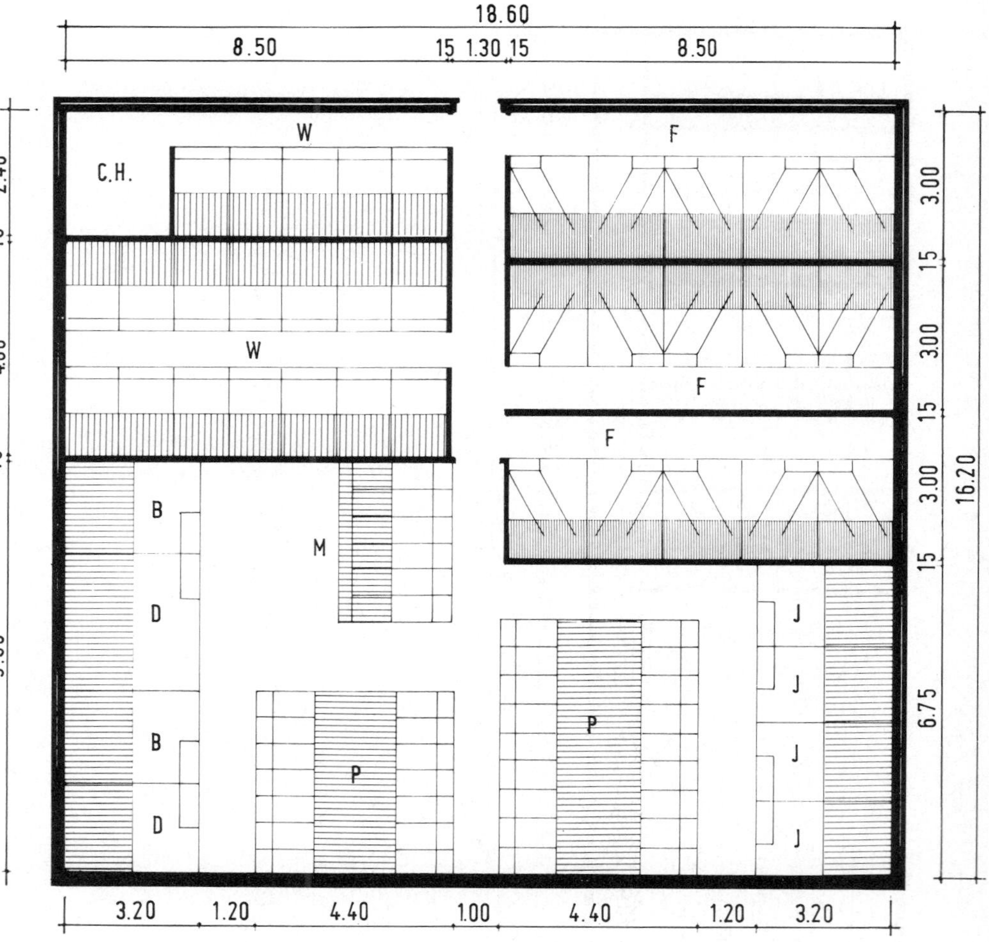

Fig. 5.26 Ground-plan of a house for 50 sows : wide design (6 m^2/sow): 15 farrowing pens, 3 sections (F); 19 weaner pens, 2 sections (W); 6 individual pen stalls for just-served sows (M) and 2 group pens (D) for dry sows; 2 boar pens (B) and 4 pens for gilts (J); 32 individual tie stalls for pregnant sows (P).
C.H. = central heating.

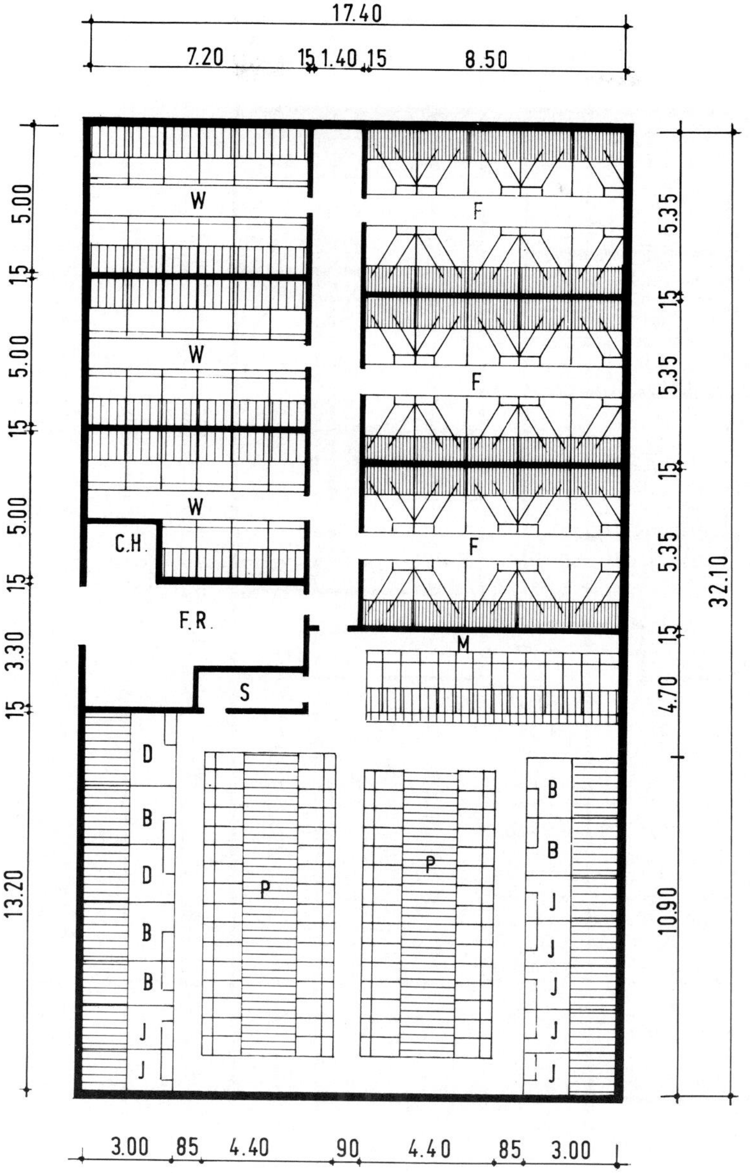

Fig. 5.27 Ground-plan of a house for 100 sows : rearing up to 20 kg (5.6 m^2 /sow) : 30 farrowing pens, 3 sections (F); 34 weaner pens, 3 sections (W); 14 individual pen stalls for just-served sows (M) and 2 group pens for dry sows (D); 62 individual tie stalls for pregnant sows (P), 5 boar pens (B) and 7 pens for gilts (J).
C.H. = central heating ; F.R. = feed room ; S = sow shower.

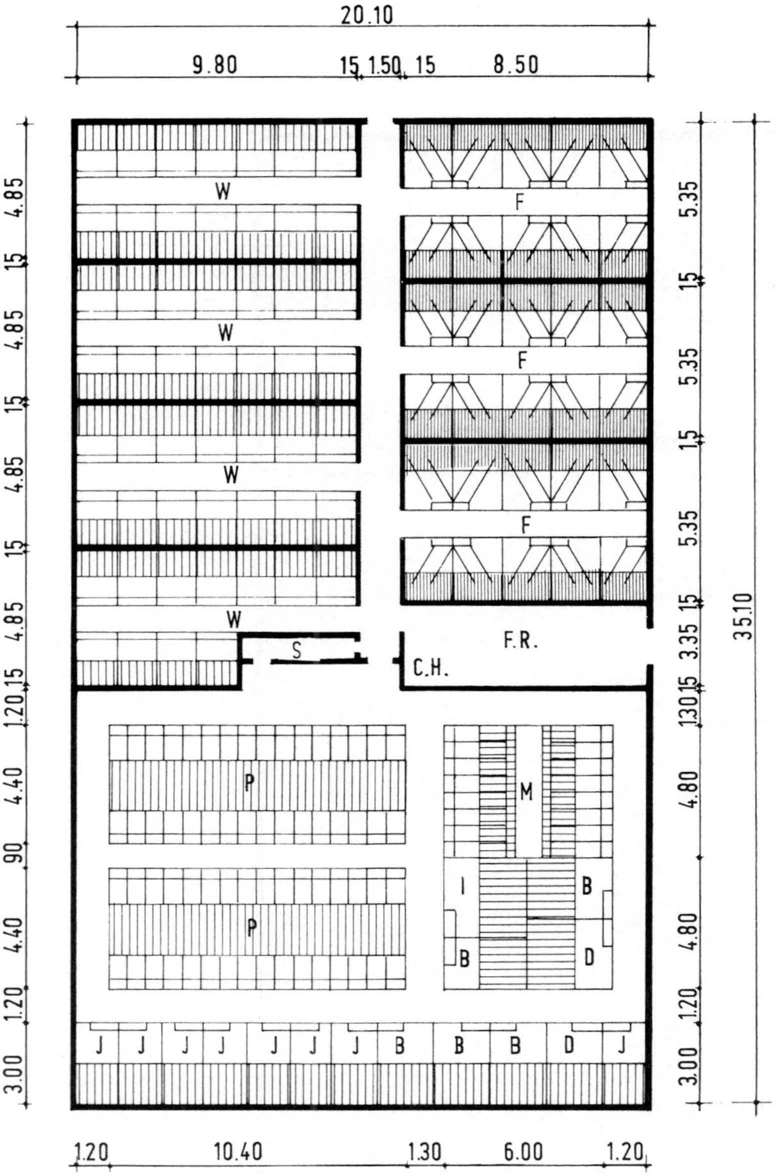

Fig. 5.28 Ground-plan of a house for 100 sows ; rearing up to 14 weeks (7 m²/sow) : 30 farrowing pens, 3 sections (F); 53 weaner pens, 4 sections (W); service pen (I); 16 individual pen stalls for just-served sows (M) and 2 group pens for dry sows (D); 5 boar pens (B) and 8 pens for breeding sows (J); 64 individual tie stalls for pregnant sows (P). C.H. = central heating ; F.R. = feed room ; S = sow shower.

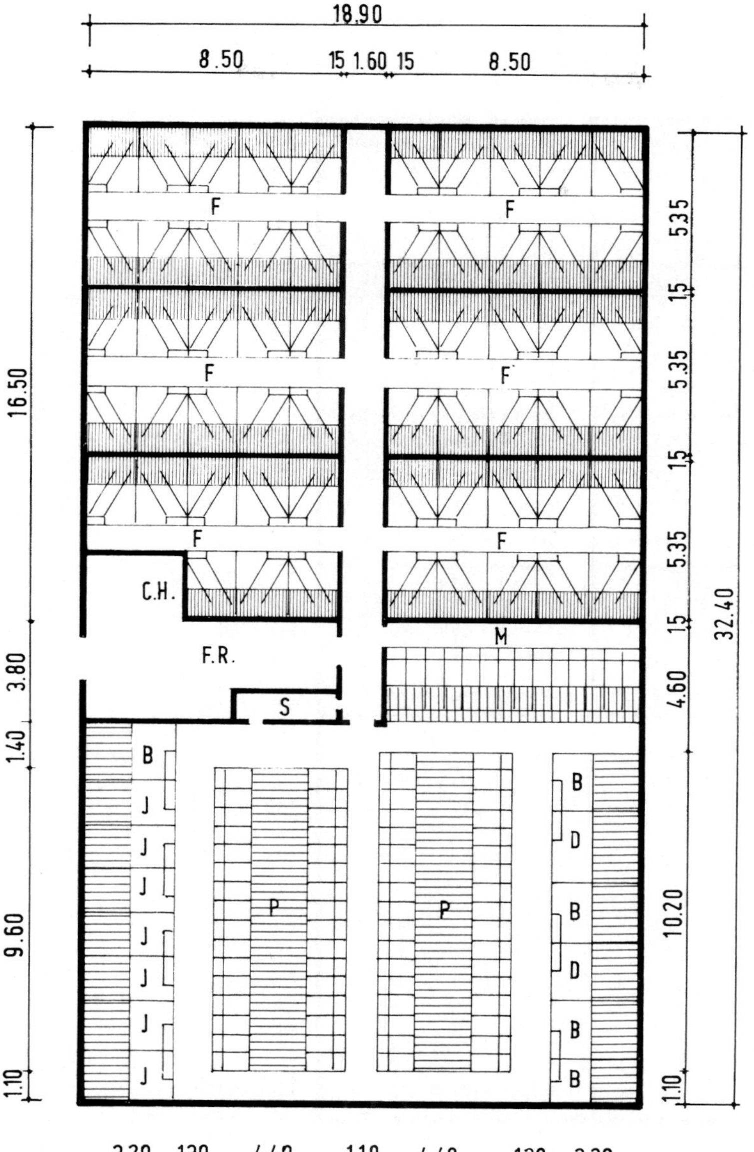

Fig. 5.29 Ground-plan of a house for 100 sows : rearing up to 20 kg in the farrowing pen (6.1 m^2 /sow) : 58 farrowing-weaner pens, 6 sections (F); 14 individual pen stalls for just-mated sows (M) and 2 group pens for dry sows (D); 66 individual tie stalls for pregnant sows (P), 5 boar pens (B) and 7 pens for gilts (J).
C.H. = central heating ; F.R. = feed room ; S = sow shower.

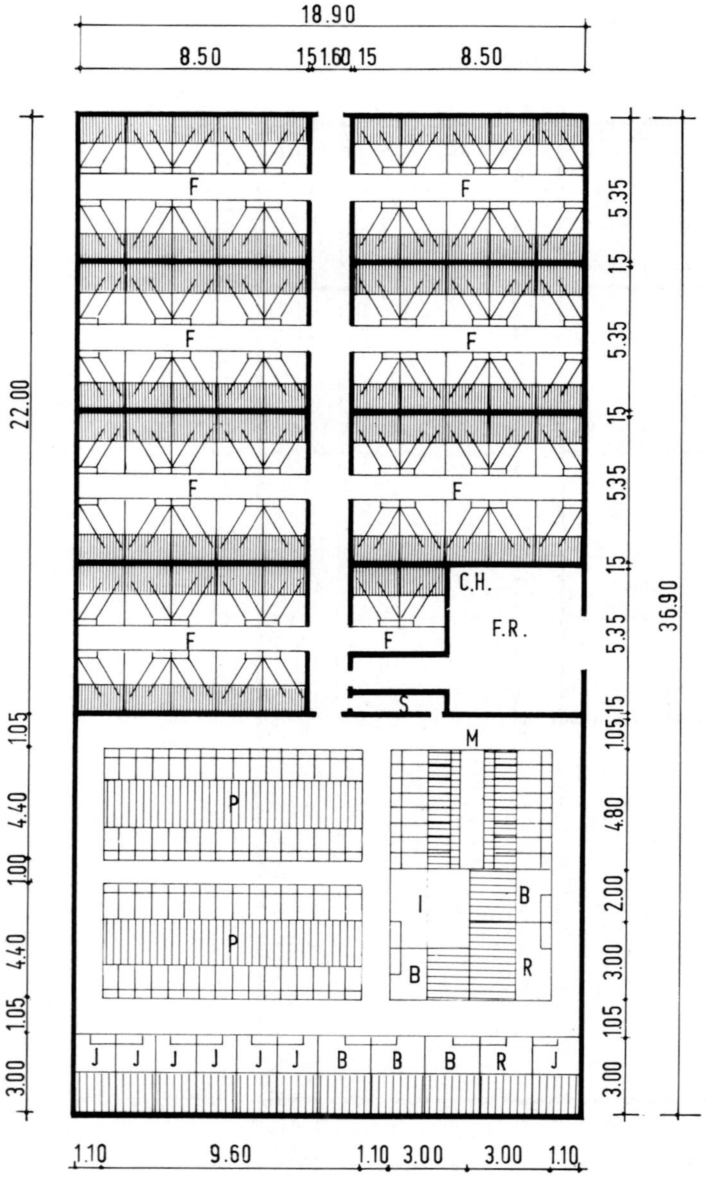

Fig. 5.30 Ground-plan of a house for 100 sows : rearing up to 14 weeks in the farrowing pen (7 m² /sow) : 72 farrowing-rearing pens, 7 sections (F); 16 individual pen stalls for just-served sows (M) and 2 group pens for dry sows (R); 5 boar pens (B), one service pen (I) and 7 pens for gilts (J); 64 individual tie stalls for pregnant sows (P).
C.H. = central heating ; F.R. = feed room ; S = sow shower.

Finally a layout is given for a house which will accommodate 200 sows, weaned at 5 weeks, where the weaners remain in weaner pens up to 14 weeks (fig. 5.31) and are then transferred to the fattening house at the farm.

5.4.2 The identification of pigs

On a large farm the stockman is always confronted with the problem of recognizing the animals, identification marks are therefore indispensable. They should be legible at a distance, easy and painless to apply, tamperproof and durable.

Of the different systems such as tattooing, ear notching and ear tagging, the last is probably the most suitable (fig. 5.32). Young breeding sows are tattooed as piglets with the mother sow or provided with ear notches at the age of 4 weeks. They are given the number of the mother sow. At the first mating they also receive an ear tag. Ear tags, on which one can inscribe any number, by means of special ink, have the distinct advantage that in the event of loss or damage the original number can again be written on the new ear tag. The ear tags can best be applied during the servicing.

5.4.3 The sow calendar

The daily activities pertaining to sow herd management are so different and individualistic that the use of a sow calendar is indispensable. This calendar shows the day-to-day activities and supervisions which have to be performed. We have successively designed :

- a disc calendar, in 1970 ;
- a small size (31 cm x 44 cm) sliding calendar, in 1972 ;
- a pocket size folding calendar, in 1977 ;
- a 3W-sow calendar, in 1978 (Daelemans, 1978).

They all offer the same possibilities. To simplify matters, only the most recent and cheapest calendar will be discussed. Copying the calendar is allowed provided that the source is clearly mentioned.

The 3W-sow calendar or 3-week-sow-calendar allows a very simple detection and control of heat because the year is divided in 3-week tables. This calendar has a fixed, built-in weaning age, viz. weaning at 5 weeks, which is nowadays often applied. The transfer to the farrowing pen takes place after 110 days of gestation.

5.4.3.1 The data-recording or data-input

Principally, recording of data is only carried out at serving or sale.

The service date

The number of each sow served, together with the letter representing the boar are recorded on the very same day under the heading of service. With a second service the previous service date is scratched and the second service is entered as a normal service but is underlined once. With a third service the same rules are followed but the new entry is underlined twice. From fig. 5.33 for instance we learn that sow 35 was served for the first time on June 17, 1984 by boar

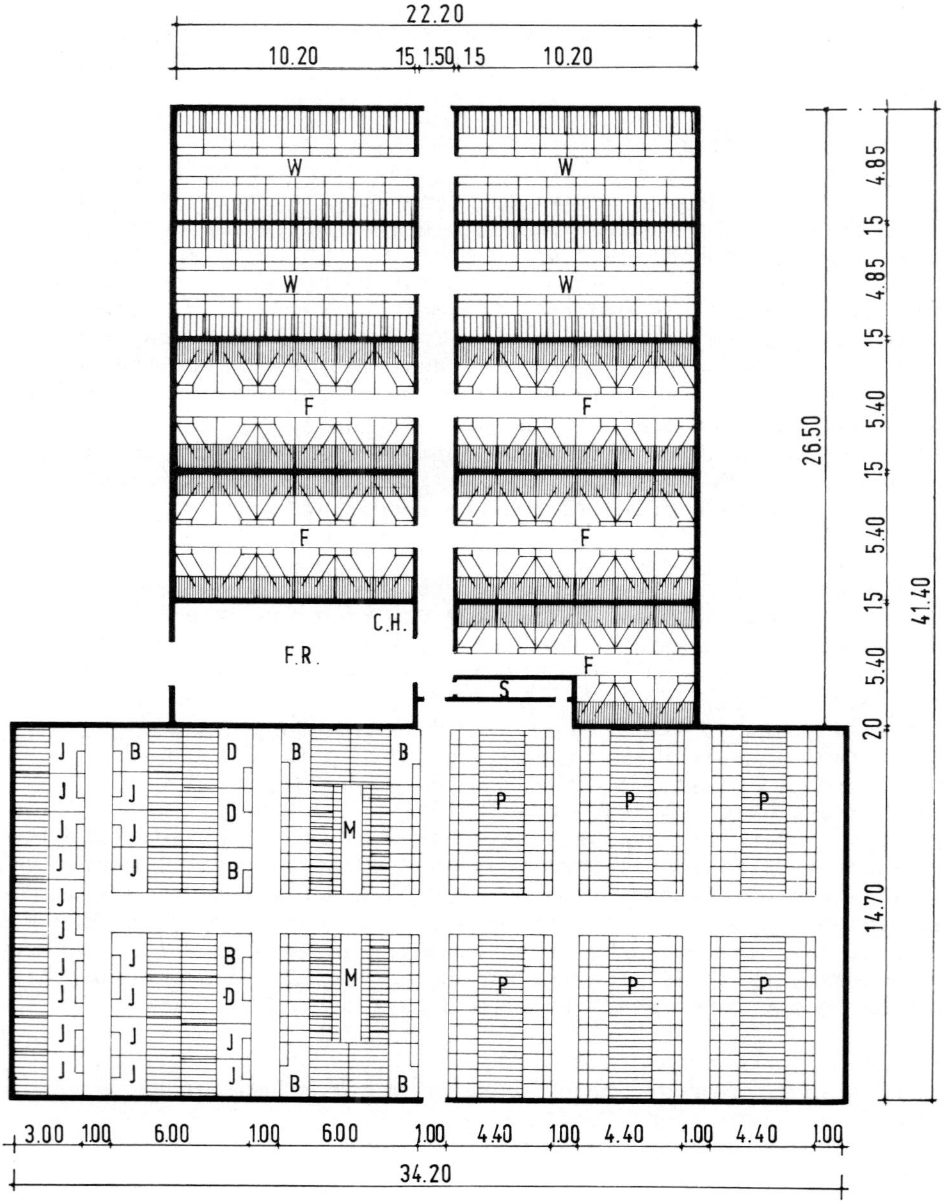

Fig. 5.31 Ground-plan of a house for 200 sows : weaning at 5 weeks, rearing up to 20 kg (5.4 m² /sow) : 57 farrowing pens, 5 sections (F); 65 weaner pens (W), 32 individual pen stalls for just-served sows (M); 3 pens for dry sows (D) and 132 individual tie stalls for pregnant sows (P), 7 boar pens (B) and 19 pens for gilts (J).
C.H. = central heating ; F.R. = feed room ; S = sow shower.

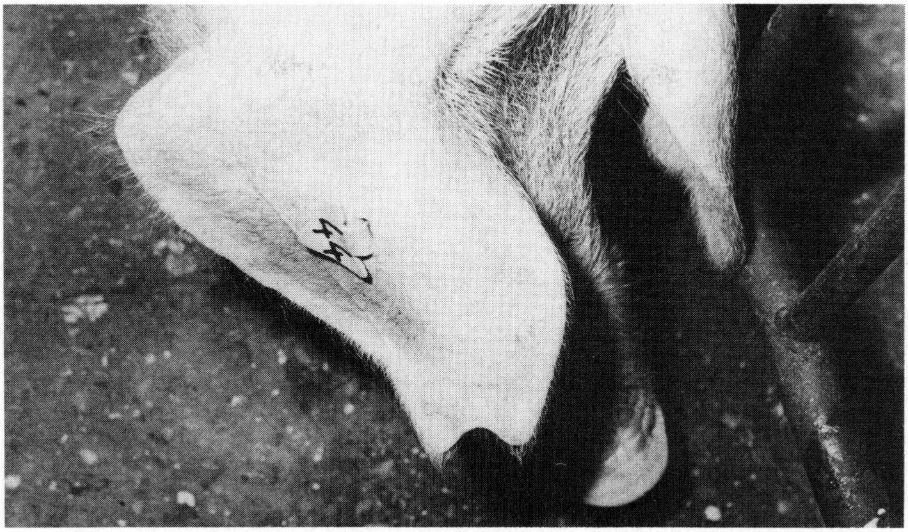

Fig. 5.32 An ear tag in plastic for marking a sow.

Morto ; sow 22 was served for the first time by boar P on January 30, she was re-served also by boar P on February 21, and served for the third time by boar M on March 15.

A sow which leaves the farm is immediately recorded and crossed on this date. In fig. 5.33, sow 58 has had 3 matings and is finally sold on June 11, sow 46 left the farm on April 9, etc.

Optional entries

When entering a first service after farrowing (thus of sows which already have produced at least one litter) the latest valid service-entry of that sow can be circled. In this way it is easy to detect from the calendar the sows which show no heat after weaning and thus are not productive. From fig. 5.33 we can deduce for example that sow 35 was entered as being served on June 17 and that her previous service-date was January 10. The circle on January 10 means that this sow was served after having farrowed. Sow 16 (January 8) and sow 29 (January 6) have also been served (viz. June 14), while sow 31 has not yet been served although this should have taken place before the others : this sow should be followed more closely, because she did not show heat in due time.

5.4.3.2 The readings or calendar output

Heat checks

Looking along the line "SERVICE" for todays date, the sow whose number is found in the column immediately left of this date has to be checked for heat (three week-cycle). A sow whose number appears

Weaning 5w.	Jun.		Jun.	Jul.	Jul.	Aug.	Aug.		Aug.	Sep.	Sep.	Oct.	Oct.		Nov.	
SERVICE	JAN.		JAN.	FEB.	FEB.	MAR.	MAR.		MAR.	APR.	APR.	MAY	MAY		JUN.	
YEAR 1984	1	21 Apr.	23	~~23R~~ 13 May	13		7	76M	28	91M	19	32R	10	92P	1	53M
	2	(5M)	24		14	14M 4 Jun.	8	26 Jun.	29	17 Jul.	20	8 Aug.	11	29 Aug.	2	20 Sep.
	3	23 Apr.	25	13M 15 May	15	63M	9	47M	30	4P	21	61P	12	~~52M~~	3	44P
	4	31M	26		16	6 Jun.	10	36P 28 Jun.	31	19 Jul.	22	10 Aug.	13	31 Aug.	4	52R 22 Sep.
	5	25 Apr.	27	10R 17 May	17	~~36M~~	11		1	~~34R~~	23	42R	14	8P	5	
	6	(29P)	28	77P ~~24~~	18	8 Jun.	12	51R 30 Jun.	2	40M 21 Jul.	24	34M 12 Aug.	15	26M 2 Sep.	6	6M 24 Sep.
	7	27 Apr.	29	19 May	19	30R · ~~39~~	13	67P	3	12R	25		16		7	20P
	8	(16M)	30	~~22P~~	20	33R 10 Jun.	14	2 Jul.	4	23 Jul.	26	59P 14 Aug.	17	38R 4 Sep.	8	26 Sep.
	9	29 Apr.	31	45M 21 May	21	~~22P~~	15	22M	5	~~58R~~	27		18		9	48R
	10	(35P)	1	73P	22	19P 12 Jun.	16	50M 4 Jul.	6	25 Jul.	28	~~58P~~ 16 Aug.	19	~~58P~~ 6 Sep.	10	57P 28 Sep.
	11	~~1R~~ 1 May	2	1R 23 May	23	83M	17	64R	7	90M	29		20	69P	11	~~58~~
	12	21R · ~~17~~	3		24	14 Jun.	18	85M 6 Jul.	8	27 Jul.	30	71R 18 Aug.	21	8 Sep.	12	5P 30 Sep.
	13	41R 3 May	4	56P 25 May	25	~~85P~~	19		9	~~46~~	1	~~89M~~	22	89M	13	27M
	14	~~25P~~ 37M	5		26	54R 16 Jun.	20	8 Jul.	10	60R 29 Jul.	2	20 Aug.	23	10 Sep.	14	16M · 29M 2 Oct.
	15	5 May	6	25M 27 May	27	81P	21	18P · 55R	11	68M	3	97P	24	~~84R~~	15	
	16	98M	7	88P	28	18 Jun.	22	10 Jul.	12	31 Jul.	4	22 Aug.	25	12 Sep.	16	84R 4 Oct.
	17	11M 7 May	8	29 May	1	87R	23	70P	13	82P	5	78P	26	65M	17	35M
	18	74R	9	9P	2	20 Jun.	24	12 Jul.	14	2 Aug.	6	80R 24 Aug.	27	14 Sep.	18	6 Oct.
	19	3P 9 May	10	31 May	3	72P	25		15	~~80R~~	7	43P	28	75R	19	
	20	2R · 28M	11	~~3M~~	4	3P 22 Jun.	26	14 Jul.	16	4 Aug.	8	79R 26 Aug.	29	62M 16 Sep.	20	8 Oct.
Transfer sow	21	11 May	12	86R 2 Jun.	5		27	93P	17	7P	9		30		21	
	22	15M			6	24 Jun.			18	6 Aug.			31	66R 18 Sep.		
Finished	Nov.		Nov.	Dec.	Dec.	Jan.	Jan.		Jan.	Feb.	Feb.	Mar.	Mar.		Apr.	

Conception rate	Boar	Jan.	Feb.	Mar.	Apr.	May	Jun.	Jul.	Aug.	Sep.	Oct.	Nov.	Dec.	Total	%	Boar
	Morto	10/10	4 ~~5~~/6	6/6	4/4	4/6								28/32	87.5	
	Revit	3/6	6/8	8/8	4/5	6/7								27/34	79.4	
	Robic	6/7	5/5	4/4	5/8	5/6								25/30	83.3	
	Herd	19/23	15/19	18/18	13/17	15/19								80/96	83.3	Herd

FIG 5.33 THE 3W - SOW - CALENDAR

in the column at left of the previous column has also to be checked for heat (six week-cycle). Assuming that today is June 17, 1984 sow 65 (three week-cycle) and sow 78 (six week-cycle) have to be checked for heat.

The transfer of sows

The dates shown at the right side of each column compose a secondary calendar in which only every second day is noted. This secondary calendar becomes useful for finding the sows which have to be transferred. The sows which are marked there are 110 days pregnant and are therefore in the prenatal stage where they have to be transferred to the farrowing pen. Assuming that today is June 17, we find that sow 81 is ready to be transferred.

Weaning

Looking along the line "WEANING-5W" (= 5 weeks) for todays date, the piglets of the sows whose numbers appear on this date have to be weaned (this is the case for the piglets of sow 11, assuming that today is June 17).

The sale of piglets

If one wishes, it is also possible to read off the sow, of which the piglets are ready to be sold. Therefore one looks on the calendar for the heading "WEANING-5W". Under this heading we look for the date of piglets to be weaned and move on to the previous column to the left (i.e. sows served three weeks earlier). The sows which are marked there have 8-weekold piglets which can be considered for sale. Assuming that today is July 1 we can read off that the piglets of sow 73 have to be weaned and that in the compartment left of this, thus in the previous column, the piglets of sow 35 are eight weeks old and ready for sale. We can also see that sow 35 has already been served since her number has been circled.

Finished pigs (ready for slaughter)

At the bottom of the calendar a line is provided marked "FINISHED", which allows the user to find the pigs which have reached the age of 190 days. In this way we can for instance see that the pigs produced by sow 35 and boar P (served on January 10, 1984) will be ready for slaughter on November 10 of the same year.

Conception rate

The records on a 3W-calendar are never rubbed out and are therefore always available, which is a great advantage. Indeed, it enables the monthly calculation of the conception rate of each boar and of the total herd. Each month one can count the number of matings per boar and the number of these matings which are not scratched i.e. : those which indicate a conception. In fig. 5.33 one can read off (and even check it on the calendar), that 10 matings by boar Morto in January 1984 resulted in 10 conceptions whereas the herd conception rate in

January was 19 out of 23 matings.

5.4.4 Pig herd recording

The use of a calendar, though very useful, gives no indication of the breeding efficiency of the farm (number of litters per sow and per year, size of the litter, replacement percentage of sows, etc.). In order to obtain this interesting data it is necessary to have a breeding herd bookkeeping and to record sets of data. We have therefore developed a simple and convenient system based on the monthly recording of data on four data-sheets which will hold the data of one year. This system is not described in detail here and therefore we refer the reader to the original publication of Daelemans (1975). This system has been updated for computer-use and the program is used by several important private companies which have made it available to their customers (Daelemans, 1984).

5.5 THE LABOUR ORGANIZATION AND THE BUILDING COSTS IN PIG BREEDING

According to our research the *labour time requirement* can be compared as functions of the weaning age and the layout of the pig house, which is represented in table 5.1. From this table it appears that the daily activities for the care of a suckling sow and her piglets amount to 2.7 to 2.9 man-hours per year (A_2), which by automatic feeding and the installation of nipple drinkers in the trough can be reduced to ca. 1.5 man-hour per sow (A_5). The daily care of sows without piglets requires ca. 2.2 man-hours per year when no free run is provided (B_2) ; the weaning age has little influence on this. The installation of an automatic feeding system for dry and pregnant sows allows only a small labour time reduction viz. ca. 0.5 man-hour per sow and per year, whereas the installation of nipple drinkers in the trough allows a labour time reduction of ca. 0.25 man-hour per sow and per year. Both measures reduce the annual time for the care of dry and pregnant sows to 1.5 man-hour per sow and per year (B_5). Besides the daily activities for the care of the sows there are also periodic or special activities. These irregular tasks include : obstetric help with the delivery of the litter, arranging the service, the care (including the injections) and the castration of the piglets, the cleaning of the houses, the transfer of sows and piglets, and administration. The labour times for the special tasks vary according to the weaning age and require on average 8.1 to 8.4 man-hours per sow and per year.

From the information given in table 5.1 it is possible to calculate the total labour time requirement in man-hours per sow and per year. These are represented in table 5.2. From this table we can conclude that the average labour time requirement in modern pig breeding varies between 11 man-hours and 16 man-hours per sow and per year. When straw was used in all phases of the cycle and wet feeding was administered, the time for the care of the sow amounted to 35 to 40 man-hours per sow and per year. The strawless housing of sows with and without piglets and of piglets, the omission of the wet feeding and its replacement by manual dry feeding together with the elimination of the free run of dry and pregnant sows allowed an important reduction

TABLE 5.1 Some labour partial-times for the care of sows and piglets, according to the phase of the cycle (in hours/sow/year).

	Weaning method	Middle-early	Normal
	Weaning age (days)	28	35
	Number of litters per sow and per annum	2.2	2.1
Suckling sows + piglets	A_1 littered farrowing pen, piglet pens with slats and feed hopper ; feeding twice daily with feed trolley and water distribution in the trough by means of a hose	5.03	5.61
	A_2 part-slatted farrowing pen, same method of feeding as A_1	2.70	2.89
	A_3 automatic water supply in the trough : time advantage compared to A_1 & A_2	-0.85	-1.00
	A_4 automatic feeding of the sows : time advantage compared to A_1 & A_2	-0.34	-0.40
	A_5 minimum labour time requirement for the care (= A_2-A_3-A_4)	1.51	1.49
Dry and pregnant sows	B_1 individual pen stalls with slats, daily run, manual feeding of meal with feed trolley twice daily, one tap for direct supply of water to 20 sows	3.22	3.09
	B_2 individual tie stalls or pen stalls with slats, without run, same feeding method	2.23	2.14
	B_3 id. as B_1 or B_2 but automatic moistening of concentrates, time advantage	-0.25	-0.24
	B_4 id. as B_1 or B_2 but automatic feeding of sows, time advantage	-0.50	-0.48
	B_5 minimum labour time requirement for the care	1.48	1.42
All	C_1 special activities (arranging mating, aid with delivery of litter, treatments of piglets, sale etc.)	8.38	8.13

of the labour time requirement per piglet produced ; middle-early weaning also contributed to this reduction. The production of a ready-to-sell piglet in this way requires 0.5 to 0.7 man-hour. Furthermore it appears from this table that, if one wants to compare the different labour time requirements or the labour productivity of management systems with different weaning ages, it is advisable to express the results per litter and not per sow since the number of litters per sow and per year is likely to vary. Furthermore we can observe that

TABLE 5.2 The total labour time requirement on pig breeding farms (based on the information given in table 5.1).

Weaning method Weaning age (days) Number of litters per sow and per annum	Middle-early 28 2.2	Normal 35 2.1
1. Sows with piglets in littered farrowing pens and weaned piglets in weaner pens ; sows without piglets in strawless individual pen stalls, with run, incl. special activities (see table 5.1, methods : $A_1 + B_1 + C_1$)		
hours/sow/annum	16.63	16.83
hours/litter	7.56	8.01
labour productivity (litters/stockman)*	238	225
labour productivity (sows/stockman)	108	107
2. As in 1, but complete strawless housing (see table 5.1, methods : $A_2 + B_2 + C_1$ for sows with and without piglets, individual pen stall or tie stall, no run)		
hours/sow/annum	13.31	13.16
hours/litter	6.05	6.27
labour productivity (litters/stockman)*	298	287
labour productivity (sows/stockman)	135	137
3. Minimum daily time for care of the sows and the piglets incl. special activities (see table 5.1, methods : $A_5 + B_5 + C_1$		
hours/sow/annum	11.37	11.04
hours/litter	5.17	5.26
labour productivity (litters/stockman)*	348	342
labour productivity (sows/stockman)	158	163

* We assume 1,800 man-hours per year

the layout of the house (the use of straw or slats for example) and the equipment (automatic watering and feeding) can greatly influence the labour time requirement. Striking, however, is the very important contribution of the periodic activities which cannot be mechanized to the total labour time requirement viz. ca. 70 %. Visual inspection of the animals remains necessary and can best be done during the manual distribution of dry feed. The pig breeder striving for labour saving methods, take most advantage from the layout of the house and the organization of the labour and in last instance think about mechanization.

The building costs of a modern-equipped sow farm amount to £ 375 to £ 500 per sow place according to the layout, the extent of compartmentalizing, the application of farrowing-rearing pens, etc. The financial repercussion of compartmentalizing the farrowing house, and the weaner house, of leaving the piglets in the farrowing pen (= farrowing-rearing house) and in some cases compartmentalizing of the latter, can be deduced from a study attributed to Debruyckere and Martens (1980). The following deductions can be made from this study :

1. compartmentalizing the farrowing house involves an additional investment of ca. £ 25 per sow place at the level of 100 sows or of 5 % of the total capital investment ;
2. compartmentalizing the weaner house requires an additional investment of £ 42 per sow place at the level of 100 sows and of £ 22 per sow place at the level of 200 sows per farm, or 9 to 5 % ;
3. the application of farrowing-rearing houses (= leaving the weaned piglets up to ca. 20 kg in the farrowing pen) requires an additional investment of ca. £ 57 per sow place or 12 to 14 % of the total investment ;
4. compartmentalizing a farrowing-rearing house requires a high additional investment, viz. £ 114 to £ 128 per sow place, which corresponds with an additional investment of not less than 25 % to 32 % ;
5. compared to a compartmentalized farrowing and weaner house (1° + 2°) the compartmentalized farrowing-rearing house (4°) requires however an additional investment of £ 48 per sow place (= +10 %) at the level of 100 sows.

The compartmentalizing of the farrowing house, although requiring an additional investment, is already widely applied and is justified by a number of advantages in hygiene, labour and energy consumption. The limited compartmentalizing of the weaner house viz. the creation of two or three compartments also leads to a reduction of the energy consumption. The pig house temperature can then also be adapted to the age of the animals. In this way the animals can be better prepared for transfer to the fattening house. The problem of the farrowing-rearing pens has to be evaluated in view of its application on a split or farrow-to-finishing farm. On the pig-selling farm, where piglets are normally weaned at 5 weeks one can take into consideration short-time rearing in the farrowing pens leaving them one more week there before transferring them to the weaner pens. The farrow-to-finish farm on the contrary can expect more of special weaner pens where the weaners are kept up to 14 weeks, because not only the construction costs are lower but also the costs involved for heating the house can be reduced. A delayed transfer to the fattening house has undoubtedly a number of advantages.

5.6 ZOOTECHNICAL AND VETERINARY ASPECTS OF THE HOUSING OF BREEDING PIGS

The efficiency of pig breeding has markedly improved in the last few years. On modern farms equipped with rational and well-climatized buildings and run by capable and dedicated stockmen, satisfactory to excellent results are obtained. A computerized herd recording performed by us at about twenty specialized farms each having about 100 sows has shown that on an average 15 weaned piglets were obtained per sow and per year in 1980 and 1981. There were however important differences between the various farms. The extreme values were respectively 12 and 19 (with a Standard Deviation (S.D.) of ca. 2). The piglet mortality (between farrowing and weaning) averaged 14 % but varied between 7 % and 21 % (S.D. ± 4 %) whereas the number of extra sows (and thus the number of non-productive sow-days) amounted on an average to 7 % (S.D. = 3.5 %). Improvement of the housing and the house climate has apparently reduced the piglet mortality, although the standard deviation indicates that not all pig breeders have solved these particular problems. In England, excellent pig breeding results were obtained in 1982 on 161 farms with an average of 146 sows per farm (Ridgeon, 1982) : 19.5 weaned piglets per sow and per year (the twenty best farms even reached a mean of 20.6 and the 20 worst farms a mean of 17.5). The piglet mortality averaged 15 % (13 % on the 20 best farms and 19 % on the twenty worst farms).

5.6.1 The housing of young sows and the breeding results

Sows have a relatively short lifetime. They are normally culled from the reproduction cycle after 3.25 litters or at the age of 3.5 years following data gathered on 300 farms with a total of 22,662 sows (Maes, 1982).

The reasons for their culling are numerous and are given below in descending sequence :

- no pregnancy or frequent return to service : 30.1 % ;
- mortality, accident, emergency slaughters : 18.8 % ;
- small litters, bad production : 12.2 % ;
- old age or worn out : 10.4 % ;
- leg weakness : 9.7 % ;
- failure to come on heat : 5.9 % ;
- bad mother characteristics : 5.9 % ;
- abortion : 4.1 % ;
- miscellaneous : 3 %.

According to these figures, it appears that annually ca. 60 % of the sows have to be replaced. In order to satisfy this demand and also to apply a sufficient selection of the gilts about the same number of maiden gilts have to be bred annually as there are productive sows present. Taking into account a breeding period of 180 days (up to the age of 250 days minus the 10 week piglet-period) a minimum of one maiden gilt place per two productive sows has to be reserved.

The gilts are kept in groups of 5 to 10 animals, in part or fully slatted floor pens with feed hopper. The pens resemble fattening pens (see further). The gilts are, from the age of 160 days, housed beside the boars in the service house, giving natural stimulation for the sows to come on heat. Paterson and Lindsay (1980) demonstrated that gilts which were kept from the age of 160 days in the vicinity of the boars, came earlier on heat than gilts kept separately. Furthermore, the same authors found no differences in production between young sows served in the first, second or third oestrous period provided that they were of the same age and had the same weight at the date of service. Kirkwood and Hughes (1980) also found the presence of the boar to be a positive influence, whereby the continuous or discontinuous presence (30 minutes per day) of the boars led to the same results. The same authors found that only "ready-to-mate" boars were able to stimulate heat. Hemsworth et al. (1981) have shown that the presence of sows had also a beneficial influence on the boars : boars, which were accommodated together with sows, made a more intensive courtship and had significantly longer ejaculation times.

There is no consensus on the influence of light. Keeping female pigs in the dark during the first two months of their life has, according to Salmon-Legagneur (1970), no influence on the prematureness and the fertility, whether or not they are provided with a free run.

Pay and Davies (1973) and Hacker et al. (1974) on the contrary came to the conclusion that an unnatural light scheme really exercised an influence on the prematureness.

5.6.2 The housing of dry sows and the breeding results

As with the housing of gilts, the housing of dry sows has to promote, in a natural way, ovulation. It is therefore necessary to keep them in the proximity of capable boars in the above-mentioned service house. Several investigators indeed found a more pronounced and more frequent ovulation when sows were kept within odour-distance and field of sight and hearing of the boar (Pay and Davies, 1973 ; Thomas, 1972). Transport-stress can lead to the same effect (Thomas, 1972 ; Kuiper and Sturm, 1975).

There is no agreement concerning the group-housing of dry sows. Some authors point to the fact that group housing leads to a better induction and distinction of the heat (Klatt et al., 1972 ; Tuinte, 1971), others obtain opposite results (Salehar, 1964 ; Glende et al., 1973). It is generally accepted that copious feeding after weaning leads to larger litters. This and other facts lead to the conclusion that dry sows have to be accommodated in groups of 5 to 10 animals, in pens with a feed hopper and a part-slatted floor. Immediately after service the sows are transferred to individual pens, which is beneficial in providing a quiet environment during the beginning of the gestation (Blendl, 1971 ; Klatt et al., 1972 ; Tuinte, 1971).

5.6.3 The housing of pregnant sows and the breeding results

In first instance the question arises whether pregnant sows have to be provided with an outdoor run or have to be kept solely indoors. Furthermore it is important to know whether strawless accommodation leads to the same good results as the traditional littered housing or not. Finally, the question can arise, which types of nonlittered houses are to be preferred.

Concerning the housing of dry and pregnant sows and their reaction upon the confinement England and Spurr (1969) ; Backstrom (1973) and Koomans and Mertens (1972) found no significant differences in the number of liveborn and stillborn piglets per litter and their average birth-weight in a research where individual and group housing without run were compared with a rearing method where dry and pregnant sows lived partially in the open air. Mennerich (1972) has found that, with pregnant sows kept on pasture during the summer, 17.9 piglets per year reached the weaning age compared to 17.5 in the case of a housing with a yard and 18.0 in the case of confinement during the whole year. The piglet-mortality amounted to 13.3 % in the case of confinement during the whole year and 15.2 % when they were kept on pasture. According to these results no better breeding results were obtained with keeping pigs part or fully in the open air. Moreover, it is generally known that pigs which are kept on pasture are more liable to worm infections (Hoorens et al., 1973).

Tuinte (1971) has investigated the culling percentages of sows as a function of their housing and came to the results given in table 5.3. According to this inquiry, which was carried out on 44 farms during a period of six months higher culling percentages were obtained in strawless houses than in houses where straw was provided. Litterless housing seems to have a negative influence on the heat occurrence of the sow. In strawless houses with individual pens, lesions to the limbs are more frequently observed than in littered houses where sows are accommodated in groups. When sows are individually housed the losses due to these lesions of the limbs are very important even when the houses have been littered.

TABLE 5.3 Culling percentages of sows in relation to their housing conditions (Tuinte, 1971).

Nature	Strawless					Littered
	Individual pen stalls			Group housing		
	with slats, abutting on slatted passages	with slats	without slats	with yard	without yard	Littered individual pen stalls
Lesions to the limbs	7.10	3.08	1.42	0.91	0.79	1.93
No oestrus	0.90	2.40	1.04	0.39	0.39	0.66
No gestation	2.58	3.54	6.79	2.09	5.51	4.17
Disorder of the uterus	0	1.09	1.56	0	0	0.64
Miscellaneous	7.72	6.99	10.39	10.51	2.81	5.10
Total	18.30	17.10	21.20	13.90	9.50	12.50

According to Hoorens et al. (1973) myopathy due to load is strongly influenced by the lack of movement. Leg weakness and arthrosis deformans are seriously accentuated by bad flooring. The presence of panaritium, necro-bacillosis of the claws and subcutaneous abscesses is accentuated by roughly or badly finished slatted floors.

An extensive comparative study of the breeding results obtained with group housing and individual housing of both littered and litterless houses was carried out by Blendl (1971). The results of this investigation are summarized in table 5.4.

TABLE 5.4 Breeding results of sows in relation to their housing (Blendl, 1971).

Nature	Group housing n = 435			Individual housing n = 490		
	$\bar{x}$	S.D.	C.V.	$\bar{x}$	S.D.	C.V.
Total pig mortality (%)	17.3	1.13	6.53	11.3	1.69	14.97
Crushed by the sow (%)	10.5	1.13	10.73	0.5	0.09	17.31
Still born (%)	5.0	1.40	28.17	0.3	0.07	22.58
Died (%)	1.8	0.14	7.73	10.4	1.85	17.72
	n = 435			n = 490		
Farrow-to-farrow interval (in days)	192.8	33.20	17.20	180.1	18.85	10.50
Not becoming pregnant (%)	28.9	3.61	12.49	24.1	1.48	6.14

where $\bar{x}$ = average ; S.D. = standard deviation ; n = number of observations ;C.V. = coefficient of variation.

Group housing gives less favourable results concerning total pig mortality than individual housing, which was confirmed by Klatt et al. (1972). When pregnant sows are individually housed they become used to lying down carefully on a limited space and so, later on, the crushing of piglets will occur less frequently. Due to the fact that individually kept pregnant sows remain quiet and on their own, still-born piglets due to traumata are rather rare. The farrow-to-farrow interval is shorter with individual housing than with group housing of pregnant sows. The number of sows which do not become pregnant is smaller when they are kept individually, although this percentage is still very high with the population of pigs which was investigated by Blendl (1971). Finally, the same author came to the conclusion that the breeding results of pregnant sows housed in individual pen stalls were better than of those kept in individual tie stalls (table 5.5). The latter type of house is however cheaper than a house with individual pen stalls.

From the zootechnical point of view individual housing of pregnant sows is certainly to be preferred. Despite the differences in the breeding results between the individual pen stalls and the tie stalls the last are to be preferred certainly when the cheaper and labour-saving facing-out arrangement (fig. 5.7) is chosen.

TABLE 5.5 The breeding results of sows respectively kept in individua pen stalls and tie stalls during gestation (Blendl, 1971).

Cause	Individual pen stalls n = 44			Individual tie stalls n = 42		
	$\bar{x}$	S.D.	C.V.	$\bar{x}$	S.D.	C.V.
Crushed by the sow (%)	3.62	1.32	36.46	5.64	1.73	30.75
Still born (%)	3.21	0.72	22.43	3.90	1.89	17.45
Died (%)	6.20	1.70	27.42	10.85	1.04	26.67
Total death-rate (%)	13.03	4.91	37.68	20.39	6.97	34.18

where $\bar{x}$ = average ; S.D. = standard deviation ; n = number of observations ; C.V. = coefficient of variation.

5.6.4 The housing of suckling sows and the breeding results

The present-day heated farrowing pens with a separate and limited sow crate and supplementary heated piglet creep without straw lead to an important reduction of the piglets'death-rate viz. to 15 % or less. In a comparative investigation into weaning at 12 days, 4 weeks and 6 weeks no significant differences could be detected concerning the death-rate and the growth of piglets which in all three cases weighed ca. 20 kg at the age of 10 weeks (Maton and Daelemans, 1974).

We have also carried out an investigation into the flooring of farrowing pens in which we compared littered farrowing pens with farrowing pens carpeted with rubber mats and others which were provided with an insulated concrete floor slightly littered with sawdust.

We were able to reach the following conclusions :

1. sole lesions were often found with piglets during the first and second week, but they are rare from the third or fourth week after birth, irrespective the type of floor used : an insulated concrete floor slightly littered with sawdust, an insulated floor with plenty of straw or an insulated floor with a soft gummi mat ;
2. piglets often receive slight injuries to their knees during the first month of life, the frequency of these injuries is higher than that of the sole lesions and they normally heal slower ; no relation was found between these lesions and the type of flooring used in the farrowing pens ;
3. serious sole and knee lesions are hardly or not found with piglets kept on one of the three types of flooring mentioned above ;
4. from a practical point of view no disadvantages concerning the incidence of sole and knee lesions occur with piglets kept on concrete floors which were slightly littered with sawdust; with this type of housing it is possible to obtain savings on both straw and labour, compared to the strawed farrowing pens, and on material

costs in comparison with the gummi mat carpeted floors.

Similar findings were published by Gravås (1977) who could not establish significant differences in the frequency of knee lesions with piglets kept on insulated concrete floors slightly littered with sawdust, on a 5 mm thick rubber mat over an insulated floor and on an insulated floor treated with epoxy resins. Piglets kept on rubber mats however showed larger and deeper wounds than those kept on a concrete floor.

In general it is possible to omit the use of straw in the farrowing pen provided that a suitable house and creep temperature are guaranteed. Part slatted floors of which the fully concreted part is insulated and littered with sawdust and the remaining slatted part made of concrete slats or punched metal sheets provide good housing for the suckling sow and her piglets.

5.7 THE LAYOUT OF HOUSES FOR FATTENING PIGS

5.7.1 Generalities

Until some twenty years ago the finishing of pigs was a small and often badly tended branch of the farm which under the pressure of economic imperatives moved to *specialization*.

Table 2.3 represented the increasing number of pigs per farm for a number of E.E.C.-countries : a number of pig farms disappeared whereas the remaining farmers increased their number of pigs.

The distribution of piggeries throughout Western Europe and in its countries is not uniform. A marked concentration exists in north-west Europe. Thiede (1983) has illustrated this concentration with a detailed table (table 5.6) showing the number of pigs in 21 European regions each having at least 1 million pigs. From this table it appears that the importance of these typical "pig regions" is still increasing. The 21 regions had a total of ca. 52.3 millions of pigs in 1982, whereas the remaining 64 European regions counted only 25.9 millions of pigs. According to the same table the highest European pig concentration was situated in West-Flanders (Belgium) with not less than 1,170 pigs per 100 ha. followed by the South of the Netherlands also with more than 1,000 pigs per 100 ha.

The fact that the production of finishing pigs is usually not related to land availability allows many stockmen to spend those labour hours on the fattening of pigs which otherwise can hardly be made productive because of the ever-growing shortage of farmland and the increasing mechanization. The *land-free character* of this specialization has however attracted a number of non-farmers to the fattening of pigs, as a secondary profession or as a guaranteed market for feed. This has led to the development of large units. From an economic point of view it seems that in many European countries, the *finishing of pigs* is carried out *on industrial scale, often even on the farms*. In fact only two production factors are involved viz. the capital investment and the labour ; the factor farmland is excluded from management because all the feed is purchased.

In this book the layout of houses for fattening pigs, intended for

TABLE 5.6 The number of pigs in 21 regions each with at least 1 million pigs.

Regions[1]	Agric. utili-zed area x1,000 ha	Number of pigs					Pigs per 100 ha[2]
		1977 x1,000	1980 x1,000	1982 x1,000	1982 to 1977 x1,000	%	
West-Denmark (DK)	2,289	6,599	7,827	7,932[3]	+1,333	+20.2	347[3]
South of the NL[4]	517	3,798	4,804	5,215	+1,417	+37.3	1,009
Brittany (F)	1,952	4,251	5,278	5,170	+ 919	+21.6	265
East of the NL[4]	563	3,428	4,075	4,146	+ 718	+20.9	736
Weser-Ems (D)	1,003		3,748	3,866			385
Münster (D)	436	2,026	2,434	2,597	+ 571	+28.2	596
West Flanders (B)	218	2,247	2,492	2,550	+ 303	+13.5	1,170
Emilia-Romagna (I)	1,376	2,303	2,222	2,297	- 6	- 0.3	167
Lombardy (I)	1,235	2,009	1,920	2,107	+ 98	+ 4.9	171
Schlesw.-Holstein (D)	1,103	1,796	1,807	1,740	- 56	- 1.1	158
Centro (I)	2,162	1,780	1,637	1,691	- 89	- 5.0	78
East-Denmark (DK)	498	1,405	1,578	1,562[3]	+ 157	+11.2	313[3]
Yorkshire (GB)	1,194	1,256	1,399	1,561	+ 305	+24.3	131
Detmold (D)	375	1,303	1,484	1,513	+ 210	+16.1	403
East Anglia (GB)	982	1,337	1,337	1,449	+ 467	+ 8.4	148
Hannover (D)	531		1,317	1,312			247
Lüneburg (D)	842		1,259	1,241			147
South-East England	1,679	1,256	1,163	1,155	- 101	- 8.0	69
Ireland (IRL)	5,650	996	1,096	1,145	+ 149	+14.5	20
Niederbayern (D)	587	966	1,030	1,028	+ 62	+ 6.4	175
Stuttgart (D)	511	878	974	1,021	+ 143	+16.3	200
21 regions	25,703		50,881	52,298			203
remaining 63 regions	66,858		26,362	25,933			39
EUR-9	92,561	72,130	77,243	78,231[3]	+6,101	+ 8.5	85[3]
Greece	9,234			1,218			13
EUR-10	101,795			79,449[3]			78[3]

where [1] = areas arranged according to the number of pigs in 1982 ; [2] = number of pigs per 100 ha of agricultural utilized area ; [3] = Denmark 1981 ; [4] = the Netherlands.

the intensive production of meat, is described. In those pig houses cereals are administered, supplemented with a feed rich in proteins (e.g. soya) and with the necessary additives (minerals, vitamins). This compound feed can be given out under different forms viz. in the form of meal, pellets or mixed with water. In Belgium these concentrates are generally administered "ad libitum" i.e. unrestricted, in specially designed feed hoppers. The Belgian Landrace pig and the Pietrain pig have for a long time been selected and bred in such a way that there is no fear of excessive fat production. With different breeds in other countries this is still possible and therefore rationed trough-feeding is the general rule. Because the diet is then restricted to reduce the fat content of the carcasses it is, in any case more accurate to speak about finishing rather than fattening.

The finishing or fattening house accommodates pigs either from an initial weight of ca. 20 kg or from the age of 9 weeks on those farms which have to buy their weaners, or from the age of 12 to 14 weeks or ca. 30 to 35 kg on farrow-to-finish farms. In both cases a slaughter weight of ca. 95 to 100 kg for Belgian Landrace pigs is reached at the age of 26 to 28 weeks.

Besides specialized breeding farms and finishing farms more and more enterprises are closing the cycle to form so-called "farrow-to-finish" units. On these farms all piglets are kept and finished. The question arises how many pens for finishing pigs have to be provided in relation to the number of sows present.

The following formula allows the calculation of the required number of pens provided that the sows are not kept in groups :

$$M = \frac{S \cdot W \cdot Y \cdot L}{C \cdot N}$$

where M = the number of pens for finishers ; S = the number of productive sows ; W = the residence time in weeks per round, viz. mostly 16 weeks, of which 15 weeks are for fattening (from 30 kg to 95 kg) and 1 week for cleaning of the house ; Y = coefficient for irregularity (= 0.95 for fatteners) ; L = the average size of a litter, expressed in weaned piglets per litter (= 7.5) ; C = the duration of one cycle which is 20, 21, 22 or 23 weeks according to weaning at 3, 4, 5 or 6 weeks ; N = the number of pigs per pen.

A farrow-to-finish farm with 100 productive sows, of which the piglets are weaned at the age of 5 weeks requires the following number of pens which normally hold 10 pigs each :

$$M = \frac{100 \times 16 \times 0.95 \times 7.5}{22 \times 10} = 52$$

This farm thus has normally 520 finishing pig places. Assuming 3 finishing rounds per year (rearing up to 30 kg in the weaner house) this farm will have an annual output of some 1,560 finishers.

The housing of finishing pigs has been subjected to radical changes in the last decades. For labour-technical and economic motives strawless housing has been introduced and the pens of the so-called Danish house have been transformed first to strawless partly slatted floor pens, later to half or fully slatted floor pens. Farms where the odour nuisance has to be minimized can however consider the use of the open and strawed finishing house. Environmental problems in animal production are mainly related to the intensive production of pigmeat. The method of feeding has also contributed to a large extent to these developments. Therefore the feeding methods will be treated in this book before the equipment of the pens.

5.7.2 The feeding systems for finishing pigs

Fig. 5.34 represents an outline of the different methods for feeding fatteners. Up to 1960 slop feeding in a trough was generally applied. The swill was prepared in the feed room and manually distributed in the troughs. Since 1960 other feeding systems gained popularity ; floor feeding was temporarily applied around 1965. Floor feeding was the first fully automatic dry feeding system. It was abandoned rather quickly because of the poorer results obtained with daily growth and feed conversion - the two main parameters in the production of fatteners. Dry ad lib feeding by means of a feed hopper is generally applied in Belgium but not widespread in other countries where restricted trough feeding is the rule. Trough feeding is sometimes done using dry meal but mostly in liquid form. In the last case we can distinguish on one hand wet feeding, where water is supplied in the trough and meal is distributed afterwards over the entire length of the trough, and on the other hand slop feeding whereby the mixture of meal and water is prepared in a tank in the feed room. This mixture can be prepared in advance viz. directly after the previous feeding and is then called pre-soaked slop in contrast to slop which is prepared immediately prior to feeding.

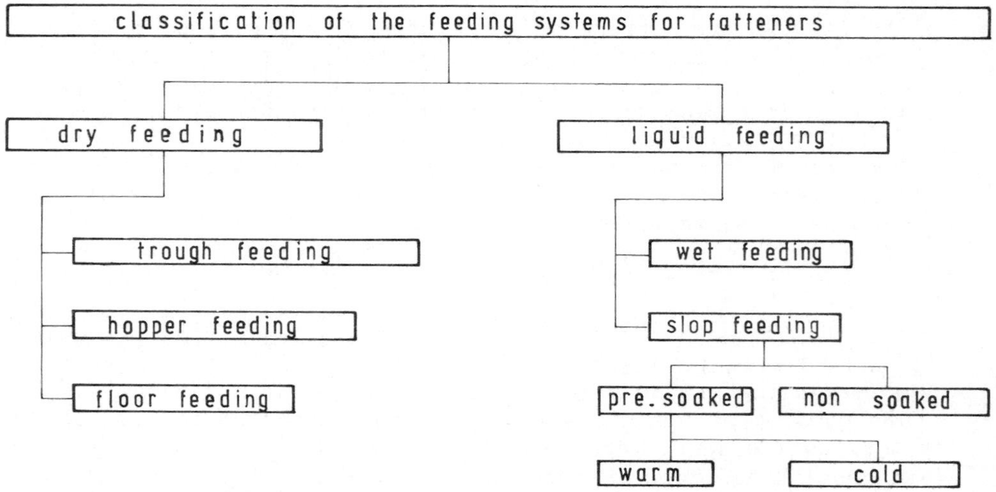

Fig. 5.34 Outline of the various feeding systems as used for finishing pigs.

5.7.2.1 The ad lib dry feeding by means of a feed hopper (automatic feeder)

The feed hopper is made of galvanized metal, asbestos cement or reinforced concrete and is ca. 120 cm long, 45 cm wide and 90 cm high (fig. 5.35). The feeder basically consists of a box, into which about 100 kg of concentrates are poured and which becomes gradually narrower towards the bottom where it forms a trough. This trough is divided into four equal parts so that four animals can eat simultaneously. It has a flat bottom and a non-spilling lip which prevents spillage of feed. Feed is poured into the box and the trough is filled by gravity. Feed removed by the animals is immediately replaced by a new supply which glides towards the trough. The box can hold more feed than is eaten in one day and so needs only to be refilled periodically. In order to regulate the amount of feed which glides towards the trough, the hopper bottom opening can be adjusted for meal or pellets. When meal is used, the stockman has to inspect the contents of the hopper daily, because blockages and "hole forming" can sometimes occur in the hopper. In fact some ingenious systems exist which are composed of a mixing installation or chain, manipulated by the animal's nose in order to prevent such blockages.

Fig. 5.35 A fattening house equipped with feed hoppers.

In earlier days the feed hoppers were manually filled with feed from sacks. Later specially designed feed trolleys with lifting dumping bins were used. Now most feeders are filled mechanically by means of an auger conveyor powered by an electric motor. This is a metal tube in which a steel jack screw pushes the meal or pellets forwards starting from the silo (fig. 5.36). This feed conveyor tube

Fig. 5.36 Automatic filling of the feed hoppers.

is branched off above each feed hopper and supplies the feed to the hoppers. The feeders are filled one after the other, starting with the one closest to the silo. In the last hopper there is a membrane activated pressure switch which stops the conveyor as soon as the required level is reached and reactivates it when no longer pressed by the feed. This implies that the last pen must always house enough pigs. For safety it is advisable to provide a second switching off device. This can be done by suspending a bucket from a spring at the end of the auger conveyor and beyond the feeder in which the first switching unit is installed. If the first switching unit fails, the bucket will be filled and subsequently the spring is pulled down thereby activating the emergency switch which stops the motor. The installation can only be switched on again after the bucket is emptied.

Until recently the number of bends in the conveyor tube had to be minimized, viz. maximum two bends, which were placed between the silo and the place where the conveyor tube entered the house. It was thus necessary to have the feed hoppers in one line or at most in two lines when a central passage was involved. The auger conveyor was normally suspended from the ceiling at a height of ca. 2.3 m above the central passage. The branched-off pipes were then directed in a sloping manner to the various feeders. Today, however, flexible bends are commercially available which allow the use of more 90° bends which simplifies the installation.

Instead of an auger conveyor which transports the meal in only one direction it is also possible to use an endless chain. This chain makes a complete round. By installing the overflow above the filling funnel the safety bucket becomes redundant. If the switch in the last automatic feeder fails to stop the installation, then the meal is simply carried along the circuit until the installation is switched off either manually or by means of a motor protection unit. The endless chain however should not start directly at the silo but from a funnel which is filled by gravity. If the endless chain starts at the silo the safety bucket has to be supplied. The automatic feeder can either be stationary in which case it is part of the fence of the pen or can be movable. Compared to the rationed feeding of finishing pigs which can be done partly or fully automatically and nowadays even computer-controlled, the dry feeding in automatic feeders requires a much lower investment.

5.7.2.2 Trough feeding

The trough is an essential element in pig houses where restricted feeding is applied. It is a long, narrow, half round (Ø 30 cm) and open concrete receptacle, which is 15 cm deep. It should be installed ca. 5 cm above the floor. The inside lip of the trough nearest to the pigs is 20 cm high whereas the outside lip, which abuts on the alley is somewhat higher viz. 25 to 30 cm. A suitable trough length should be chosen to enable all the pigs of a pen to eat simultaneously. The required length of the trough depends on the number of pigs in the pen and on their final weight. If the pigs remain in the pen until they are finished, a trough length of 33 cm per pig is required and there are no partitions per pig but only per pen. The trough is generally situated along the feeding passage as shown in fig. 5.37.

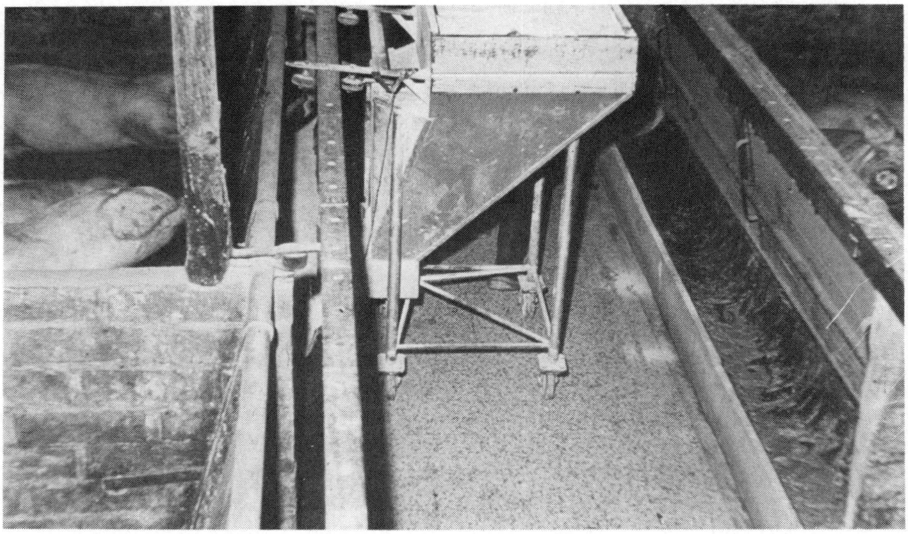

Fig. 5.37 The trough situated along the feeding passage.

Consequently the longest side of the pens abut on the feed passage and this implies more passage area in the house and hence an increase in the building costs. With liquid trough feeding the trough can be positioned perpendicularly to the feed passage whereby the length and thus the area of the passage can be reduced. Above the trough a pen partition is required to keep the pigs in the pen. If the longest pen side adjoins the feeding passage, as is mostly the case, these partitions must allow unobstructed eating : they comprise a framework of bars.

In some cases swing-partitions are applied. They swing under a horizontal tube and can be set in two positions. With the frame set above the outside trough lip the pigs have free access to the trough. With the partition set above the inside lip the pigs are refused access to the trough and the trough can be filled without the stockman being hindered by the pigs. An additional advantage is that new feed can be distributed immediately after the current feeding : the trough partition is set in the filling position and feed is distributed, and at the next feeding time the partition is set in the feeding position and pigs are allowed to the trough. This feeding method reduces the time during which the pigs are "shouting" for feed - a very disturbing and disagreeable noise.

A special form of trough is the circular trough which has two major advantages. Firstly, the pen width is no longer related to the trough length and secondly the feed is now delivered at one single point per pen. The problem of an unequal distribution of the feed over the entire trough length is thereby eliminated and mechanization made less expensive. Fig. 5.38 shows a round trough design.

5.7.2.3 Various trough feeding systems

Handfeeding i.e. shovelling feed from a wheel barrow or feed cart into the trough is outdated. Nowadays mechanical trough feeding prevails. Trough feeding often implies restricted feeding and mechanical feeding systems are therefore basically volumetric or gravimetric dosing units.

5.7.2.3.1 Volumetric trough feeding devices

All the treated feed systems can be applied for both dry feeding in the trough or, after supplying the necessary quantity of water in the trough, for wet feeding.

Feed trolley

There are two types of feed carriers viz. the auger type feed trolley and the trolley with metering rail.

The *auger type feed trolley* is an easily manoeuvrable wheeled hopper. A battery activates an auger for a preset time, dosing each trough with feed. The driver pushes the trolley along the trough, adapting its speed to the delivered quantity in order to obtain a regular distribution of feed over the entire trough length.

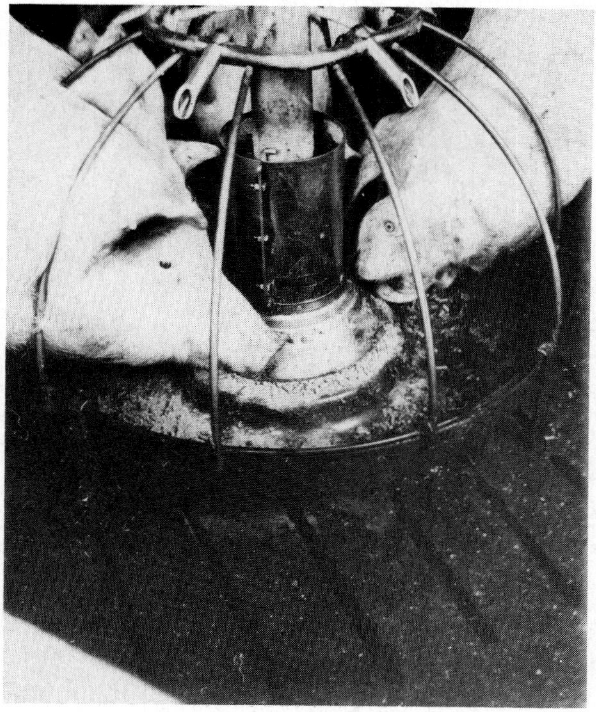

Fig. 5.38 Circular trough equipped with nipple drinkers.

The *trolley with metering rail or the so-called pin and rail trolley* (fig. 5.39). A rail is installed under or above the trough. This rail can contain a variable number of pins. The trolley is pushed by hand along the trough. A linking device holds the trolley at a fixed distance from the trough. Each time the linking device encounters a pin the dosing unit tips a fixed volume of feed. The pig farmer rations the discharged feed by installing the required number of pins and spreads them regularly over the entire length of the trough in order to obtain a regular distribution of feed.

Adjustable feed dispensers

One or more feed dispensers are installed above each trough (fig. 5.40). They are automatically filled from the silo by a feed conveyor of the auger or chain type. The feed volume in each dispenser can be regulated by either adjusting a movable base or the telescopic

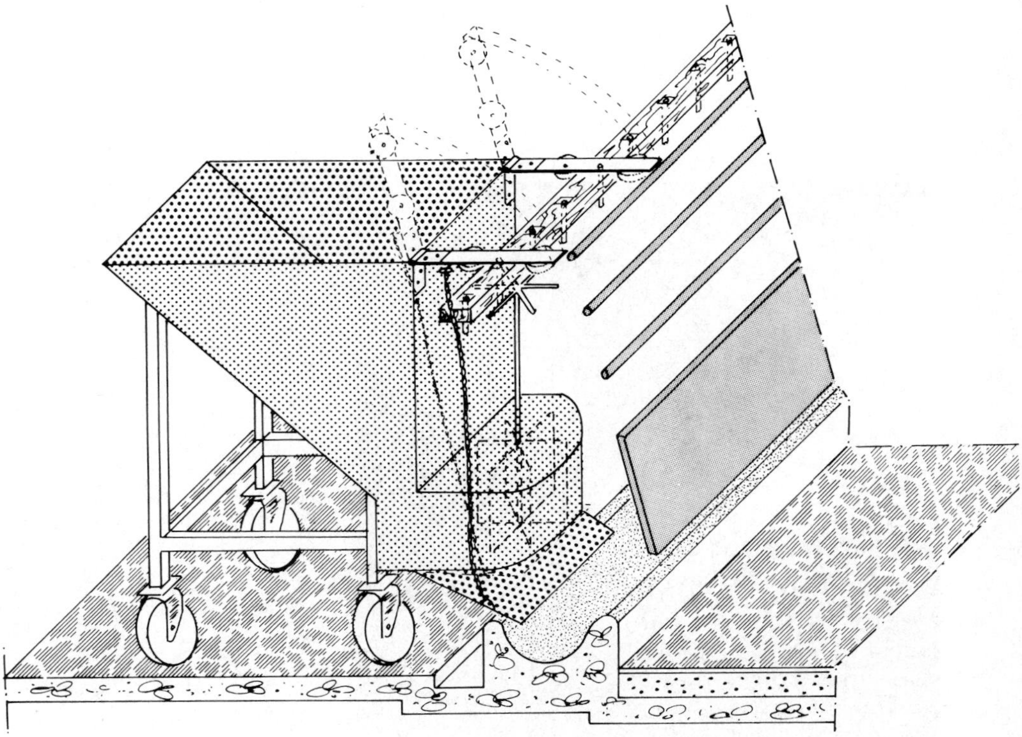

Fig. 5.39 The feed trolley with metering rail as used in fattening houses.

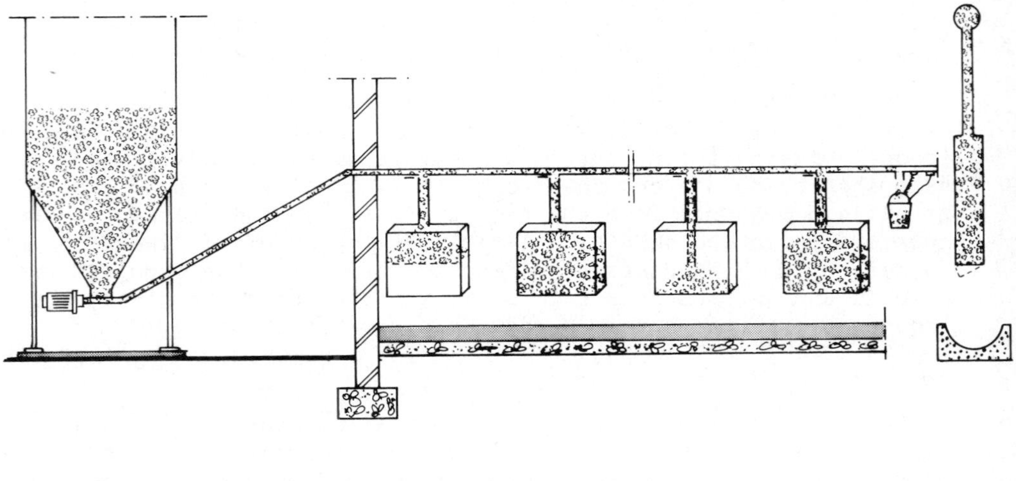

Fig. 5.40 Trough feeding with feed dispensers.

delivery tube of each dispenser box. Fig. 5.40 shows both types of adjustments used with the feed box dispensers. At feeding time an electrically operated cable pulls all the bottom locks open and the contents are discharged into the troughs. A second cable movement closes all the dispensers again. The conveying system is then activated and the dispensers are filled for the next meal. A pressure contact in the last dispenser box stops the conveyor. The adjustment of the ration is in both cases carried out manually by positioning the movable bottom or telescopic pipe in the desired position.

Dosing pipe system

Above the trough a frame of vertical tubes is installed (fig. 5.41), linked at the top by a horizontal tube coming from the silo and which contains the feed conveyor. A sliding lath is provided underneath the frame and can be set in two extreme positions : in one of the positions it will block all the outlets of the pipes whereas in the other position all pipes will be emptied. A chain or auger conveyor fills the pipes, one by one, starting with those nearest to the silo. Furthermore each pipe can be closed by a slide under the upper tube thereby enabling the rationing of feed in each trough, as a function of the number and the weight of the pigs in the pen. At feeding the sliding lath is moved so that all pipes are opened and their contents discharged. Immediately after the lath has returned to its original position the feed conveyor is activated and the pipes filled ready for the next feeding. A pressure contact installed in the last pipe stops the conveyor automatically. A second contact (security bucket) completes the installation.

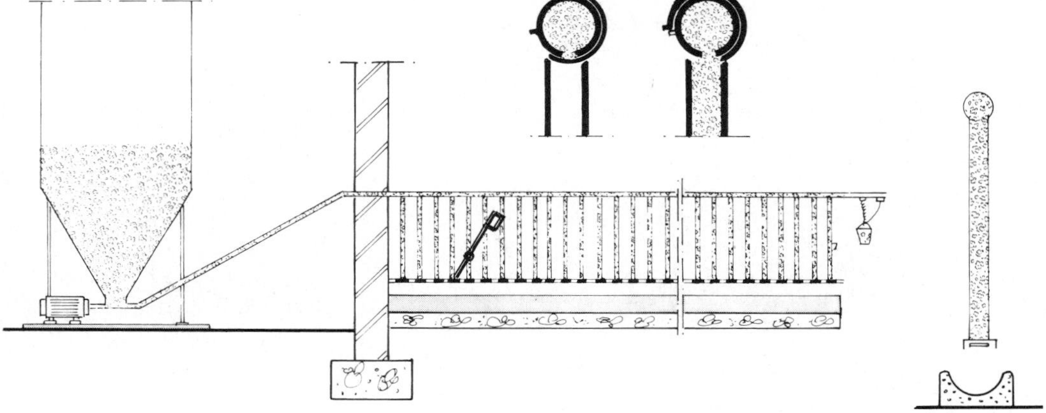

Fig. 5.41 Trough feeding with dosing pipes.

5.7.2.3.2 The gravimetric trough feeding system

This system is more expensive than the volumetric systems but is more accurate as it is based on the measurement of a weight. With the gravimetric system a number of feed dispenser boxes hang on balances (fig. 5.42). The feed supply is interrupted the moment the ration is reached. At that moment the weight of the dispenser box (with its contents) equals that of the counterweight and forces the shutter over the inlet. At feeding all the boxes are tipped or the bottoms of the boxes are opened by pulling a rope. Directly after feeding the boxes are refilled for the next meal. The last dispenser box is equipped with a safety device for stopping the conveying system. A second switch (security bucket) completes the installation. The feed to be dispensed, can be adjusted by sliding the counterweight over a scaled and notched beam. In this way the content of the dispenser box can be reset.

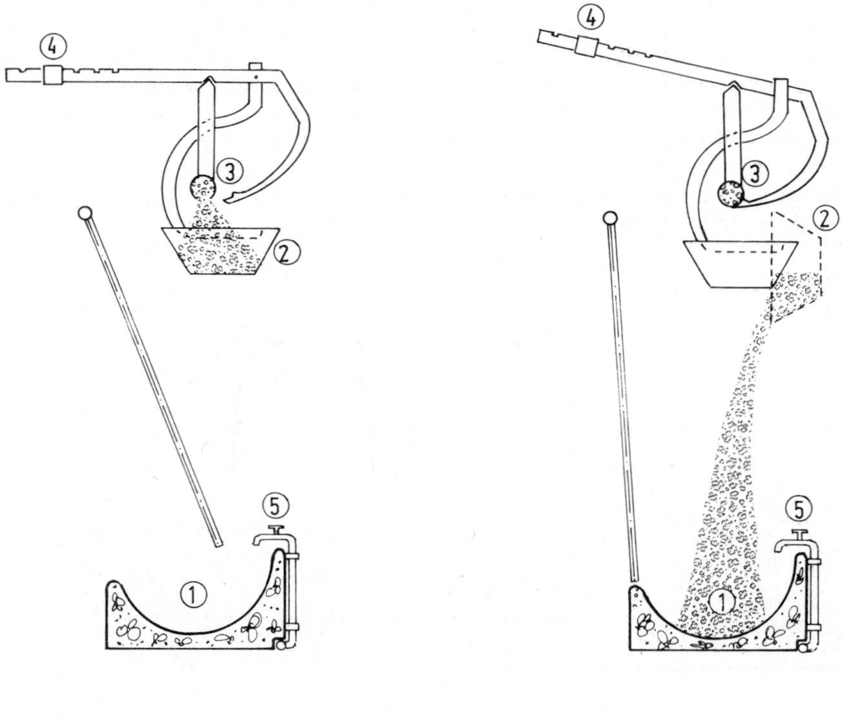

Fig. 5.42 Mechanical gravimetric feeding system, at left : the weighing and supply pan ; at right : the feeding proper.
Legend : 1 = trough ; 2 = dispenser box ; 3 = feed supply tube ; 4 = adjustable weight ; 5 = water supply.

5.7.2.4 Wet feeding

Wet feeding, i.e. the separate delivery of water and feed into the trough at a rate of 2 l of water per kg of concentrates, requires that water is poured into the trough before the addition of the concentrates. This was, until some years ago, done with a hose pipe and was time-consuming and inaccurate. Later on, a water distribution system was developed consisting of a number of water dispensing boxes, installed above the troughs (fig. 5.43). At the top, the boxes are connected to a filling pipe. In the first box the water supply is automatically shut off by a float valve, but an overflow protection is installed at the other end of the system. Each box is provided with a flexible outlet pipe inside the box. Resetting the height of this pipe enables the adjustment of the required water quantity. A water valve is provided underneath each box and all the valves of one trough line can simultaneously be opened or closed.

After distributing the water in the trough the feed is delivered in the trough by one of the above-mentioned mechanical trough feeding systems.

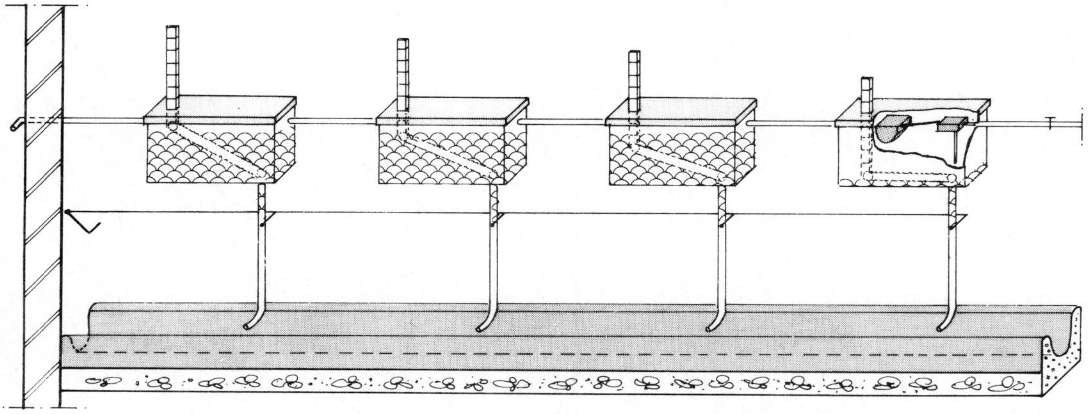

Fig. 5.43 A water dispensing system for wet feeding.

5.7.2.5 Slop feeding

Slop feeding (fig. 5.44) implies the distribution of liquid feed prepared in a tank outside the pig house. The slop is a mixture of concentrates and water in a ratio of 2.5 to 3 litres of water per kg of concentrates. This high water content is required in order to obtain a mixture which can be pumped to the pig house. This mixture is prepared by pouring the required quantity of water into the tank and then adding the necessary quantity of concentrates. A mixer agitates the feed until a homogeneous mixture is obtained. At feeding time the liquid is pumped around the pig house. A cock above each trough

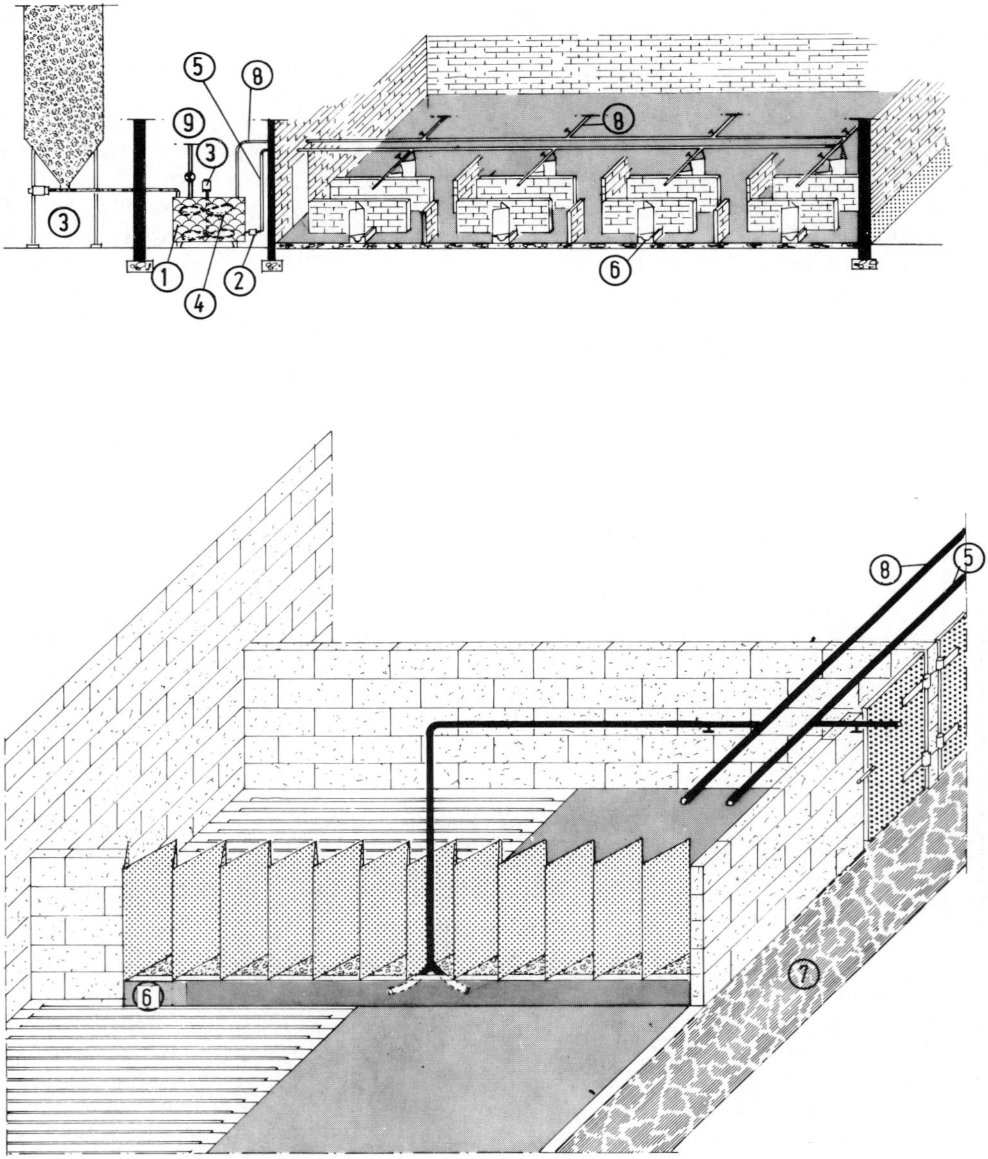

Fig. 5.44 A slop feeding system.
Legend : 1 = feed preparation tank ; 2 = pump ; 3 = electric motors ; 4 = stirring equipment ; 5 = pipeline for slop ; 6 = trough ; 7 = passage ; 8 = return pipe ; 9 = water supply.

allows the delivery of the desired quantity of slop into each trough. The cock can be manipulated by hand or can be activated electromagnetically for a preset time. In order to reduce the building costs the trough is often placed perpendicularly to the passage and between two pens. A zig-zag separation allows the pigs of two neighbouring pens to eat from one and the same trough.

5.7.3 The different types of housing for finishing pigs

Until some twenty years ago the *littered Danish fattening house* was probably the most familiar type of housing. It was characterized by the fact that along the feeding passage a trough was installed, behind which a littered lying area was found. This area gave access to a 15 cm lower dunging passage through an opening with door in the wall between the pen and the dunging passage (fig. 5.45). The water bowl was installed in the dunging passage. The doors of the dunging passage were mounted in such a way that in one position the pigs were fenced-in in the lying area during mucking-out whereas in the other position the doors formed a partition in the dunging passage between the different pens. In order to save space, pens of various sizes were made and pigs had to be moved twice or three times as they grew to provide them with a more spacious pen (fig. 5.45). In an attempt to reach an even greater saving in space the *Swedish fattening house* was developed. This new housing type comprised only one central dunging passage and two feeding passages along the side walls. The pigs were allowed in half of the dunging passage behind their pens. The doors were installed in such a way that the whole dunging passage could be freed by locking the pigs in their lying area during the mucking-out. Trough feeding was generally applied which led to wide, shallow and therefore expensive pens.

The installation of slatted floors in the dunging passage of this type of house made mucking-out superfluous, whereby the dwarf wall (with door) between the pen and the dunging passage could be eliminated and the doors between the different pens replaced by walls or horizontal bars. Soon it became apparent that, in order to keep the pigs clean, the slatted area had to be increased and therefore part slatted (about half the area) or fully slatted floors were chosen. Besides this type of house a number of special house types were developed such as the partly lidded pen house and the open fronted house.

The housing of fattening pigs in individual fully slatted pens was abandoned after a short period because it was neither economical norpractical. Moving pigs from one pen to another in order to save space and building costs was also abandoned or at least reduced to a minimum (only once) because of the labour it requires and for zootechnical reasons : the performance is depressed resulting in a standstill period whilst the pigs are subjected to stress as some time is required in order to let them settle down in their new accommodation. Finally a new tendency has arisen viz. the compartmentalizing of the house, whereby the transversal layout is chosen for the farrow-to-finishing houses. The most important and most frequently built house types will now be discussed.

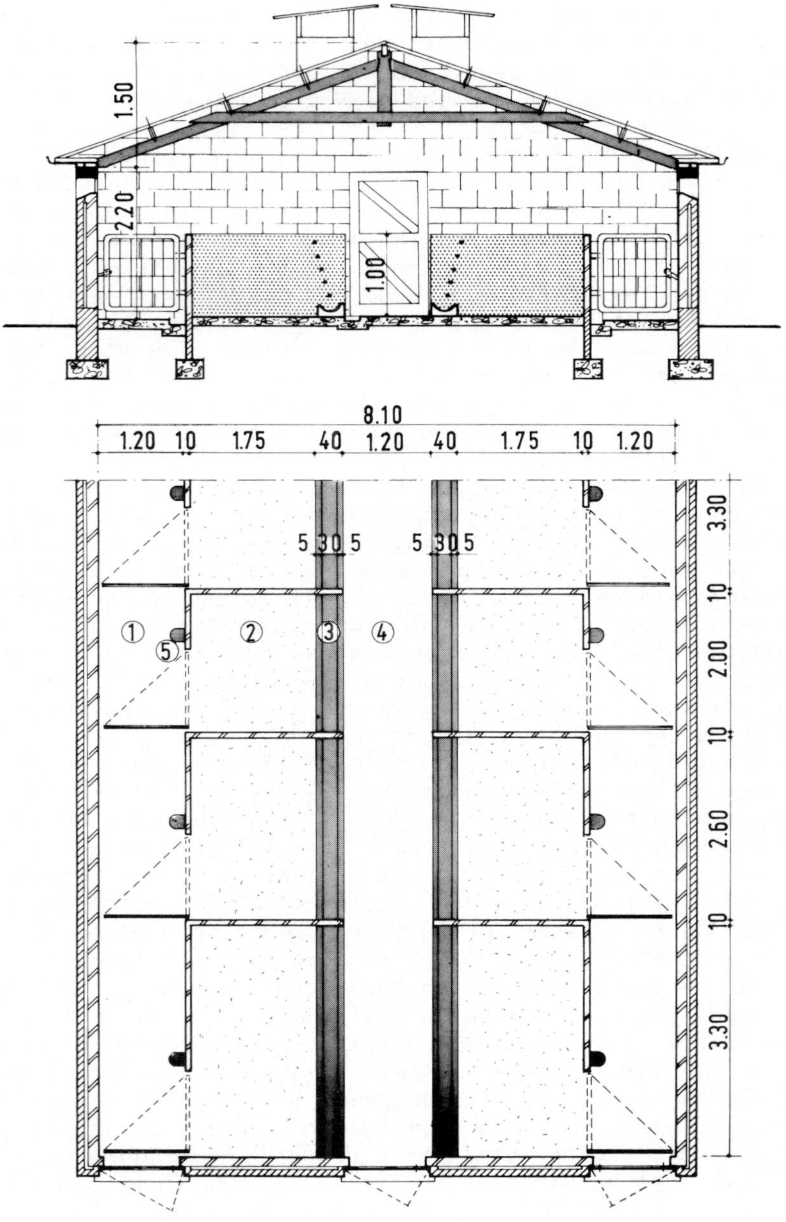

Fig. 5.45 Ground-plan and cross section of a two-row Danish fattening house with trough feeding.
Legend : 1 = dunging passage ; 2 = lying area ; 3 = trough ; 4 = feeding passage ; 5 = water bowl.

5.7.3.1 The part-slatted house with feed hoppers

This house comprises two or more rows of pens which are positioned along the feeding passages. Each pen consists of a fully concreted lying place with the feed hopper in the front along the feed pass and a dunging area with slatted floors at the rear. The lying area is 2.2 m long and the dunging area 2 m, the total length is thus 4.2 m. The width of the pen has to be calculated in such a way that a pen area of 0.84 m^2/animal is obtained : if 10 pigs are to be accommodated in this pen until their finishing weight, then the pen ought to be 2 m wide. If the pigs are moved once then the pen, which holds 10 pigs, must be 1.40 m wide and 4.2 m deep (0.60 m^2/animal) in the initial phase. More and more 12 and sometimes 15 fattening pigs are accommodated in one pen

The shortest side of the pen is positioned along the passageway and with the same number of rows as with trough feeding wider and hence cheaper pig houses can be obtained. The passage is also reduced to about 80 % of its original area which enables an additional saving. Fig. 5.46 represents the layout of a partly slatted pig house with feed hoppers, while fig. 5.35 gives an overall view of this house.

The *lying place* should be composed of a well-insulated, impervious and durable top layer which has a fall of ca. 4 % towards the dunging area. Such a lying place is not littered and this enables a substantial labour saving. The top layer can either be a screed of 2 to 3 cm made of cement and sand, or Bernit or Stallit, on a 15 cm layer of argex or expanded clay concrete (argex beads 10 - 20 mm Ø) supported by a concrete layer of ca. 5 cm.

The *dunging area* consists of a reinforced concrete slatted floor (beams or slat panels). When using beams the slat width should be 10 cm (fig. 5.47). The bottom face of the slat holds one main steel bar with a diameter of 12 mm whereas the top face is made up of two secondary steel bars each with a diameter of 8 mm. The reinforcement bars are connected to each other by braces with a diameter of 5 mm, which are provided every 20 cm in a beam with a length of 1.6 m. Such a beam can bear a load of 250 kg/m and is supported over ca. 10 cm at both sides by dwarf walls. A slot width of 2 cm is provided while the slats are tapered to facilitate the passage of faecal material. Slotted panels, sometimes also called "waffle slats" (fig. 5.48) are reinforced flat concrete panels of 6 cm thickness which are commercially available in different sizes e.g. 1.25 m x 0.50 m, 1.50 m x 0.50 m, 2.00 m x 0.40 m. They are characterized by a number of rows of short slots. They are more "pig-friendly" and the "normal" slats therefore have been gradually ousted by this new type of slotted panels. A liquid manure channel is provided below the slats which at the same time acts as a cellar. Some water is put into the channel preventing the droppings from adhering to the channel walls and bottom. The channels have a fall of 0.5 cm/m towards the end of the house where they drain into a pit of which the bottom is ca. 1 m lower than that of the channel. In this way it is always possible to remove a volume of one slurry tanker from the pit. During the trip to and from the land and the spreading, the pit gets filled again. This pit is often installed in the house itself.

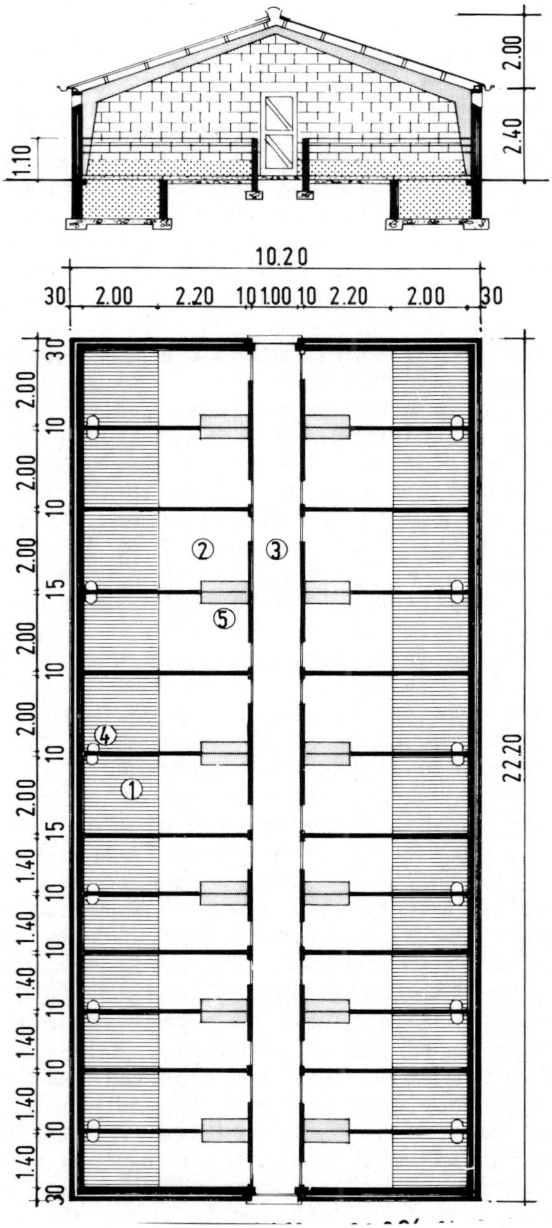

Fig. 5.46 Ground-plan and cross-section of a two-row partly slatted fattening house for 240 (24 x 10) fatteners, which are once transferred.
Legend : 1 = slats ; 2 = lying place ; 3 = passage ; 4 = drinking bowl ; 5 = feed hopper.

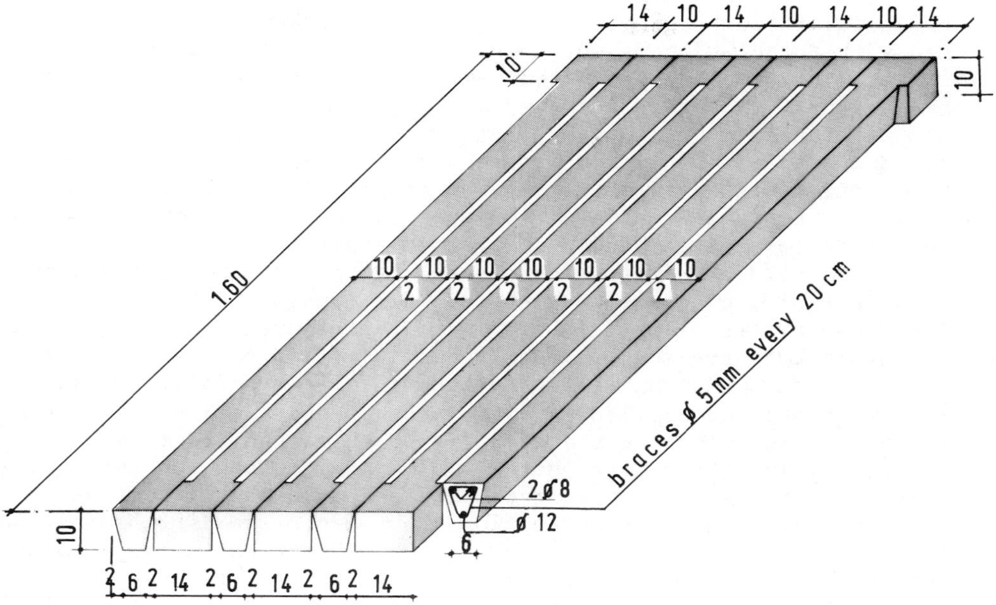

Fig. 5.47 Concrete beams forming a slatted floor in a fattening house.

Fig. 5.48 Slotted panels, also called "waffle-panels" in a fattening house.

By installing a draw-off it becomes possible to glide the suction pipe into the pit without entering the house. The draw-off is made of a concrete sewer pipe with a diameter of 0.5 m which is placed obliquely through the foundation wall. The draw-off should be provided with a lid which is only removed for extracting manure.

The liquid manure production amounts to 5 litres per fattener and per day or 0.15 m^3 per fattener place and per month or 0.5 m^3 per pig and per fattening period ; channel depths are adapted to these rated quantities. The liquid manure should always remain at least 0.2 m below the slats. The removal of the liquid manure and the spraying of it on the field can be performed with the same equipment as is described for cattle housed on slatted floors.

The *partitions* between the pens are only 1 m high, provided that only fattening pigs are housed. They can consist of "full" materials (concrete plates or walls) which will contribute in keeping the animals calm but have the disadvantage of limiting the survey of the pigs by the pig keeper. A combination of full materials and galvanized tubes can also be applied. A satisfactory construction consists of ca. 50 cm high concrete plates above which galvanized tubes are horizontally installed over another 50 cm. Reinforced concrete plates which have to be slid in concrete posts are generally applied.

In another type of construction the lower plate is 0.5 m high and full whereas the upper has an equivalent height but shows a number of openings.

In these ways the animals are protected against draught while sufficient observation of the animals is possible.

The drinking bowl or nipple drinker is installed above the dunging area to avoid wetting of the lying bed with spilled water. The watering facilities are described later in this chapter.

5.7.3.2 The part-slatted house with trough along the passage

The trough is installed along the feeding passage and consequently determines the width of the pen. Abutting against the trough is the insulated lying place (see higher) which measures 1.35 m and behind that a slatted floor of 1.2 m. Fig. 5.49 gives the ground-plan of two half-slatted houses with trough ; the first layout has only one size of pens (no moving) whereas the second layout has two different sizes of pens (moving once). Table 5.7 gives the main pen dimensions.

5.7.3.3 The fully slatted house with feed hoppers

This type of house comprises two or more rows of pens situated along feeding passages (fig. 5.50). The feed hoppers are placed on the slatted floor of the pen and along the feeding passage. The entire floor area of all pens and even of the whole house (see further) is either made of reinforced concrete slats or of the more "pig-friendly" slatted panels of reinforced concrete which are also called "waffle-panels". They will form both the lying as well as the dunging area. Two types of "waffle-panels" are applied as flooring for the

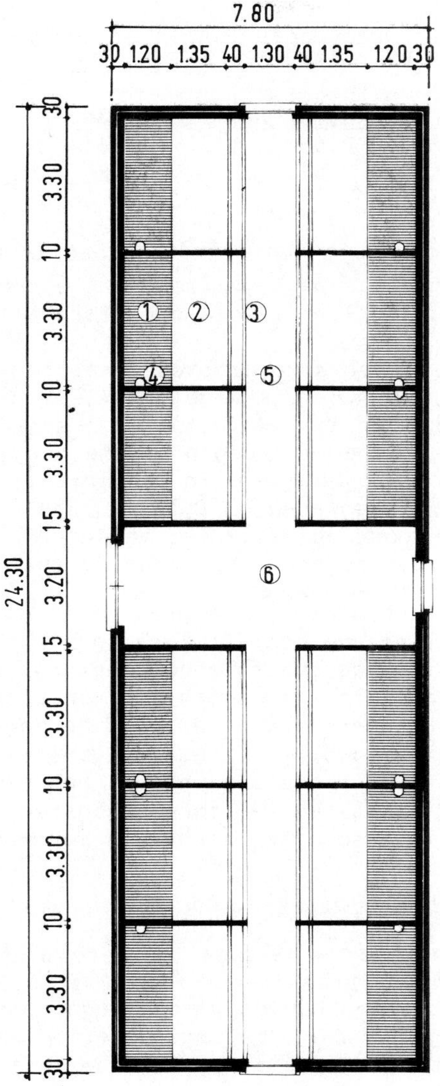

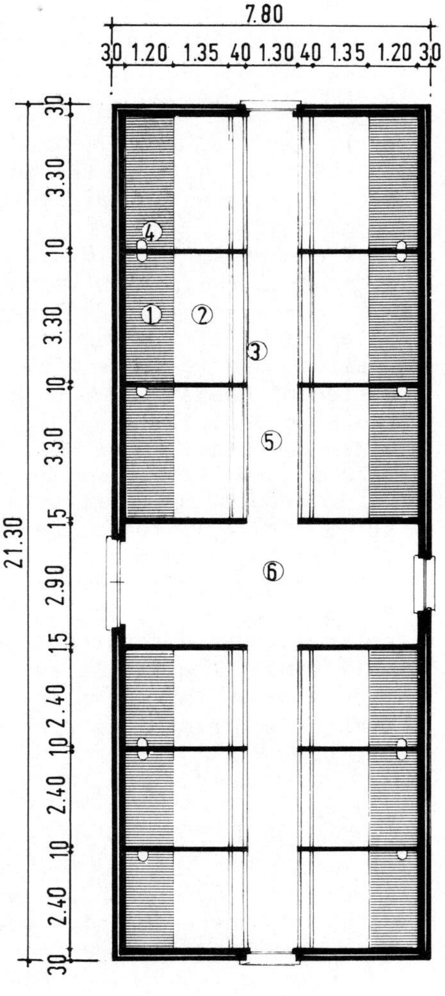

Fig. 5.49 Ground-plan of two half-slatted houses with trough feeding for fatteners.
Legend : 1 = slats ; 2 = lying area ; 3 = trough ; 4 = water bowl ; 5 = passage ; 6 = feed room.

TABLE 5.7 The main pen dimensions for a half-slatted house with trough along the passage.

Number of movings	Weight category (kg)	Trough length (= pen width) (m)	Pen depth (m)	Area of the pen (m^2)
0	20 - 100	3.30	2.55	0.84
1	20 - 40	2.40	2.55	0.61
	40 - 100	3.30	2.55	0.84

pen : the ones with a small number of slots (void : 10 %) are installed in the front half of the pen near the feed passage whereas the ones with a large number of slots (void : 17 %) will form the rear half (fig. 5.48).

A pen measuring 2 m wide and 3.5 m deep accommodates 10 fattening pigs up to their final weight while a pen of 2 m wide and 4.2 m deep will normally hold 12 fatteners. Each pig is then allotted a space of 0.70 m^2. If pigs are moved once and when the depth of the pen, remains identical the width ought then to be 1.4 m in the initial fattening phase when each pig will have an area of 0.50 m^2. The stocking density in a fully slatted house is maximal : the area provided per pig is 0.84 m^2 in a part-slatted house compared to only 0.70 m^2, or 17 % less, in a fully slatted house.

To reduce the construction costs the whole house area is often slatted which facilitates the mechanical excavation. It is also possible to lay the slats on dwarf walls or to include concrete columns in the cellar floor on which prefabricated beams are laid which will support the slats (fig. 5.51). The concrete wall panels are slid into the concrete columns. Such a construction method is ideally suited for do-it-yourself since it allows a reduction of labour costs as the farmer can employ his own labourers instead of specialized and more expensive contractor workers. This construction is also often sold as a turn-key project.

5.7.3.4 The fully slatted house with trough along the passage

The trough is installed along the feeding passage and determines the width of the pen. The whole flooring area is slatted. Fig. 5.52 shows the ground-plan of two fully slatted houses with trough, one with pens where the pigs stay during the whole fattening period and one where the pigs are transferred once. Table 5.8 gives the most important dimensions of the house.

5.7.3.5 The partly or fully slatted floor house with perpendicularly placed trough

Slop feeding systems allow the homogeneous distribution of the slop over the entire trough and this from one central delivery point. This implies that in these cases the trough can be placed perpendicularly to the feeding passage.

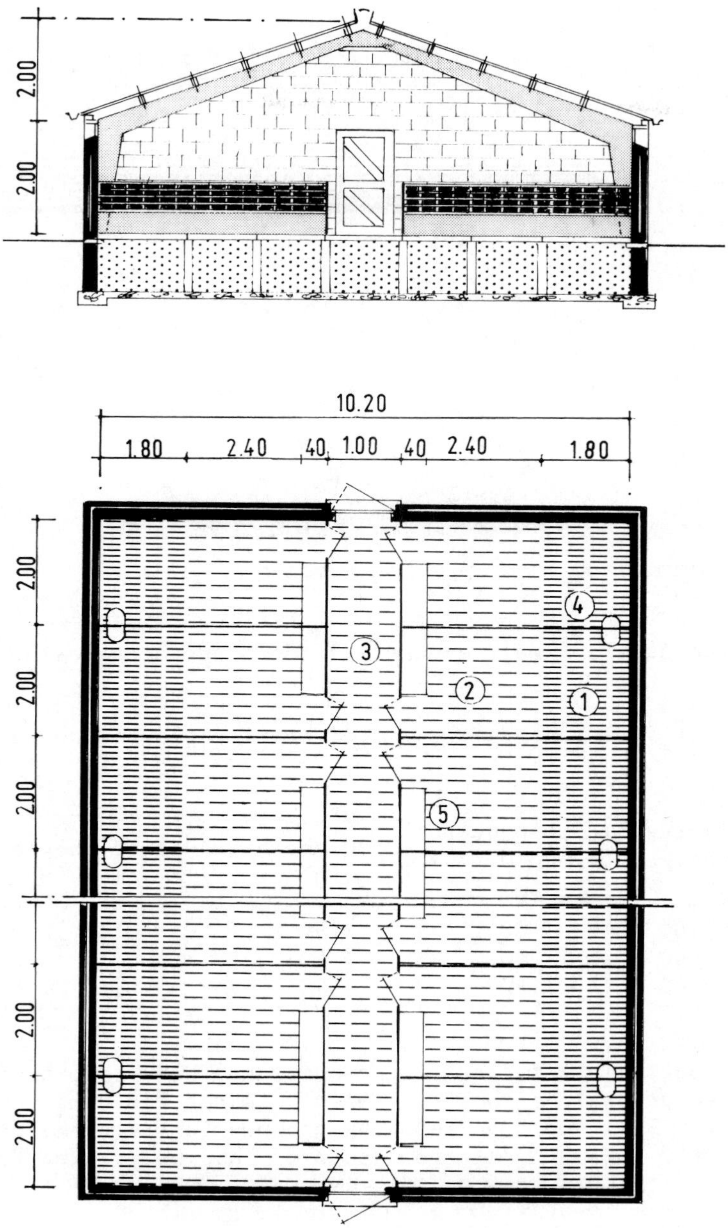

Fig. 5.50 Ground-plan and cross-section of a two-row fully slatted house with 12 pigs per pen.
Legend : 1 = slats in the dunging area ; 2 = slats in the lying area ; 3 = passage ; 4 = drinking bowl ; 5 = feed hopper.

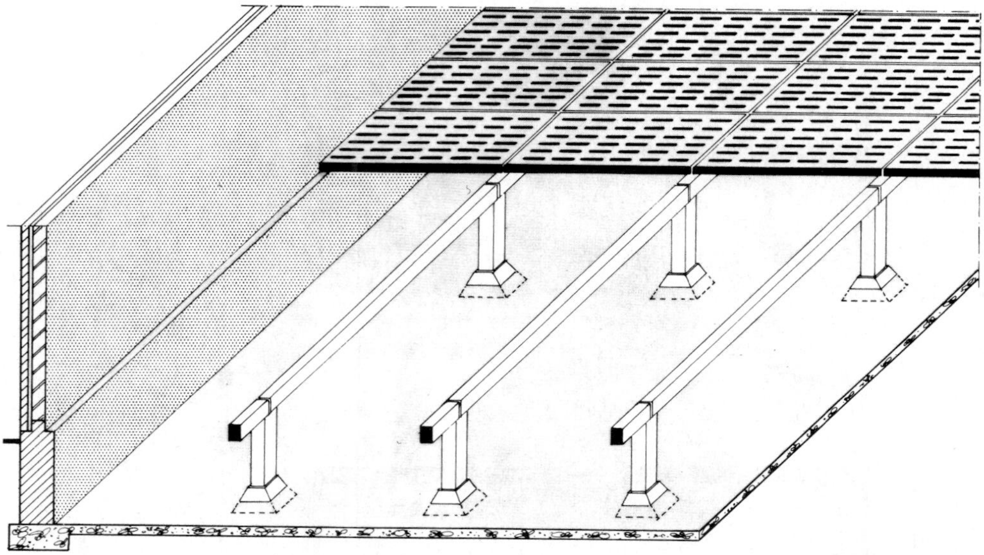

Fig. 5.51 Slotted panels ("waffle-panels") supported by a prefabricated foundation.

TABLE 5.8 The most important dimensions of a fully slatted house with trough.

Number of movings	Weight category (kg)	Trough length (= pen width) (m)	Pen depth (m)	Area of the pen (m^2)
0	20 - 100	3.30	1.90	0.63
1	20 - 40	2.40	1.90	0.46
	40 - 100	3.30	1.90	0.63

This layout results in a substantial reduction of the length of the passageway and allows the construction of a wider and hence cheaper pig house. Figs 5.53 and 5.54 show the ground-plans and the cross-sections of a partly and a fully slatted house of this particular type.

5.7.3.6 The compartmentalized finishing house

Nowadays compartmentalized and heated fattening houses are also built according to a Dutch design based on the results of an investigation carried out in the Netherlands. Compartmentalizing involves the division of a house into a number of small and separate rooms or compartments. The lengthwise arrangement of the pens is unsuitable since this would involve traversing a number of rooms to reach the desired one and this would annul the idea of separation.

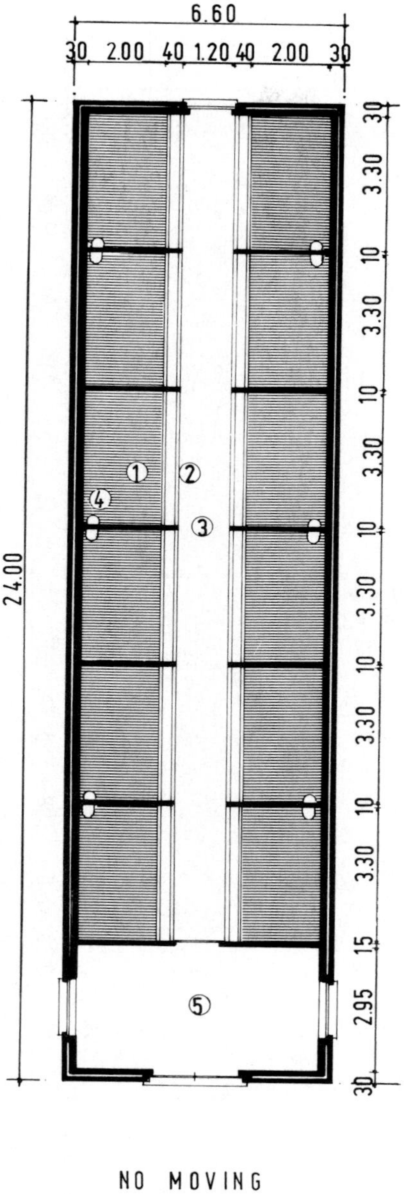

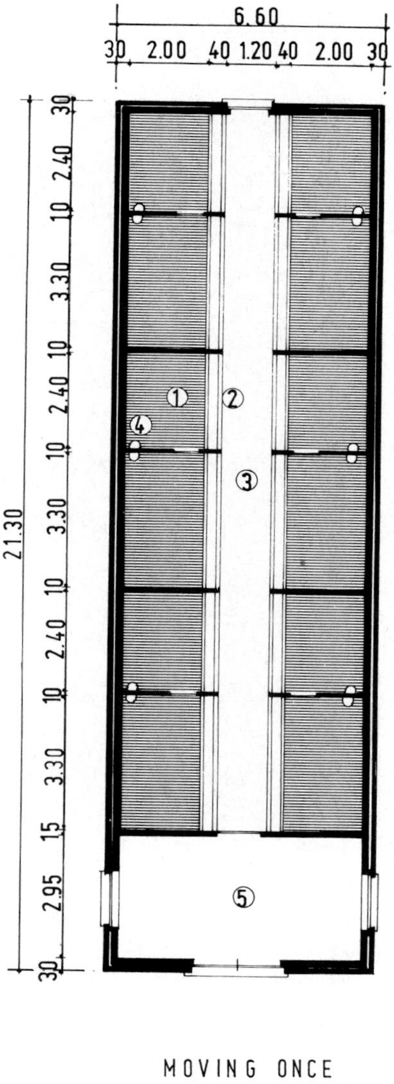

Fig. 5.52 Ground-plan of two fully slatted houses with trough feeding for fatteners, with 10 pigs per pen (120 fatteners per house). Legend : 1 = slats ; 2 = trough ; 3 = passage ; 4 = water bowl ; 5 = feed room.

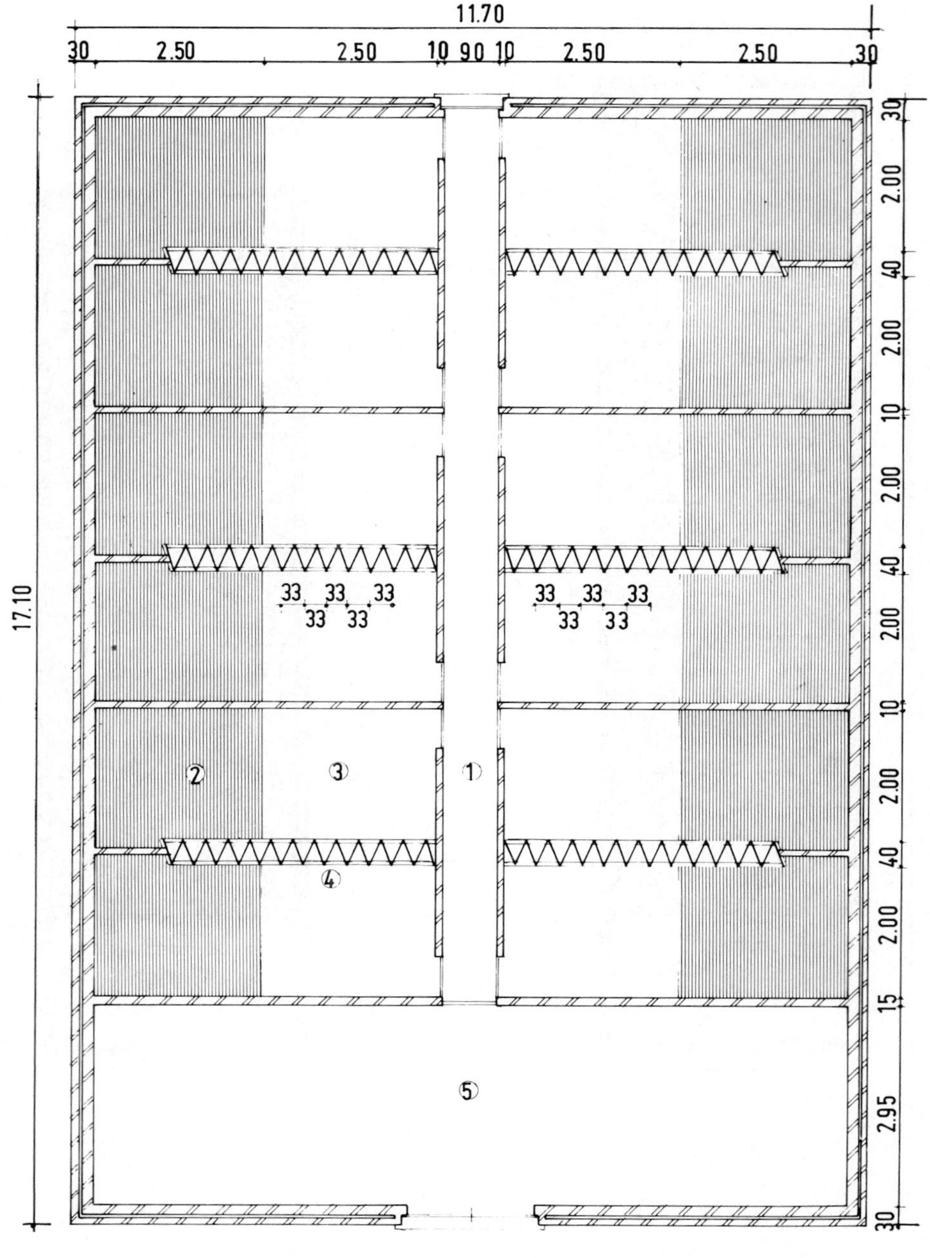

Fig. 5.53 A two-row partly slatted house with perpendicularly placed trough and intended to accommodate 144 pigs, (12 pigs per pen). Legend : 1 = passage ; 2 = slatted dunging area ; 3 = lying area ; 4 = zig-zag partitions above the trough ; 5 = feed preparation room.

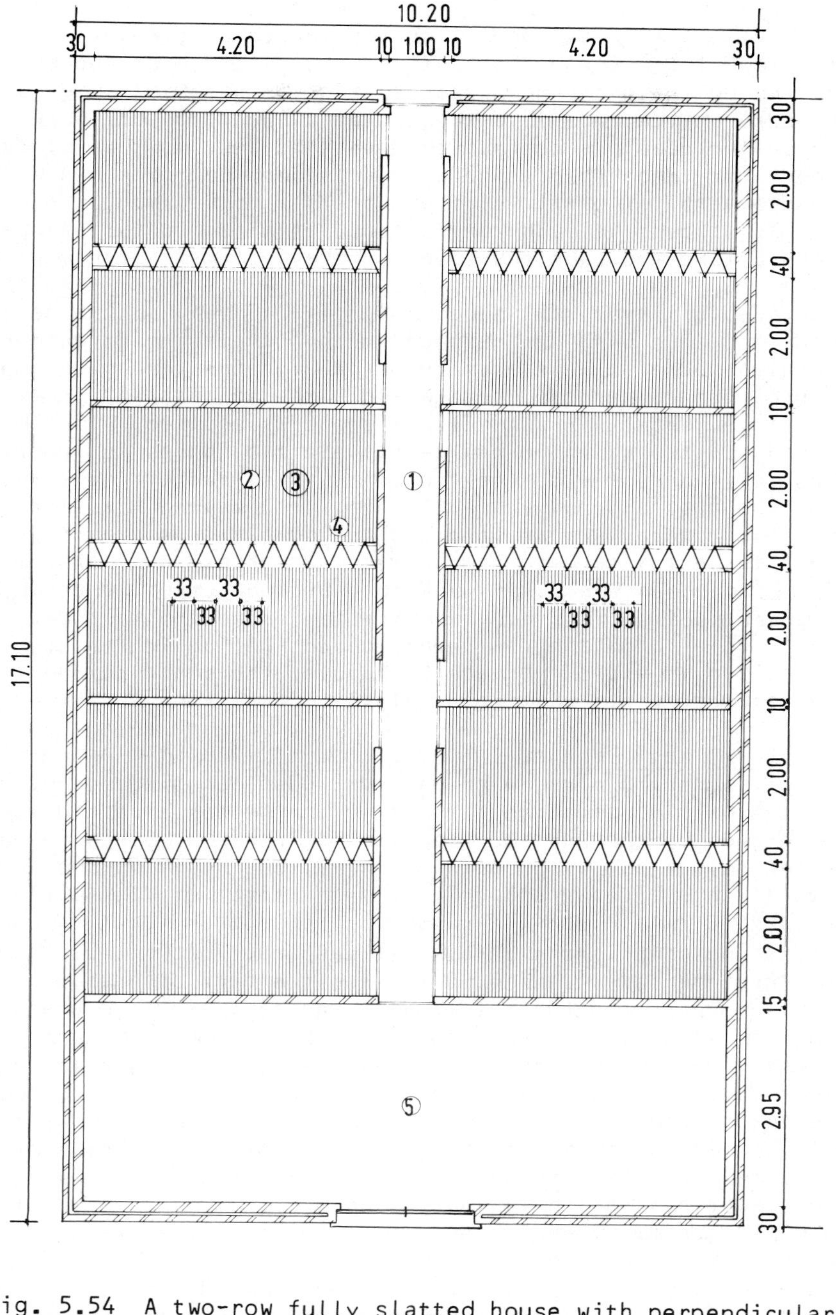

Fig. 5.54 A two-row fully slatted house with perpendicularly placed trough for 144 pigs, (12 pigs per pen).
Legend : 1 = passage ; 2 + 3 = fully slatted dunging and lying area
4 = zig-zag partitions above the trough ; 5 = feed preparation room.

The *parallel arrangement,* also sometimes called the *comb-shaped arrangement,* is ideally suited for compartmentalizing. Along one of the longitudinal walls a passage is provided over the entire length of the building giving access to each compartment. Figs 5.55 and 5.56 represent two possible layouts of such a type of house. The first design includes one central feed passage and two rows of partly-slatted pens per compartment, whereas the second design shows one feed passage and one row of partly slatted pens per compartment.

The number of pens in each row is mostly limited to 6, maximum 7, to keep the volume and thus the speed of the inlet air stream within certain limits.

In practice, compartmentalizing always corresponds with a parallel arrangement and thus with one passage in the lengthwise direction which is completely shut-off. This acts as a lock for the air inlet and one can then speak of mechanical ventilation with indirect air inlet. Furthermore, since the all-in all-out principle is applied, supplementary heating is installed which in the partly slatted houses is generally carried out by means of floor heating. All these factors make this type of housing the most expensive one. Finally it appeared necessary, in order to avoid countercurrents and sucking of air from one compartment to another, to ventilate each compartment separately by means of an air inlet channel through the outer wall to the compartment. One can in fact no longer speak of indirect ventilation. However, in that case only, one is able to obtain the required ventilation in the desired air stream direction in all compartments whatever the stocking density and hence the required ventilation quantity.

5.7.3.7 The finishing house with partly-lidded pens

The first finishing house with partly-lidded pens intended for 240 fatteners was put in operation in 1971 in the Netherlands and was a design developed by the Advisory Service for Farm Buildings in Wageningen. The second was put in operation in 1975 and by 1979 some 40 finishing houses of this type were operative with a total of 16,000 pig places (Freriks, 1979). The finishing house with partly-lidded pens (fig. 5.57) is strawless and comprises two rows. It has ample natural ventilation and has no supplementary heating. Seen from the central passage, each pen is slatted over a length of 2 m while the remaining 3 m are covered with a full concrete insulated floor. The feed hopper and water bowl are installed next to the central passage on the slatted area. When on warm summer days the pigs are lying down on the "fresher" slatted floor they have the tendency to defecate on the full concrete floor. Therefore, a 60 cm wide slatted area is often provided at the rear of the lying area, next to the outside wall. This construction tends to improve the cleanliness of the animals (fig. 5.58). The width of each pen which normally accommodates 15 fatteners to their final weight, amounts to 2.2 m, so that each animal has 0.84 m².

In order to furnish a favourable microclimate during the colder periods, the lying place is covered with insulation panels. The name of this particular type of house is derived from this construction-detail. The partitions between the various lying areas must be closed, those between the dunging areas not. The feed hopper can also

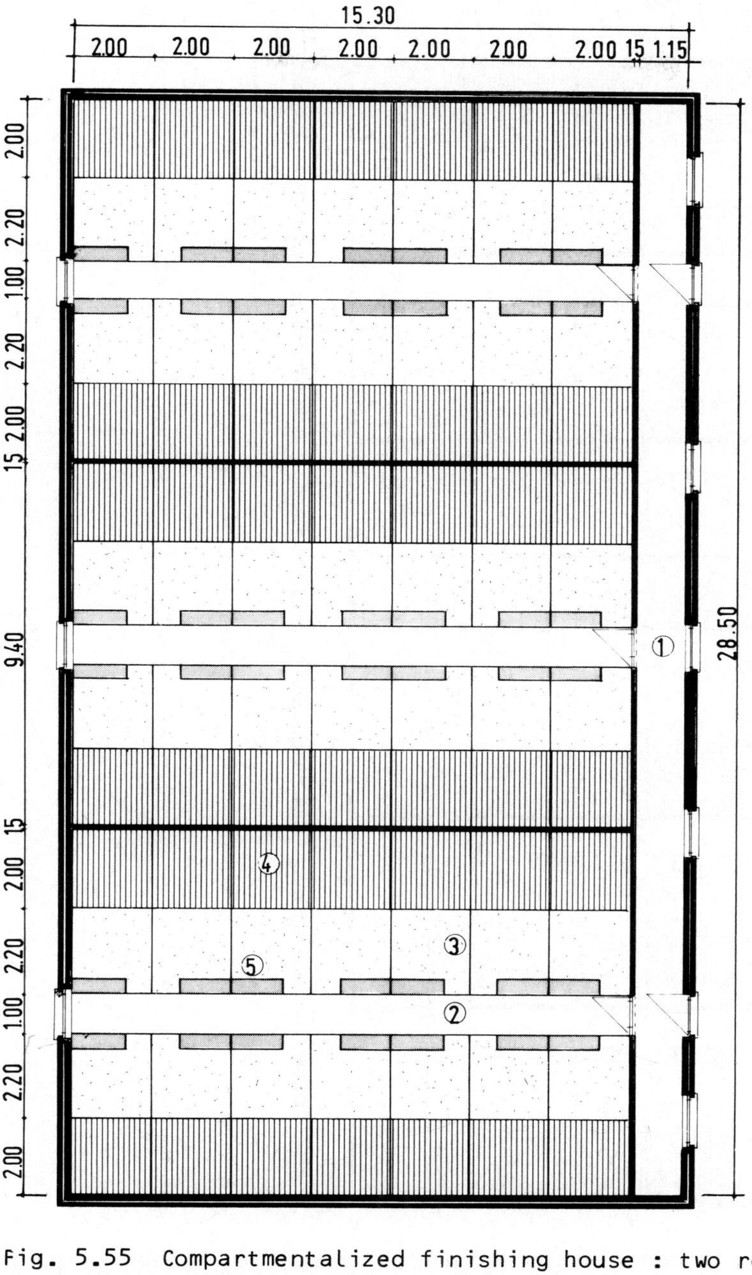

Fig. 5.55 Compartmentalized finishing house : two rows of partly slatted pens per compartment, 10 fatteners per pen, 140 pigs per compartment.
Legend : 1 = passage in the lengthwise direction ; 2 = feed passage ; 3 = lying area ; 4 = dunging area ; 5 = feed hopper.

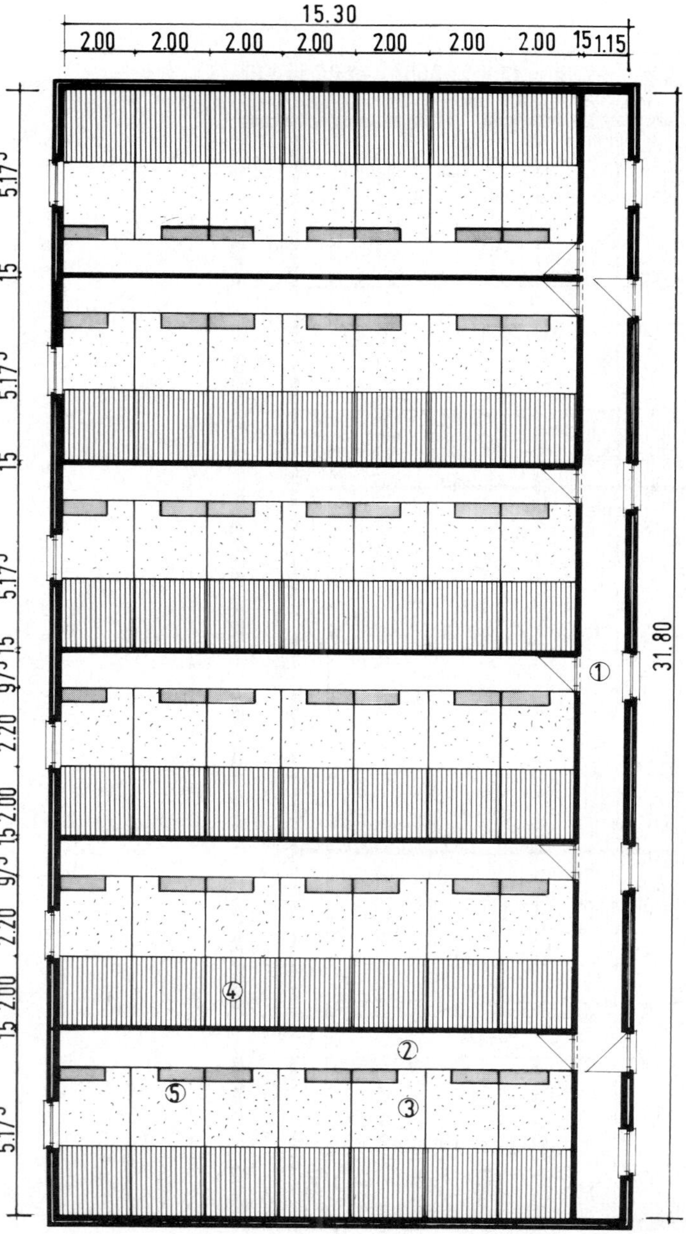

Fig. 5.56 Compartmentalized finishing house, one row per compartment, 70 fatteners per compartment, with partly slatted pens, 10 pigs per pen.
Legend : 1 = passage in the lengthwise direction ; 2 = feeding passage ; 3 = lying area ; 4 = dunging area ; 5 = automatic feeder.

Fig. 5.57 A finishing house equipped with partly-lidded pens.

be placed between two dunging areas, perpendicular to the feeding passage (fig. 5.58).

The insulated cover has to be kept out of reach of the pigs and is therefore placed at a height of 1.2 m. During the winter it has to cover the whole lying area but in other seasons it has to be partly or fully removable, otherwise the pigs will tend to lie down on the slatted floor during warm periods. The temperature of the lying area can easily be 25°C higher than the outside temperature during the winter ; the relative humidity in the pen varies between 60 % and 80 % while no harmful concentrations of noxious gases could be detected (Freriks, 1979).

The natural ventilation is by means of an ample sized open ridge and adjustable air inlets or an adjustable curtain under the eaves. The roof is insulated. The finishing house with partly lidded pens is easy to construct and is also a relatively cheap type of housing.

5.7.3.8 Other types of houses for finishing pigs

The continuous striving towards improvement regularly leads to more or less radical changes to the existing houses. In practice a number of variations of the above-mentioned types of houses and their dimensions are found. In 1970 attempts have been made to accommodate the pigs individually in one or two tier cages with restricted space (0.4 m x 1.4 m) and slatted floor. These attempts have failed. In special cases (feed experiments, selection...) individual accommodation might be indicated. The pens are then 1 m wide and 1.6 m long of which the rear half is slatted. In this particular case the water supply is provided by means of nipple drinkers, while according to the

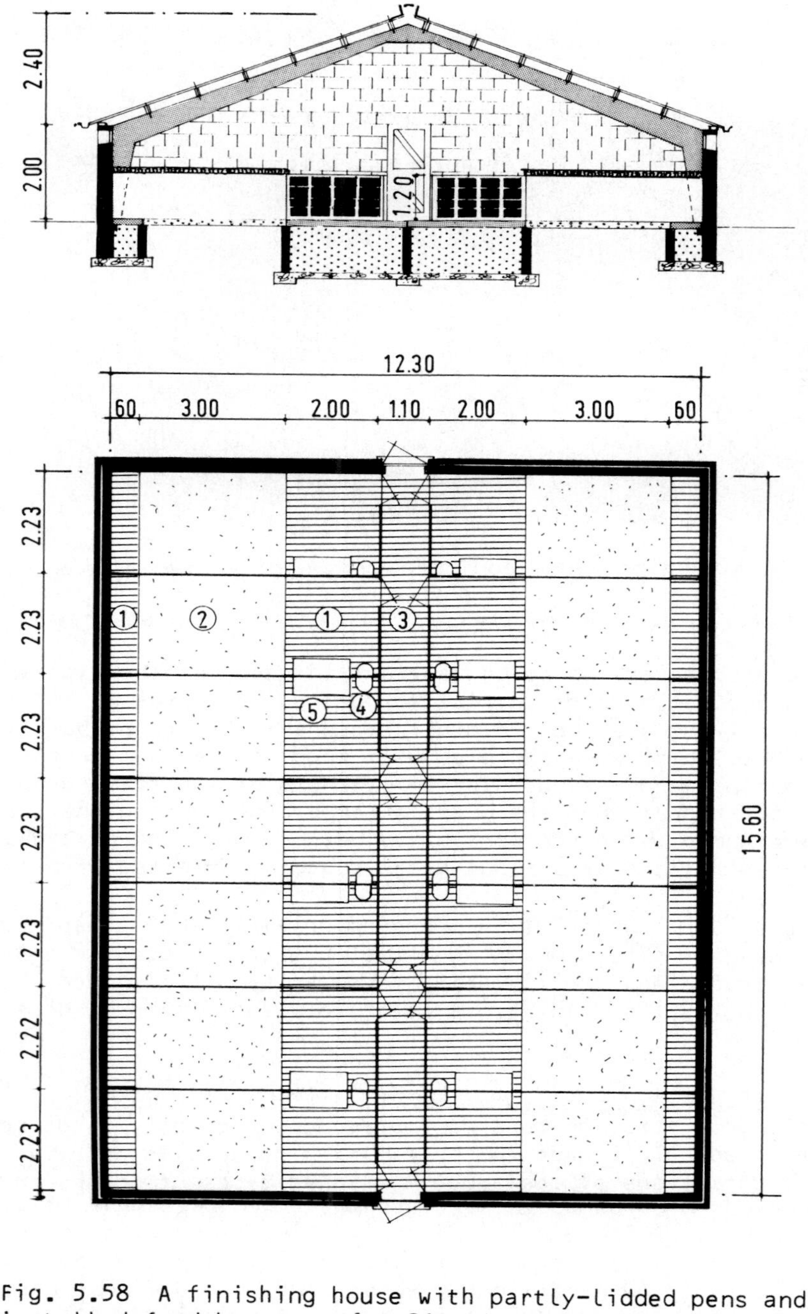

Fig. 5.58 A finishing house with partly-lidded pens and transversely installed feed hoppers, for 210 pigs (15 pigs per pen).
Legend : 1 = slats ; 2 = lying area ; 3 = passageway ; 4 = water bowl ; 5 = feed hopper.

type of feed trial, a trough or feed hopper is installed in the front of the pen.

The littered open fronted house which had limited application developed in the Seventies to a strawless open fronted house. This open front can be closed by means of a flap. The floor of this house is partly or fully slatted. With partly slatted houses the slatted area is situated along the feeding passage (fig. 5.59). The pen dimensions are 2 m x 5 m (of which 2 m are slatted, while the remainder forms the lying bed) for 12 pigs and 2 m x 6 m (of which 3 m are slatted and 3 m are reserved for the lying bed) for 15 pigs per pen.

In the Netherlands the transversely compartmentalized finishing house is often naturally ventilated. This type of house is known as the "Veluwe" house for fatteners.

Two other variations of the fully slatted house exist but are rarely used. In the first modification the rear half of the pen remains slatted, while the front half is characterized by a strip in the middle which is fully concreted. The concrete slatted panels along the pen sides remain. Another modification has a partly slatted pen with lowered lying area. The crossing between the slats and the 15 cm lower lying bed is achieved by a short declivity of about 40 cm long. The first results show that the pigs are making use of the lying bed. In summer time however they often lie on the slats and soil the lying bed and themselves.

In general, insufficient experience has been acquired with these types of houses to decide upon their usefulness.

5.7.4 Special equipment in the finishing house

5.7.4.1 The ventilation and lighting

The ventilation of finishing houses, which are nowadays equipped with feed hoppers and fully or partly slatted floors, is carried out either naturally or by means of extraction fans. The air inlet is formed by gaps under the eaves over the entire length of both side walls and controlled by long and adjustable hopper flaps which are intended to direct the entering air over a sufficient distance in the house. The air outlet takes place through an adjustable open ridge or through a number of insulated stacks. In the latter case an outlet area of 1 m^2 per 100 m^2 of floor area is required while the air inlet is taken twice as large. The areas for inlets and outlets are given in Chapter 3. It is possible to automatically adjust the natural ventilation using thermostatically controlled hopper flaps at the air in- and outlets, whereby the airflow pattern is less affected by the influence of the wind. Mechanical ventilation is another possibility and the fans are normally installed in the ridge. The house is normally windowless so that only background light is allowed to penetrate through the air in- and/or outlets. This facilitates the insulation, reduces cannibalism and quietens the animals. A waterproof electrical lighting is required with an intensity of ca. 30 lux per m^2, which is ca. 0.8 W per m^2 when striplight is used.

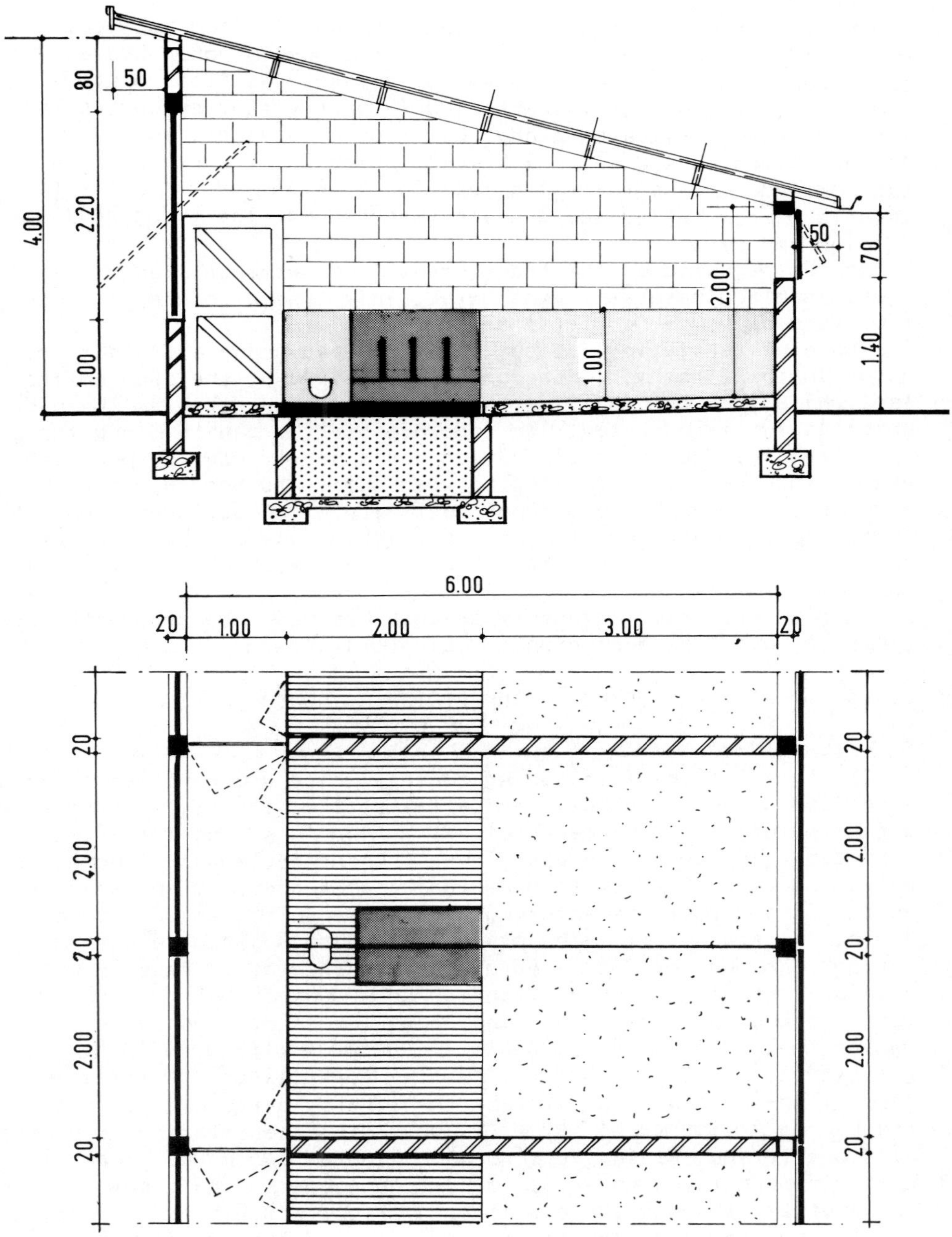

Fig. 5.59 The single row part-slatted open fronted finishing house with ample natural cross ventilation.

5.7.4.2 Watering facilities

The distribution of drinking water is carried out by automatic drinking bowls or nipple drinkers at a rate of one per pen. Spillage of drinking water is generally higher with nipple drinkers than with automatic water bowls and amounts to 18 % for bite drinkers and 46 % for nipple drinkers of the push-type. In order to reduce this important spillage a rimmed plate is installed under the drinker. The water which is spilled is collected on this rimmed plate and can be drunk by the pigs so that it is no longer wasted (fig. 5.60).

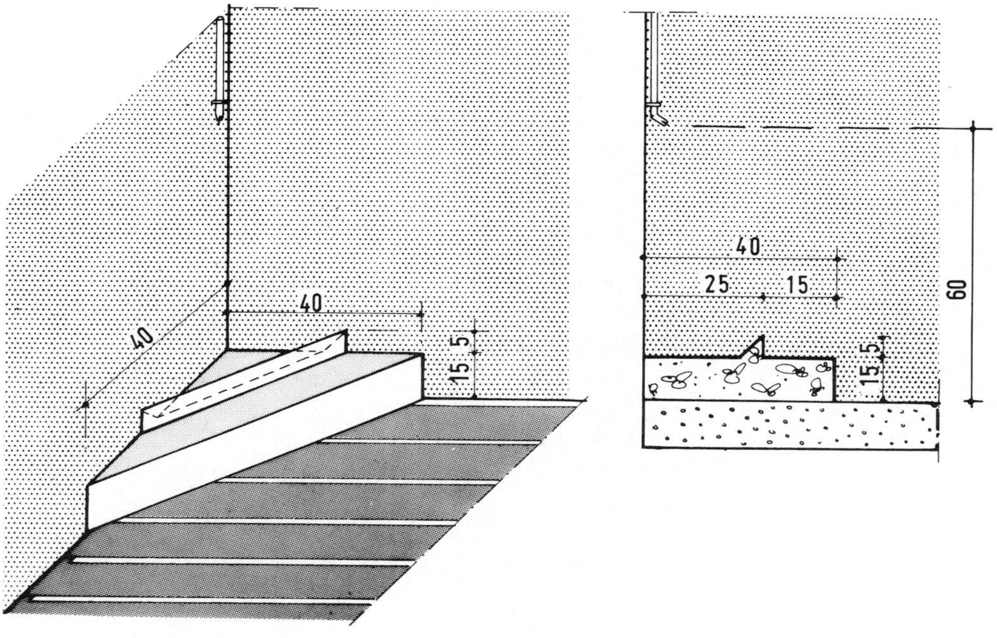

Fig. 5.60 Collecting-pan for spilled water installed under the drinker.

5.8 LABOUR-TECHNICAL AND ECONOMIC ASPECTS OF THE HOUSING OF FINISHING PIGS

The comparison of the labour time requirement for the care of finishing pigs in various types of houses and with the application of different methods of feeding is based on our own research (table 5.9). The times indicated represent both the daily activities viz. : feeding (twice daily), littering, mucking-out (only for nr. 1) and cleaning of the lying area as well as periodic activities viz. delivery of the pigs in the different pens, perhaps moving of the pigs (1x), veterinary care, cleaning and disinfecting the pens, weighing and delivery of the finished pigs and the time involved for making notes. The pe-

TABLE 5.9 The labour time requirement for the care of fattening pigs.

Type of house	Labour method	Time in man-min per 10 pigs and per day	Total time in man-hours per delivered fattener (120 days)	Labour productivity in number of pigs per stockman (8 h/day)
1 Littered Danish house, without slats in the dunging area	Manual wet feeding with feed trolley, filled with 250 kg of meal at the silo and with feed scoop ; twice daily trough feeding ; supplying water in the trough by means of a hose ; automatic water bowls, manual littering and mucking out	5.7	1.14	842
2 Partly slatted house	Wet feeding and drinking facilities as in nr. 1	3.2	0.64	1,500
3 Partly slatted house	Automatic dry feeding at feed hopper ; automatic drinking bowls	1.3	0.26	3,692
4 Fully slatted house	idem	1.25	0.25	3,840

riodic activities require 0.95 min per 10 pigs and per day and thus cover a large part (up to 3/4) of the total time for the care of fattening pigs in partly and fully slatted houses. In the last mentioned houses, equipped with automatically filled feed hoppers, the daily "activities" are limited to supervision.

From table 5.9 we are able to draw the following conclusions. The strawless housing, by applying partly slatted floors, allows an important (44 %) reduction of the labour time (compare nr. 2 with nr.1).

The use of an auger for filling the feed hoppers in houses with partly or fully slatted floors allows the reduction of the total labour time requirement to ca. 15 minutes per finished pig (nrs 3 and 4). This is indeed a remarkable result of modern technology. Such a house with up to 3,840 fatteners can be run by one stockman. In practice, however he shall have to call in a colleague for performing certain activities such as weighing or to replace him when he is ill or on leave etc.

The *construction costs of a finishing house* differ according to the type of housing. We have calculated the building costs for various types of houses for finishing pigs for the year 1980 (Daelemans et al., 1980). We were able to make following deductions :

- a fully slatted floor finishing house requires a smaller investment than a partly slatted house (fig. 5.61) viz. £ 7.50 less per place for a finishing house with 500 pigs ;
- the lengthwise arrangement of the pens for partly slatted houses (fig. 5.61) requires the smallest investment with a two-row layout but from 400 places onwards the three-row design turns out to be less expensive ;
- the lengthwise arrangement of the pens involves a smaller investment than the transversal arrangement : compartmentalizing requires supplementary internal walls and gives rise to important additional costs which will be much higher as the degree of compartmentalizing increases viz. with the single row compartmentalizing (fig. 5.62). A three-row house with lengthwise arrangement, housing 500 pigs costs £ 67/place ; a house with transversal arrangement and six compartments of 1 x 7 pens costs not less than £ 88/place.

5.9 ZOOTECHNICAL AND VETERINARY ASPECTS OF THE HOUSING OF FINISHING PIGS

Several factors influence the results of the production of fatteners (Castryck, 1981) such as : the house-climate, the group size, the transferring, the application of the all-in all-out system, the method of feeding, the type of pen. All these factors will be dealt with in detail.

The house-climate is determined by a number of parameters of which the temperature is the most important.

The important role played by the house-temperature is described by Mateman et al., (1982a) as follows : "If pigs are housed in a temperature zone which can be called optimum (*thermoneutral zone*) the pigs will not be required to generate extra heat for the heating of their own body. If the house temperature drops below this temperature zone (below the critical temperature) then the pig is obliged to utilize part of its feed which was originally intended for its growth, for the extra heating of its own body".

The same authors (Mateman et al., 1982a) investigated the relation between the intake of feed and the temperature with finishing pigs and have therefore divided the finishing period in two parts viz. : from 25 kg up to 60 kg and from 60 kg to 100 kg. They reached the following conclusions :

- too low a house temperature will cause a reduced growth of 9 g/day/°C in the first fattening period and of 18 g/day/°C in the

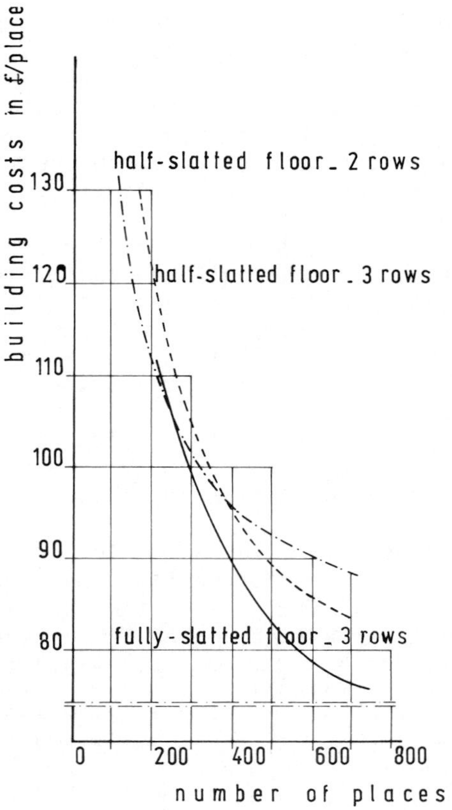

Fig. 5.61 Investment-costs for finishing houses with partly and fully slatted floors

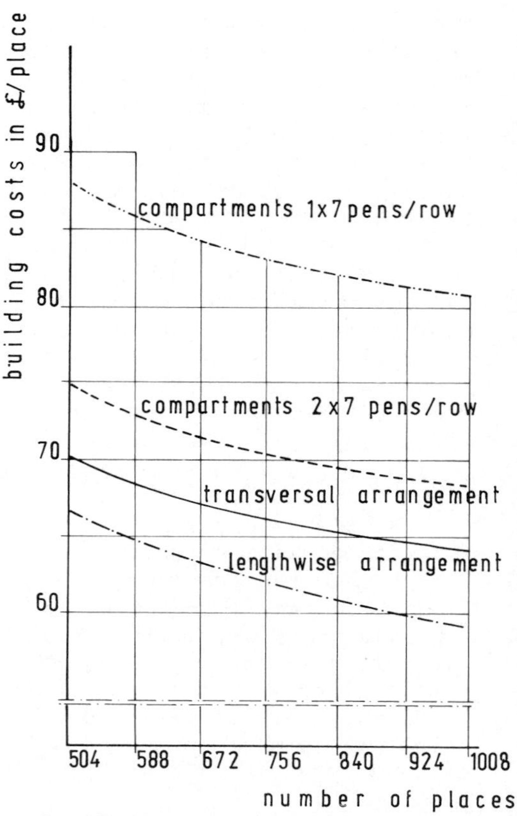

Fig. 5.62 Investment-costs for finishing houses with lengthwise and transversal arrangement (compartments = transversal arrangement).

second period ;

- the above-mentioned growth depression can be prevented by supplying extra feed ; from the research it seemed that each °C below the optimum temperature required a supplementary intake of 20 g of feed per day in the first period and of 40 g/day in the second period ;
- rationed feeding led to other conclusions than ad lib feeding : ad lib fed pigs tolerate lower temperatures without harm or growth depression owing to their higher feed intake than ration-fed pigs.

The determination of the critical temperature is not so easy, as it is influenced by a number of factors :

- the feed regime : with ad lib feeding the critical temperature is a few degrees lower ;
- the air velocity in the vicinity of the animals at low ambient temperatures : an increase in the air velocity results in an increase of the critical temperature ;

- the size of the group : the critical temperature will be the highest with individually accommodated animals, but will decrease as the size of the group increases ;
- the flooring of the house : littered houses have a lower critical temperature than partly slatted floors which in turn have a lower one than fully slatted floors ;
- the weight of the animal (or the age) : as animals grow older their critical temperature decreases ;
- the time of day : the pigs normally rest during the night, they then have a lower skin temperature and the critical temperature is therefore also lower ;
- the constitution of the animal : not all animals react in the same way to the critical temperature.

The lower critical temperature can, within certain limits, be influenced by for instance the use of partly slatted floors, by a restriction of the air velocity and by feeding the animals ad lib, which is often practised in Belgium. The question whether supplementary heating is justified during the winter remains open. There is some disagreement on this subject in the literature. Following some sources (Anon., 1982b) supplementary heating of the house is economically justified with ad lib feeding and fully slatted floors. On the other hand Mateman et al. (1982b) calculated that the administration of extra feed often outweighs the advantages of supplementary heating. Concerning the optimum temperature as a function of the age and the type of housing we refer to table 3.4. From what is mentioned above it appears that it is not so easy to determine the lower critical temperature for each situation and this makes the setting of the desired temperature when mechanical or mechanically controlled natural ventilation is used rather problematic. Based on the above-mentioned factors Bruce and Clark (1979) developed a simulation model to enable the calculation of the lowest critical temperature (LCT) in each situation. Boon (1981) has developed another method for its approximation and this is based on the fact that pigs tend to huddle in cold conditions whereas otherwise they prefer to lie alone in a warm environment. He therefore studied the postural behaviour of pigs (12 pigs per pen) between 23 kg and 75 kg by means of time lapse photography and evaluated their lying behaviour by an "huddling index" which represented the degree of huddling. From this it appeared that pigs try to adapt themselves to the surroundings by crowding together when it is cold (social warmth) and in these conditions they even occupy 20 % less lying area than the minimum required. To be complete we must also mention that the upper temperature of the thermoneutral zone is also limited. This temperature limit is called the upper critical temperature (UCT). If the temperature in the piggery exceeds the upper critical temperature it will also be disadvantageous because the heat production of the pig rises due to the increasing evaporation (increased respiratory rate) while its body temperature will also increase. Pigs cannot sweat and they adapt themselves to the extreme temperature by wallowing in the dunging area or by urinating on the lying place with the intention of moistening their skin when lying down so as to give off the heat through the evaporation of this moisture.

It is advisable during the hottest days of the summer to apply maximum ventilation to prevent the house temperature from rising too much above the outside temperature. In such circumstances the air speed in the neighbourhood of the animals is likely to exceed 0.2 m/s but this is not regarded as harmful under these conditions as it will increase the cooling effect. The upper critical temperature is influenced by the same factors as the lower (see higher) and decreases as the animals become heavier. Houses accommodating heavy pigs will reach the upper critical temperature faster and the stockman will therefore be obliged to take additional measures earlier (opening of doors and windows, switching on a movable fan, etc).

A relative humidity of 60 to 80 % is believed to be optimum, whereas the air speed in the vicinity of the animals should in "normal" circumstances be kept below 0.2 m/s (table 3.4). A sufficient supply of oxygen must be provided while the concentration of noxious gases in the piggery must be kept as low as possible.

Several researchers have studied the *cubic capacity or air space* of the pig house in relation to the production results. Lindqvist (1974) and Haaring (1976) found significantly less lesions of the lungs as the air space in the house increased. From both investigations it appears that the cubic capacity should not be less than 3 m^3 per pig. Tielen (1974) however found no direct correlation between the cubic capacity and the production results.

Concerning the *group size*, several authors have found that with restricted feeding the daily growth decreased and the feed conversion increased with an increase of the group size (Pechert, 1970 ; Ober, 1972). Petersen and Nielsen (1977) compared group sizes of 8, 16 and 32 pigs with restricted and ad lib feeding. For restricted feeding the authors reached significantly better results as the group size was smaller. For ad lib feeding however they were not able to detect significant differences. Papp (1984) compared groups of 5, 10, 15 and 20 pigs and found that with self-feeding and a constant individual area (0.84 m^2/pig) an increased group size decreased the average daily feed intake and growth of the fattening pigs. The feed conversion on the other hand improved. Ethological observations give some explanations about it. Group size affects the feeding behaviour. Increased group size decreases the eating time and the intensity of eating, consequently feed intake decreases. The most important factor in eating behaviour is the access to the feed hopper. Group size has hardly any effect on the duration of resting and activity of fattening pigs. As smaller group sizes are accommodated per pen the number of partitions will inevitably increase leading to an increased expenditure. Furthermore, bad results (irregular growth) have been experienced in practice with larger groups (30 and more) and therefore the size of the group should be limited to 10 to 12 pigs when ration feeding is applied and 12 to 15 for ad lib feeding.

The *pen shape* can be varied (square, rectangular) with self-feeding. Applying this feeding system and allowing a constant individual area of 0.84 m^2/pig, Papp (1984) found no significant effect of the pen shape on the fattening performance.

For economic reasons there is a tendency to reduce the *area per pig*. This leads to crowded conditions at the end of the finishing period. Papp (1984) experimented with 4 different pen sizes using partly slatted floors (0.75 m^2, 0.76 - 0.85 m^2, 0.86 - 0.95 m^2 and 0.96 - 1.05 m^2). Pigs allowed less than 0.75 m^2 individual area show a significantly decreased daily growth compared to those with a larger area. Trends in average daily feed intake are similar. The feed conversion is also worst with the smallest area whereas there is a substantially smaller difference between the other sizes.

Moving of pigs ought to be avoided or at least kept to a minimum. Moving pigs will temporarily affect the performances and will upset the animals. Ober (1972) found that the growth rate associated with moving pigs 0, 1, 2 and more than twice amounted on average to 606 g, 597 g, 589 g and 583 g per day respectively and hence decreased slightly with an increased number of transfers. Hottelmann (1969) came to a similar conclusion in a comparative survey carried out in Westfalen (GFR) by means of a computer and involving 100,000 finishing pigs. In practice, finishing pigs will therefore be moved once at the maximum : after they have been housed in a first pen from 20 to 25 kg up to a weight of 40 to 45 kg they are moved to a second pen for finishing. At a farrow-to-finish farm they remain in the weaner pen up to a weight of ca. 35 kg and are then accommodated in the finishing house where they remain until they have reached their final weight. In such a way moving is omitted but the pigs will possess too large a space in the initial phase. This is a waste of space which inevitably increases the building costs per pig place and can also lead to a fouling of the young pigs.

Noise, visitors, interference by dogs and rats etc. will all contribute in upsetting the pigs and must therefore be avoided.

At farms where only fattening of purchased weaners is carried out, the "all-in all-out" principle is generally applied : the finishing house is populated within a short time and the pigs are practically all at their final weight, ready to be slaughtered at the same time. An appraisal of this particular system can be summarized as follows :

Advantages

- the likelihood of infiltration of infections is smaller, since drivers and salesmen, who are in fact a great risk, come less frequently to the farm ;
- an outbreak of disease can be restricted to one group of animals ;
- it is possible to interrupt continuous infections on a farm whereby the infectious agent is not given a chance to manifest again in young and healthy animals ;
- certain preventive measures can be taken, for larger groups of animals and simultaneously, such as for the prevention of mange and worms. The animals are all treated via the drinking water as soon as they enter the house. A special container, connected to the water distribution system, is provided in each house by which medicaments are added to the drinking water ;

- entire animal houses and compartments can now be cleaned and disinfected at regular intervals with products which for safety reasons cannot otherwise be applied in populated houses (e.g. the dispersal of formalin) ;
- moving animals is restricted to a minimum and stress is hence avoided ;
- the climate in the house can be adapted to the age of the animals, which is beneficial for the health, growth and feed conversion of the animals ;
- the likelihood of making mistakes with feeding is minimized, the feed norms are easier to handle and the same feed is used throughout one section.

Disadvantages

- it is often difficult to purchase all the animals from one farm ; when weaners have to be purchased from several farms the likelihood of diseases increases ;
- the likelihood of diseases is also greater, since a larger group of young and vulnerable animals is now present in the house.

In conclusion we can state that the advantages, mainly of veterinary nature, largely outweigh the disadvantages.

There is no consensus on whether or not compartmentalizing and hence the all-in all-out system should be applied on the "farrow-to-finishing"-farms. The advantage of the continuous system mainly lies in the fact that the installation of expensive heating and ventilation equipment becomes unnecessary since young and older fattening pigs are accommodated together. Compartmentalized and heated finishing houses don't generally lead to better production results and incur higher production costs making them economically difficult to justify.

Concerning the *ventilation*, natural and mechanical ventilation can be distinguished. Finishing houses have to be amply ventilated whereby with natural ventilation often an adjustable, covered open ridge is chosen instead of stacks. A disadvantage however is the rather difficult daily fine setting of the ventilation related to the climatic conditions outside and this might result in daily temperature fluctuations. Recently developed thermostatically controlled hopper flaps can solve this problem. With mechanical ventilation the quantity of extracted air is controlled by a thermostat. A misconception exists about this system : farmers often think that it can proceed without supervision and this therefore often leads to less satisfying production results. The control apparatus does not always function without failure. Mechanical ventilation is basically calculated for summer conditions and the minimum winter ventilation quantities may therefore be too high. In small compartments mainly, this can result in too high an air speed at pig level whereby the costs of heating can rise considerably in houses where supplementary heating is provided. Granier (1978) compared natural ventilation (stack) with three different methods of mechanical ventilation viz. extraction-ventilation with respectively low and high extraction volumes and pressure ventilation. The comparison was carried out with restricted feeding. He found no differences except for pressure ventilation which gave less satisfactory results. Pressure ventilation is also to be discouraged since it leads to a considerably higher electricity consumption (Nilsson, 1981). Geers et al. (1984) made a survey in 137 pig fattening units concerning the influence of ventilation on the feed conversion. They found in this connection no significant differences between natural and mechanical ventilation. Satisfactory results can be obtained with both mechanical and natural ventilation, but mechanical ventilation requires higher installation and operational costs. There is much diversity in the electricity consumption figures mentioned by some authors concerning mechanical ventilation : they vary between 7.5 - 10 kWh (Debruyckere et al., 1982) and 18 kWh per delivered pig (Legters, 1982). Legters (1982) has calculated the total saving (i.e. the saving on both energy costs as well as housing costs) obtained with natural ventilation as compared to mechanical ventilation under Dutch conditions and reached a saving of ca. £ 1.25 per delivered pig.

The *feeding* of finishing pigs can be carried out ad libitum or otherwise rationed feeding can be applied at the trough or on the floor. Which system is chosen depends mainly on the breed. Hoorens et al.(1973) mentions from a research carried out in Flanders (Belgium) that a slight feed restriction is only advantageous for fat breeds of pigs but that it is unfavourable with the typical meat pigs or intermediary types. For both the last mentioned types of pigs, which predominate in Belgium, restricted feeding as compared to ad lib feeding, leads not only to a lower growth rate but also to a reduced slaughter quality due to a less favourable development of the so-called lean parts whereas the feed conversion remains the same in both cases. Restricted trough feeding leads to a prolonged fattening period and hence to a less efficient utilization of the pig house. Moreover it requires expensive installations for the mechanical distribution of the feed and is therefore economically not justified in Belgium.

Koomans and Mertens (1974) also mention the importance of inheritance for the possibility of ad lib feeding. In a research carried out in the Netherlands they obtained a lower slaughter quality with ad lib feeding than with rationed feeding : the fat thickness was 34.7 and 30.8 mm respectively, the percentage of Extra and 1st Grade pigs was 48 % and 74 % respectively and these differences were significant ($P < 0.05$). The Dutch Landrace pig appears therefore less suitable for ad lib feeding than the Belgian. Furthermore Hoorens et al. (1973) demonstrated that floor feeding as compared to trough feeding leads to a reduction of the growth of 6 % and 3.3 % respectively when meal and pellets were given. The feed conversion was 3.9 % and 1.5 % higher respectively with the administration of meal and pellets. Floor feeding has therefore been abandoned in Western-Europe. Concerning dry feeding versus wet feeding, Hoorens et al. (1973) found a better growth rate in the second case, and an equivalent feed conversion and slaughter quality with ad lib feeding, but he throws doubt upon his own conclusions. Wet feeding leads to a reduction of the dressing percentage which is due to an increased filling of the gastro-enteric tract and lower losses due to spillage and dust. The same authors (Hoorens et al., 1973) prefer concentrates in the form of pellets rather than meal because a number of advantages are attributed to the use of pellets : an increase of ca. 7 % in the growth rate, an amelioration of the feed conversion (about 9 %) and a reduction of the spillage. He finds that the advantages associated with the use of pellets outweigh the disadvantages viz. the higher cost of the pellets and the increased incidence of stomach ulcers.

Let us finally compare the influence of the *pen type* on the production results. In the Netherlands, Freriks (1979) found in a comparative research between an open littered house and a closed insulated Danish fattening house, both with ad lib feeding and with 98 pigs per house, an average feed conversion of 3.24 and 3.22 respectively and a daily growth of 866 g and 825 g respectively. The slaughter quality of the pigs from the open fronted house was however lower than that of the pigs from the closed house : the number of IA quotations was 20 % lower in the open house and the fat thickness increased from an average of 32.6 mm to 36.7 mm. Similar results were obtained by Koomans (1977). Taking into account the rather unfavourable results and the fact that the littered open house involves more labour and that from 500 animals and more, the investment becomes larger (concrete ways around the building for mechanical mucking-out) it is obvious that this type of housing cannot be recommended. Koomans (1977) also compared the littered open house with the partly slatted open house. The latter type appeared to lead to a lower growth rate but to a more favourable feed conversion and a better slaughter quality. The open house derived from this design and which can be closed with a flap during colder weather therefore has better prospects.

Eventual differences and the extent of the differences in the production results between the littered Danish house, the partly slatted house and the fully slatted house is yet another question which has to be answered. Klein-Hessling (1969) found in these three types of finishing pig houses a feed conversion of 3.46 (-), 3.63 (+5 %) and 3.85 (+11 %) respectively whilst the daily growth was 606 g (-), 593 g (-2.2 %) and 564 g (-7 %) respectively. Ober (1972) found a daily growth of 618 g (-), 607 g (-2.2 %) and 586 g (-5.2 %) respectively in the same three house types. Results published in the Yearbook of the Zootechnical Institute of Copenhagen, (Thomsen and Pedersen, 1969) mentioned, for the three types of pig houses, a feed conversion of 3.32 (-), 3.48 (+5 %) and 3.75 (+13 %) respectively and a daily growth of 586 g (-), 541 g (-8 %) and 522 g (-11 %) respectively. Both the feed conversion as well as the daily growth thus appeared to be better in a littered house than in a strawless partly slatted house and these in turn appeared to be better than in a fully slatted house. Other researchers (Bells et al., 1967 ; Danielson, 1968 ; Jones et al., 1967) however found no differences in the feed conversion and the growth rate in a comparison between the three different designs.

Our research (Maton et al., 1978), conducted with pigs fed ad lib up to a weight of 45 kg receiving afterwards rationed wet feed at a trough, confirmed the conclusions mentioned earlier viz. that partly slatted houses lead to significantly better production results than fully slatted houses. Recent research (Maton and Daelemans, 1984) carried out in fattening houses where exclusively ad lib feeding was applied, again confirmed the above-mentioned differences. In an experiment with five replicates and with each time 12 pigs, all brothers and sisters, in a partly slatted pen and a fully slatted pen, a significantly better feed conversion was obtained in the partly slatted pen viz. 3,065 versus 3,196 (-131 g/kg) while the daily growth was also better (+11 g) although not significantly.

About 55 % of the fattening houses built in Belgium in the period 1979 - 1980 (Anon., 1981) were of the fully slatted type. In 1980 - 1981 this percentage decreased to 28 % (Anon., 1982a). Hence an important reduction in the number of fully slatted houses can be observed. The principal reason why a number of farmers still choose fully slatted houses can probably be found in the fact that, mainly with higher temperatures and hence in the summer, pigs housed in partly slatted houses are often seriously fouled because they then prefer to lie on the slats. In order to avoid this disadvantage a number of intermediate solutions are sometimes introduced (the installation of a short second slatted area along the outer wall of the partly lidded pen or a fully slatted house with a large sized solid plate behind the trough).

Cannibalism (tail biting) is a serious problem with the strawless housing of pigs. Outbreaks are spontaneous, sometimes severe and causes can be different. Haaring (1976) investigated 92 farms and observed tail biting in 50 % of the partly slatted houses and in 80 % of the fully slatted houses. Installing a play-chain and docking of the tails might contribute to an improvement. Darkening the house (background light through air in- and outlets) has also a beneficial influence on this particular problem (Van Putten, 1980). As the stocking density increases the incidence of tail biting will become greater (Lindqvist, 1974). The regular administering of a handful of straw in half slatted houses also leads, according to Van Putten (1980), to an important improvement. It keeps the pigs busy for 1.5 hours per day (an increase of 100 %) while the number of conflict-situations and bullying, which indicate a lesser welfare, are reduced to one third.

In summary, the partly slatted house with a 50 % lying area and provided with a top layer of volcanic stones (Bernit, Stallit) offers more advantages than the fully slatted house. Naturally ventilated and unheated group pig houses (i.e. the type of housing not practising the all-in all-out principle) on "farrow-to-finishing"-farms can lead to equally economically satisfactory results as the mechanically ventilated and heated houses. Automation of the natural ventilation can then be recommended. For farms which produce fattening pigs from purchased weaners, the same conclusions are valid for ventilation and heating but here the all-in all-out principle should be the general rule.

5.10 THE HOUSING OF WILD BOARS

The wild boar, which was widely found in many European countries, was considered as "harmful" and has nowadays largely disappeared from the forests.

The high commercial value of its meat has led to the creation of wild boar farms where the animals are bred in confinement (in special houses) or partially in the wild.

The *breeding of wild boars in a restricted natural environment* requires a large fenced, mainly wooded area, (300 ha per 100 adult animals) which has to be provided with a 1.20 m high wire fence. This fence is completed with a double barbed wire installed at 10 cm above the ground which must prevent burrowing. It is recommended to build some littered shelters at dry places and to provide a few fodder places. It must be possible to feed the piglets separately and this can be carried out by providing a few pop holes allowing only access for piglets to their fodder place (Juhel, 1981).

The *breeding of wild boars in confinement* takes place in houses of which the open side gives access to an outside run (30 m^2 per animal) provided with a wire fence as described above. As for their tame relatives, the house is also adapted to the stage of reproduction : houses for dry and pregnant sows and for boars, farrowing pens, weaner and rearing pens. Dry and pregnant sows are kept in batches of 10 with one boar in a collective pen which is closed at two sides and connects to an outside run of 360 m^2 . It is recommended to have more outside runs than strictly necessary and each of the runs can in turn be left unused. The individual farrowing pens are covered and insulated constructions (walls made of straw stacks, protected by a strong wiremesh with small meshes) of 2 m x 2 m, with a concrete floor and giving access to a covered concrete apron in the open air with similar dimensions and equipped with a feed area for the sows and a feed hopper for the piglets. The abovementioned apron is in turn connected to two outside runs of 3 m x 10 m which are alternately used for half a month. The sows remain in the individual farrowing pens until the piglets have reached the age of one month. After this period, batches of 3 to 5 sows, together with their litters, are moved to a collective pen, which can be similar to the house for pregnant sows, and hence includes an outside run with the same dimensions. Separate feeding places for piglets and individual sow feeding places are desirable. Weaning takes place at the age of 3 months, whereby the piglets remain in the weaner house and the sows are transferred to a house for dry and pregnant sows.

REFERENCES

Anon., 1981. Landbouw- en tuinbouwtelling op 15 mei 1980, Nationaal Instituut voor de Statistiek, Brussel, Belgium, 221 pp.

Anon., 1982a. Landbouw- en tuinbouwtelling op 15 mei 1981, Nationaal Instituut voor de Statistiek, Brussel, Belgium, 255 pp.

Anon., 1982b. Wel of geen verwarming in volledig roostervloerstallen, Bedrijfsontwikkeling, 13 : 913-914.

Aumaître A., 1983. Le confort de la truie face aux nécessités d'un élevage rationnel, L'élevage Porcin, nr. 133, pp. 43-48.

Backstrom L., 1973. Environment and animal health in piglet production, Acta. Vet. Scand. Suppl., 41, 240 pp.

Baxter S.H., 1974. Movable heat pads for young pigs, Farm Buildings Progress, 38 : 19-22.

Báxter S., 1984. Intensive Pig Production, Environmental Management and Design. Granada Technical Books, London, G. Britain, 588 pp.

Bekaert H. and Daelemans J., 1970. De drinkwatervoorziening bij zuigende biggen, Landbouwtijdschrift, 23 : 925-939.

Bells E.S., Neil Marshall E.Mc., Stanley J. and Thomas H.R., 1967. Studies of slatted-floor swine housing in controlled, semicontrolled and uncontrolled environments, Transactions ASAE, 10 : 561-563.

Blendl H., 1971. Haltungsverfahren in der Ferkelerzeugung, Landtechnik, 26 : 231-236.

Boon C.R., 1981. The effect of departures from lower critical temperature on the group postural behaviour of pigs, Animal Prod. 33 : 71-79.

Bruce J.M. and Clark J.J., 1979. Models of heat production and critical temperature for growing pigs, Animal Production, 28 : 353-369.

Castryck F., 1981. De invloed van de omgevingsfaktoren op de gezondheidstoestand van het varken, Licenciaatsthesis, Fakulteit Diergeneeskunde, Rijksuniversiteit Gent, Belgium, 97 pp.

Daelemans J., 1975. Een zoötechnisch kontrolesysteem voor de zeugenhouderij, Landbouwtijdschrift, 28 : 99-116.

Daelemans J., 1978. De 3W-zeugenkalender, De Belgische Veefokkerij, 32 : 14-15.

Daelemans J., 1984. Foktechnische zeugenboekhouding met computer, Rijksstation voor Landbouwtechniek, Merelbeke, Belgium, 19 pp.

Daelemans J., Martens L. and Maton A., 1980. De investeringen in de mestvarkensstal en zijn uitrusting, Landbouwtijdschrift, 33 : 477-505.

Danielson M., 1968. Slatted floors for growing finishing pigs. Nebraska swine progress report, 129, 10 pp.

Debruyckere M. and Martens L., 1980. De rendabiliteit van de verdere mechanisering in de zeugenhouderij, Landbouwtijdschrift, 33 : 457-475.

Debruyckere M., Van Laken J., Christiaens J. and Van Der Biest W., 1982. Klimaatregeling in stallen, Ventilatie en Verwarming, Vereniging der Elektriciteitsbedrijven in België, Brussel, Belgium, 115 pp.

England D.C. and Spurr D.T., 1969. Litter size of swine confined during gestation, Journal of Animal Science, 28 : 220-223.

Freriks J.H., 1979. Nieuw staltype voor mestvarkens, Consulentschap in Algemene Dienst voor Boerderijbouw en -inrichting, Wageningen, the Netherlands, 5 pp.

Geers R., Goedseels V., Berckmans D., Wijnhoven J. and Merckx J., 1984. Feed conversion ratio of pigs in relation to season and the engineering and control of the pig house environment, Proceedings C.I.G.R., Budapest, Hungary, Sept. 3-7, 1984, Tome 2 : 130-134.

Glende P., Klatt G. and Richter H., 1973. Zweckmässige Möglichkeiten der Haltung guster und tragender Sauen zur Intensivierung der Produktion, Tierzucht, 27 : 223-227.

Granier R., 1978. La climatisation des porcheries, Station expérimentale, Institut Technique du Porc, Rapport annuel, pp. 29-55.

Gravås L., 1977. The effect of the floor on behaviour and knee damages of piglets and sows, C.I.G.R.-Section II, Seminar on Agricultural Buildings, As, Norway 8-12 August, pp. 85-100.

Haaring H., 1976. Stalklimaat en gezondheid op varkensmestbedrijven, Doktoraalskriptie Gezondheidsdienst Vakgroep Veehouderij, Landbouwhogeschool, Wageningen, the Netherlands.

Hacker R.R., King G.J. and Bears W.H., 1974. Abstract nr.76 : Effect of complete darkness on growth and reproduction in gilts, Journal of Animal Science, 39, pp. 155.

Hemsworth P.H., Winfield C.G. and Chamley W.A., 1981. The influence of the presence of the female on the sexual behaviour and plasma testosterone levels of the mature male pig, Anim. Prod., 32 : 61-65.

Hoorens J., Debruyckere M., De Moor A., Maton A., Oyaert W., Pensaert M., Vandeplassche H. and Vanschoubroek F., 1973. Huisvesting, voeding en ziekten van het varken, Story Scientia Uitg., Gent, Belgium, 496 pp.

Hottelmann F., 1969. Hunderdtausend Schweine im Komputer, Landw. Wochenblatt Westfalen, W. Germany, nr. 6, pp. 26-29.

Jones H., Conrad J. and Kadlec J., 1967. A comparison of swine finishing houses, Research Progress Report, 134, Purdue University, Sept.1967.

Juhel L., 1981. L'élevage rationnel du sanglier, L'élevage porcin, Juin, nr. 107, pp. 47-52.

Kirkwood R.N. and Hughes P.E., 1980. A note on the efficacy of continuous v. limited boar exposure on puberty attainment in the gilt, Animal Production, 31 : 205-207.

Klatt G., Richter H. and Glende P., 1972. Zweckmässige Lösungen für die Schweinehaltung, ein Beitrag für die Rationalisierung und die industriemässige Produktion, Tierzucht, 26 : 146-149.

Klein-Hessling P., 1969. Bericht über Untersuchungen im Versuchsgut der Landwirtschaftkammer Westfalen, Landw. Wochenblatt Westfalen, W. Germany, nr. 6, pp. 23-25.

Koomans P., 1977. Varkens mesten in goedkope stallen, Landbouwmechanisatie, 28 : 1001-1003.

Koomans P. and Mertens J.A.M., 1972. Proeven met vastgebonden zeugen, Mededelingen Instituut voor Landbouwbedrijfsgebouwen, nr. 50, 4 pp.

Koomans P. and Mertens J., 1974. Beperkt en onbeperkt voederen van varkens in verband met de huisvesting, Mededelingen Instituut voor Mechanisatie, Arbeid en Gebouwen, Wageningen, the Netherlands, nr. 5, 27 pp.

Kuiper C.J. and Sturm J.M.J., 1975. Anaphrodisie bij gelten en zeugen, Tijdschrift Diergeneeskunde, 100 : 824-835.

Legters J.W., 1982. Natuurlijke of mechanische ventilatie in mestvarkensstallen, Bedrijfsontwikkeling, 13 : 740-742.

Lindqvist J.O., 1974. Animal health and environment in the production of fattening pigs, Agria. Vet. Scand. Supplementum, 51 : 1-78.

Maes F., 1982. Uitval bij zeugen, De Boer en de Tuinder, 88, nr. 19, pp. 15-17.

Mateman G., Brandsma H.A. and Verstegen M.W.A., 1982a. Temperatuur en voerniveau in relatie tot slachtkwaliteit van mestvarkens, Bedrijfsontwikkeling, 13 : 259-266.

Mateman G., Brandsma H.A. and Verstegen M.W.A., 1982b. Meer voeren in de koude betekent een betere voederkonversie, Bedrijfsontwikkeling, 13 : 993-997.

Maton A. and Daelemans J., 1974. Zoötechnische aspekten van de huisvesting van kweekvarkens, Landbouwtijdschrift, 27 : 1-19.

Maton A., Daelemans J. and Lambrecht J., 1978. De bevloering van mestvarkenshokken : haar technische eigenschappen en zoötechnische effekten, Landbouwtijdschrift, 31 : 309-318.

Maton A. and Daelemans J., 1984. De bevloering van vleesvarkensstallen, Landbouwtijdschrift, 37 : 1035-1049.

Maton A., Daelemans J. and Hoorens J., 1978. De invloed van de huisvestingswijze op zool- en carpusletsels bij zuigende biggen, Vlaams Diergeneeskundig Tijdschrift, 47 : 16-21.

Mennerich H., 1972. Weide, Auslauf, ganzjährige Stallhaltung für tragende Sauen, Der Tierzüchter, 24 : 408-409.

Mitchell D. and Smith W.J., 1978. Piglet foot dimensions for design of slotted floors, Farm Buildings Progress, 51 : 7-9.

Nilsson N.C., 1981. Control and design of mechanical ventilation systems, Communications of the C.I.G.R.-Section II, Seminar : Modelling Design and Evaluation of Agricultural Buildings, Aberdeen, Scotland, pp. 143-147.

Ober J., 1972. Die Mastleistung im Schweinestall, Bauen auf dem Lande, 23 : 307-309.

Papp J., 1984. The effect of group size, pen shape and space allowance on fattening performance of pigs, Proceedings C.I.G.R., Budapest, Hungary, Sept 3-7, 1984, Tome 2 : 104-108.

Paterson A.M. and Lindsay D.R., 1980. Induction of Puberty in Gilts, Anim. Prod., 31 : 291-297.

Pay M.C. and Davies T.E., 1973. Growth, food consumption and litter production of female pigs mated at puberty and at low body weights, Animal Production, 17 : 85-91.

Pechert H., 1970. Verhaltensforschung bei Schweinen, Bauen auf dem Lande, 21 : 235-237.

Petersen S. and Nielsen K., 1977. Flokstørrelsens indfly delse pä slagtesvins produktionsevne, Statens Byggeforsknings Institut, Landbrugsbyggeri, Denmark, nr. 49, 11 pp.

Ridgeon R.F., 1982. Pig Management Scheme Results for Agricultural Economics Unit, Department of Land Economy, Cambridge, G. Britain, Economic Report nr. 91, 36 pp.

Salehar A., 1964. Belegen der Sauen nach dem Absetzen der Ferkel, Ve Congr. Intern. Riprod. Anim. Fecond. Artif., Trento IV, 441.

Salmon-Legagneur E., 1970. Etude de quelques facteurs de variations de l'âge et du poids des truies au premier oestrus, Journées Rech. Porcine en France, Paris, France, pp. 41-46.

Thiede G., 1983. Regionale Konzentration der Schweinehaltung in der EG, Deutsche Geflügelwirtschaft und Schweineproduktion, 35 : 1154-1155.

Thomas P., 1972. Factors affecting sow fertility, Agriculture, 79 : 395-398.

Thomsen R.N. and Pedersen O.K., 1969. De sammenlignede forsøg med svin fra stasanerkendte avlscentre, In : Afdelingen for forsøg med svin, Yearbook of the Zootechnical Institute of Copenhagen, Denmark, pp. 85-106

Tielen M.J.M., 1974. De frekwentie en de zoötechnische preventie van long- en leveraandoeningen bij varkens, Mededelingen Landbouwhogeschool Wageningen, the Netherlands, nr. 7, 141 pp.

Tuinte J.H.G., 1971. Uitval van zeugen bij verschillende huisvestingsmethodes, Maandblad voor de Varkenshouderij, 33 : 201-203.

Van Putten G., 1980. Analyse und Vorbeugen des Schwanzbeissens beim Mastschwein, Deutsche tierärztliche Wochenschrift, 77 : 134-135.

Chapter 6

THE HOUSING OF POULTRY

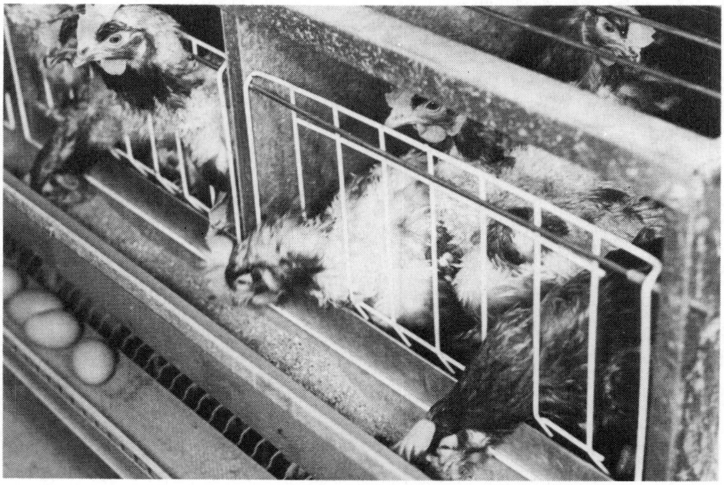

Chapter 6

THE HOUSING OF POULTRY

6.1 GENERALITIES

As in previous chapters it is useful to give briefly some information of a general nature which may be important for the housing of poultry.

Poultry is generally kept for the production of eggs (layers, geese) or for the production of meat (broilers, turkeys, guinea fowls, slaughter pigeons, geese, ducks) or as a pastime (ornamental fowls, pigeons). Egg farming and broiler production are by far the major branches of poultry production.

Through intensive selection, based on scientific research, a number of so-called lines in the layer have been developed. Eggs originating from valuable parents are hatched in an incubator and the chicks are delivered by the hatcheries as day-old chicks to the poultry keeper. The point-of-lay is reached after five months and on average 275 eggs are laid during the first year (with a weight of 55 to 65 g). After 12 to 15 months of laying the layer is slaughtered (spent fowl) ; she weighs ca. 2 kg and has consumed ca. 50 kg of feed. With low egg prices the layers are sometimes artificially brought to moulting (by fasting) and kept for a second laying period of 6 to 10 months.

Chicks intended for broilers are kept for ca. 8 weeks, they weigh after this period ca. 1.6 kg and have consumed ca. 3.5 kg of feed. This period is more and more shortened to 6 weeks whereby the animals then weigh ca. 1.3 kg and have consumed ca. 2.5 kg of meal.

Turkey-hens are either kept as small frying turkeys, which weigh 3 to 4 kg after 10 to 12 weeks and which are mainly sold as Christmas turkeys or exported live (especially to Germany), or as medium frying turkeys of 6 to 7 kg after ca. 15 weeks, and either sold as turkey-parts or used as Christmas turkeys. Turkey-cocks are kept until 18 - 20 weeks of age, weighing 10 - 12 kg and are sold in parts.

The guinea fowl originates from Central-Africa, where it still lives in the wild, it was imported into Western-Europe by Portuguese seafarers and especially by the legions of Julius Caesar. The guinea fowl is kept for 10 to 13 weeks of age until a final liveweight of 1.0 to 1.4 kg. The feed conversion amounts to 3.0 to 3.5. It gives a high meat-efficiency viz. ca. 80 % compared to only 65 % for broilers and the low-fat meat (3 to 4 %) has a "half-wild" taste. Guinea fowls are therefore considered as a luxury product.

Pheasants, quails, pigeons, ducks and geese are also intensively bred.

In the past poultry breeding was generally carried out at the farmyard. Since outside runs have become superfluous and grains have been replaced by high quality concentrates, the production of eggs and poultry can now take place away from the traditional farmyard.

Intensive housing of poultry in large units for the production of eggs and meat no longer needs to be part of a farm. Since the 1950's large specialized units for the production of eggs and poultry have developed.

6.2 THE HOUSING OF LAYERS

6.2.1 The construction and equipment of laying houses

Houses for laying hens are fully closed low profile houses or hangars built of insulated, lightweight construction materials (fig. 6.1). In recent years prefabricated elements have often been applied for the construction of trusses, walls and roofs.

Fig. 6.1 A modern laying house.

The following types of houses can be distinguished :

- the deep-litter houses ;
- the slatted floor houses ;
- the cage houses, incl. deep pit houses.

The first mentioned type of laying house is the oldest : in the 1950's they formed the transitional stage between extensive layer keeping (roosting- and laying house with outside run) and intensive layer keeping with the omission of the outside run. In the Sixties the different types of slatted floor houses knew a certain popularity but they were soon substituted by the cage houses. The building activity in the last decade in the field of laying houses was rather limited owing to an economic crisis caused by an overproduction of eggs. Practically all new built laying houses were equipped with cages and accommodate in general a minimum of 10,000 birds. We will however also describe in detail the deep-litter and slatted floor houses for layers. On one hand, there are still a number of (older) houses of this type in Western Europe. On the other hand this type of house is still irreplaceable for the keeping of breeding hens (layers, broilers, etc.).

6.2.1.1 The deep-litter houses

Description

The deep-litter house consists of a large exercise area, provided with peat, wood shavings, sawdust, chopped straw or another littering depending largely on what is locally available, and a roosting area over the droppings pit. The roosting area, where the flocks rest during the night, is equipped with perches at a height of 0.8 to 1.0 m above the ground. The ability to sleep on perches has to be acquired. The perches are 5 cm wide and 1.5 cm thick ; they are 25 cm to 33 cm apart (centre to centre) and mounted on top of a wire-mesh netting (2.5 cm x 5 cm). Most of the droppings fall through the netting into the pit. The area for roosting measures 1 m^2 per 18 birds. The litter layer which is about 0.3 to 0.4 m thick must be kept dry and turning of the litter may be necessary. Its depth must be maintained, if necessary by adding some new litter. The stocking density is 5 to 6 birds per m^2. The house is generally not cleaned out during the laying period. In earlier days it was customary to provide the birds with an exercise area but in really intensive poultry keeping this method has been abandoned completely.

The equipment of the deep-litter house is nowadays largely directed to labour time reduction and labour relief. Taking care of the birds is almost completely mechanized. The deep-litter house is preferably provided with a central service passage when egg collection is done by hand (fig. 6.2). If egg collection belts are installed this central service passage can be omitted.

The feeding systems

Two main feeding systems are currently in use : the feed dishes or feed pans and the feed trough. The feed pans can be filled either by a conveyor or an auger while the feed trough is normally filled by a trolley or a chain with flat links.

- feed pans filled by means of a cable conveyor or tube feeder : the hopper which supplies the conveyor with meal is either filled by means of an auger or is placed below the outlet-pipe of the feed silo whereby it is gravity-filled with meal (fig. 6.3). An endless chain, driven by an electric motor and provided with disc-shaped carriers, which fit exactly in a tube, is pulled through this tube. The chain carries the meal to and around the poultry house. At the underside of the horizontal tube a number of vertical tubes connect with the feed pans. The disc-shaped carriers take the feed from the hopper, through a carry-up tube to a horizontal tube from which the meal drops in the first feed pan.

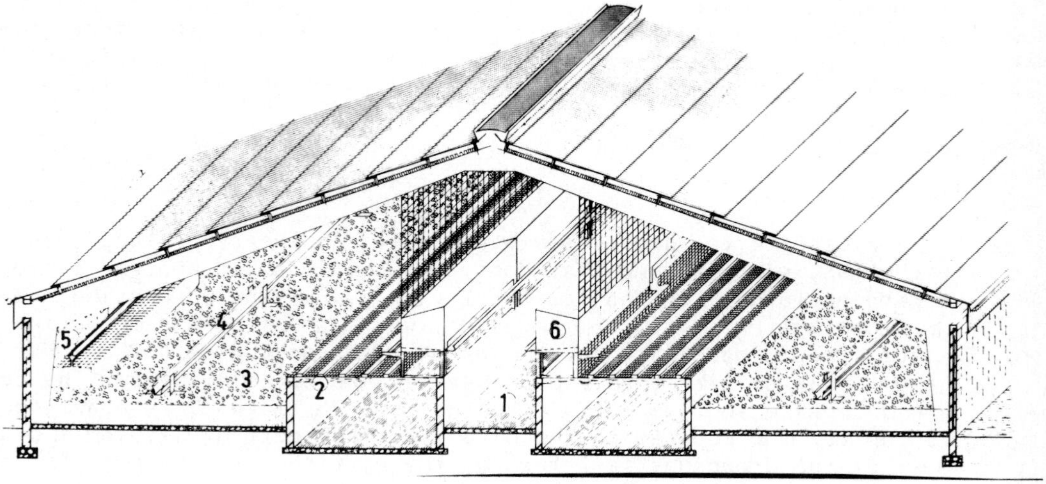

Fig. 6.2 The deep-litter house with central service passage. Legend : 1 = central passage ; 2 = perches on a wire-mesh netting above the droppings pit ; 3 = built-up litter ; 4 = feed trough ; 5 = waterer above spillage box ; 6 = laying nests.

When the first feed pan and its vertical pipe are completely filled (up to the connection with the horizontal tube) the feed will be conveyed to the second feed pan until this is also completely filled, etc. As soon as all feed pans are filled, the meal is returned to the hopper. At this moment the electric motor is switched off by means of an automatic or hand operated switch.

- feed pans filled by means of an auger : the feed pans or dishes are placed in one or several rows, above which an auger is installed (fig. 6.4). The complete installation is often suspended from the ceiling in order to facilitate height adjustment related to rearing ; it is also advantageous for cleaning the poultry house. At the front of the tube a hopper and a drive mechanism with an electric motor are provided. The hopper is automatically filled with meal by means of an auger. The feed dishes or pans are aligned along one row and are connected to the hopper by means of a tube in which the auger is installed. The tube has a number of outlets through which the feed can be distributed to the feed pans. The feed pans are filled in turn. At the end the tube is provided with a feed pan which is suspended in such a way that when it is full, it will stop the motor by means of a switch. With this feed system all pans are filled, one after the other, by means of an auger, the motor is switched off until more feed is required.

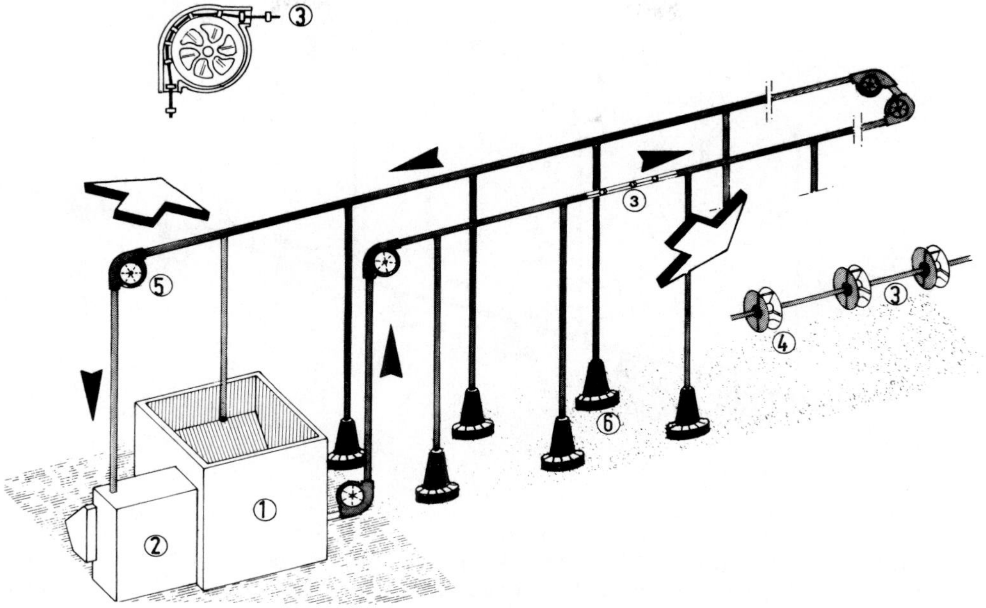

Fig. 6.3 The cable conveyor feeder.
Legend : 1 = hopper ; 2 = drive unit ; 3 = endless chain ; 4 = disc-shaped carriers ; 5 = corner wheel ; 6 = feed pan or dish.

- feed trough filled by trolley (fig. 6.5) : a feed trough is installed along the entire length of the poultry house. Two rails, capable of carrying the feed trolley are placed next to the feed trough. The trolley consists of a hopper, mounted on wheels, which travels over the rails. Another possibility is that the edges of the feed trough are reinforced and that the feed trolley rides on the feed trough itself. The feed trolley is equipped with two outlet pipes which allow the filling of two parallel feed troughs with one pass. Less feed will flow out as the pipes debouch closer to the bottom of the trough. The adjustment of the feed quantity can easily be carried out either by lowering or raising the feed trolley with respect to the wheels, which run on the rails or by making the outlet pipes telescopic. The feed trolley is filled either by moving it under the silo or by means of an auger which transports the feed from the silo to the trolley. The trolley is mostly moved by means of an electric motor. Limit switches will automatically stop the feed trolley at the end of the feed trough. A stirring mechanism can be driven by the wheels and assures a smooth and regular flow of meal.

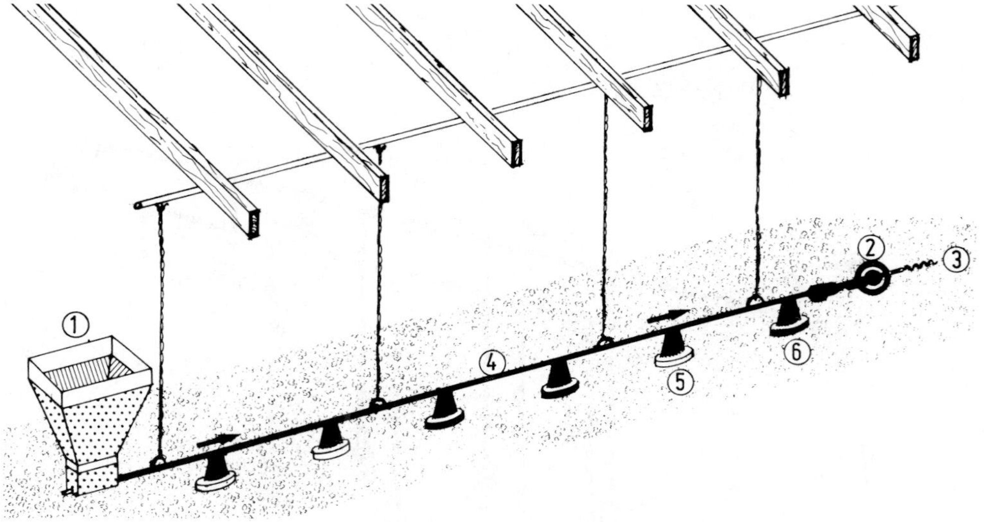

Fig. 6.4 A feeding system employing an auger.
Legend : 1 = hopper ; 2 = drive unit ; 3 = auger ; 4 = tube in which the auger is installed ; 5 = feed pan ; 6 = feed pan with switching mechanism.

- feed trough filled by a chain with carriers (fig. 6.6) : an endless feed trough is installed in the poultry house and runs under and through a funnel-shaped meal container. This hopper is preferably situated under the outlet of the silo and is automatically and continuously filled by gravity, although it can also be filled by means of an auger. The chain is provided with special flat links which transport the feed through the feed trough as soon as it is activated by the electric motor. The carriers differ according to the brand. This chain can be activated regularly by means of a timer and has the advantage of stimulating the feed consumption by its sound. Furthermore a "just-around-stop" switch can restrict the movement of the chain to one complete round. The feed trough is normally mounted at a height of 20 cm above the exercise area. A feeding space of 10 cm per bird is provided in order to allow all the birds to feed simultaneously which is highly desirable. If both sides of the feed trough can be used, only 1 m of feed length is provided per 20 birds. Since feathers and other impurities can befoul the feed during its crossing, it is sometimes advisable to include a feed cleaner at the end of the chain. This cleaner consists of a turning screen through which the feed passes ; the rotary movement sieves the meal and it is collected in the feed trough. The impurities are removed by the slope of the mechanism.

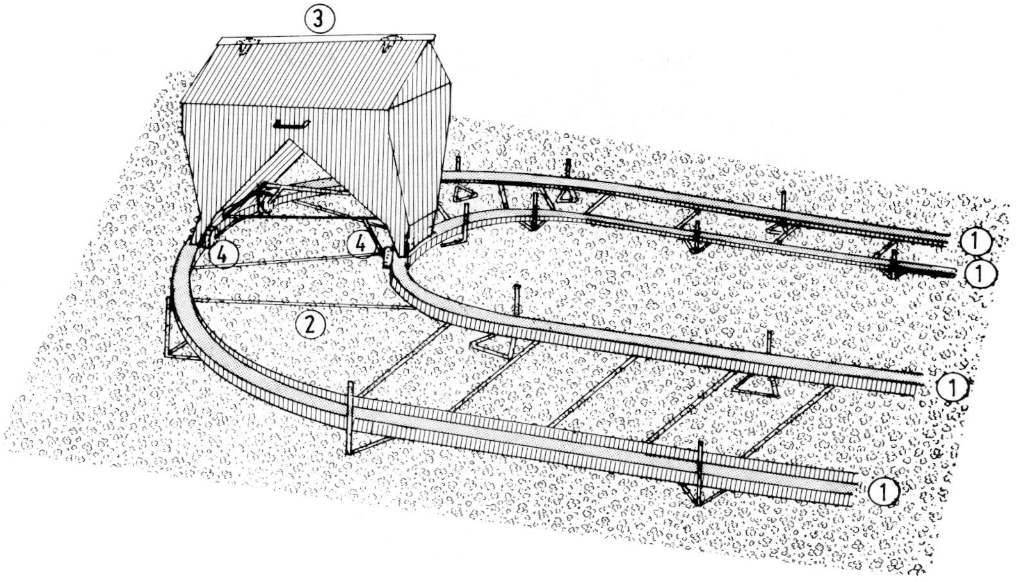

Fig. 6.5 The feed trolley.
Legend : 1 = feed trough ; 2 = curve ; 3 = trolley ; 4 = outlets of the trolley.

Watering devices

The drinking water distribution is mostly carried out automatically. The poultry drinkers are usually installed above the droppings pit or above a platform equipped with a sewer which prevents the litter from being wetted by spilled water. The water troughs are preferably reachable from both sides. All currently used systems are based on two principles viz. : a water trough or a suspended drinking pan, both equipped with a mechanism for maintaining a constant level :

- the water trough with weight adjustment (fig. 6.7) : water troughs of ca. 2 m long are placed at regular distances from each other, a constant level is maintained by means of a weight-controlled mechanism. Sometimes one continuous water trough is applied but this is usually unnecessarily long. The length of the water trough is reckoned at 2.5 cm per bird and if the water trough is reachable from both sides, 1 m of trough is satisfactory for 80 layers.

- the suspended round drinkers : fig. 6.8 gives a view of the arrangement with round suspended drinkers. A float-controlled reservoir with a content of ca. 80 l is connected to the main and supplies water through a horizontal pipe to a row of round suspended drinkers. This pipe is attached to a cable or a rail by means of a number of

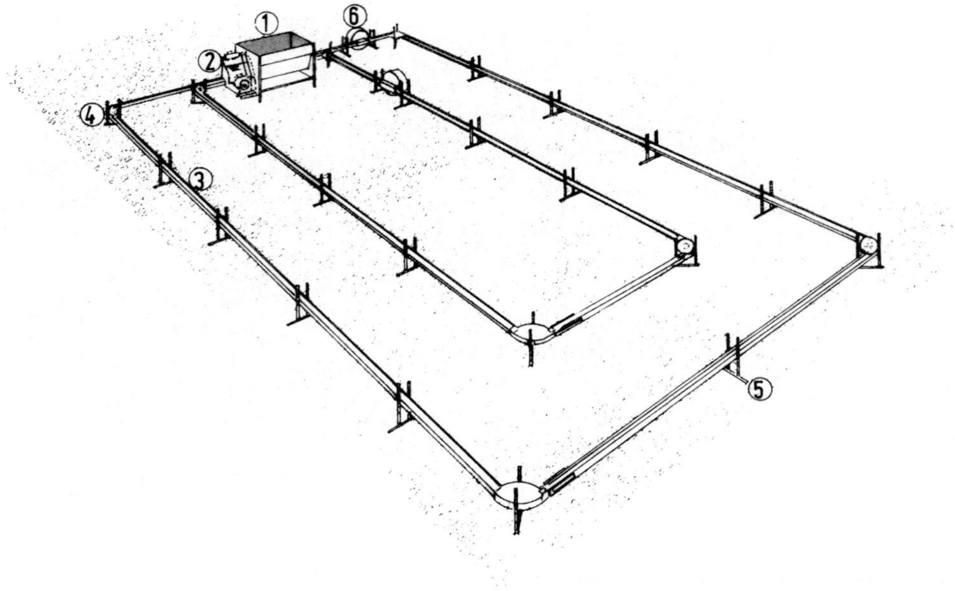

Fig. 6.6 The feed chain.
Legend : 1 = hopper ; 2 = drive unit ; 3 = feed trough ; 4 = curves ; 5 = supports for the height adjustment of the trough with accessory equipment ; 6 = feed cleaner.

clips and the cable or rail itself is suspended from the ceiling. The drinking pans are fed by the pipeline through vertical plastic tubes. The round drinkers are at regular distances from each other suspended from the cable or the rail by a wire. The different parts of the round drinker are represented in fig. 6.8.

The drinking water is supplied by the main (high-pressure) to the float reservoir where water becomes available at a low pressure and is supplied to the various drinkers via a pipeline. The water enters the drinker through a long tube provided with a filter and runs then through a short tube equipped with a double acting valve. This valve shuts-off the supply of water, firstly when sufficient water is available in the round water trough and secondly when the drinking pan is removed for cleaning. The water, after passing the double acting valve, runs down along the outer surface of the pan into a water gutter from which the birds take water. Different sizes of drinkers are commercially available according to the size of the birds. The water height is controlled by the double acting valve which is itself activated by a pressure spring and the weight of the pan and can be adjusted by an adjustment screw and a set screw. The drinker

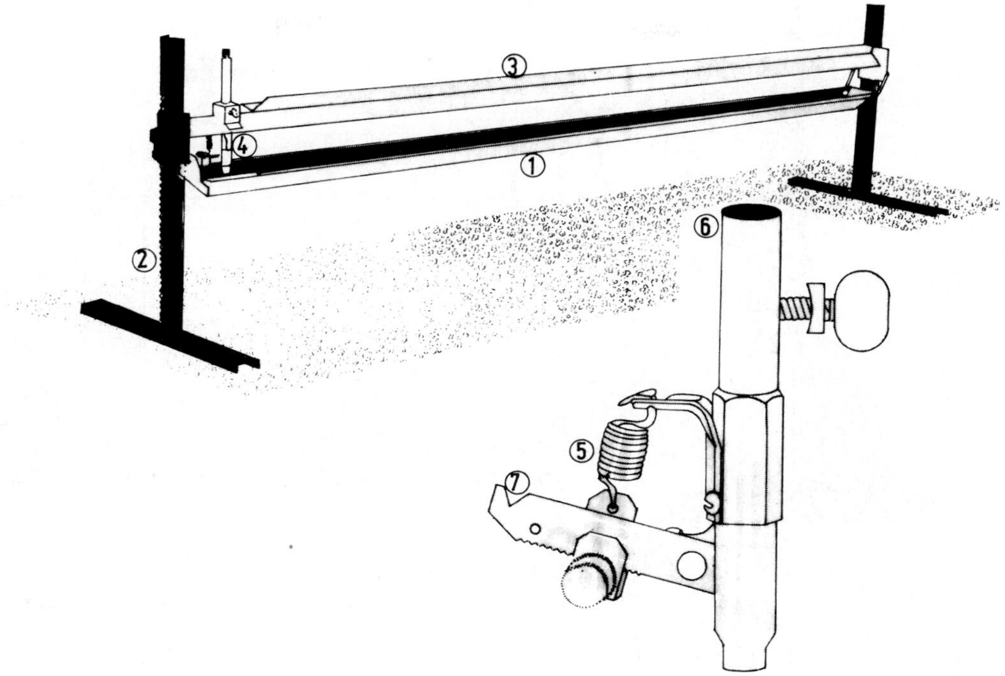

Fig. 6.7 The water trough with weight adjustment.
Legend : 1 = water trough ; 2 = supports with height control ; 3 = non-roost wire or rotating lath ; 4 = float-mechanism ; 5 = spring ; 6 = water supply duct ; 7 = notch for keeping the water trough in place.

is provided with stabilization baffles to minimize swinging.

The cleaning out

The litter must be loose and so occasional stirring may be needed. The litter can be kept relatively dry by providing suitable ventilation, but should not be too dry in order to avoid dust problems. If necessary additional litter may be added on the exercise area.

Cleaning out consists of removing the manure from the droppings pit and the litter from the exercise area. Cleaning out is only done after the laying period.

The laying nests

Birds and thus laying hens lay their eggs instinctively in nests.

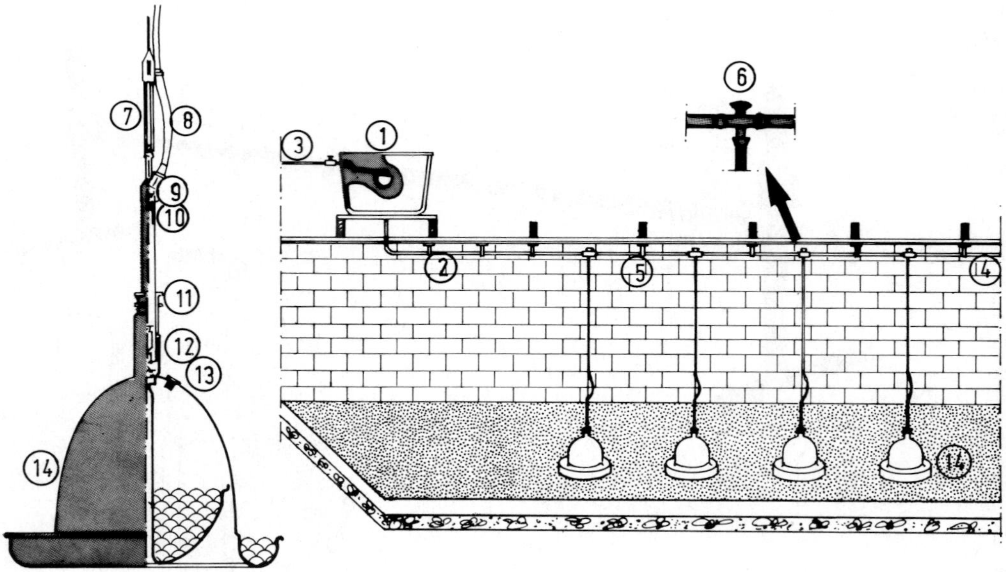

Fig. 6.8 The drinking system with round drinkers.
Legend : 1 = break pressure reservoir with float ; 2 = low pressure pipe ; 3 = main ; 4 = free end of the low pressure pipe for adjustment of the pressure height ; 5 = fixing clips ; 6 = T-connections ; 7 = suspending cable for height adjustment ; 8 = supply tube to drinker ; 9 = mounting cap ; 10 = filter ; 11 = adjustment screw + set screw ; 12 = double-acting valve ; 13 = filling hole for inside water reservoir (stabilization baffles) ; 14 = round water trough.

Consequently houses for layers have to be provided with nests. Their form and location are important to stimulate the hens to lay the eggs in the nests. In practice, eggs are sometimes laid on the floor of the laying house and are therefore called "floor eggs". Floor eggs appear frequently in deep litter houses. Their number normally amounts to ca. 2 % although figures between 10 % and 30 % may be found. Dark nests are preferred because the hen likes seclusion and a raised area for laying. Dark nests also reduce the likelihood of egg eating. It is therefore not recommended to locate the nests in front of a window or directly on the ground. Hens roosting in the nests should be removed in the evening since eggs may then be fouled. The nests should also not be placed too high. Kraggerud (1963) has observed that when layers have the choice between nests at 0.5 m, 1 m and 1.5 m of height, approximately an equal number of hens will deposit their eggs in the nests at 0.5 m and 1 m high but remarkably less in the highest nests. The last only represented ca. 60 to 70 % of the number counted in the middle nest.

Eggs can either be collected by hand or mechanically and the latter system is increasingly applied. According to the kind of nests we can distinguish : individual nests and community nests either without or with a roll-away system.

Individual laying nests

- Ordinary single nests
  The space in an individual nest is restricted so that only one hen can occupy the laying nest. Usually nests are stacked in 3 to 4 tiers along one or more walls of the poultry house (fig. 6.9).

Fig. 6.9 Individual laying nests along the walls of the poultry house.

  The floors of the laying nests are littered with oat hulls, chaff, short straw, shavings, ground corn cobs, shredded corn stalks, peat moss or sawdust. Twenty-five individual nests are provided per 100 hens if no roll-away nests are installed. In order to reduce breakage and dirty eggs and to avoid excessive accumulation it is recommended to collect the eggs several times a day. The eggs can then be cooled earlier after laying and are stored in a room at a temperature of 10 to 12°C and with a relative humidity of 75 %. The eggs are stored large end up to prevent displacement of the yolk towards the shell which might encourage deterioration or decay.
- Individual laying nests with a roll-away system for the eggs.
  Individual nests with a roll-away system are provided with a sloping bottom made of wire-mesh. The sloping floor of the nest is installed in such a way that the eggs, as soon as they are laid, will roll to a gathering area at the side of the nest and out of reach of the hens.

In recent years the floors of the nests have often been provided with plastic nest trays which feel softer and are more attractive to the hens. Twenty individual nests with roll-away system are usually provided per 100 hens. Eggs can be collected by hand from a central service passage. Nests of this type can also be mounted in such a way that the eggs roll directly on a jute belt. This jute belt, driven by an electric motor, conveys the eggs to a collection table once or twice a day.

Community nests

Community nests can either be of the ordinary type or can be equipped with a roll-away system. Community nests (or box-nests) consist of a rectangular box, the floor area of which amounts to 1.2 m² per 100 hens if no egg roll-away system is provided and to 1 m² if the roll-away system is installed. If such a system is applied an egg conveyor belt can be included.
Table 6.1 lists the most important specifications for the equipment of deep litter houses.

6.2.1.2 The slatted floor houses

This type of house (figs 6.10 and 6.11) differs from the previous in that the floor is part or fully slatted with a grating of laths (2.5 cm thick, 3 cm wide, with a slot of 2.7 to 3 cm) or with a wire floor

Fig. 6.10 The fully slatted floor house for laying hens. Note the feeding system.

with meshes (1" x 2", 2.5 cm x 5 cm), in the roosting area, 5 cm wide laths are placed every 25 to 33 cm which function as roosts. A central service passage with roll-away nests at both sides is recommended (figs 6.10 and 6.11).

Table 6.1 lists the most important specifications for the slatted floor houses.

TABLE 6.1 Summary of the most important specifications of deep litter houses and slatted floor houses with central service passage.

House type	Deep litter house		Slatted floor house	
Number of birds per house	2,500	5,000	5,000	10,000
Number of birds per m² of floor area	5	5	8	8
Required floor area in m²	500	1,000	625	1,250
Width of the central service passage in m	1.2	1.2	1.2	1.2
Width of the front hall in m	4	4	4	4
Trough length per bird in cm	5	5	5	5
Number of feed pans	100	200	200	400
Water trough length in cm per bird	1.25	1.25	1.25	1.25
Number of birds per m² of roosting area	18	18	18	18
Number of nests per 100 birds				
- individual ordinary nests	25	25	25	25
- individual roll-away nests	20	20	20	20
Floor area of nests per 100 birds in m²				
- ordinary community nests	1.2	1.2	1.2	1.2
- community nests with roll-away system	1	1	1	1
With a house width of 12 m				
- total house length in m	50.3	96.6	61.9	119.7
- total floor area in m²/bird	0.24	0.23	0.15	0.14
- total house area in m²	604	1,159	743	1,436
With a house width of 16 m				
- total house length in m	37.8	71.6	46.2	88.5
- total floor area in m²/bird	0.24	0.23	0.15	0.14
- total house area in m²	605	1,146	740	1,415

6.2.1.3 The cage houses

The use of wire cages for the housing of laying hens has become standard practice in recent years. Here the laying hens are restricted to the cage and are not allowed to run free contrary to the abovementioned houses. The group of cages, each connected to others, is referred to as a battery (fig. 6.12). The batteries are placed in rows between which are service passages each with a width of 0.8 to 1.20 m. The cages are either supported by or suspended from a steel frame and the bottom of the lowest row of cages is at a height of 0.30 to 0.50 m above the floor.

A cage measures normally 40 to 50 cm in the front, is 45 cm "deep"

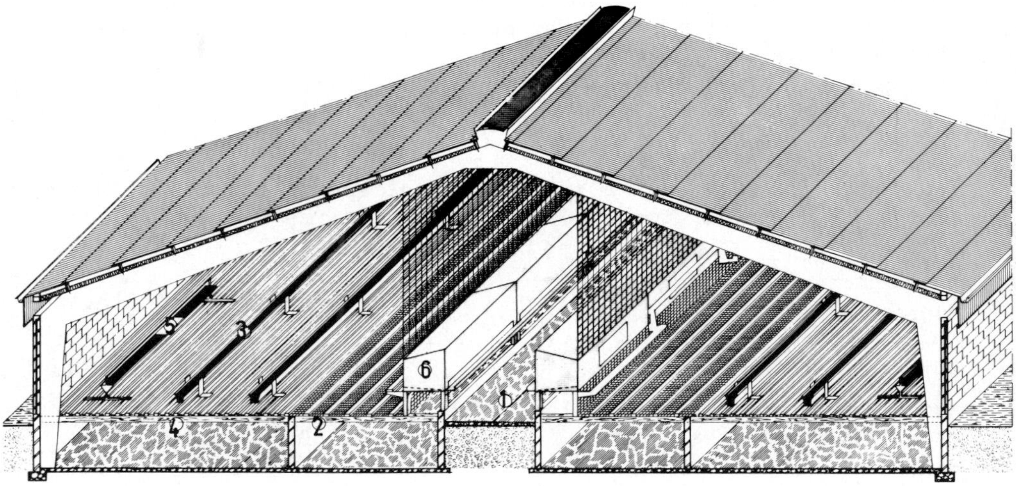

Fig. 6.11 The fully slatted floor house with central passage. Legend : 1 = central service passage ; 2 = roosts on a wire-mesh above the droppings pit ; 3 = feed trough ; 4 = grating of laths above the droppings pit ; 5 = water trough ; 6 = laying nests.

Fig. 6.12 The cages of a battery for laying hens.

and ca. 45 cm high and houses 4 to 5 laying hens. Each laying hen has a feeder length of 10 cm and a minimum area of 450 cm^2 at her disposal. A number of groups for animal protection and welfare are putting heavy pressure on the E.E.C.-Governments to adapt the cages in order to improve the welfare of the birds (Anon., 1975a). This particular problem is discussed in detail in "Zootechnical and veterinary aspects of the housing of layers".

The cages are either fully fabricated from galvanized wire or sometimes consist of side walls made of solid galvanized metal plate or perforated plastic sheet. Wire of 2.4 mm thickness was often used in the past but for economic reasons it is now mostly 1.8 mm thick. The wires usually form rectangular openings of 2 to 2.5 cm wide and 5 to 7 cm long. The bottom is sloping to allow the eggs to roll away and extends forwards to bring the eggs out of reach of the laying hens. The end of the wire-floor is bent upwards to retain the rolling eggs. The front of the cage is made of special wirework which allows the birds to stick their head through and to take feed from the feed trough. The wirework normally consists of 6 to 8 rectangles of galvanized metal wire each having a width of 6 to 7 cm. Each rectangle is reinforced with a horizontal wire. The purpose of this horizontal wire is to prevent the loosening of the spotwelded connections. The front side is indeed subjected to the pressure exercised by the birds when taking their feed. The frontage is removable in order to put birds in or out. Cage houses can be divided into three main groups depending on their model :

- tier cages ;
- Californian cages incl. deep-pit cages ;
- one tier cages or flat deck.

This last type is rarely built when erecting new houses and has been gradually superseded by the two other types of batteries. Compared to the three tier battery the Californian battery has the distinct advantage of allowing an easier removal of the manure (e.g. deep-pit). The fact that one can start with a half-mechanized and hence cheaper Californian battery, which can be mechanized later is one attraction for this type of battery.

6.2.1.3.1 Description of the various types of batteries

Tier battery

The cages are installed in two rows back-to-back. Two or more rows are placed above each other. According to the number of double rows of cages which are placed one above the other we can distinguish the 2-, 3-, 4- and even 5-tier battery. The three-tier battery (fig. 6.13), whereby an already high density can be noted, (15 to 20 birds/m^2) was the most popular type of battery in the Sixties. Fig. 6.14 shows the construction and the most important dimensions of this often sold three-tier battery, in which each cage accommodates four hens.

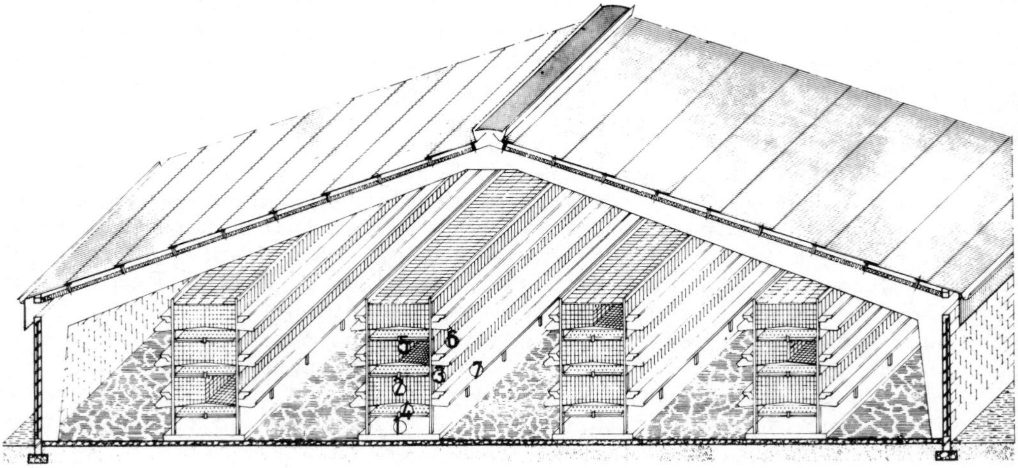

Fig. 6.13 The four-row three-tier battery house.
Legend : 1 = battery ; 2 = sides of the cages ; 3 = frontage of the cages ; 4 = manure platform ; 5 = nipples ; 6 = feed trough ; 7 = egg roll-away area.

The Californian battery

The ordinary Californian battery consists of six rows of cages of which the upper two rows are mounted back-to-back, while the four lower rows are situated apart, forming a staircase-like structure (fig. 6.15). The cages have the same dimensions as those of a tier battery and are supported by a steel frame. The manure of all the laying hens accumulates directly on the floor of the house, in a liquid manure channel or in a cellar underneath, hence deep pit (see further). Fig. 6.16 shows the construction and the most important dimensions of an ordinary Californian battery with three tiers.

In recent years a number of other forms of batteries have been derived from the common Californian battery viz. : the pyramid battery (figs 6.17 and 6.18) and the compact Californian battery (figs 6.19 and 6.20). The idea behind these constructions is to increase the density or the number of hens per m^2 of floor area. With the pyramid battery the cages are in fact moved away from each other and the droppings of each bird fall directly on the floor. With the compact battery the rear sides of the lower - and with a three-tier battery also those of the middle rows of cages - are brought close to each other (at a distance of 20 to 30 cm) (fig. 6.19).

In order to prevent droppings of some birds falling on the birds beneath, platforms are installed under the upper and the middle rows of cages.

The removal of the manure from those platforms, is carried out by means of a scraper which is mounted on the feed trolley and which makes the droppings fall into a manure channel behind the cages.

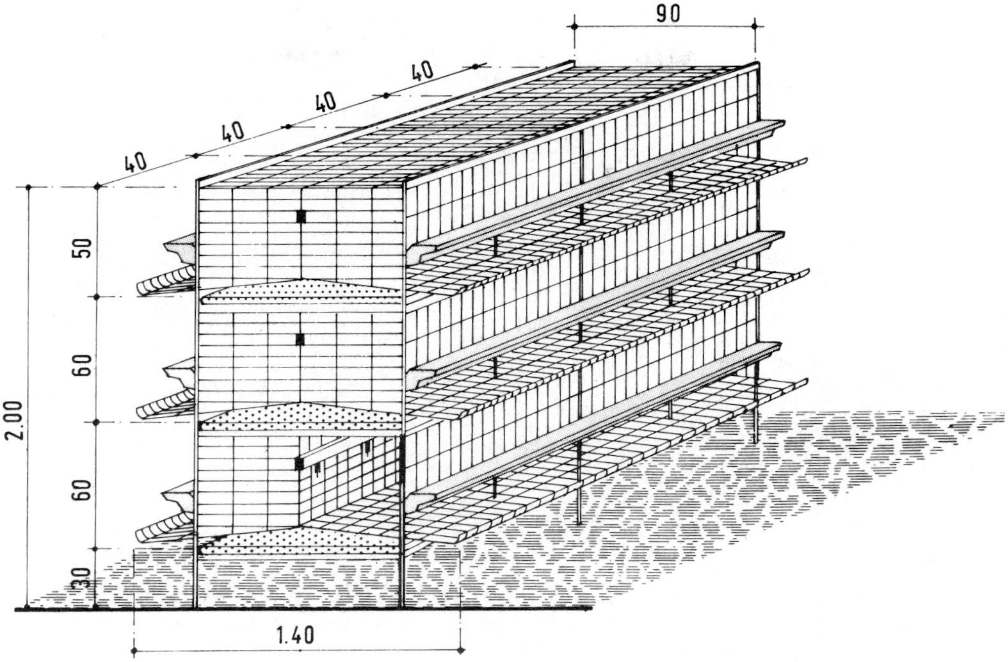

Fig. 6.14 Two elements of a three-tier battery.

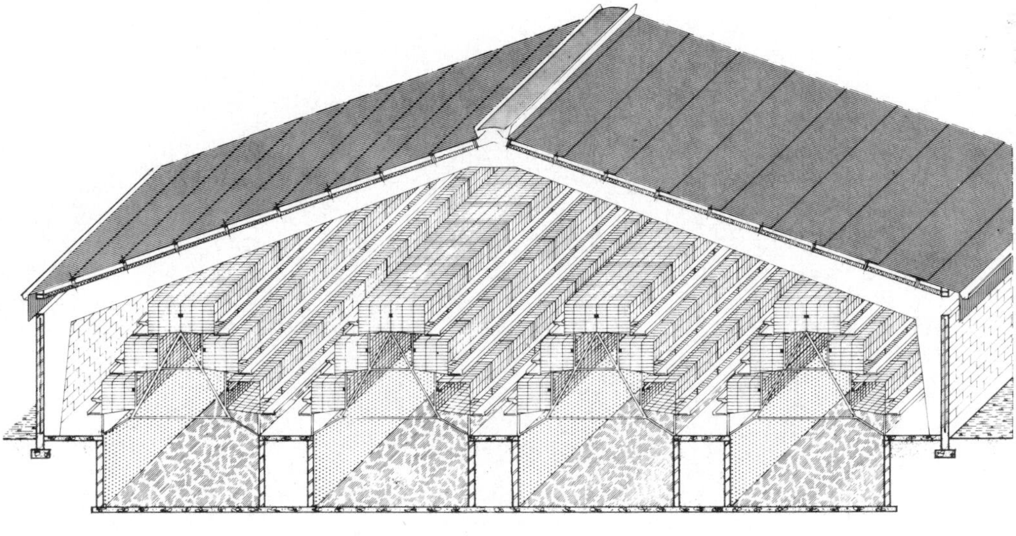

Fig. 6.15 The four-row Californian battery house with three tiers.

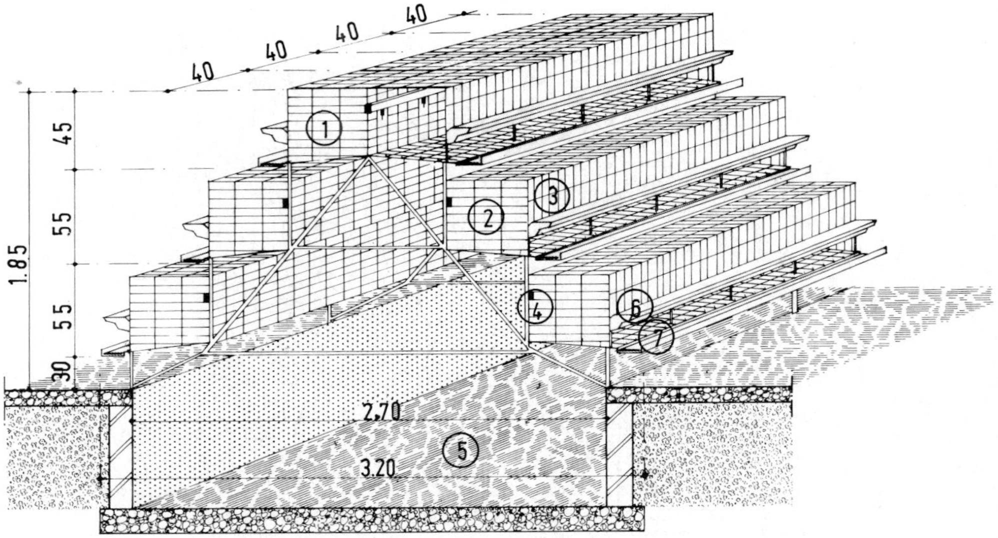

Fig. 6.16 The common Californian battery with three tiers.
Legend : 1 = Californian battery ; 2 = sides of the cages ; 3 = frontages of the cages ; 4 = nipples ; 5 = liquid manure channel ; 6 = feed trough ; 7 = egg roll-away system.

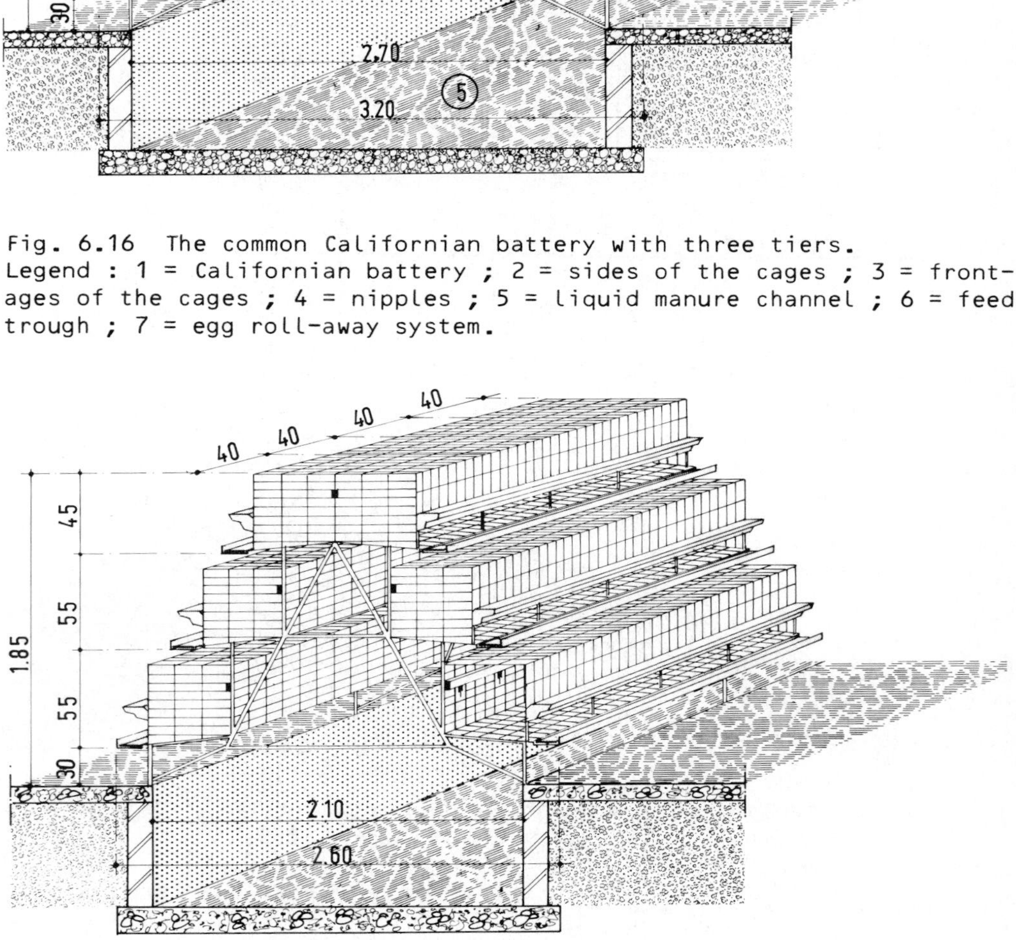

Fig. 6.17 The pyramid battery with three tiers.

Fig. 6.18 Poultry house for laying hens, equipped with three-tier pyramid batteries.

Fig. 6.19 The compact Californian battery with three tiers.

Fig. 6.20 A laying house with three-tier compact batteries.

Four-tier Californian batteries are nowadays popular.

One-tier battery or flat deck

The flat deck battery consists of four horizontally arranged rows of cages each next to the other and supported by a steel frame. The bottom of the cages is ca. 60 cm above the floor. The batteries are separated by passages. Each battery is split up in a number of sections each containing 4 rows of 6 cages or a total of 24 cages per section. Each cage is 40 cm long and 45 cm "deep" and accommodates 4 hens. In this way a stock density of 12 hens per m^2 is obtained. A section can for instance have a total width of 205 cm and a length of 240 cm (= 6 x 40 cm). Two openings are provided in the battery : one between the first and the second row and another one between the third and the fourth row. Each opening has a width of 12.5 cm and holds, from top to bottom, a water trough, a feed trough with feed chain and an egg conveying belt. Flat deck batteries are nowadays seldom applied because of their low stocking density which leads to high building costs. Another disadvantage is that eggs cannot be collected by hand in case of a breakdown (fig. 6.21).

Table 6.2 summarizes the main specifications for the equipment of the various types of battery houses.

6.2.1.3.2 The feeding systems in battery houses

The feed troughs are situated outside the cages over the entire length of the row of cages (fig. 6.12). They are manufactured from strong and inflexible metal plate and are preferably V-shaped to enable

TABLE 6.2 Summary of the most important specifications of battery houses for laying hens.

Type of battery	Three-tier and compact three-tier Californian battery		Pyramid-three tier Californian battery	
Width of the house in metres	15		15	
Number of hens per house	14,400	21,600	11,520	17,280
Number of hens per cage	4	4	4	4
Cage-width in cm per hen	10	10	10	10
Cage-width in cm	40	40	40	40
Number of battery-rows	5	5	4	4
Number of hens per length of battery	2,880	4,320	2,880	4,320
Length of the battery in metres	48	72	48	72
Required length of the house in m	51	75	51	75
Required house area in m^2	765	1,125	765	1,125
Number of hens per m^2	18.8	19.2	15.1	15.4
Width of the front hall in m	4	4	4	4
Total length of the house in m	55	79	55	79
Total area of the house in m^2	825	1,185	825	1,185
Length of the feed trough, in m	1,440	2,160	1,152	1,728
Length of the roll-away system in m	1,440	2,160	1,152	1,728

the debeaked hens to eat easily. V-shaped feed troughs also have the distinct advantage of a lower feed spillage compared to the deep and open feed troughs. Feeding is done twice a day by means of a feed trolley or by a chain or auger feeder.

The feed trolley (fig. 6.22) is installed in such a manner that it can fill the feed troughs of a complete battery (2 x 3 rows of cages for instance) over its entire length in only one pass. The feed trolley is therefore equipped with the same number of hoppers as there are feed troughs. Each hopper empties through a pipe into the feed trough. By adjusting the height of the outlet pipe the feed quantity delivered in the feed trough can be controlled. The different hoppers on the feed trolley are filled from the silo by means of an auger, mounted in a horizontal tube, which dumps the feed in a number of vertical pipes attached to it. The trolley is mostly driven by an electric motor in both directions in order to allow the automation of both the feed distribution as well as the filling. Limit switches automatically stop the feed trolley at the end of the feed trough. The chain and auger feeders are also important feeding systems (fig. 6.23). The chain or auger travels through the feed trough ; their construction and working are similar to the ones described in the deep litter house (p. 324, 326).

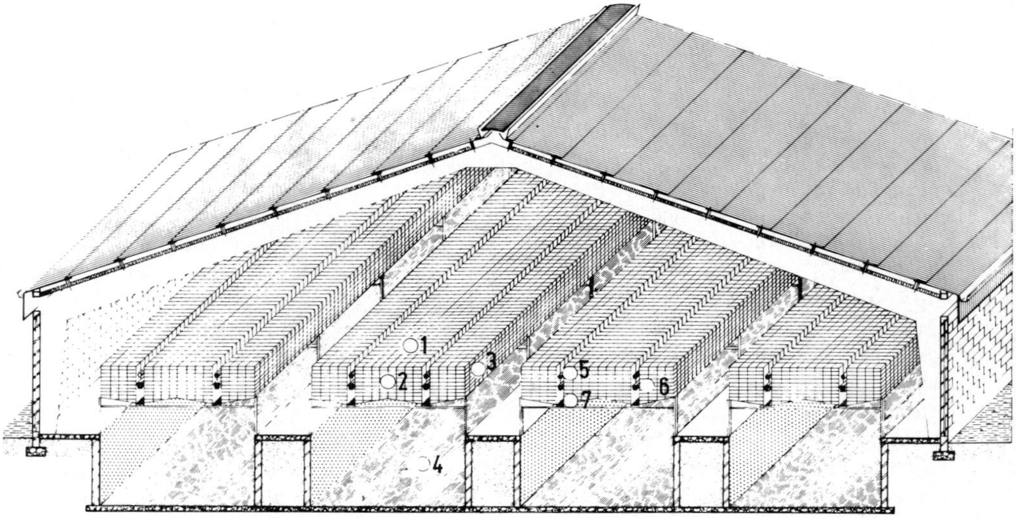

Fig. 6.21 A poultry house equipped with four-row flat deck batteries. Legend : 1 = battery ; 2 = sides of the cages ; 3 = frontages of the cages ; 4 = liquid manure channel ; 5 = water trough ; 6 = feed trough ; 7 = egg conveyor belt.

Fig. 6.22 The feed trolley for a three-tier Californian battery.

Fig. 6.23 The chain feeder in a battery house.

6.2.1.3.3 The drinking devices in battery houses

The water trough

The water trough is usually filled via a float-system as described for other types of houses. The water trough can be placed at the outer side of each row of cages, preferably above the feed trough and along the entire length of the battery. Another possibility, where applicable, is the installation of a common water trough per tier, in the middle of the rows of cages. Such a water trough can then be cleaned by pulling a cylindrical scraper through it. The scraper should fit in the water trough and can be moved by a cable and winch system. A central water trough has the advantage that spilled water accumulates on the manure platform and never in the feed trough.

The nipple drinker

Nipple drinkers are nowadays generally used (fig. 6.24). Each cage has its own nipple drinker.

When a hen pushes the nipple drinker, the valve is opened and water can flow into her beak. As soon as she releases the valve, it closes

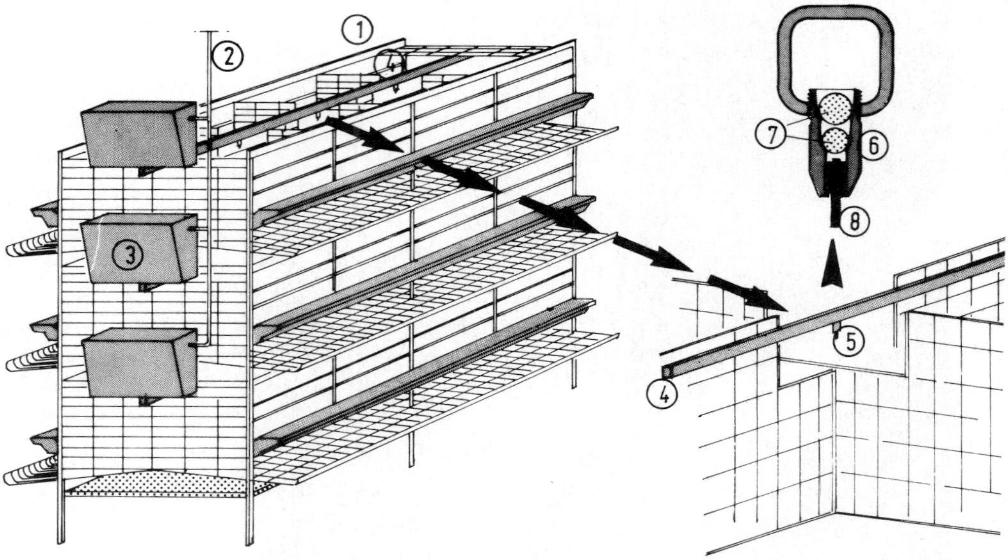

Fig. 6.24 The nipple drinker system.
Legend : 1 = three tier battery ; 2 = water supply duct ; 3 = float-box with float ; 4 = low pressure pipe with nipple drinkers ; 5 = nipple drinker ; 6 = nipple body ; 7 = second valve (double ball) ; 8 = first valve with valve-stem.

by gravity and the water supply is shut off. The nipple drinkers are installed in such a way that they are reachable from two or four cages according to the type of battery. Hens from one cage can therefore always reach two nipple drinkers. A break pressure reservoir with float control is necessary since the drinkers are designed to work from a low pressure pipe. The piping is made of non-translucent tubing to prevent the growth of algae.

A number of advantages are attributed to the nipple drinkers. Unlike the water trough, nipple drinkers always provide clean water, the likelihood of infections is minimized since the transfer via a common water trough is avoided, the cleaning labour can be omitted, cannibalism occurs less frequently and the consistency of the manure is better. Wetting of the feed is excluded because the nipple drinkers are not installed above the feed trough. Several investigations have clearly demonstrated that the feed consumption per egg is lower when nipple drinkers are provided than with the application of a water trough (Langeveld, 1970).

6.2.1.3.4 Egg collection

Nests are not provided. The hens lay their eggs on the wire floor of the cage in which they are accommodated (fig. 6.12). The sloping wire mesh floor allows the eggs to roll out of the cage and they are then stopped by the bent-over wire floor. The width of the collecting "gutter" which is created in this way has to be at least 20 cm to facilitate collection of the eggs by hand. The eggs accumulate in a row in front of the cage and it is recommended to provide some kind of protection to the eggs whereby it becomes impossible for the hens to touch or eat them once they have rolled out of the cage. This can easily be done by installing a metal plate in front of the cage. The wire mesh bottom of the cage under the feed trough must be regularly brushed off in order to prevent the formation of dust-rings on the eggs. This is often done by a nylon brush mounted on the feed trolley. Jute or plastic belts are common nowadays (fig. 6.12). They are ca. 10 cm wide and are fitted in the lowest part of the sloping wire floor. The eggs roll away straight onto the conveying belt. Once or twice a day the egg collecting belt is activated by an electric motor and the eggs are transferred to an egg collection table or to a collecting and packing machine (fig. 6.25) by means of a transveyor (fig. 6.26). This machine can also sort the eggs according to their weight.

Fig. 6.25 The egg collecting and packing machine.

6.2.1.3.5 The cleaning-out systems

Different types of mucking-out systems are currently in use.

Fig. 6.26 The egg collecting belts deliver the eggs to a transveyor.

The manure scraper

Tier batteries are provided with a manure platform, made of glass or asbestos cement, underneath each tier of cages. The manure falls through the wire mesh floor of the cage and accumulates on the platform beneath the cages. At regular times (every two or three hours) the manure scrapers of one battery are put into action by means of a timer. They are driven by an electric motor together with a winch and cable system. The manure is collected and deposited in a manure pit or a channel at one end of the battery. The scrapers are lifted to return to their initial position.

Unlike this system, the manure from all rows of cages and hence of the battery from the common and pyramid Californian batteries as well as from the flat deck battery accumulates directly in a manure channel beneath the batteries. With compact Californian batteries, the scraper is mounted slantwise on the feed trolley. It pushes the manure away from the platform into a manure channel situated in the middle beneath the battery. The manure is removed from this channel by another scraper (fig. 6.27) which brings it to a manure cellar or transversal channel :

- direct evacuation : with this system the scraper directs the manure straight to the manure cellar, water is added if necessary ;
- mechanical evacuation : a mechanical manure removal system is installed in a channel at the end and across the house (endless chain or automatic dung channel cleaner of the push-type), it may be provided with an elevator belt outside the house and the manure is stored in an above-ground silo.

Fig. 6.27 The manure scraper underneath a battery.

If the manure is to be stored in a manure cellar its content is rated at 0.18 to 0.20 l per hen and per day. A storage of 1 m^3 will suffice for 55 birds and for a period of 3 months. The manure is lifted, transported and spread by means of a slurry tanker.

In the last decade the "dry-manure system" has gained popularity. The manure is left for ca. 3 weeks on the platform or belt and dries by means of ample ventilation of the house (final moisture content = 60 - 65 %). Sometimes, air is blown through a perforated duct installed above or at both sides of the manure platform of each tier. Each manure platform is equipped with a special scraper which pushes the manure per section (adjustable between 10 and 80 cm) towards one end of the house and has thus a forward and backward movement. In this case the scraper will be less susceptible to wear. If the manure has accumulated on a belt it can be evacuated by the belt itself. The "dry" manure is easier to stack and to handle and has a less offensive odour than fresh poultry manure.

The belt cleaner

With tier batteries, the manure of each tier falls on a pre-stretched nylon or plastic belt beneath the cages. A winch equipped with a non-movable scraper is therefore installed at one end of the battery. The plastic belt is regularly turned around the drum of the winch by means of an electric motor (e.g. once a week). The fixed scraper removes the manure from the belt during its turning. Breakdowns sometimes occur with this type of cleaning-out system. The manure is later removed from the house by one of the above-mentioned methods. Either a single or double belt can be used. In the first case the belt has to return to its original position after the manure has been removed. Sometimes one motor is used to turn all the belts of one battery. Removal of the manure from the house can be carried out as mentioned before.

The deep pit system

This cleaning-out method forms the foundation of a new type of housing. This house consists of two parts viz. : the house itself and the deep pit (figs 6.28, 6.29, 6.30). The latter is ca. 2.5 m deep, is mostly built completely above-ground and is provided with a concrete floor. The house itself contains several rows of Californian batteries between which passages of ca. 80 cm wide are installed. Similar passages are also provided along the walls. In this way a building is obtained which is about 4 to 4.5 m high at the outside measured from ground level to the eaves. The walls are often made of cellular concrete and are ca. 20 cm thick. The saddle roof consists of corrugated asbestos cement sheets which are insulated at the inside. The manure produced by the hens accumulates in the deep pit. This pit has a great depth first of all to allow the storage of the manure during a number of years (5 to 6 years) and secondly because mucking-out has to be carried out by tractor and front loader.

Fig. 6.28 The deep pit of a laying house equipped with Californian batteries.

Fig. 6.29 Ventilation slots in one of the longitudinal walls of the deep pit.

Fig. 6.30 Double door in the gable-wall of a deep pit poultry house. Note the access ramp leading to this door and enabling the loading of the eggs on a truck.

The basic idea behind all this is to obtain a dry manure which is easy to handle. This can be carried out by passing air over the manure either by natural ventilation e.g. through a number of slots in the longitudinal walls of the deep pit or by artificial ventilation by means of perforated plastic ducts installed beneath the passages between the batteries. The duct has two openings with a diameter of 1 cm, every 50 cm, which are directed towards the manure. It is capped at one of its ends whilst the other end is connected to a ventilator which blows air into the duct. These ducts are suspended from the ceiling by a number of ropes which can be pulled up so that the duct is always ca. 50 cm above the manure. The house itself is mostly provided with natural ventilation whereby the air enters through an adjustable curtain and is extracted via an adjustable open ridge. Besides the continuous withdrawal of moisture by the natural ventilation the manure is also dried by a microbiological fermentation which causes the temperature to rise to ca. 35°C. Spillage of drinking water on the manure has to be avoided. Spillage gutters are therefore provided beneath the nipple drinkers to evacuate the spilled water. The volume of the manure thereby remains practically constant. The final product is reasonably dry (> 50 % D.M.) and is easy to handle (Hoogerkamp, 1974).

A high and wide door in the rear wall allows mucking-out with tractor and front loader, when the deep pit has to be emptied after 5 to 6 years.

No important odour development takes place during the drying process, the storage or the spreading of the manure. The deep pit house has obtained a good reputation with respect to the environment. On the other hand there are some disadvantages connected with this particular type of house viz. the higher capital lay-out, problems with flies and the less attractive view of such a high type of house (it is necessary to install stairs at the entrance of the battery house and a double door in the gable-wall). The latter allows the loading of cartoned eggs on a truck which is driven up an access ramp in front of the double door. The deep pit house, which has been popular in Great Britain for a long time (Anon., 1975b) has also found its way on the Continent.

6.2.1.4 Alternatives for the laying batteries

In the European Community the keeping of laying hens in batteries is increasingly opposed by the public (see further). This has even led to the introduction of bills in some E.E.C.-countries leading to a ban on the housing of laying hens in battery cages. All these problems have led to the development of new types of battery cages and alternative forms for housing laying hens. Although most of these developments are still in the experimental stage it is useful to describe them briefly.

Improving existing types of batteries

In order to meet the demands for an increased welfare of the hen in the existing housing systems the cages can be made larger or less hens can be kept in a cage. The use of more suitable materials might lead to a reduction of the number of lesions. The cages could be made large enough to include some perches for the birds. A further evolution has led to the development of a totally new type of cage, the so-called get-away cage.

The get-away cage

The get-away cage was originally a cage measuring 1 m x 1 m x 0.7 m, provided with perches, laying nests and sand boxes and intended for 20 birds (Kuit, 1983). Fig. 6.31 represents a get-away cage of 1 m x 0.65 m x 0.80 m for 16 to 20 hens (Wegner, 1980) with two tiers. A number of alternative cages have been developed from this original type, such as a cage of 2.0 m x 0.7 m x 0.55 m, installed in three tiers which allows a better supervision of the birds (Kuit, 1983).

The storeyed house

A traditional littered house accommodates 5 to 6 birds per m^2. By constructing such a littered house with several storeys e.g. three, it is possible to increase the stocking density threefold. The IMAG in Wageningen developed such a storeyed house of which the cross section is given in fig. 6.32.

A litter-layer is applied in the house and several rows of storeys are constructed each having a width of ca. 1.70 m and equipped with perches, feed troughs and nipples. One row of storeys comprises the laying nests. A small passageway of ca. 0.65 m wide is included between the rows and allows the poultry keeper access for supervision.
A manure conveying belt is provided per storey and per row. The eggs are collected with a belt and conveyed to the egg collection station.

A number of problems remain unsolved with this type of housing viz.:
- the distribution of the hens over the entire house and over the different storeys ;
- floor eggs ;
- hygiene ;
- cannibalism ;
- coccidiosis and helminths due to the presence of the litter.

The results obtained with this new type of housing were not encouraging and until now it has not found practical application.

6.2.1.5 The lighting

Lighting can be carried out either artificially or combined with natural lighting. The first system assumes a windowless building whereby heat losses through the walls are reduced to a minimum and a strict application of the light pattern is possible. On the contrary, it is not always possible to apply certain lighting patterns in houses with windows or a ventilation curtain. As a matter of fact one cannot shorten the natural day length, but, a lengthening of the day remains possible with the use of artificial lighting.

The lamps are automatically switched on and off by means of a timer. Furthermore the light intensity can be adjusted with artificial lighting. High light intensities (sun beams through the windows of the house) stimulate hens to peck their cagemates (cannibalism) and can

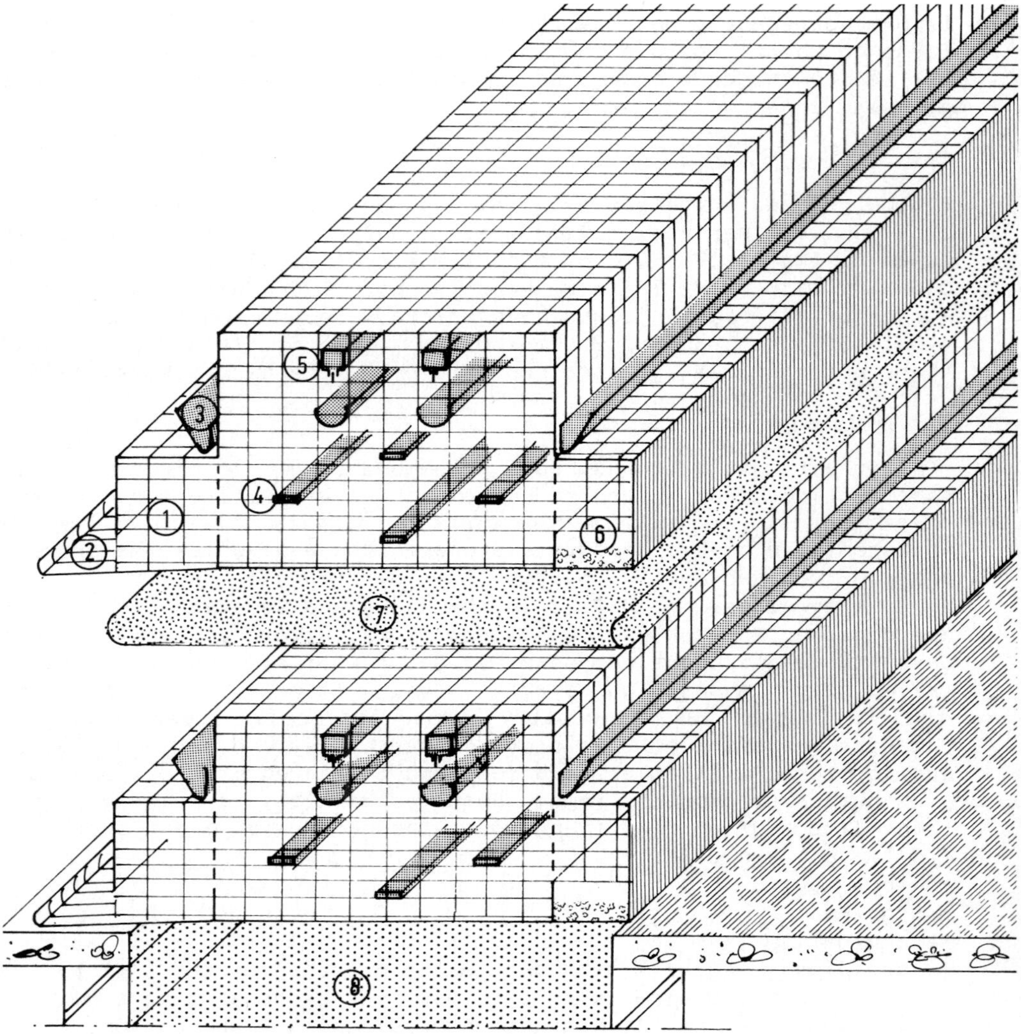

Fig. 6.31 The get-away cage for laying hens (Wegner, 1980).
Legend : 1 = laying nest ; 2 = egg collecting belt ; 3 = feed trough ; 4 = perches ; 5 = nipple drinkers ; 6 = sand boxes ; 7 = manure platform ; 8 = manure channel.

be avoided by a reduction of the light intensity. A light intensity of ca. 15 lux or 0.4 W/m² with fluorescent lamps appears to give the best results (Comberg and Hinrichsen, 1974).

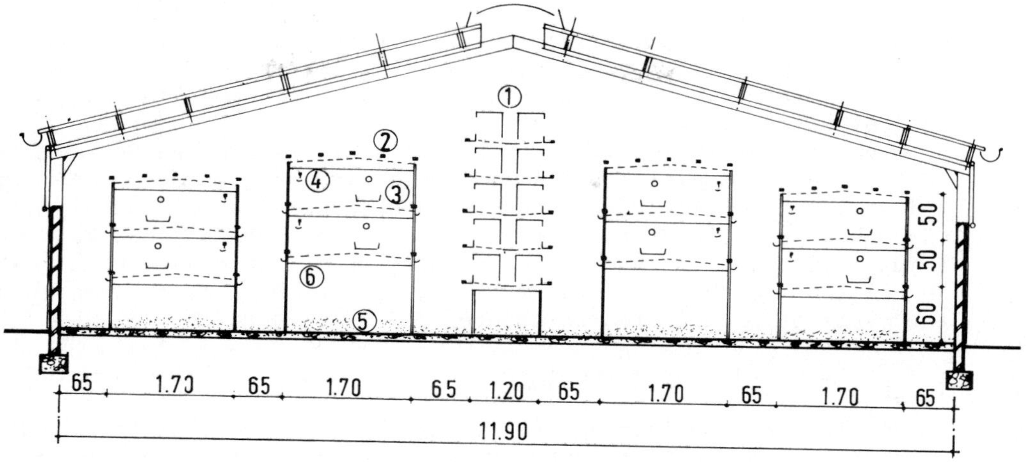

Fig. 6.32 The storeyed house for laying hens (IMAG-design).
Legend : 1 = laying nests ; 2 = perches ; 3 = feed troughs ; 4 = nipple drinkers ; 5 = litter ; 6 = manure belt.

The light pattern (day length adjustment)

The day length is a major environmental parameter for layers and affects growth, sexual maturity, egg production, weight of the eggs and mortality. The research in this connection has led to widely scattered conclusions. A lot of poultry keepers apply the following method. The chicks are initially provided with light for 24 hours a day. Light is then reduced by one hour per day and after seven days the day length is abruptly reduced to 8 hours per day until the pullets reach 18 weeks of age. After 18 weeks the light can be increased by one hour per week to reach a day length of 16 hours at the age of 26 weeks and this day length is maintained until an important drop in the egg production occurs. One or a couple of hours more light are then suddenly given to finally reach a day length of 20 to 22 hours just prior to the end of the laying period. Others provide light for 20 hrs a day for the first three weeks, 9 - 10 hrs a day from the 4th to the 20th week and then stepping up to 17 hrs a day (Petersen, 1984).

Simons and Zegwaard (1983) point to the advantages of the intermittent lighting programme. The hens are thereby exposed, when they reach 36 weeks of age, to a lighting programme consisting of alternately 15 min of light and 45 min of darkness each hour and this for a period of 14 hours, followed by a period of complete darkness lasting 10 hours (biomittent lighting programme). The egg production remains the same but feed consumption, electricity consumption, egg breakage and culling are lower.

6.2.1.6 The ventilation

Ventilation can either be carried out dynamically or statically. With ventilators in the ridge of the building a maximum ventilation of

5 m^3/kg liveweight and per hour is reckoned. The minimum ventilation is 5 to 10 times smaller and hence amounts to 0.5 to 1 m^3/kg liveweight and per hour. The ventilators are controlled by a thermostat and their speed is continuously adjustable between 10 and 100 %. The air inlet takes place through adjustable openings in the longitudinal walls of the building.

Dynamic ventilation is however less and less applied and natural ventilation is preferred. The air outlet is an adjustable open ridge whilst the air inlet takes place through a ventilation curtain or through adjustable slots in the upper part of the longitudinal walls (fig. 6.29). The in- and outlet regulation is increasingly automated in relation to the house temperature. For further information about the ventilation we refer to Chapter 3.

6.2.2 Labour organizational and economic aspects of the housing of layers

The labour organization and labour requirement of poultry keeping were studied by our Institute for a number of years by carrying out time studies in a large number of specialized farms. From this research we were able to calculate for a number of different types of houses (Daelemans, 1967), the labour time requirement for daily care and for periodic activities.

Table 6.3 represents the labour time requirement for the periodical activities in the layer house. Especially the cleaning-out of deep litter and slatted floor houses requires a high labour input.

TABLE 6.3 The labour time requirement for the periodic activities in a layer house, expressed in man-hours per 1,000 laying hens.

Activity	Type of house		
	Deep litter	Slatted floor	Battery
1. Moving-in the hens	5.00	5.00	8.33
2. Bringing in the litter	5.83	-	-
3. Removing the hens	8.33	8.33	5.83
4. Cleaning-out the house (a)	37.50	25.00	-
5. Disinfecting the house	1.00	1.00	1.67
6. Disinfecting slats and batteries	-	5.83	6.67
Total (1 - 6)	57.66	45.16	22.50
Total without item 4	20.16	20.16	22.50

(a) This work is not always performed by the poultry keeper himself.

The total labour time requirement in the various types of layer houses is given in table 6.4. The daily labour time requirement in the layer houses depends on several factors but mainly on the degree of mechanization and varies from one - to fourfold. We can therefore state that the daily activities will require roughly between

TABLE 6.4 The daily labour time requirement, expressed in minutes per 1,000 laying hens, at 75 % production, for several types of laying houses.

Type of house	Deep litter		Slatted floor	Three tier battery		Californian battery 3-tier	Californian battery 4-tier
	(a)	(b)	(b)	(a)	(b)	(a)	(b)
Feeding							
- with hoppers and by hand	8.8	-	-	-	-	-	-
- with trolley, filled by auger and pushed by hand	-	-	-	4.8	-	4.8	-
- with automatic chain feeder	-	0	0	-	0	-	0
Cleaning of the water troughs	3.8	3.8	3.8	1.4	-	3.2	-
Egg collection							
- with trolley and by hand	26.4	-	-	21.6	-	21.6	-
- with collection belt	-	12.0	12.0	-	12.0	-	-
- with collection belt and packing machine	-	-	-	-	-	-	6.0
- collection of floor eggs	2.0	2.0	1.0	-	-	-	-
Packing eggs	4.0	4.0	4.0	4.0	4.0	4.0	1.0
Miscellaneous (c)	7.5	7.5	7.5	6.0	6.0	6.0	6.0
Total in minutes per 1,000 hens and per day	52.5	29.3	28.3	37.8	22.0	39.6	13.0
Number of hens in one 8-hour workday (excl. periodic activities)	9,143	16,382	16,961	12,698	21,818	12,121	36,923
Number of hours per 1,000 hens and per year							
- without periodic activities	319.4	178.2	172.2	230.0	133.8	240.9	79.1
- with periodic activities	377.0	235.8	217.3	252.3	156.1	263.2	101.4

where (a) = low degree of mechanization ; (b) = high degree of mechanization ; (c) = supervision, incl. bird health control, checking of nipple drinkers.

13 and 52 minutes per 1,000 layers. The labour time requirement was drastically reduced for each type of housing with the introduction of mechanization. In a four-tier step-deck battery one person can perform the daily activities involved with ca. 37,000 laying hens excluding the periodic activities, compared to ca. 9,000 layers in a deep litter house with a low degree of mechanization. A high level of labour productivity is already obtained in the management of laying hens in a four-tier battery house with a maximum degree of mechanization : a mere 5 minutes per year are required for the care of one hen ! Such a house has quasi-attained the ultimate automation : nipple drinkers, automatic feeding with chain feeder, egg collection belt combined with a grading and packing machine. Nearly 50 % of the labour time requirement in such a house is taken by supervisory activities.

Concerning the investment required for the construction and equipment of a laying house the following can be said. Building activity has been reduced during the last five years owing to the crisis situation in the egg production sector. Almost no new deep litter houses or slatted floor houses have been built thus making it impossible to put forward a construction price for them. We assume that the annual costs of slatted floor houses will be approximately the same as those for battery houses and that those of deep litter houses will be higher than both of these. The construction and layout of a battery house for 10,000 layers and equipped with Californian batteries, mechanical trolley, nipple drinkers, liquid manure removal, egg collection belt and natural ventilation requires an investment of £ 4.40 - £ 5.00 per layer. The described equipment of a laying house is now customary when building a new house. The construction of a deep-pit house, as described earlier, will require an additional investment of ca. £ 0.65 per laying hen.

6.2.3 Zootechnical and veterinary aspects of the housing of layers

6.2.3.1 Advantages and disadvantages of battery houses with respect to other types of housing

The "Institut für Kleintierzucht" in Celle (GFR) (Wegner, 1982) has carried out an extensive comparative research during three successive laying periods whereby about 2,300 hens were involved at the beginning of each period. The investigated types of houses were poultry houses with a free range, deep litter houses and battery houses. For each type of house the hens were divided into four groups of 192 laying hens each. The stocking density in the free range house and in the deep litter house amounted to 6 birds per m² whilst in the battery house it amounted to 9 birds per m². Each cage of 480 cm² in the battery house accommodated 4 hens with 12 cm of feed length per hen. The following parameters of production were determined : number and weight of the eggs, feed consumption, body weight and the quality characteristics of the final products viz. the eggs and the boiling hens. No distinct differences were found in the number of eggs per hen. There was a slight tendency to an increase in the number of eggs per beginning hen in the battery house because the culling rate is here clearly lower than in other types of houses. The higher percentages of cullings in the deep litter houses were mainly caused by cannibalism, while those of the free range houses were attributed to attacks by birds of prey. A distinctly lower feed consumption and hence a better feed conversion i.e. a lower feed consumption per kg of eggs was obtained in the battery house. The weights of the boiling hens on the battery were ca. 50 g lower than of those kept in deep litter houses or with a free range. The eggs of the battery hens had a better shell quality and a darker yolk colour. The latter can possibly be attributed to a better absorption of the carotenes by the battery hens. The battery house also produced a lower number of stained or dirty eggs. A similar conclusion was reached in an investigation concerning infections in eggs. Quality control also confirmed that eggs from hens kept in batteries, showed a higher vitamin-A content. No differences were found for the other quality parameters viz. : the levels of amino acids and minerals, the albumen height (measure of freshness), the number of eggs with foreign substances inside, the odour and flavour. The meat quality of the boiling hens from battery houses was lower (lower weight of breasts) than those of the two other types of houses. Clinical examination of the birds revealed no differences in the quantitative and qualitative composition of the blood (number of blood cells per cubic millimetre, concentrations of haemoglobin, glucose, plasma lipids, serum lipids, triglycerides, cholesterol, total proteins, creatinkinase) with respect to the type of housing. The same is valid for the concentrations of calcium, inorganic phosphates and magnesium in the blood. Post-mortem examination also revealed no differences in the incidence of diseases with respect to the type of housing. Hens housed on deep litter and especially those with a free range were however more infested with parasites. The hens housed in batteries showed no roundworms or other intestinal parasites. The likelihood of infections is however larger in houses with a free range or in deep litter houses than in slatted floor houses and cages. No differences could be found in the working of the endocrine glands (suprarenal gland, thyroid, pancreas) and the hormonal physiology with respect to the type of housing.

The air in animal houses contains both dust and micro-organisms. According to Batel (1977) aerial dust concentrations in poultry houses do not appear to exceed the maximum allowable concentrations of 8 mg per m^3. The air of poultry houses contains large amounts of micro-organisms viz. bacteria, fungi and sometimes viruses either free or bound to dust particles or water drops.

Matthes (1979) collected the results of several researchers. These results are represented in table 6.5. From these figures it appears that especially the air in poultry houses has an abundance of micro-organisms and this is even more pronounced in littered houses.

TABLE 6.5 The concentrations of airborne micro-organisms in animal houses, dwelling-houses and in the open air (Matthes, 1979).

Place	Concentrations of airborne micro-organisms per l of air	Literature
Cattle houses	31 - 562	Various
Piggeries	4 - 11,400	authors
Laying hens, floor	16,730 - 48,461	Hurtienne (1967)
	2,200 - 16,000	Hilliger (1969)
	50,228 - 160,956	Kösters & Müller (1970)
	185 - 3,595	Gebhardt (1973)
	9,368 - 22,456	Gebhardt (1973)
	1,920	Sarikas (1976)
Laying hens, battery	680 - 5,860	Kösters & Müller (1970)
	342 - 2,003	Kösters & Müller (1970)
	90 - 366	Gebhardt (1973)
	50 - 200	Gärtner (1975)
	200 - 300	Gärtner (1975)
Dwellings	maximum 1	Kösters & Müller (1970)
Open air	0.01 - 0.1	Rüden et al. (1978)

From the above-mentioned research of Wegner (1982) we can conclude that poultry can be housed more economically and hygienically in cages than in deep litter houses or with free range. However the same research stresses a number of behavioural problems associated with the housing of hens in cages.

6.2.3.2 Welfare of the hens in connection with their housing in cages

Certain pressure groups and some public opinion have in recent years expressed their opposition to cage-housing of hens. They find that the housing of laying hens in cages cannot be associated with the welfare of the animals. This opposition however is not backed-up by irrefutable scientific findings. The welfare of the bird is indeed sometimes difficult to judge and if its behaviour is taken as a measure of its welfare a number of difficulties arise. Concerning an important activity of the chicken viz. its feeding behaviour, several researchers have reached contradictory results. Some find a uniformly distributed feed intake during the entire day length period, others find one peak, two peaks or even three peaks (Hughes, 1972 ; Fujita, 1973 ; Anon., 1981). Certain birds which do not have the possibility to feed during the busy-hours due to a shortage of feed length will keep themselves busy in a rather bizarre way by preening and moving around in a highly nervous state (Duncan and Wood-Gush, 1972).

The rationing of feed particularly creates frustration which is made visible by the pecking of feathers and cage (Hammer, 1973). Anyway, it is generally accepted that a feed length of 10 cm per bird is sufficient for production purposes, without making a judgement about the welfare of the animal (Lee and Bolton, 1976).

Chickens don't need to drink simultaneously and they will even not contest for access when they are thirsty (Banks et al., 1980). They will take more time to drink from the nipple drinkers than from the water trough and they prefer the latter. Problems concerning the drinking by hens in cages seem to be very limited. Some lines of white hens make unusual movements, show an increased aggression and hence appear to be frustrated in the pre-laying period (Wood-Gush, 1972). By providing them with trap nests they will return to a normal laying behaviour.

On the other hand Hughes and Wood-Gush (1973) have demonstrated that laying hens are less aggressive when they are housed in cages than on litter, if they are kept in fixed groups of small size. Increasing the available space for the hens from 400 cm^2/bird to 800 cm^2/bird greatly increases the aggression but improves the production : the optimum would probably be reached between 450 cm^2 to 500 cm^2 of space per bird. According to Hughes (1975) a certain relation exists between mortality and group size (mortality is lower in a group of 7 than in a group of 10) while Wells (1973) found a relation between mortality and cage density (mortality is lower at 500 cm^2/bird than with 465 cm^2/bird, and is the highest with 350 cm^2/bird).

There is abundant evidence that cage-housed birds have reduced bone mass and reduced bone breaking strength compared to birds in littered houses and the same is valid for the wings (Simonsen et al., 1980). With certain breeds pecking also occurs more frequently (Simonsen et al., 1980). Another explanation for the increased frequency of pecking might be the search for missing nutrients (Ca, Na).

There are thus certain indications that cage-rearing of laying hens might have detrimental effects to the welfare of the animal but there is yet no irrefutable proof.

Already proven are the economic consequences of the increased space which is provided to the cage-housed birds (Anon., 1981). An increase of the area per bird from 400 cm^2 to 500 cm^2 corresponds with an increase of the production costs of 4 % and from 400 cm^2 to 600 cm^2 per bird it results in an increase of 9 %. A decrease of the area per bird causes an increase in the mortality : 8.6 % and 14 % with an area of respectively 440 cm^2 and 360 cm^2 per bird. The economic optimum space per bird amounts to 440 cm^2 when these two factors are taken into consideration. Jongenburger (1982) also mentions an economic optimum area of ca. 450 cm^2 per bird.

The obligation for our producers to increase the area allotted to each bird (by removing one or even two laying hens, for example, from the existing cages) would allow other countries to offer eggs at lower prices, which would only aggravate the already heavy competition battle. To prevent this unfair competition the E.E.C. would be obliged to increase the import duties at the border of the Community and to award premiums for the export. The whole of these measures and factors would lead to an increase of the egg price for the consumer and this increase is not likely to be lower than the increase in the production costs viz. at least 5 % which would result in a decrease in egg consumption of about 1.5 %. We can also add that if such E.E.C.-regulations become statutory the producer would be obliged to invest if he wants to keep his initial flock. From this we can easily understand the egg producers' fear of the possible application of new E.E.C.-regulations.

At the time of writing this book the Board of Ministers of Agriculture of the E.E.C. had not reached an agreement concerning the precise standards which must become statutory in the E.E.C. member countries. The European Commission proposes a minimum available cage area of 500 cm^2/bird as per July, 1, 1995. This proposal was discussed by the Economic and Social Committee of the E.E.C. on 28 and 29 October 1981 and was rejected. This Committee has not found that protection of the laying hen is a matter of priority for the E.E.C. as long as there are no guarantees for the financial situation of the producer. The Committee has also objected to the fact that the Commission draws proposals and standards without any scientific basis. In spite of this negative advice, the European Parliament accepted the standard of 500 cm^2 per laying hen and has even advised that the transition period be 1 July, 1990 instead of 1 July 1995 as previously proposed by the Commission. On 11 June, 1982, the Commission changed its proposal concerning the transition period and brought it into accordance with this of the European Parliament. The Board of Ministers of Agriculture of the E.E.C. however rejected this proposal. A number of alternatives have been introduced since then such as a minimum cage area of 450 cm^2 per layer as from July 1, 1984 and of 550 cm^2 as from July 1, 1993. None of the proposals were accepted. The latest proposal dates from March, 1983 and advocates an area of 450 cm^2 per layer as from 1 January, 1984. This proposal was not agreed by the Board of Ministers. The Animal Protection Societies are shocked by the irresolution of the Board of Ministers of Agriculture but it appears logical and understandable to us, in the light of what has been mentioned above, that the Board does not decide upon new regulations with undue haste.

6.3 THE HOUSING OF BROILERS

Broilers (slaughter-ready chickens of only 6 to 7 weeks old) are nowadays also housed in hangars which are either prefabricated or traditionally built. Portal framed buildings are mostly used, whereby the whole inside space remains free ; the most frequent width is 12 m. Walls are either made of light concrete, perhaps combined with brickwork or of wood and, together with the roof which is made of corrugated asbestos cement, must be well-insulated. The lower part of the inner side of the wall is rendered to facilitate cleaning.

6.3.1 The littered house for broilers

The chickens are running free in this house which has a concrete floor covered with a layer of litter. Since a thorough disinfection is necessary after the flock has left the house in order to prevent problems with coccidiosis, helminths, etc., a concrete floor is essential. It is 6 to 8 cm thick over a layer of ca. 25 cm of coarse sand and is covered with a layer of litter made of wood shavings, chopped straw and peat. Additional wood shavings can only be applied after the first week, otherwise they might be picked up by chicks of less than 1 week old and plugs might then be formed in the gizzard which cause nutritional disorders often followed by death. The layer of litter is 10 cm (winter) to 5 cm (summer) thick : a thicker layer reduces the heat losses during the winter while a thinner layer allows the loss of superfluous heat by older chicks during hot summer weather.

The density normally amounts to 17 or 18 chicks per m^2 taking into account that moving is no longer practised. If space heating is applied only a third of the house will be put at the chicks'disposal during the first two or three weeks after which they are admitted to the whole house. If brooders are used the same principle is applied, be it in another way. Both heating systems will be discussed later.

The houses are windowless and normal white light is provided during the first two weeks after which it is replaced by dim light (0.5 W/m^2) which is beneficial to the tranquillity of the birds and which prevents cannibalism. There is nowadays a tendency towards the application of an intermittent lighting pattern i.e. an alternation of 1 h of light and 2 h of darkness. In this way a better feed conversion and a substantial energy-saving can be achieved. For the first couple of days carton plates, upon which pellets are laid, are at the chicks'disposal. The idea behind this is to stimulate the chicks to eat by the noise made by their feed-picking housemates. The chain feeder which was customary in earlier days has gradually been replaced by round feed pans and low level feed pipes : they are automatically filled with an auger or with a nylon rope with disc-shaped carriers. One feed pan is provided for every 75 chicks. The feed pipes which are connected to the feederpans are equipped at their upper side with a non-roost system i.e. a wire which prevents the chicks from sitting on the tube. After the transfer of the broilers to the poultry slaughter plant the feed installation is lifted in a very simple way and screwed to the ceiling in order to enable the cleaning of the house. Compared to dismantling and removing the chain feeder, the lifting of the feed installation results in a substantial saving in labour. Moreover, the use of feed pans, compared to a chain feeder, reduces the incidence of breast blisters (fig. 6.33).

The supply of drinking water is carried out automatically by means of suspended round drinkers - one for every 150 chicks (fig. 6.8). These round drinkers are also lifted to the ceiling for mucking-out and cleaning of the poultry house.

A suitable temperature is of the utmost importance for the health and growth of the chicks. The temperature shall be kept between 32 and 33°C during the first week and can then decrease weekly by 2.5°C to reach a final temperature between 18 to 21°C after 5 to 6 weeks and this temperature is then maintained until the end. These temperature requirements imply the installation of a heating device, which will

Fig. 6.33 The littered house for broilers. Note the suspended round drinkers and the feed pans.

certainly need to function during the first few weeks. An automatic, thermostatically controlled heating installation is therefore employed. Besides the traditional energy sources it is also possible to use pellets made from the compressed manure of the broilers for fuelling the heating installation. Either total heating of the broiler house or local heating of limited areas in the house can be applied. In the first case indirect fired air heaters, evenly spread over the entire house, are used. In the second case gas fired brooders are used. These consist of gas burners sometimes equipped with a large reflector. The hover space of one such brooder heats sufficient space for 300 - 500 broilers since the brooders are normally suspended at a certain height above the floor. The regulation of the heat production is controlled by a thermostat. The chicktrays are placed in the vicinity of the hovers and a guard of wire mesh is placed around them during the first two weeks to prevent the chicks of getting lost. Forcing warm air into the house by means of air heaters results in a rather uniform temperature in the entire house and encourages an even spreading of the chicks and hence a better growth. Such a uniform temperature is not easy to realize with gas brooders and good ventilation is thus essential. Extraction ventilation is normally applied and the stale air is extracted from the house by means of a number of fans. The ventilation is rated at an air replacement of max. 1 m^3/h/kg live-weight and at a maximum air velocity of 0.15 to 0.30 m/s according to the age of the flock.

The extraction of the foul air can be carried out in different ways. Ridge ventilation - the best system in wide houses - employs a number of fans in the ridge where the air is extracted and fresh air is supplied through adjustable inlets under the eaves in the side walls. With cross ventilation the fans are installed in one side wall whilst the air-inlets are situated in the opposite side wall. A modified version of cross ventilation allows the air to enter through a perforated ceiling whereby the likelihood of draughts is reduced.

Mucking-out of the broiler house is carried out by means of a tractor equipped with front loader and is necessary after the delivery of a flock : housing a new flock on this old litter cannot be tolerated as it is conducive to the development of many disease organisms such as coccidiosis, helminths etc. The presence of a concrete floor is required in order to enable a thorough disinfection after removal of the litter. Normally six flocks of chicks can annually be produced (6 to 7 weeks of rearing + 1.5 weeks of idle time per house). Concerning the *labour time requirement for the keeping of broilers* it can be said that one person is able to care for 50,000 birds provided that the farm is well-organized, mechanized and professionally run. Some activities however will require additional labourers viz. at the arrival of a new flock of chicks or the delivery of a finished flock. When all the birds of the flock are of the same age which is for hygienic reasons usually the case, a serious "labour peak" is created at the end of a rearing period (catching the birds, preparing them ready for transport to the poultry slaughter plant). In order to avoid this labour peak birds of different ages could be kept at the poultry farm as these will be ready at different dates. This however, has to be discouraged. The uniformity of age of the flock has a number of important advantages from a veterinary point of view. With an outbreak of a contagious disease it is possible to rid the farm of the disease quickly. The presence of flocks of different ages would only promote and prolong the infection and may finally result in a permanent infection.

The *capital investment* for raising broilers is high because relatively high costs are involved with the construction and equipment of broiler houses. An investment of ca. £ 2.50 per broiler place is required at the level of a flock of 15,000 broilers.

The price-risk with the raising of broilers is important and for this reason a full or partial integration of the cycle is realized : breeder, hatchery, broiler raiser, producer of feeds, poultry slaughter plant.

6.3.2 The cage house for broilers

A number of attempts have been made to raise broilers, similarly to layers, in one or multiple tier cage houses. This idea has found no acceptance because cage-housed broilers often showed breast blisters which negatively influenced the slaughter quality. In some cases 30 % of the finished broilers, which were raised in cage houses, showed enlarged breastbone follicles (Brantas, 1973). The incidence of perosis is also higher in cage houses than in littered houses. Cage-raising of broilers is thus yet not ready for practice, although it would offer many advantages. The risk of coccidiosis could nearly be excluded and hence the addition of coccidiostats would no longer be necessary. The delivery of finished broilers could be simplified by transporting the cages with the broilers in them to the slaughter plant thus eliminating the labour for catching the broilers and excluding the risks of lesions involved with the catching. Further research into these possibilities is therefore justified.

6.4 THE HOUSING OF TURKEYS

The raising of turkeys is a successful and quickly growing branch of poultry. In the traditional production areas such as the United States of America, Canada, Mexico and Great Britain the growth was steady but rather slow (Stigter, 1982) during the last decade (1970 - 1980). The production in the U.S.A. increased 29 % while in G. Britain it rose from 68,500 tonnes to 111,000 tonnes or an increase of ca. 61 %. A spectacular increase was noted in Italy and France, 315 % and 485 % respectively. In Italy the production rose from 65,000 tonnes to 205,000 tonnes while in France it increased from 35,000 tonnes to 170,000 tonnes.

Whilst the U.K.-production in 1970 amounted to one third of the total E.E.C.-production it decreased in 1980 to only one fifth of the total E.E.C.-production. The consumption of turkey in kg per head of the population in the different E.E.C.-countries in 1979 was as follows : France 2.6 kg ; Italy 3.6 kg ; United Kingdom 2.0 kg ; the Netherlands 0.55 kg (Stigter, 1982) and in Belgium, in 1981, it amounted to 1.3 kg per person (Anon., 1982). Israel is probably the largest consumer of turkey with not less than 11 kg per person (Stigter, 1982). Another point of interest is the ever-increasing demand for turkey parts and this requires the raising of heavy turkeys (6 - 16 kg).

The housing of turkeys is closely related to the production method, which can either be extensive or intensive. In the first case the poults are reared in confinement for a period of 8 to 12 weeks. After this period they are fattened on dry pasture provided with a number of stationary or movable sheds. They are kept until the age of 16 weeks (light turkeys) or 26 weeks (heavy turkeys). With the intensive method the poults are kept in brooder houses for about five weeks after which they are transferred to a finishing house, which provides for maximum 1,500 birds where they are fattened until the age of 10 to 20 weeks (respectively light turkey hens and heavy turkey cocks).

The extensive method offers a healthy environment to the turkeys and leads to a relatively low feed consumption and a good meat quality. Drawbacks of the extensive method are the increased likelihood of contagious diseases (especially the much feared black head disease), the difficulties for reaching a rational labour organization and a possible increased mortality with unfavourable weather conditions. It can only be applied in spring and summer in our Western European climate. A number of large turkey farms in the U.S.A. still practise this extensive method (fig. 6.34).

The *brooding of poults* is carried out in well-insulated ($k < 0.7$ W/(m² .K)) windowless hangars, similar to those for broilers. The flooring must here also allow a thorough disinfection and should there-

Fig. 6.34 A large turkey farm in the U.S.A.

fore be made of concrete on which a 5 to 10 cm thick layer of wood shavings (6 kg/m^2) or peat is applied. This layer of litter is compacted to prevent picking. Turkeys are in fact cursores which unlike chickens do not scratch. An especially strong ventilation is required viz. 7.5 m^3/kg liveweight/h compared to only 5 m^3/kg liveweight/h for broilers and is mostly realized by a number of fans. Some turkey growers have nevertheless installed natural ventilation and their recently built houses are provided with curtains and an open ridge. The side walls of these houses are between 2.75 and 3.00 m high and the houses are also employed for growing. The temperature required for day-old turkey poults is also higher than for broiler chicks and is situated between 37 - 38°C and is gradually lowered to reach 20°C after 4 weeks. A local heating by means of gas brooders, equipped with reflectors and sufficient for 300 - 500 poults seems to give better results than the heating of the complete house by means of air heaters. The latter system is also insupportable to the poultry keeper during the first weeks. Therefore local heating with a temperature of 38°C in the hover space and 25°C in the house itself is preferred. A guard made of metal gauze (Ø 4 to 4.5 m) is installed around the reflector for the first ten days. This ring must allow a sufficient ventilation at the level of the poults and must prevent, by its elasticity, the crushing of poults (fig. 6.35). The feeding and drinking facilities are installed inside the guard in a star-fashioned position and within easy reach of the poults. Feeders and drinkers are preferably red as this colour is best seen by the turkeys. In a later stage the guards are removed and wire separations are installed which will hold 1,000 to 1,500 turkeys. The poults are

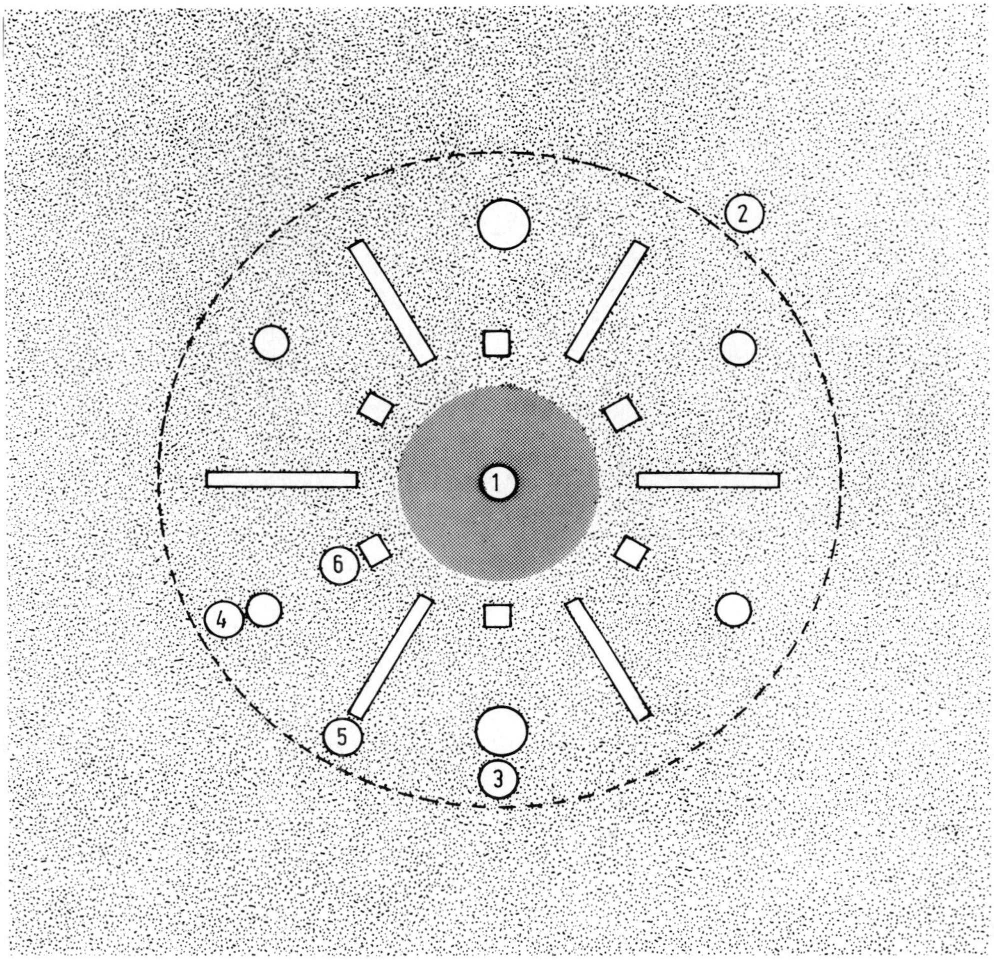

Fig. 6.35 The rearing guard for poults.
Legend : 1 = dull emitter ; 2 = guard of metal gauze ; 3 = automatic drinkers ; 4 = poults' fount ; 5 = feed trough ; 6 = egg carton filled with feed for the first four days.

debeaked at the age of 10 days in order to avoid cannibalism (Bossuyt, 1979). Flocks are sometimes smaller viz. when cocks and hens are kept apart to facilitate the loading of fattened turkeys as the hens are the first to be removed. Tube feeding systems are more and more applied.

Feeder pans, filled by a tube feeding system, can be applied from the fifth week onwards. The feeder pan can simultaneously be used by

60 to 70 turkeys. Nowadays round suspended drinkers are often used to supply drinking water. One drinker will suffice for 100 turkeys.

The poults are provided with ample white light (min. 4 W/m²) during the first few days to facilitate their search for the feeders and founts but the light intensity is then gradually decreased to reduce the incidence of cannibalism. The poults are debeaked for the same reason at the age of 10 days. The stocking rate is ca. 10 poults/m².

The *fattening of turkeys* is carried out according to the intensive method in insulated and closed houses whilst it is done in open air for the extensive method. In the latter case (fig. 6.34) the birds, after spending 8 weeks in a brooder house, are transferred to a dry pasture which is provided with a number of fixed or removable sheds together with automatic feeders and water troughs. A pasture of 8,000 m² is sufficient for 1,000 turkeys and the birds are regularly moved to another pasture (rotational grazing).

With the intensive breeding of turkeys the birds, after a period of 5 weeks in a brooder house, are transferred to a finishing or fattening house where they are kept until they have reached their final weight. This fattening house resembles the brooder house in terms of construction and equipment and is also littered. It is divided into a number of sections each accommodating 1,000 to 1,500 birds. The temperature is preferably set between 18 and 20°C while the birds are nowadays permanently kept inside and hence have no free run (sun porch or solarium) mainly in view of labour saving. Natural ventilation can be successfully applied viz. through a number of adjustable air inlets in the upper side of the longitudinal walls and an adjustable open ridge which functions as an air outlet. It is ofthe utmost importance to keep the litter dry to prevent the much feared black head disease. Needless to say that the finishing house, like the brooder house, is mucked-out, cleaned and disinfected each time a flock is removed. The fattening house is windowless and some light is only allowed into the house through the ventilation openings : this will prevent cannibalism. Feed and water distribution are carried out as described above. The stocking density will vary according to the production of small, medium or heavy turkeys whilst cocks and hens are always kept in separate sections. With the production of heavy turkeys an initial occupancy of 8 birds/m² is maintained ; the hens are sold at the age of 10 - 12 weeks at a weight of 3 to 4 kg (small turkeys) ; the cocks then receive an additional area, previously occupied by the hens, and the stocking density is thereby reduced to ca. 3.5 birds/m² taking into account the culled ones. They are sold at the age of 18 to 20 weeks at a weight of 10 to 12 kg (heavy turkeys). For the production of medium turkeys, hens are kept until the age of 15 weeks, corresponding to a weight of 6 to 7 kg. The delivery of a smaller number of heavier turkeys reduces the costs of heating, especially in the brooding period, and the labour requirement and increases the income of the turkey keeper by the higher price per kg obtained with the sale of heavy turkeys (Stigter, 1982). The investment required for the intensive production of turkeys is high and exceeds that for broilers (higher degree of insulation and ventilation) the investment in buildings can be rated at £ 45/m² and £ 0.75 per day-old poult. The production of turkeys is even more speculative than that of broilers and this explains the extended vertical integration in this branch.

6.5 THE HOUSING OF LAYER PULLETS

Raising of pullets destined for layers is nowadays carried out by the intensive method and in closed houses. The chicks can be brooded on the floor or, as is now generally done, on batteries.

In the first case the chicks are kept in a limited part of the house during a period of 6 to 8 weeks after which the partitions are removed and the chicks allowed access to the whole house where they are then kept up to the age of 20 weeks. The construction and equipment of this littered house can be compared with the broiler house and it accommodates 25 chicks per m² up to the age of 4 weeks and then 8 chicks per m² to 20 weeks. Nowadays the chicks are generally brooded on batteries viz. on batteries of the three-tier type. One tier of cages, mostly the upper one, is populated with day-old chicks (ca. 80 chicks/m² of cage area) and they remain there until they are 4 weeks old. This upper tier is heated with gas brooders or infra-red lamps to 30° - 35°C in the first week and the temperature is then gradually lowered to reach a temperature of 20°C after 4 weeks. The chicks are then distributed over the three tiers of the battery to obtain a stocking rate of 25 to 30 chicks per m² at an ambient temperature of 20°C. It is also possible to accommodate the chicks of up to 4 weeks in one complete three-tier battery in a separate part of the house kept at a suitable temperature after which they are transferred to other batteries without supplementary heating. Although chicks have then to be transferred over a longer distance it allows a substantial reduction of the energy requirement. Fig. 6.36 shows one type of three-tier battery with its construction and principal dimensions. The front, rear and floor are made of plastic covered wire. The meshes of the floor measure 18 mm by 18 mm. However, in the initial phase of raising a perforated plastic sheet with circular or square openings of 100 mm² is laid on the cage floor. The partitions consist of full plates (asbestos cement or galvanized metal boards) which contribute to a quiet environment for the chicks. The front is often movable. In the first phase of brooding the front is tilted forwards, just over the feed trough, while sides of plastic covered wire prevent the escape of the birds ; the feed trough is covered with plastic coated wire mesh through which the birds have access to their feed and in which a chain feeder supplies the feed. As soon as the chicks are 4 weeks old the front of the cage is put vertically whilst the special partitions and the covering of the feed trough are removed. The chicks can take their feed directly from the feed trough, in which a feed chain is moving, controlled by a timer which guarantees an automatic and rationed feeding. The front of the cage can be tilted upwards for easy handling of the birds.

Similar to the layer houses, a feed trolley instead of a feed chain is used in some of the cage houses for pullets. The drinking water supply is by means of nipple drinkers adjustable in height, of

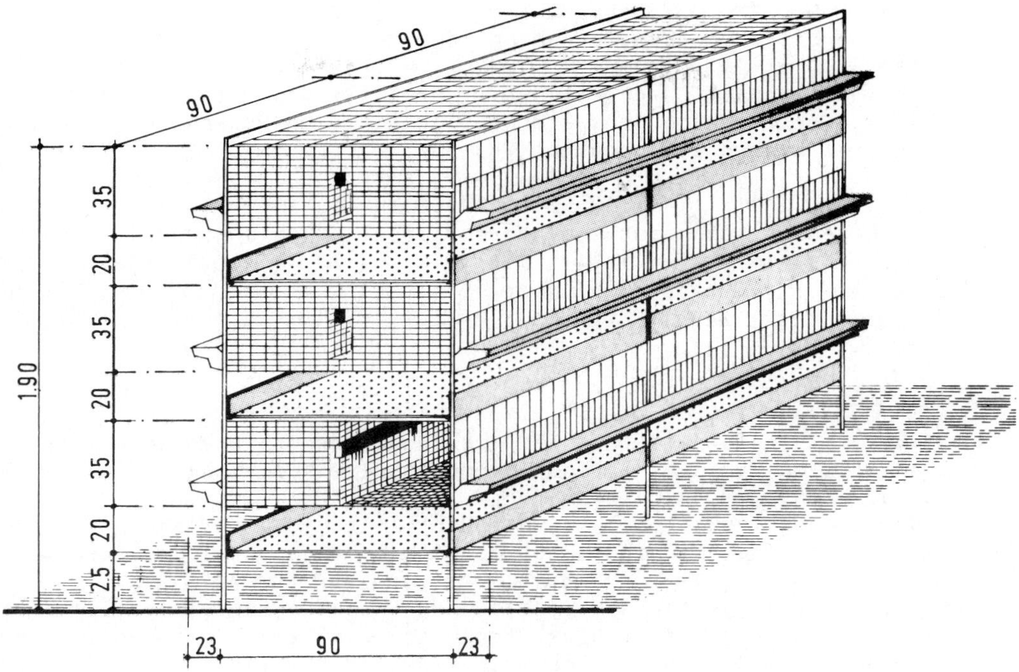

Fig. 6.36 The three-tier battery for the housing of growing young hens.

which six are provided per tier and placed between two rows of cages. The spilled water is collected either by a gutter installed under a row of nipple drinkers or by a cup placed under each of them.

In tier batteries the droppings are evacuated in two ways : either scrapers or a synthetic belt can be used, placed beneath each tier. When a belt is preferred, a fixed scraper at the end of the belt removes the manure. With a Californian or step deck battery, the droppings fall on the floor between two rows of cages.

One strives towards the application of the all-in all-out principle in each section or in the entire house whereby a thorough cleaning and disinfection is possible after each flock is removed. A lighting pattern, as described above, is applied.

Cage-rearing of pullets has a number of distinct advantages compared to the littered rearing house : the transfer to the battery for layers creates less stress, a better supervision of the flock is possible, the incidence of parasitic diseases is much lower, a substantial saving is possible on the coccidiostats, the number of crushed hens is lower and less space is required per bird. However, a higher capital investment is required for the batteries but nevertheless cage-housing of pullets is now generally applied.

6.6 THE HOUSING OF BREEDING HENS

The broiler or layer breeding hens deliver the *hatching eggs*. After incubation in the *incubator* chicks emerge from the fertile eggs. These chicks are destined as broilers or, after rearing, as layers. The breeding hens are kept in littered houses, or, especially the layer breeders, in batteries.

6.6.1 The deep litter house for breeding hens

The general construction and equipment of this house is similar to this described for layers. We will therefore treat only those aspects which are of special interest for the housing of breeding hens in deep litter houses. A number of basic rules must be observed.

Brooding

Brooding refers to the early period of growth when chicks are unable to maintain their body temperature without the aid of supplementary heat.

- The brooder house must be thoroughly cleaned and disinfected prior to populating the house. Not more than 10 chicks are placed per m^2 and the flocks preferably do not exceed 500 birds.
- The house is heated at least 12 hours before the arrival of the chicks and a constant house temperature of 25°C and a hover temperature of 35°C are to be maintained in the first week. The temperature can then be lowered by 4°C per week until a final temperature of 15°C is reached in the house. The guard encircling the hover space is made larger as the chicks grow older.
- Easy access to waterers must be provided to the chicks to avoid an increase in the mortality : one water fount per 100 chicks will give good service during the first few days. Later on, 1 cm of water trough length per bird is needed. Water troughs must of course be kept clean.
- During the first days the feed is administered on chick trays or egg cartons. Afterwards a feed trough is used. Up to the age of 8 weeks a feed length of 4 to 6 cm per chick (= 2 to 3 cm of trough length) suffices. Grit is provided from the first day.
- Sufficient fresh air must be provided, but draught must be avoided ;
- A high degree of sanitation is maintained and the programmed vaccination is closely followed.

Rearing

- The following lighting pattern can be recommended for a light-tight and mechanically ventilated house :

Age		Lighting per day
0 - 1 week	:	24 hours (1st day), decreased by 1 h per day, to reach 18 hours on the 7th day
1 - 18 weeks	:	8 hours

from 18 weeks on to point-of-lay (26 weeks) : 8 - 9 hours.

The "short day rearing" results in a retarded point-of-lay but a compensatory egg production is obtained. It can only be done in a light-tight house which is therefore of great importance in the rearing period.

- If restricted feeding is practised all hens should get their share of the feed. This can be accomplished by providing enough feed length so that all birds can eat simultaneously. Depending on the age a feed length of 10 to 15 cm is therefore required (5 to 7.5 cm of trough length). The feed ration must be administered in one or two portions.
- With restricted lighting and rationed feeding, the cocks have to be housed separately since they must not be subjected to these treatments. It is advisable to install a number of escape-perches in the male section to allow frightened birds to find a safe place.
- Overcrowding should be avoided : a stocking density of 7 or 8 hens. per m^2 is adequate for hens at the age of 8 weeks.

The laying period

- One of the following two lighting patterns can be followed :
  - the day length can be changed over to 16 hours at the beginning of the lay (26 weeks) and this level can be maintained throughout the laying cycle ;
  - a step up lighting can be applied : from an initial day length of 8 - 9 hours at the beginning of the lay, the day length can be increased by weekly increments of 18 minutes (not with 3 min per day but with 18 min per week) to a minimum of 14 hours.
- Fluctuations in the house climate should be avoided or at least reduced to a minimum. A good insulation ($k < 0.7$ W/(m^2.K)) contributes to achieve this suitable house climate. An ample, draught-free ventilation promotes the health of the breeders. Excellent results have been obtained with an open ridge ventilation and mechanical ventilation is therefore only indispensable in connection with the strict application of a lighting pattern. Supplementary heating during the winter is, according to the latest insights, not strictly necessary. If any stray light enters the house it is recommended to adjust the light pattern viz. starting in time with additional lighting.
- The stocking density is different to the one used for layers : 6 white leghorn layers per m^2 result in a liveweight of ca. 12 kg/m^2 ; 4 broiler-breeders per m^2 reach together a weight of ca. 14 kg.
- In order to obtain a good fertilization it is necessary to keep one cock with every 10 hens. The cocks are supplementary fed with breeder pellets in troughs which are out of reach for the hens. To avoid lesions to the hens the nails of the two inner digits of the cocks are removed.
- Generally the broiler breeders are kept no longer than 9 to 10

months since after this period a drop in production is observed.
- Individual laying nests are preferred viz. 1 per 4 hens and eggs are collected 4 or 5 times a day.

6.6.2 Cage houses for breeding hens

A rather recent development in the housing of breeding hens and more in particular of layer breeders is the application of batteries. A three-tier battery resembling the one used for layers is hereby applied. There are however a number of minor differences. The breeding hens are individually housed in cages measuring 900 to 1000 cm², or twice the space reserved for layers which are accommodated in groups of 4 birds. Cage housing became possible by the fact that more and more dwarf breeding hens are raised which weigh ca. 2.5 kg and which are artificially inseminated.

6.7 THE HOUSING OF TURKEY BREEDING HENS

Turkey breeding hens can be housed in either deep litter houses or in cage houses.

6.7.1 Turkey breeding hens in deep litter houses

The birds are kept in closed deep litter houses after being raised as described above for turkey poults. The lighting pattern is also of considerable importance. During the rearing the birds receive 6 hours of light per day. Mechanical ventilation is necessary to enable the darkening of the house. Turkey hens are forced into lay after 32 weeks by abruptly changing the day length from 6 hours to 15 hours (a day length of 24 hours produces soft-shelled eggs). Forced moulting is carried out after a laying period of 4 months and is followed by a rest period of 3 months. Fig. 6.37 illustrates the relation between the rate of lay and the progressing laying period. As can be seen in the laying graph, the optimum rate of lay is reached after 3 weeks and then drops quickly by steps. This is attributed to the frequent occurrence of broody hens which go out of lay immediately.

Preventing broodiness as much as possible is a very heavy task. This will be discussed later on. Turkeys are kept on a thick layer of litter which needs regular and frequent stirring. Indeed, on one hand turkeys (cursores) don't scratch like hens and on the other hand the litter must be kept dry to prevent any outbreak of the much feared black head disease. There are no perches present. The stocking rate is 1.75 to 2 turkey breeding hens per m².

Round feed troughs are commonly used and are mechanically filled. A continuous water trough at the level of the turkey's back is installed along one of the side walls of the house : turkeys run free over the entire area of the house, unlike hens which stay in a limited space. The width of the house can amount to 14 metres. Individual nests are better than community nests. Laying nests equipped with a tumbler lock are preferred since such a device prevents two turkey hens from being in the same nest. Sometimes one laying nest is provided for every four turkeys, although in practice one nest per seven turkey hens is frequently found and appears to give good results. Wood shavings or peat are used as litter in the nests. The

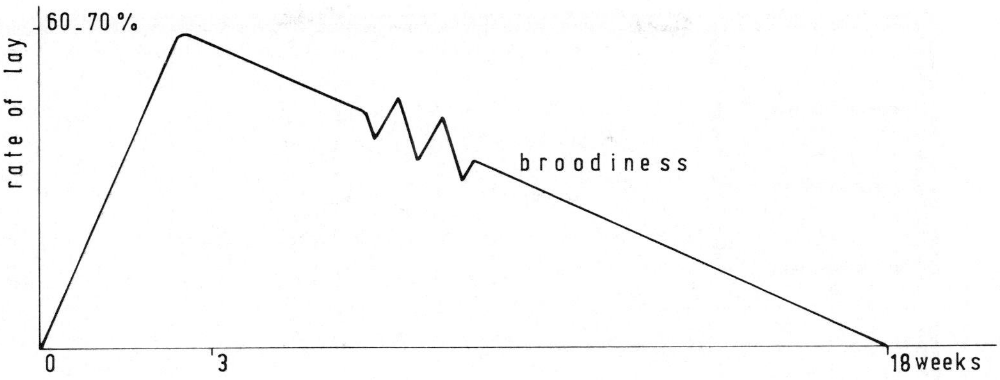

Fig. 6.37 Laying graph of turkey breeding hens.

nest width is between 35 and 40 cm.

Eggs are normally collected five or six times a day, although in England it is often done hourly. The eggs, after being collected, are cleaned and disinfected in a fumigation cabinet. The egg storage room is disinfected weekly. Eggs are stored at an ambient temperature of 12°C and a relative humidity of 70 to 80 % in boxes which are regularly turned. In contrast to layers, which normally receive a light intensity of 11 lux or 2 to 4 W/m², the turkeys are given very intensive light viz. 20 to 30 lux or 4 to 10 W/m². Fluorescent lighting is preferred because of its lower electricity consumption compared to incandescent lamps. Strip lighting however tends to flicker sometimes thereby disturbing the tranquillity in the house.

As mentioned above the suppression of broodiness is a serious task. Turkeys easily become broody for a period of 3 to 4 weeks if this situation is not thwarted. Broodiness can be established through a number of signs viz. the birds become warmer, they get a "snake's neck", they parade and the pubic bones are narrowing but the most accurate means of establishing broodiness is by the so-called "egg-feeling". The following possibilities exist for suppressing broodiness:

- A hormone treatment with testosterone, oestrogens or the intramuscular administration of copper sulphate ; all these means are either too expensive or demand too much labour or both.
- The isolation of the hens in separate sections, which requires too much labour.
- A shift system, whereby the house comprises two sections (A and B), where the laying nests are only installed in one of the two sections and where the same section contains a number of small compartments (1, 2 and 3) for broody hens. A ground-plan of a house suitable for the shift system is schematically given in fig. 6.38. A partition, with a number of openings which can be closed, is installed between the sections A and B of the house. Turkeys are normally allowed in both sections A and B. The eggs are for the last

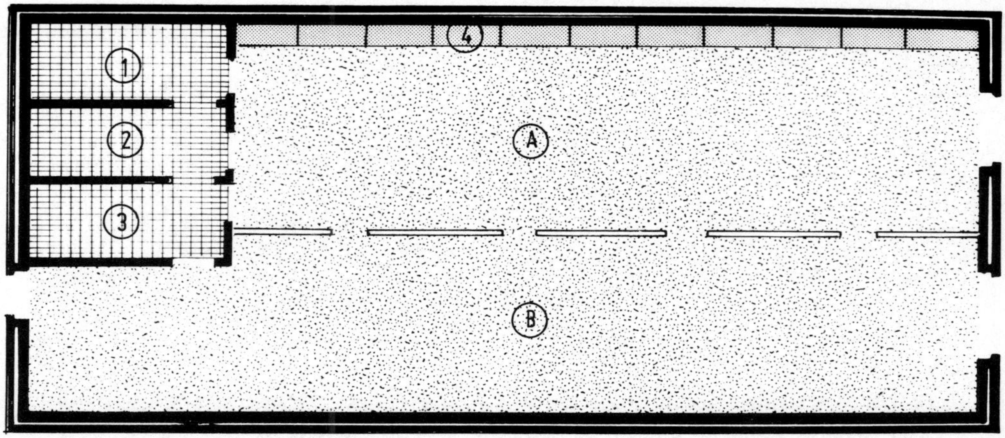

Fig. 6.38 Ground-plan of a house for turkey breeding hens suitable for the application of the shift principle.
Legend : A and B = compartments ; 1, 2 and 3 = compartments for broody hens ; 4 = laying nests.

time collected one hour before switching off the lights and the turkeys are then chased away from the nests. When the lights go out the different gates in the partition between A and B are closed. All birds which are then found in section A or in the laying nests are considered as broody and are chased in the first compartment for broody hens (1). The same treatment is repeated the next day after the broody hens from compartment 1 have been driven to compartment 2. Broody hens from compartment 2 are chased on the third day to compartment 3, those of compartment 1 to compartment 2 while compartment 1 can be filled up again. At the end of the fourth day the first broody hens which are kept in compartment 3 are released into the house. Broody hens thus remain in the compartments reserved for them for three days ; they are moved every day and receive 24 hours of light per day. Compartments for broody turkey hens must therefore be well-shielded from the other sections of the house. A high intensity of light is maintained in the broodiness compartments while the hens are cooled down by means of a sufficiently-rated ventilation. Feed and water are provided in the broodiness compartments but laying nests are of course lacking.

- The rotation-system or transfer-system. The house for turkey breeding hens is divided into a number of compartments (6 to 8). Each compartment is fully equipped and ready to house the birds. One of the compartments however is left free. A ground-plan of such a type of house is given in fig. 6.39.
  During the period in which broodiness can occur (from the fourth week of lay onwards) the hens are moved once per three weeks from

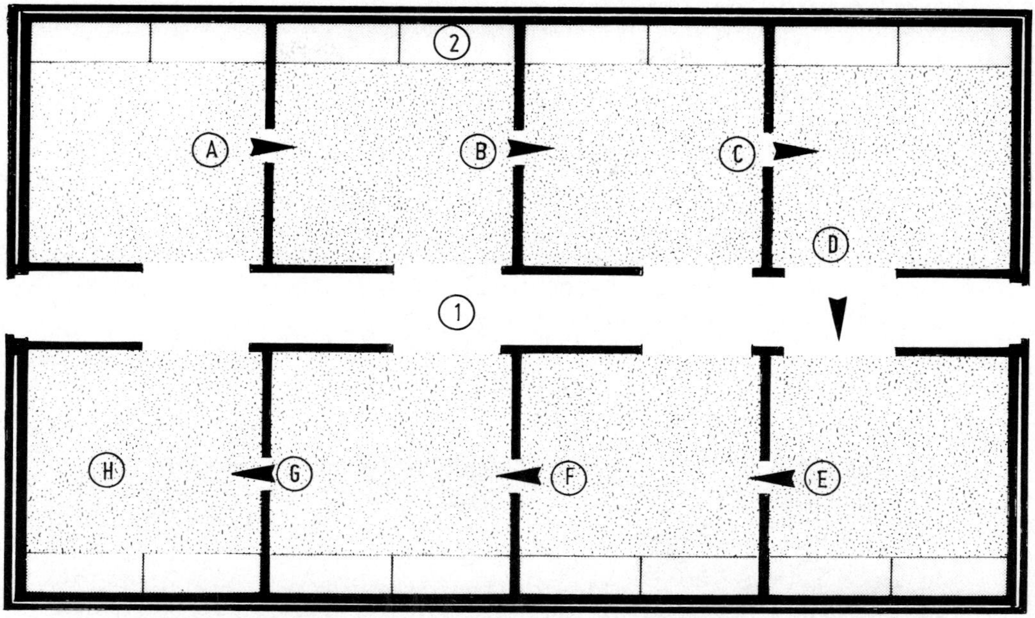

Fig. 6.39 The rotation or transfer system for turkey breeding hens. Legend : A, B, C, D, E, F, G, H = strawed compartments ; 1 = central alley ; 2 = laying nests.

one compartment to the next. It is recommended to combine this transfer with the artificial insemination. Compartment H (fig. 6.39) is unoccupied. The hens from G are transferred to H thereby freeing compartment G etc. In this way compartment A becomes unoccupied. With the next move compartment B becomes available etc. This moving of all hens once per three weeks (and not of only a part of the population as is carried out with the shift system) appears to suppress broodiness in practice.

- The noise : noise phenomena of 110 to 135 dB appear also to be effective in the suppression of broodiness.
- The electric shocks : electric shocks seem to influence broodiness but the results are no better than those obtained with the transfer system.
- Breeding-technical measures : the problem of broodiness can be solved through selection but this is only possible in the long term.
- Recently a device against broodiness was developed in Israel : above a row of laying nests a bar is installed on which one flap per nest is fixed ; at regular intervals, commanded by a timer the bar revolves and the flaps sweep over the nests, thus chasing the breeding hens.

In general the collection of eggs several times a day, together with intensive lighting contribute largely in suppressing broodiness.

6.7.2 Cage housing of turkey breeding hens

Cage housing (fig. 6.40) has a number of advantages :

- a beneficial influence on broodiness ;
- the lack of litter and subsequently no need for disinfection of the litter ;
- a smaller risk for an outbreak of the black head disease ;
- no necessity for adding testostats to the feed ;
- a substantial reduction of the labour requirement : one labourer can take care of ca. 5,000 turkey breeding hens compared to ca. 2,000 if they are kept on litter.

Fig. 6.40 Cage housing of turkey breeding hens.

The egg production is certainly not increased. Two turkeys are housed in each cage of which the dimensions are : 43 - 45 cm wide, 50 cm deep and 55 cm high. Turkeys housed in batteries are sometimes suffering from lameness and the only remedy for this is the immediate release of the animal. The keeping of turkey cocks (10 to 20 % of the flock of hens) in individual cages (fig. 6.41) is desirable since the production of semen appears to be 20 to 30 % higher in the case of cage housing. The dimensions of the floor of the individual cage for turkey cocks are 60 cm x 60 cm with a 5 cm-wide lath, acting as a perch, in the middle.

The egg production of a turkey hen amounts to 110 hatching eggs (varying between 90 and 120) per 11 months (two production periods

Fig. 6.41 Individual cages for turkey-cocks.

of 4 months with a 3-month rest period in between). The production in the winter half-year is lower than in the summer half-year.

6.8 THE HOUSING OF GUINEA FOWLS

6.8.1 Generalities

Guinea fowls are intensively bred in France where they form an important branch in poultry breeding but in many other countries the production is limited and forms only a minor part of the total poultry production (e.g. in Belgium only 438 tonnes in 1982 or a mere 0.3 % of the total poultry production).

Guinea fowls are extremely frightened and semi-wild and those facts should be taken into account when housing them. Guinea fowls are monogamous and this limited progress in their intensive breeding. It meant a serious financial load on the production of hatching eggs. Experiments performed by Dr. Petitjean at the I.N.R.A. in France have resulted in the development of a method for artificial insemination whereby only one cock is needed for every three or four hens. Artificial insemination together with the cage housing of the breeding hens of guinea fowls has been a breakthrough for guinea fowl breeding in France, and particularly in Brittany, since the early Seventies (Kuit, 1973).

We will discuss the housing of the breeding hens of guinea fowls and of guinea fowls intended for the production of meat.

6.8.2 The housing of guinea fowl breeding hens

The raising of day-old chicks hatching from the eggs and intended as hens or cocks is carried out in well-insulated (k < 0.7 W/(m^2.K)), windowless and mechanically ventilated houses of the low profile hangar type which resemble those applied for the rearing of other poultry (Kuit, 1973). The floor of the house is concreted (to allow a thorough disinfection and cleaning after each breeding period) and provided with a 15 cm thick layer of litter. This litter preferably consists of short straw or peat. Wood shavings often give less satisfactory results because of too much dust which makes the chicks nervous and of the danger of the communication of fungal diseases to which guinea fowls are very sensitive. About 3 kg of straw are given per m^2 of floor area which normally holds 8 to 12 chicks.

The house is equipped with a number of brooders under which a temperature of ca. 38°C is maintained with space for a flock of ca. 500 chicks. The hover space is surrounded by a guard with a diameter of 3 m and a height of 60 cm. The temperature is gradually lowered with increments of 2°C per week, for two weeks, followed by weekly increments of 3 to 4°C and this until a temperature of 20 - 22°C is reached at the age of 8 weeks. A temperature below 18°C has to be avoided as guinea fowls tend to huddle at this lower temperature and the death rate might thereby increase considerably due to suffocation.

The birds are de-winged after birth to prevent them flying in the house. It is customary to de-wing the cocks at the right side and the hens at the left. Raising both sexes separately under different brooders and successively in different houses is required because each sex is subjected to a different lighting pattern. Inside the guard one fount is provided for every fifty chicks and this is filled with water a few hours before the chicks are moved into the area. During the first few days the feed is supplied on carton trays which being noisy stimulate the chicks to eat. From the fourth day small boxes with feed are supplied and from the sixteenth or seventeenth day the chicks are able to take their feed from the feed trough which is provided with an automatic feed chain. The rim of the feed trough must be at the level of the birds'back otherwise excessive spillage might occur. Guinea fowls scratch with their beaks, unlike chickens which use their feet and for this reason considerable spillage can occur if the height of the feed trough is incorrectly adjusted. Cocks and hens are raised in different compartments. The day length is gradually shortened in the females'compartment from 20 hours a day to 8 hours a day at the age of 28 weeks. Darkening must be carried out gradually e.g. by means of a dimmer so as not to startle the birds. Similar to the rearing of other poultry short day rearing is also applied with guinea fowls. A constant day length of 14 hours is however maintained in the compartment housing the cocks. Dynamic ventilation is applied throughout the house.

As soon as the birds reach the age of 28 weeks they are moved to a well insulated (k < 0.7 W/(m^2.K)), windowless house equipped with three-tier batteries. The cocks are accommodated in the lower tier of cages and hens are moved into the upper cages (fig. 6.42)..

Fig. 6.42 A three-tier battery house for guinea fowl breeding hens.

The battery comprises six rows of cages, spread over three tiers, each of them containing two rows of cages back-to-back. Each cage measures 50 cm along the front, is 45 cm deep and 50 cm high and houses five guinea fowls. The partitions of the cages, together with the covers of the upper cages are made of full galvanized metal plates whereas the front, the rear and the floor of the cages are made of wire mesh. The floor of the cage slopes towards a folded lip outside the cage where the eggs are collected twice daily by hand. Underneath each tier of cages there are synthetic belts to hold and transport the droppings which pass through the wire cage floors, as described above. Scrapers to remove the droppings from platforms under each tier can also be used. The feed is supplied in feed troughs filled by means of an electric powered feed trolley as described above. Two nipple drinkers are supplied per cage. Passages of ca. 1.20 m wide are provided between the batteries and are used for the execution of the artificial insemination (five times per month) and for routine supervision. The ventilation is carried out by extraction fans installed in the ridge of the roof while the air inlets are located in the longitudinal walls, above the radiators of the central heating.

Throughout the entire laying period a day length of 14 hours is maintained by means of lamps, suspended at half the height of the battery. In order to allow a trouble-free execution of the artificial insemination the lamps can be pulled up either by means of a cable running over a number of wheels or by a winch and controlled by a central lever. The temperature must be maintained at 20°C and this requires a supplementary heating e.g. a central heating system with radiators installed under the air inlets. After a stay of 3 weeks in the batteries, hence at the age of 31 weeks, the guinea fowl hens start to lay. The laying percentage reaches 50 % at 34 weeks and increases further to 80 - 90 %. The lay normally extends over a period of ca. 40 weeks. Although the guinea fowl hen produces up to 170 eggs, the production of chicks per hen only amounts to between 90 and 100.

The daily care of guinea fowls requires two persons since artificial insemination cannot be carried out by one single person. They can supervise 2,000 to 2,500 breeding hens including other chores and especially egg collection. The labour conditions are not favourable, mainly because of the sharp and unpleasant noise produced by the birds which in large houses can be deafening.

6.8.3 The housing of guinea fowls intended for meat production

The fattening of guinea fowls, delivered at the farm as day-old chicks, is carried out in a well-insulated ($k < 0.7$ W/(m^2.K)), windowless house, which is mechanically ventilated and provided with a concrete floor covered with a thick layer of litter. The housing of guinea fowls intended for meat production is with a few exceptions similar to that described for the rearing of guinea fowl breeding hens. The stocking rate is somewhat higher with guinea fowls for fattening than for guinea fowls intended as breeding hens viz. 12 to 15 chicks per m^2. Flocks should not exceed 2,500 birds. Guinea fowl cocks and hens are not separately housed. They are provided with 20 hours of intense light a day during the first week (ca. 2 W/m^2) but from the second week of life the day length is reduced to 16 hours with a decreased light intensity (ca. 0.3 W/m^2). The lights are operated with a dimmer device preventing abrupt changes in light intensity and hence startling of the birds. A few nightlights remain on. Guinea fowls grow more slowly than broilers. They are fattened either up to the age of 10 - 11 weeks and a liveweight of ca. 1.050 kg or up to the age of 13 weeks when they weigh ca. 1.450 kg. Heavier birds have a more pronounced gamy taste than lighter ones.

6.9 THE HOUSING OF PHEASANTS

The pheasant has lived for many centuries in the wild in Western-Europe and has been a prime game bird. The natural stock of game has declined considerably since World War II and this can be attributed to the continuous urbanization, intensive application of insecticides and pesticides in agriculture, environmental pollution, etc. In order to replenish the stock of game, pheasants are now widely bred in specialized farms and then released in the shoot. The Mongolian game pheasant is widely found throughout Western Europe.

Pheasant breeding hens are kept in large aviaries which are 2 to 3 metres high, screened off with wire mesh incl. the top. A natural vegetation develops within the aviary and birds are able to fly around and live in a quasi-natural environment although limited in area. The breeding hens winter in the aviary, which is provided with a rudimentary shelter protecting the birds during severe weather conditions (mainly snow). The pheasants receive a compound feed of meal or pellets in a trough. Founts are also provided. One cock is kept for every 6 to 8 hens. The hens come into lay at the beginning of April and the lay extends up to the end of June. Egg production amounts to 40 to 45 eggs per hen. The eggs are set and remain in the "setter" for 21 days at e.g. a temperature of 37.6°C and a R.H. of 56 % after which they are moved to the "hatcher" for 4 days at e.g. 37.2°C and 58 % R.H. progressing to 73 % R.H. as hatching begins. Just-hatched chicks are removed every two to three hours and put in carton boxes with perforated sides. The boxes are stored in the hatchery, having a temperature of 30°C, where the fluff of the chicks can dry.

The chicks are then moved to a well-insulated ($k < 0.7$ W/(m^2.K)), windowless house which is constantly heated and mechanically ventilated and which resembles the broiler house. The floor is concreted and covered with a layer of coarse sand and white wood shavings. The chicks are placed under a brooder in flocks of max. 500 birds and surrounded with a guard, which can be removed after two weeks. An initial temperature of 37°C under the brooder and ca. 25°C in the house is maintained during the first week. The temperature is then gradually lowered by 4°C per week to reach 20°C in the fifth week and this temperature is maintained for the remaining time. The chicks are provided with light day and night. If the weather is good the chicks are allowed to the aviary on the fifth day. The aviary, which is 2 to 3 metres high, abuts against the house and is completely fenced (incl. the top) with wire netting. The stocking rate amounts to 1 bird per square metre. The aviary is provided with a layer of dry sand and is annually ploughed and harrowed by means of a two-wheel tractor (fig. 6.43). During the first few days the chicks are allowed in only a limited part of the house to help them find the feed trough and fount as soon as possible. Kraft paper is put under the feed trough to prevent the mixing of spilled feed with the wood shavings. The feed trough length is 4 m per hundred chicks if both sides can be used. The water trough length is 2 m per hundred chicks and is equipped with a float-system. After an initial period of a few days the birds are admitted to the entire house and the stocking density is then max. 12 birds per square metre. Quietness must be maintained in both the rearing house as well as the aviary since pheasants are very shy and tend to fly against the wire netting if they are disturbed e.g. in case of tumult. If this happens it is likely that some pheasants may get hurt or even killed. Debeaking or putting anti-pecking rings in the beak is a must to avoid cannibalism. Besides floor rearing, rearing in warm batteries is sometimes practised up to the age of 3 to 4 weeks : this is not only expensive but is, from our point of view, too artificial a rearing for birds which are in fact destined for game. The pheasants are kept in a collective aviary up to 10 weeks and are then released in the shoot in small groups. Prior to their release they are housed in a run, for 3 or 4 days, which is provided with a number of feed boxes and founts. In this way the pheasants will know where to find feed and water if they are finally released in the shoot. In practice, however, a number of pheasants viz. 30 to 40 % are kept longer in the aviary and are

Fig. 6.43 An aviary for raising pheasants.

only released when they are 12 to 18 weeks old . These birds often have problems adapting themselves to their new environment and a number of them perish. Furthermore, shooting pheasants which have only been released shortly before can hardly be called game and the hunt for such birds is not worth this name.

A few good pheasants are kept in the aviary as breeding hens while some wild pheasant cocks are either bought or caught and used to carry out an upgrading of the race. These pheasant breeding hens winter in the aviary and will produce new game pheasants in the coming year.

6.10 THE HOUSING OF QUAILS

The intensive breeding of quails is rather limited (e.g. in Belgium, in 1982, only 115 tonnes of meat, or a mere 0.1 % of the total poultry production). The production of meat and eggs (as a cold first course) are still the main orientations, but other markets are the re-population of shoots and especially their use as relatively cheap laboratory animals in medical, veterinary and pharmaceutical laboratories. It are mainly selected, less aggressive varieties of the Japanese quail (Coturnix sp.) which are used for breeding. These rather small birds are accommodated in a well-insulated ($k < 0.7$ W/(m² .K)), windowless and statically or dynamically ventilated house. Day-old chicks are placed in a layer of litter or most often in mini-batteries where they are *reared and fattened*. Three or more tier batteries are therefore used and the cages are 1 m long, 50 cm wide and 20 cm high and house 60 - 65 quails. The cages are made of galvanized wire mesh. Underneath each tier of cages a synthetic belt or a scraper on a platform is installed for the removal of the droppings which functions similar to the one described for layers. Each cage is equipped with a feed trough, filled by means of a feed chain, and a water trough. The temperature during the first week is 35 - 36°C, the second week 31°C, the third week 27 to 23°C, the fourth week 23 to 21°C and the fifth week ca. 20°C. Central heating in the house is therefore necessary. The day length is initially 20 hours and is after a few days gradually lowered to reach 14 hours per day in the sixth week.

Quails not intended for the production of hatching eggs are sold at the age of six weeks, whereas those destined to become breeding hens are transferred to another battery house. At the age of six weeks the cocks weigh ca. 110 g, the hens ca. 130 g, heavier varieties reach a liveweight of ca. 150 g or a net weight of 130 g.

Quail breeding hens are housed in batteries. Each battery has 3 or 4 tiers and is equipped as described above. The cages are 50 cm long, 50 cm wide and 20 cm high and accommodate 4 cocks and 12 hens. The cage has a sloping floor made of wire mesh and an upstanding rim to catch the eggs. An ambient temperature of ca. 20°C is maintained by means of a central heating system. Continuous lighting of 24 hours per day is commonly applied at industrial breeding plants. The hens come into lay from the age of seven weeks and produce 80 to 100 eggs in a period of four months, after which they are sold. At other farms a less intensive lighting pattern is sometimes applied whereby the hens are kept for a period of 50 to 60 weeks in which they produce ca. 300 eggs. The eggs are relatively heavy (1/20th of the body weight, compared to 1/40th for layers) and are brooded in the incubator (Geurden, 1974 ; Wilson, 1972).

6.11 THE HOUSING OF TABLE PIGEONS

Since the Middle-Ages pigeons have been a delicacy for the gastronome, the flesh is tasty and easy digestible but its consumption is limited (e.g. Belgium consumed ca. 1,000 tonnes in 1982 or only 0.8 % of the total poultry production).

The housing of pigeons has to be adapted to their specific way of life (Reyntens, 1971). These monogamous birds always live in brace, the hen pigeon normally produces two eggs which are brooded by the male pigeon as well as the hen pigeon. The poults are fed by both parents. The pigeon hen normally produces new eggs when the poults are 2 to 3 weeks old. Concerning the feed, pigeons are either fed ad libitum at a compartmentalized feed box where they have the choice between maize, peas, wheat and sorghum (the last two mainly as bait) and where they also receive some grit, ground oystershells etc. or they are fed ad libitum all-mash pellets. Finally pigeons must be able to fly around. Table pigeons are kept in a house, provided with a central feed passage and cages along both sides. The cages measure e.g. 3.7 m x 2.2 m and are ca. 3 m high and accommodate 8 to 10 brace of pigeons. The front of the cage, which abuts against the feed passage, is made of solid material (e.g. asbestos cement board) up to a height of ca. 1.5 m while the remainder is of glass ; it also includes a door. The rear of the cage, which in fact forms the outer wall, consists of blockwork up to a height of 1.5 m and the remainder of glass. The partitions between the cages are made of chicken wire. The floor is concreted and covered with a 2 cm thick layer of dry sand and straw. The latter must remain dry and has to be replaced regularly. A wooden grating with a manure "tray" below is also sometimes used as a floor. A perch of ca. 2 m length must always be provided. The feeding and watering facilities are similar to those used in deep litter houses for laying hens. Automatic feeders are installed at breast-height of the birds : a round feeder with a diameter of 35 cm suffices for 30 pigeons whereas when a feed trough is used a frontage of 2.5 cm per pigeon is satisfactory. An automatic drinker, with a content of 10 litres is sufficient for 30 pigeons if it is filled daily : the gutter must be designed in such a way that the bird can only put its beak in the gutter as this prevents water spillage.

A water bath is put at the pigeons'disposal once a week and is avidly used by them.

A double nest must be provided for each brace of pigeons : it comprises two parts which each measure 30 cm x 30 cm x 30 cm. These wooden nests hang in three rows above each other and each nest has an open front provided with a walk. Each box contains a nesting dish in which the hen pigeon builds her nest. A rack, filled with straw, is placed on the floor : the pigeon collects the straws and brings them to the hen pigeon which builds the nest.

Each cage can be connected to an aviary in the open air through an opening of 60 cm x 30 cm in the outer wall. A cage with the aforesaid dimensions accommodates 8 to 10 brace of pigeons . The size of the accompanying aviary is half that of the cage (Wriessnig, 1979), it is often provided with a wire mesh floor and a perch is also available. The temperature in the cage should not drop below 15°C and this necessitates supplementary heating during the winter. Natural lighting is indicated. Light should however be provided in the house for 14 to 16 hours a day to assure that the poults receive enough bait (Wriessnig, 1979) and this requires additional artificial lighting for 6 to 7 months per year.

The young pigeons are ready to be slaughtered after 28 days, thus before they leave the nest ; they weigh at that moment 450 to 500 g (Dubourg, 1973). The production of table pigeons continues throughout the year with the exception of a moulting period which lasts about six weeks.

Labour time requirement is around 75 minutes per brace of pigeons and per year. An additional chore, which is however useful to prevent unproductive brooding, is the weekly candling of the laid eggs. This can readily be done by a torch and an infertile egg is removed while the remaining egg can be put in another nest where there is also only one egg. The hen pigeon on the nest from where the eggs were removed will usually produce two new eggs eight or ten days later (Wriessnig, 1979). A kind of management recording system is required on the larger farms. It can for instance include the following data : number of the brace and of the corresponding nests, the laying dates, the hatching dates - 18 days after the second egg - the delivery dates and the number of delivered pigeons.

In order to obtain a profitable business, each brace of pigeons should produce at least 15 ready-to-slaughter poults per year. The production is usually at its maximum in the second and third year of life and drops rapidly from the fifth year of life of the parent-pigeons. The procreation of breeding hen pigeons takes place by rearing selected birds up to ca. 5 months of age. They are then coupled viz. by locking a couple in "living quarters" for some time after which they can be released in the cage as a new couple.

6.12 THE HOUSING OF DUCKS

The breeding of ducks is particularly in France a significant branch of poultry production as the production of table ducks there represents ca. 5 % of the total poultry meat production. In Belgium, however, the production of table ducks amounted to only 396 tonnes in 1982, or a mere 0.3 % of the poultry meat production and is there unimportant. The main varieties developed for meat production or as egg layers, are Muscovy, Domestic Mallard types, Rouen, the white Aylesbury and the Peking. The meat of the duck is rather rich in fat (a fat content of ca. 17 % compared to ca. 5.5 % for broilers) and is therefore not desired by everybody ; although no one calling himself a gastronome would disapprove of a "magret de canard", a "caneton à l'orange" or a "caneton aux navets". Some varieties of ducks are suited for the preparation of "pâté de foie gras" : about 65 % of this French delicacy is in fact made of ducks' livers.

With the *extensive breeding* the birds are provided with a shelter, an exercise run and a pond for brood ducks and table breeds, although the latter are sometimes provided with only just a water basin. The brood ducks are always locked in a deep litter house during the night and artificial lighting is applied to encourage egg laying in wooden nests. Such large extensive breeding farms are mainly found in Central Europe where for instance the State Farm in Tata, in the West of Hungary produces annually ca. 450,000 meat ducks for export to Western Europe and the U.S.A. In France there are also a number of large semi-extensive farms such as the l'Elevage du Moulin du Lée (Indre et Loire), which keeps ca. 25,000 breeding ducks and produces some 2 million meat ducks per year.

The *intensive breeding of ducks* by the all-in all-out system is becoming increasingly important, especially in France. The day-old chicks are housed in well-insulated ($k < 0.7\ W/(m^2.K)$) and mechanically ventilated buildings. They are kept in groups of max. 300 chicks under a brooder where a temperature of ca. 35°C is maintained for several days. The hover space is at a distance of 1 metre surrounded by a guard, made of wire mesh, with a height of 40 cm. The temperature under the hover is gradually lowered e.g. from the 3rd day to the 8th day to 30°C and further to 25°C at the age of 13 days while the temperature in the house is maintained at 17°C. In this way the chicks are accustomed to the lower temperature of the fattening house to where they will be transferred at the age of two weeks.

Some other breeders prefer to keep them in the starting house up to the age of 3 to 4 weeks before moving them to the fattening house and they lower the temperature in the hover space more gradually.

A few night lamps remain on during the night. In the first week the feed is served on carton plates which are placed inside the guard. Later on the feed is supplied in wooden troughs (3 m of feed space per 100 ducklings). After about 10 days hanging trough feeders can be used. Also the first week water is provided in so-called siphon founts viz. one with a content of 2 litres and this suffices for 50 ducklings. After the house is emptied, careful disinfection and cleaning is carried out.

The *production of meat* is also carried out in an intensive way. The ducklings are shifted at the age of 2 to 4 weeks to a naturally ventilated house provided with windows. The floor of the house is either completely littered with straw, or 50 % of straw with the remainder a wire mesh or wooden grating, or all wire mesh or a grating. Ducks tend to spill a lot of water and this creates problems for keeping the litter dry : the ducks befoul themselves and are more easily infested with parasites. This problem can be solved by placing the water troughs and founts on a slatted floor. Research carried out by Tüller (1979a) has proved that the growth and feed conversion of ducks housed in fully slatted houses are as good as of those kept on half slatted floors or on litter. The same author (Tüller, 1979a) demonstrated furthermore that fully slatted floors are cheaper than littered ones ; the litter is in fact expensive to purchase and demands much labour for its distribution and removal. The fully slatted floor can either be a wire floor or a grating made of laths. The wire floor is normally made of 4 mm thick wire and the meshes measure 60 mm x 25 mm. The manure is deposited in a pit beneath the floor and is normally ca. 40 cm deep. Ducks normally utilize the whole floor area and are very quiet. The slatted floor in fig. 6.44 consists of wooden laths (20 to 25 mm wide with slots of 20 to 25 mm) with underneath a large manure pit, which is ca. 40 cm deep and slopes towards a pumping pit. The birds are kept in groups of 500 ducks and at a rate of 6 to 7 per square metre. The feed is supplied by means of (mechanically filled) hanging feeder troughs (1 per 50 ducks). Chain feeders are an obstacle for these clumsy birds and are therefore not recommended. Short water troughs of ca. 2 m length are used for the drinking water distribution. They are suited for 250 ducks. The ducks are often allowed to an outside run, which is also provided with a slatted floor. During the summer the ducks may be allowed to the outside run after one week, but only after two weeks during the winter. They have access to the outside run through a number of openings which can be closed by means of small doors. The run is fenced with wire mesh. The entire house is properly disinfected and cleaned out after each lot of ducks is sold. It is sufficient to keep the house free of frost.

The ducks are often slaughtered when they reach the age of 7 weeks and they have then a liveweight of up to 2 kg (Tüller, 1979a). Drakes are sometimes kept up to 10 weeks when they have a liveweight of up to 3.4 kg (Tüller, 1979a).

The *keeping of breeding ducks* takes place in similar houses. One drake is kept with every five ducks and the latter produce a to-

Fig. 6.44 The fully slatted house for table ducks.

tal of ca. 80 eggs per duck and per laying period from the age of 6 or 7 months. The laying period extends over 5 to 6 months and is followed by respectively a moulting period of ca. 3 months and a new laying period in which each duck produces 60 to 70 eggs. The eggs are brooded in an incubator. Experiments have been carried out to raise ducks in cages (Rebaud, 1973), but this housing has as yet not been applied in practice.

6.13 THE HOUSING OF GEESE

Geese, from time immemorial have been bred by man and frescoes have been found in Egyptian tombs, dating back to 3,000 B.C. representing geese breeding farms (Willems and Brandt, 1971). Even now extensive geese keeping is widely practised in some countries such as Poland. In many countries its production is limited (e.g. in Belgium only 270 tonnes in 1982). In recent years there has been an increased and lively interest for table geese in Germany (Rheinland). They are kept on pasture at the rate of 50 to 150 birds per hectare. This of course leads to extensive damage to the pasture and to parasitic diseases of the geese. The table geese are slaughtered at about 30 weeks old (Tüller , 1979b).

Intensive geese keeping has only developed in the last decade and mainly in France (Rousselot-Pailley, 1975). The geese are in the first place not bred for their meat, which is juicy and tasty although very rich in fat (fat content of goose-flesh : 31 % compared to ca. 5.5 % for broilers) but, paradoxically enough, for the preparation of fat liver (foie gras), which in fact is even fatter than the flesh, but is "the" outstanding gastronomic delicacy.

6.13.1 The rearing of geese

The rearing of goslings is carried out in well-insulated (k < 0.7 $W/(m^2.K)$) buildings, with or without windows and naturally or artificially ventilated. The concreted floor is covered with a layer of litter. The chicks are placed under a brooder where a temperature of 32 to 35°C is maintained during the first two weeks, whereas the temperature in the house is 20°C. The temperature in the hover space is gradually lowered to between 26°C and 30°C in the third and fourth week and to 25°C in the fifth and sixth week while the temperature in the house is reduced to 18°C. The brooders are removed in the seventh week and the temperature in the house is maintained at 18°C. The stocking rate is reduced from 10 chicks per m^2 during the first two weeks to 5 per m^2 during the third and fourth week and finally to 2.5 goslings per m^2 in the fifth week. The goosery is therefore divided into several compartments to which the chicks are allowed in three steps. From the fifth week onwards and when weather conditions permit, the goslings are allowed to an outside run which is provided with a wooden grating or with boulders and which measures 0.5 m^2 per gosling. Feed and water facilities are similar to those described for ducks.

It is however necessary to provide a number of racks which must frequently be filled with fresh grass. Rearing geese without provision of grass will inevitably lead to cannibalism. The first rearing phase is terminated when the geese are 8 weeks old and at that time they normally weigh 4 to 5 kg.

The geese are sometimes kept longer until the age of 10 weeks when they have reached a liveweight of 5 to 6 kg and are then slaughtered as young table geese.

Further raising extends however over a period of ca. 7 months and is preferably carried out in buildings furnished with a wooden slatted floor (2 geese per m^2) and a slatted outside run (0.5 m^2 per goose). The slatted floor is 50 cm above a concrete floor and is in Germany often referred to as "Gänsebalkon" (geese balcony) (Tüller, 1979b). These slatted floors are a must if one wants to prevent parasitic diseases which are mainly carried over by contact between the birds and the excreta. The feeding equipment and watering devices are the same as described above and also include the provision of fresh grass (fig. 6.45). The fact that it is still impossible to raise geese solely with concentrates, although intensive research has been performed on this matter, largely prevents the industrial raising of these birds. The growing and harvesting, together with the daily distribution of fresh grass and the removal of wilted grass refused by the birds, forms a heavy burden for the intensive rearing

Fig. 6.45 The slatted floor house for the raising of geese. Note the installation for the distribution of fresh grass.

of geese. The grass consumption amounts to 1 kg per bird and per day. After rearing, the birds are either intended as breeding geese or kept for the production of "foie gras". Sometimes they are slaughtered as table geese at the age of 15 weeks.

6.13.2 The keeping of breeding geese

Breeding geese are kept in a half open house provided with a wooden slatted floor (3 geese per 2 m^2). The feeding equipment and watering devices are the same as described above. The daily distribution of fresh grass is also necessary, otherwise egg production may drop considerably (Rousselot-Pailley, 1975). In order to achieve a good copulation it is necessary to provide a cemented pond (0.5 m^2 per goose) which abuts against the house proper : this pond needs only to be 20 to 30 cm deep, must be easy to reach and is fenced with a galvanized wire netting (fig. 6.46). The natural laying period commences at the end of January and extends to the end of June. Geese produce eggs from ca. 8 months and the maximum laying rate is reached in the 2nd to 5th year of life. Each goose produces annually ca. 20 to 30 chicks which are obtained by brooding the eggs in an incubator. Two laying periods can be obtained per year by allowing the breeding geese to an outside run during part of the day and by subjecting them for the remainder of the time to a lighting pattern in a closed, windowless and well-ventilated house. The first laying period then commences in November and extends to March whereas the second lay begins at the end of May and finishes at the end of August. Breeding geese can be kept for up to 10 years.

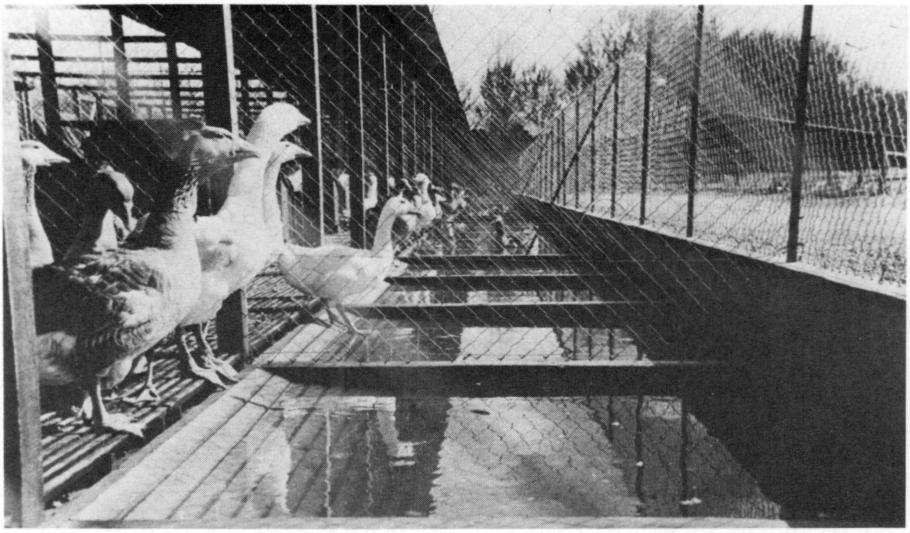

Fig. 6.46 The slatted floor house with pond for the breeding geese.

6.13.3 The keeping of geese for the production of fat liver

Geese kept for the production of fat liver are also accommodated in fully slatted houses. Adult birds are fed by literally pushing the feed into the throat of the bird (le gavage). White maize is normally used since this stimulates the storage of fats in the liver and results in a nice pink-white colouring of the liver. Finger-thick plugs of feed are pushed down the throat by means of a stick or syringe. This is done twice or three times a day for 3 to 4 weeks and leads to a total feed intake of 25 kg of maize for a Landes-goose and of even 45 kg for a Toulouse-goose. "Le gavage" requires patience and skill, not to mention the injuries to the hands. This "technique" leads to an excessive development and fattening of the liver which can be considered as pathological since it is mostly accompanied with icterus. Needless to say that this cruel treatment, which can be brought back to the gastronomic demands of the consumer, has received strong opposition from the Societies for Animal Welfare. It should be possible by surgical removal of the saturation centre in the hypothalamus of the brains of the goose to bring the geese to an intensive feed intake and thereby replacing the technique of "gavage" by "autogavage" (Castaing, 1975). This is economically not justified because the livers obtained in this way are of inferior quality while the pathological and repellent character of the production of fat liver still persists.

REFERENCES

Anon., 1975a. Dierenbescherming en legbatterijen, Bedrijfsontwikkeling, 6 : 502.

Anon., 1975b. Het deep-pit-systeem in Engeland, Bedrijfsontwikkeling, 6 : 424.

Anon., 1981. Rapport nr. SEC (81) 1283 de la C.E.E. du 31.7.81 au sujet de "L'élevage des poules pondeuses en cages : aspects économiques", Brussel, Belgium, 45 pp.

Anon., 1982. Gezinsverbruik van gevogeltevlees in België, (1975 - 1981), Agricontact nr. 132 : VII b 1-7, Sept. 1982.

Banks E.M., Wood-Gush D.G.M., Hughes B.O. and Mankovich N.J., 1980. Social rank and priority of access to resources, Behav. Processes, 4 : 197-209.

Batel W., 1977. Geruchstoff-, Staub- und Lärmebelastung in Anlagen der Tierproduktion, gemessen im Verlauf eines Jahres - zweiter Bericht, Grundlagen Landtechnik, 27 : 83-87.

Bossuyt H., 1979. Kalkoenen moet men in tomen opkweken. Pluimvee, 14 : 379-382.

Brantas G., 1973. Onderzoek naar de oorzaken van grondeieren e.a. Jaarverslag Instituut voor Pluimveeonderzoek, Spelderholt, the Netherlands, nr. 149, pp. 27.

Castaing J., 1975. Le foie gras, bien le produire, et bien le vendre, L'élevage, nr. 42, pp. 85-87.

Claeys N., 1966. Arbeidstechnische en ekonomische studie bij de inrichting van legkippenstallen, Thesis Rijkshogere Technische School voor Landbouw en Landbouwindustrieën, Gent, Belgium, 93 pp.

Comberg G. and Hinrichsen K., 1974. Tierhaltungslehre, Ulmer Verlag, Stuttgart, W. Germany, 464 pp.

Daelemans J., 1967. Gebouwen voor en arbeidsorganisatie in de legkippenhouderij, Mededelingen van het Rijksstation voor Boerderijbouwkunde te Merelbeke/Gent, Belgium, nr. 23, 146 pp.

Dubourg A., 1973. Elever des pigeons, Maison Rustique, Paris, France, 60 pp.

Duncan I.J.H. and Wood-Gush D.G.M., 1972. Thwarting of feeding behaviour in the domestic fowl, Animal Behaviour, 20 : 444-451.

Fujita H., 1973. The effect of length of daily light periods on diurnal feeding activity of laying hens, Jap. Poult. Sci., 10 : 123-127.

Geurden A., 1974. Kwartels, Pluimvee, 9 : 286.

Hammer P.R., 1973. The feeding habits and general behaviour of caged layers under different feeding regimes, Dissertation submitted for BSc, University of Nottingham, School of Agriculture, Sutton Bonington, G. Britain.

Hoogerkamp D., 1974. Deep-pit stal, Bedrijfsontwikkeling, 5 : 727-729.

Hughes B.O., 1972. A circardian rhythm of calcium intake in the domestic fowl, Br. Poult. Sci., 13 : 485-493.

Hughes B.O., 1975. Spatial preference of the domestic hen, Br. Vet. J., 131 : 560-564.

Hughes B.O. and Wood-Gush D.G.M., 1973. An increase in activity of domestic fowls produced by nutritional deficiency, Anim. Behav., 21 : 10-17.

Jongenburger H., 1982. Bezettingsdichtheid van leghennen op de batterij, Bedrijfsontwikkeling, 13 : 447-452.

Kraggerud H., 1963. Research on housing and equipments for layers, Report nr. 28 of the Institute of Agricultural Structures, Agricultural College of Norway, Vollebekk, 18 pp.

Kuit A., 1973. De parelhoenderteelt in Frankrijk, Instituut voor Pluimveeonderzoek, Het Spelderholt, the Netherlands, 5 pp.

Kuit A., 1983. Aangepaste huisvestingsvormen voor leghennen, Bedrijfsontwikkeling, 14 : 30-32.

Langeveld H., 1970. Verschillende bedrijfssystemen in de pluimveehouderij, Onze bedrijfspluimveehouderij, 5 : 35-39.

Lee D.J.W. and Bolton W., 1976. Battery cage shape, The laying performance of medium and light body-weight strains of hens, Brit. Poult. Sci., 17 : 321-326.

Litjens C., 1965. Diergeneeskunde en Slachtkuikens, Landbouwgids, Utrecht, the Netherlands, pp. 373-375.

Matthes S., 1979. Art und Zusammensetzung der Luftverunreinigungen in der Nutztierhaltung und ihre Wirkung in der Stallumgebung, Dtsch. tierärtzl. Wochenschr., 86 : 262-265

Petersen J., 1984. Jahrbuch für die Geflügelwirtschaft 1984, Ulmer Verlag, Stuttgart, W. Germany, 216 pp.

Rebaud M.L., 1973. Les méthodes d'élevage du canard à rotir, Nouvelles de L'Aviculture, 12 (188) 7-9.

Reyntens N., 1971. Het kweken van vleesduiven kan renderend zijn, Bedrijfspluimveehouderij, 6 : 355-357.

Rousselot-Pailley D.,1975. La reproduction des oies, L'élevage, numéro spécial : "Les jeunes", Paris, France, pp. 157-162.

Simons R.C.M. and Zegwaard A., 1983. Verlichting in verband met produktiviteit en energiebesparing van leghennen, Bedrijfsontwikkeling 14 : 785-789.

Simonsen H.B., Vestergaard K. and Willeberg P., 1980. Effect of floor type and density on the integument of egg-layers, Poult. Sci., 59 : 2202-2206.

Smith R. and Noles R.K., 1963. Effects of varying day lengths on laying hen production rates and annual eggs, Poultry Science, 42 : 973-982.

Stigter E., 1982. De kalkoenhouderij in Nederland, Bedrijfsontwikkeling, 13 : 269 -271.

Tüller R., 1979a. Zur Haltung von Flugenten auf Drahtrosten, Deutsche Geflügelwirtschaft und Schweineproduktion, 31 : 792-794.

Tüller R., 1979b. Balkonhaltung von Gänsen, Deutsche Geflügelwirtschaft und Schweineproduktion, 31 : 648-649.

Visser A., 1965. Slachtkuiken-moederdieren, Landbouwgids, Utrecht, the Netherlands, pp. 370-371.

Wegner R.M., 1980. Evaluation of various maintenance conditions for laying hens. In : Animal Regulation Studies (3) 73-82, Elsevier Science Publishers, Amsterdam, the Netherlands.

Wegner R.M., 1982. Ergebnisse der qualitativen und quantitativen Untersuchungen zum Verhalten, zur Leistung und physiologisch-anatomischen Status von Legehennen in unterschiedlichen Haltungssystemen (Auslauf-, Boden- und Käfighaltung), Der Tierzuchter, 34 : 416-418.

Wells R.G., 1973. Stocking density and colony size for caged layers, Proc. IV Europ. Poult. Conf. London, G. Britain, pp. 617-622.

Wilson W.O., 1972. A review of the Physiology of Coturnix (Japanese Quail) World's Poultry Science Journal, 28 (4) 413-429.

Willems R. and Brandt E., 1971. Hoenderachtigen en Watervogels, Uitg. K.M. Het Neerhof, Gent, Belgium, 362 pp.

Wood-Gush D.G.M., 1972. Strain differences in response to sub-optimal stimuli in the fowl, Animal Behaviour, 20 : 72-76.

Wriessnig E., 1979. Taubenmast, wirtschaftlich betrieben, Deutsche Geflügelwirtschaft und Schweineproduktion, 31 : 1142-1146.

Chapter 7

THE HOUSING OF HORSES

Chapter 7

THE HOUSING OF HORSES

7.1 GENERALITIES

Horses, throughout the ages, have been used for a large variety of purposes. Although horses were used for food by prehistoric people there is no evidence of domestication until the 4th or 5th millennium B.C. The horse was probably domesticated by the Mongols. The Greeks and Egyptians utilized the horse by 2,000 B.C. The Arabian breed is a modern horse with the longest known history.

Draught horses used in modern times grew from the horses selected by medieval knights, for their strength. As a draught horse it has proved invaluable for the transport of persons (diligence, horse tram, etc.) and goods (transport along the roads, in mines, harbours, etc.). It has also played an important role in many wars throughout the centuries and this until World War I where many thousands of horses were killed. In farming it was praised for its tractive power. It is nowadays largely replaced on the farm by the tractor and for the transport of persons and goods by motor vehicles. Horses are mainly attributed a recreational and sporting function although they are also still used for display (Horse-Guards, a cavalry of British Household troops i.e. Life-Guards and Royal Horse Guards) and as an aid for maintaining law and order.

In its recreational functions (riding, jumping) equestrian sport has a large interest which can be related to the striving of the city dweller to return to nature while on the other hand it continues to offer the rural youth a traditional recreation. Horse racing is the leading gambling sport in many countries and is one of the oldest pastimes dating back at least to Greek chariot racing. The first public racecourse was the Smithfield Track in London in the late 12th century. Polo and jumping are other well-known equestrian sports. The high costs involved with the keeping of riding and racing horses combined with the economic recession have contributed to a marked decline of their number. The number of draught horses used in agriculture also continues to decrease. Table 7.1 shows the evolution of the number of horses in 9 E.E.C.-countries. Table 7.2 illustrates the separate evolution of the number of draught and riding horses in Belgium.

The housing of horses must be designed in such a way that it complies with the requirements set by the specific nature of the horse. Labour technical and economic aspects of the housing must also be considered.

The horse in its natural state is an animal of the steppe which lives in herds (Franke and Nicolay, 1973). It is an outstanding example of an animal of movement : on one hand the poor vegetation of the steppe forces the horse to traverse long distances in order to find its food, on the other hand escape is the main defence of the horse against its natural predators. In this connection the horse shows a

TABLE 7.1 The number of horses in the different E.E.C.-countries (x 1,000) (Anon., 1983).

Country	Year			
	1978	1979	1980	1981
W. Germany	371	378	380	383
France	328	308	293	254
Italy	524	510	500	483
the Netherlands	76	75	70	68
BLEU	46	43	40	35
G. Britain	205	205	205	205
Ireland	115	100	105	100
Denmark	61	60	56	50
EUR 9	1,726	1,679	1,649	1,578

TABLE 7.2 The number of horses in Belgium (Anon., 1960-1982).

Year	Draught horses used in agriculture	Other horses	Total
1959	169,745	3,242	172,987
1964	119,546	4,823	124,369
1969	50,955	29,817	80,772
1975	35,498	16,402	51,900
1980	19,393	13,640	33,033
1981	18,028	13,461	31,489

continuous attentiveness for its environment through eyes and ears. The horse has a strong respiratory system which is however very sensitive and ill-kept horses therefore often suffer from incurable damage of the respiratory organs (short-windedness). As an animal of the steppe the horse tolerates low temperatures and wide temperature fluctuations. It takes its feed (mainly grass) from the ground in a standing position with bent-down head.

Although horses have been domesticated for thousands of years the specific ethological characteristics of a horse should still be taken into consideration.

With the so-called "Robusthaltung" of horses the housing is limited to a minimum. The horses are allowed on pasture or at least on an outside run (min. 30 m^2/head) during the whole year and have only access to an hangar which consists of two walls made of wood or concrete plates and two open sides directed opposite the prevailing wind direction. The roof is made of corrugated asbestos cement or metal sheets. The hangar is often divided into two compartments viz. the littered lying area (10 m^2/horse, 7 m^2/pony) and the feeding area with manger, rack and feed supplies and a hardened floor. This so-called "Robusthaltung" gives good results (Marten, 1978), even for Arabian Thoroughbreds (Marten, 1982) but requires rather large areas of land. However, in many climate zones, housing of horses is in winter-time (snow) necessary. Growing foals and certainly ponies or small horses are preferably housed in group (loose stables). Large horses can hardly be housed in groups. The establishment and the maintenance of the hierarchy within the herd would undoubtedly lead to unrest and lesions of the animals which in a closed space possess little means of escape. Furthermore, group stabling is inconvenient for the individual care of the horses. In view of the high financial value of many sport and draught horses it is irresponsible to take the aforesaid risks. Large horses should therefore be individually stabled and one then has the choice between a tie stall and a loose box. A comparison between the stabling types is therefore useful (Schnitzer, 1971).

We will describe the following types of housing : tying stalls, box stalls and loose house for horses. We will also discuss the construction and equipment of a riding school.

7.2 THE CONSTRUCTION AND EQUIPMENT OF STABLES

7.2.1 The tying stall for the individual housing of horses

The stable is often a portal framed building. The walls of the stable have a k-value of 1.1 W/(m^2.K) consisting for example, from the outside to the inside, of a 1/2 brick + cavity + 14 cm of hollow blockwork. The walls are 3 m high. The roof is made of red or black corrugated asbestos cement or metal sheets with an insulation layer, e.g. 5 cm thick chipboard. The roof has a pitch of ca. 20° (ca. 36 cm/m). The floor consists of clinkers on a layer of light concrete (8 cm) and rubble (10 cm) or of a layer of concrete (15 cm). The concrete must show a ribbed finish in the lengthwise direction of the stall to prevent the animals slipping. The stable space has to be rated at a minimum of 30 m^2 per animal. A one-row layout is used for up to twelve horses. If one wants to house more than 12 horses a two-row arrangement is preferable as this facilitates the supervision and prevents long travelling distances. In this case the facing-out arrangement is always chosen and a feeding passage of at least 1 m wide is provided in front of each row of stalls. This allows the horses to have a view over the surroundings and will quieten them. Another advantage of the feeding passage is that it allows the rationalization of the feeding work.

The stable block abuts at one side on the feed room, where hay, straw and concentrates are stored. The feed storage requirement is rated at ca. 24 m^3 per horse and per annum.

The distribution of feed can be carried out by means of a trolley, i.e. a four-wheeled cart with low platform, on which bales of hay or straw can be placed together with the receptacles containing feed and which is pulled through the feeding passage. Hay and straw are often also stored in the loft above the boxes or stalls. The loft normally has a concrete floor. This solution is less interesting from the labour-technical point of view and requires a much higher investment than ground-floor storage ; it is therefore not recommended. The other side of the stable block gives access to the outside and the manure stack. The size of the concreted apron is calculated for a manure storage of 0.6 m^3 per animal and per week (with daily mucking-out of the stable). The manure can also be stored in large containers, sometimes supplied and removed by firms with the intention of delivering it to mushroom producers. The littered stall is at least 2.5 m long, as measured from the base of the trough to the gutter, it has a slope of 2 % and is 1.5 m wide (fig. 7.1) : the horses have then sufficient freedom of movement, even when lying down or sleeping. They cannot turn in the stall and befoul it or hinder their neighbours. A concrete trough with a length of 80 cm, a width of 50 cm and a height of 1 m is situated in the front of the stall. The width of the trough gradually increases from the top, where it is 50 cm wide, towards the floor, where it is 70 cm wide. This particular design prevents traumata to the carpal joints of the horses. The hay rack is placed above or beside the trough. In the latter case a box, with a width of 40 cm and a depth of 5 cm is put under the rack and the larger part of the spilled hay is collected in it. Especially the residues of lucerne hay are completely eaten. A brick wall completes

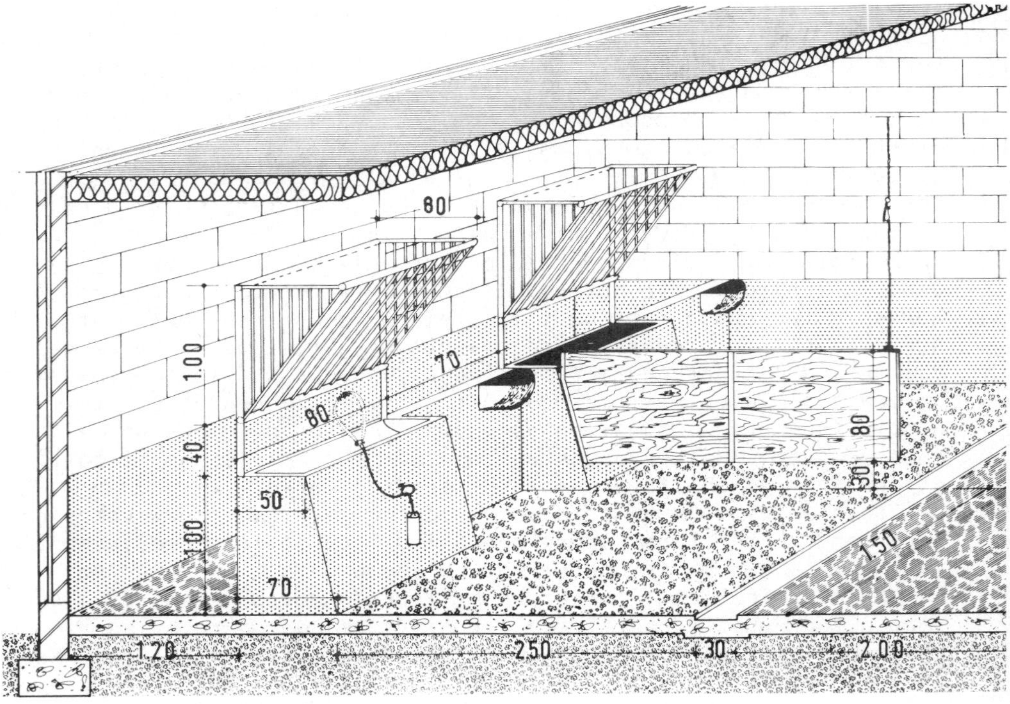

Fig. 7.1 The tying stable for horses.

the front of the stall, viz. the remaining 70 cm. This is an ideal location for the automatic drinking water bowl. Automatic water bowls equipped with an easily movable lever are to be preferred since the horse has a very sensitive muzzle. Each of the drinkers must be provided with a cock viz. to prevent a sweating horse from drinking. The horse is tethered by a tying system, which can consist of a metal ring, cemented in the trough and through which an iron chain glides ; this chain is connected to a leather strap which is put around the head (halter) or the neck (neck-collar) of the horse. A counterweight is attached to the end of the chain and keeps the chain stretched. This tying system gives the horses considerable freedom of movement.

The individual stalls can be separated from its neighbours by a stable bar i.e. a wooden beam of 10 cm x 15 cm, held in strip iron and suspended at 90 cm above the floor. It is fixed in the front on a concrete pillar (20 cm x 20 cm) by means of rings and at the rear on an iron pole (Ø 20 cm) or on the ceiling by means of a ring and chain or cable.

The partition can also be made of a planking suspended from the ceiling ; or it can consist of a gummi mat (3 cm thick) or of a fixed wall (fig. 7.2).

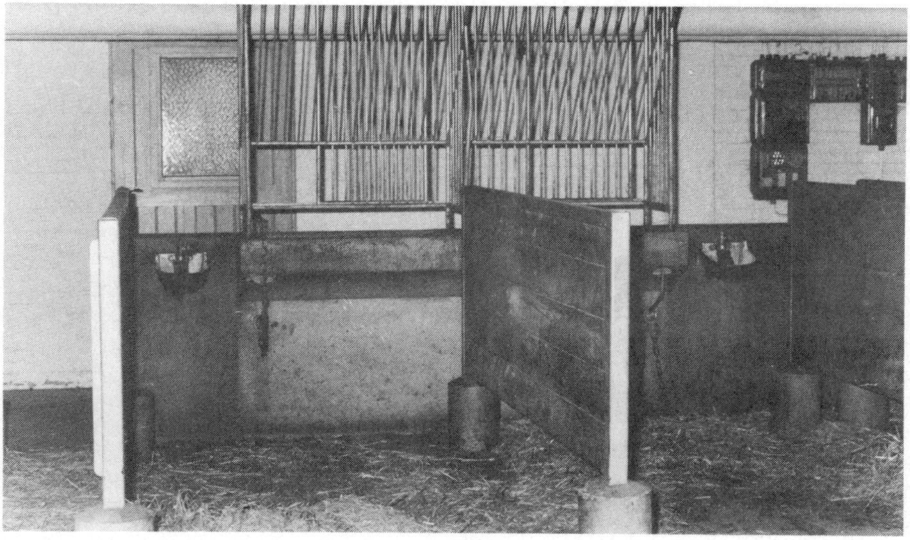

Fig. 7.2 The tying stall with fixed partition for the horses.

The wall along the "last" stall is protected by a strong cement-sand rendering or by a kicking board, i.e. a planking nailed to clamps attached to the wall, and of which the intermediate space between the clamps and the planking is filled up with concrete. A flat gutter is provided at the rear of the stall, its depth should not exceed 3 cm to prevent the horse stumbling which could result in injuries. This gutter is mainly intended for the collection of the cleaning water and its conveyance to a drainage system. The quantity of urine produced by the horse is rather small (ca. 3 to 6 l per day), and is mostly absorbed by the litter.

A wide service passage behind the row of stalls or between two rows of stalls is necessary. It must be wide enough to allow one handler with two horses to pass while it also serves for mucking-out, littering and in the absence of a feeding passage for the feeding of the horses. A wide enough service passage will also reduce the likelihood of injury to the handler and passing horses if the tied horses are frisky. The service passage must therefore be at least 2 m wide in a single-row stable and at least 3 m wide in a two-row stable.

7.2.2 The loose box for the individual housing of horses

The frames, walls, roof and flooring of the loose box stable are similar to those used for the tying stable.

For up to 8 - 10 horses the stable can be of the one-row type, whilst a larger number of horses requires a two-row arrangement with preferably a central service passage between two rows of boxes ; this arrangement is largely instrumental in obtaining a better view of the horses and in the reduction of the travelling distances. The stable abuts at one end on a storage where hay and straw supplies are stored. The other side gives access to the outside and includes a concreted apron on which the manure is stored. We refer for these items to the tying stable.

Each box accommodates one horse and has a floor area of 10 m^2 for a riding or race horse (3 m x 3.5 m) and 12 m^2 for a brood mare or stallion (3 m x 4 m) (Hardmuth, 1972). The shortest side of the box lies along the feeding passage (figs 7.3 and 7.4) and the boxes are littered. The lower part of the partition walls between the boxes and the service passage is made of a brick wall (20 cm thick) or of planks (4 cm thick) of 1.45 m high. The upper part of these partition walls consists of a wire mesh of 1 m height of which the mesh wires (Ø 4 mm) form squares of 5 cm x 5 cm. It can also consist of vertical bars (Ø 15 mm) which are 5 cm apart. This open partition helps with the herd instinct of the horse. If one of two neighbouring horses becomes aggressive the wire mesh can be screened by means of planks or other means or the horse can be shifted to another box. The meshes or gaps in the partitions should not be larger than 5 cm in order to prevent the horse getting its hoofs stuck in the gaps or biting the partition material which might result in the loss of teeth. The partitions have a total height of 2.45 m. The side walls of some of the boxes shall be provided with a wooden sloping (15°) belt (cfr. description of the manège) in order to prevent injury to the horses, which often stand close to the walls. The partitions of the boxes accommodating stallions are made of bricks up to their full height (3 m). Nervous horses can often be

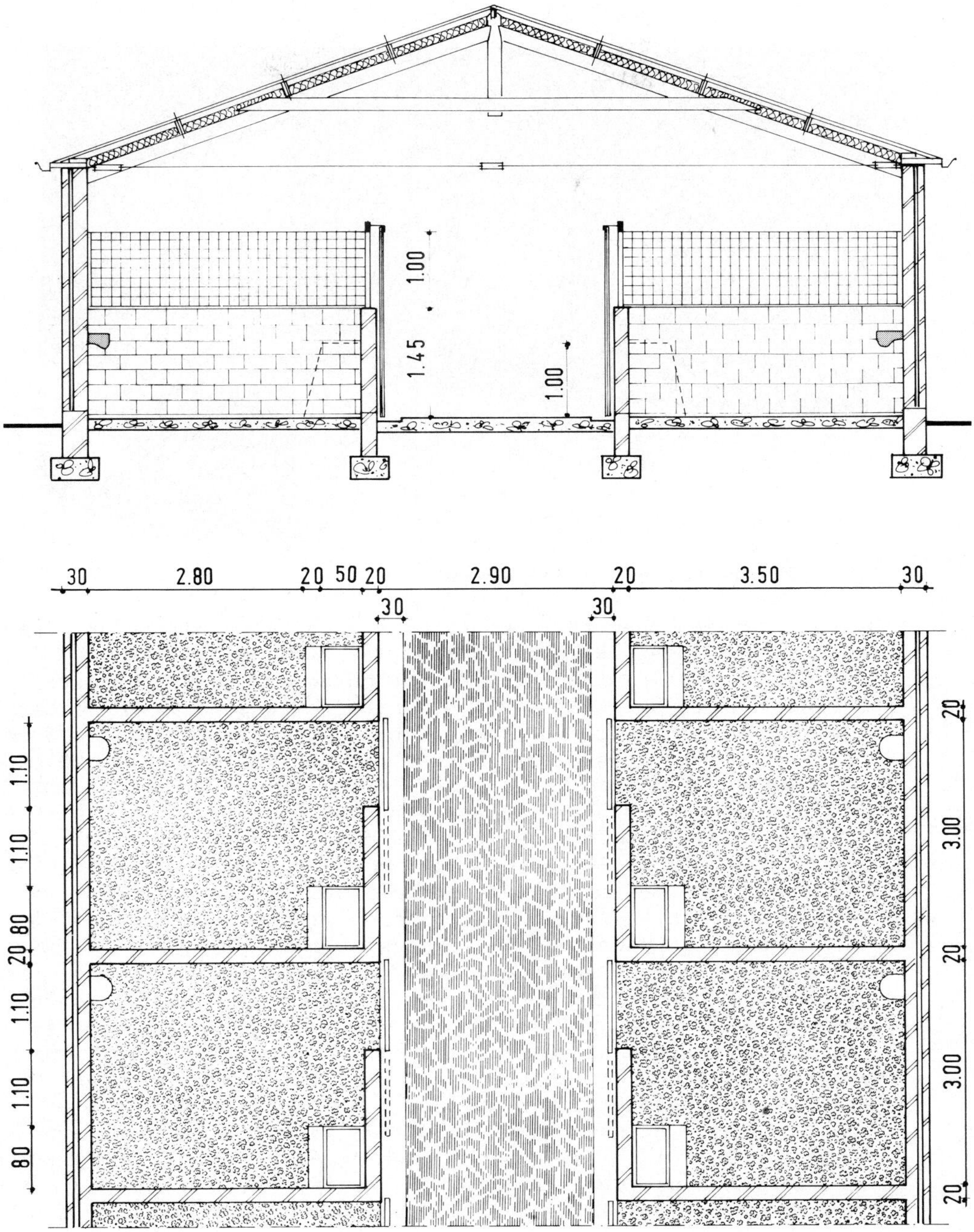

Fig. 7.3 The box stalls for horses.

Fig. 7.4 The loose box for horses.

calmed by giving them the company of a goat. The goat is provided with its own manger where it is tied for feeding.

In one of the front corners of the box a trough is installed, the construction and dimensions of which correspond to those described above for the tying stable. The hay can be administered in a plastic net which is closed at the top. It has meshes of 10 cm x 10 cm through which the horse can take its feed. The net is hung beside the trough once a day. The hay can also be thrown on the floor (Marten, 1982). In this case all the feed is given to the horse from the service passage. It is not recommended to supply the hay in a rack, mainly in view of preventing deformations of the spinal column. A sliding door is installed in the front wall and is 1.10 m wide and 2.40 m high. A threshold of 10 cm high at the inner side of the doorway prevents the door getting blocked by the accumulation of litter or manure. A sliding door, compared to a door on hinges, has the advantage of not obstructing the service passage. A flat gutter is sited in the service passage, along each row of boxes and fulfils the same function as in the tying stable. The service passage is at least 2.5 m but preferably 3 m wide.

The automatic drinking bowl is placed in such a way that the horse is unable to drink whilst eating which otherwise could lead to disorders of the digestive system by an insufficient secretion of saliva and to befouling of the drinking bowl. The service passage can also be sited under a roof overhang which is at least 1 m wide (fig. 7.5).

The front of each box is similar to the one described above but the sliding door is now substituted by a hinged door and split

Fig. 7.5 A stable block with covered service passage.

at a height of 1.2 m. The door is 1.10 m wide. The upper door can be left open when the weather conditions are favourable. The boxes are constructed and equipped as described above. This type of stable has a pleasant appearance but tends to lead to a typical and undesirable position of the horse whereby it puts the weight of the front part of its body alternately on the left and the right front leg. It is therefore advisable to install a wire mesh in the upper doorway. The open side of the stable must be orientated away from the prevailing winds and preferably to the sun.

A separate box (1 per 10 mares) is provided where foaling can take place. This foaling box measures 4 m x 5 m. The corners inside the box are curved to prevent injuries and a beam and tackle are preferably installed in this box at a height of ca. 2.5 m. A double door, which opens outwards, is provided. The construction and equipment of this box is otherwise similar to the one described above.

The mare is transferred to the foaling box a few days before the expected date of foaling and remains there after the delivery with her foal for a few more days. Constant monitoring for several days and sometimes weeks can be necessary. Foaling can start quite suddenly and progresses sometimes quickly while in 90 % of the cases the foal is delivered during the night. Special but expensive apparatus can make this continuous monitoring redundant (Marten, 1980). The principle of automatic monitoring is as follows : as soon as the mare lies down beams of infra-red light produced by a series of infra-red emitters are reflected by reflectors and reach the different receivers. The latter activate an alarm placed in the bedroom of the horse keeper. This system can be complemented by a closed circuit T.V.-installation which allows the horse keeper to "monitor" the mare (fig. 7.6) from his bedroom as soon as the alarm sounds.

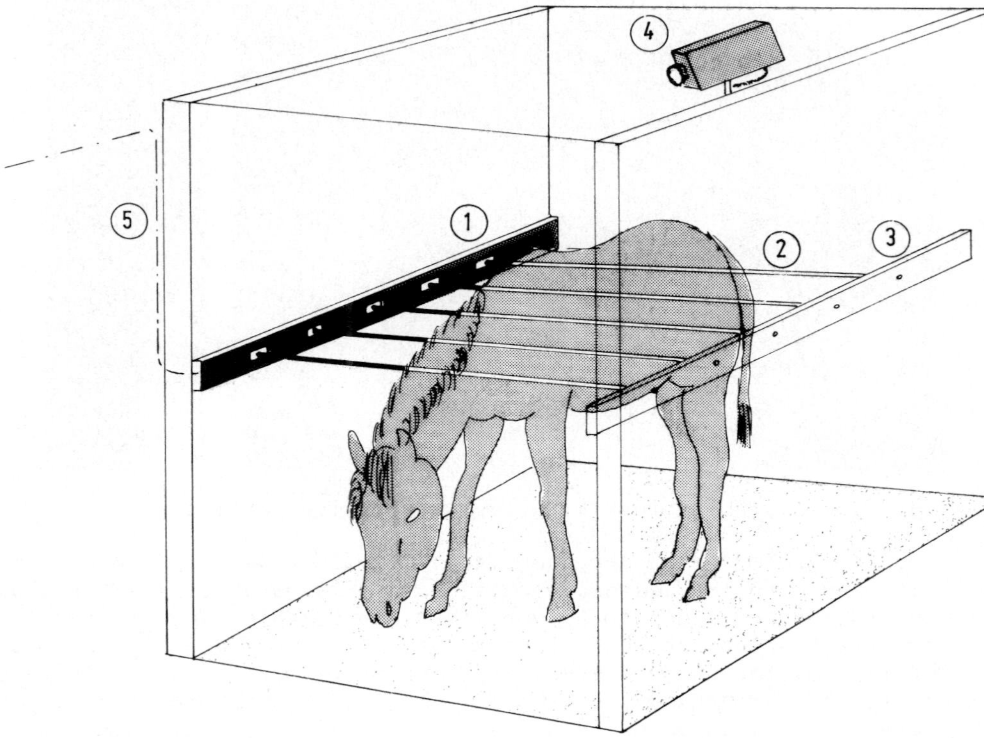

Fig. 7.6 Automatic surveillance of the mare prior to foaling. Legend : 1 = infra-red source and receiver ; 2 = infra-red light beams ; 3 = reflector ; 4 = T.V.-camera ; 5 = cable to bedroom.

7.2.3 The loose stable for the group-housing of horses

The stable frame comprises a number of portal frames. The walls consist of a concrete dwarf wall, ca. 80 cm high, against which the manure accumulates and above which a wall of brickwork (19 cm) or timber (4 cm) is built. The roof is made of corrugated asbestos cement or metal sheets. The floor consists of compacted earth covered with straw. The layer of litter (manure + straw) accumulates over several months. One of the longitudinal walls of the stable abuts against the feed room which holds the supplies of hay and straw (fig. 7.7). The trough and hay rack are filled from this room.

The horses are tethered by a short rope to the trough during feeding : movable stable bars may be provided between the horses (feeding stalls). Tying allows not only the individual feeding but

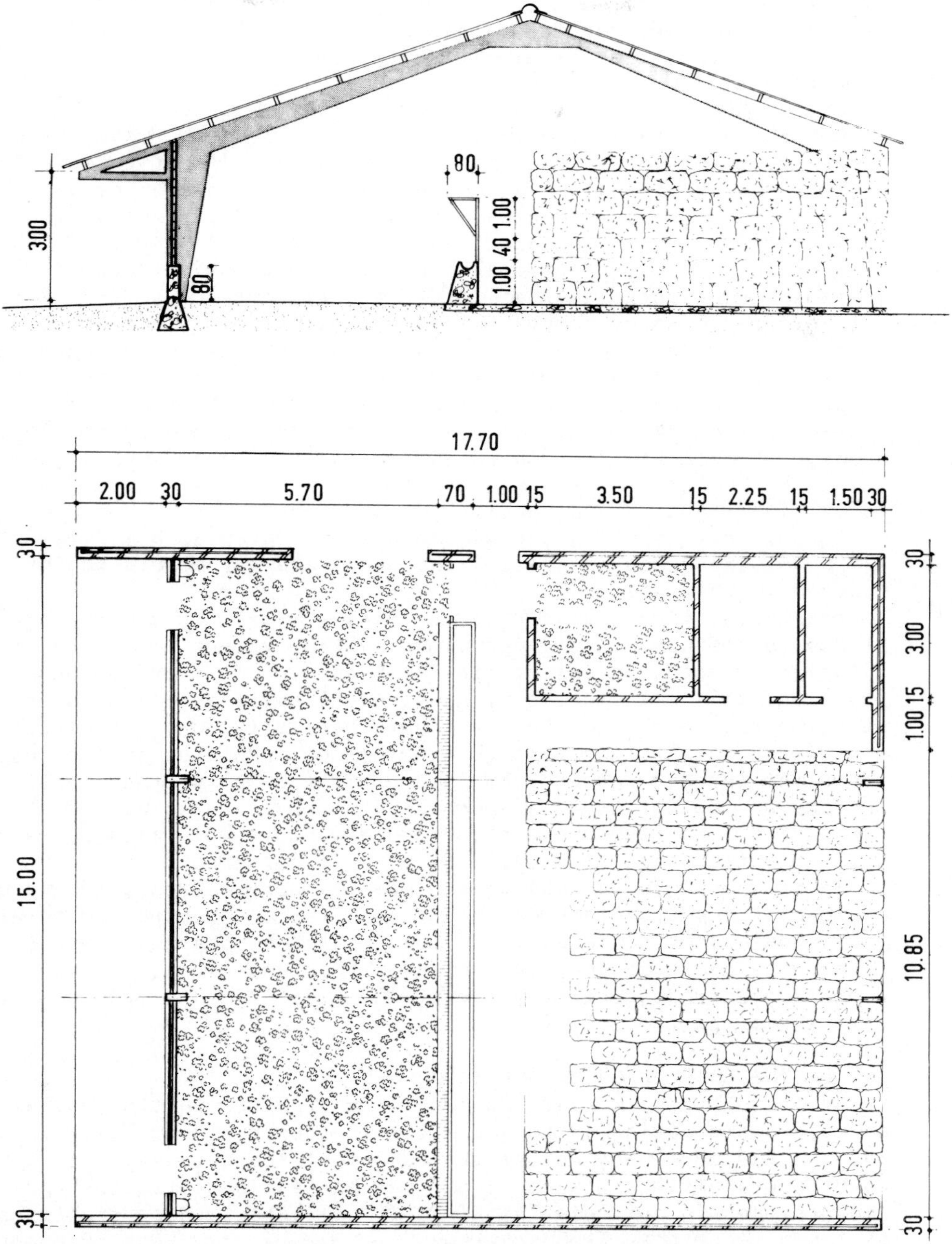

Fig. 7.7 The loose stable for horses.

also enables regular contact between man and animal, which is especially important for young horses (Marten, 1980). One automatic water bowl, which can be electrically heated, is installed for every five horses.

The stable has three doors : two of 1.1 m wide through which the horses enter or leave the stable and one of 3 m wide which is used for the mucking-out of the stable by means of a front loader. The roof of the stable can be extended (up to 2 m) to provide an overhang which covers the service passage (fig. 7.7). The stable abuts on an outside run, orientated away from the prevailing winds. The outside run consists of grass or sand. The following space requirements are considered : 12 m^2 for a mare and her foal, 10 m^2 for a full-grown and large horse, 8 m^2 for a full-grown small horse (Haflinger), 5 m^2 for a pony. The area of the outside run is equal to or larger than the area of the stable. The loose stable for group stabling is especially suited for young horses (Marten, 1980). Foaling often takes place in the stable itself and presents no problems, but separate foaling boxes can also be included.

7.2.4 The lighting and ventilation of stables

For natural lighting the window-area is rated at a minimum of 1/15th of the flooring area. Translucent plastic panels can be installed in the roof.

Artificial lighting is rated at min. 40 lux (2 W/m^2) but preferably at 100 lux (5 W/m^2) and is provided by fluorescent lamps. Natural ventilation is used. An adjustable inlet opening is provided under the eaves and this both in the tying stable as well as in the stable with loose boxes. Cold incoming air should not fall onto the horses. The air outlet takes place through a number of insulated shafts installed above the service passage and in the ridge of a two-row stable. The outlets are adjustable. Maximum ventilation is rated at 240 $m^3/h/$ horse of 500 kg (Debruyckere and Neukermans, 1973) which corresponds with an outlet size of 0.5 $m^2/100$ m^2 of floor area.

Boxes which give access to the open air or loose stables require no special installations for their ventilation. Horses can tolerate low temperatures quite well provided that they possess a well-littered lying area and that they are well-fed (Marten, 1982). The k-value of a horse stable is generally around 1.1 $W/(m^2.K)$. The relative humidity should be 60 to 80 %.

7.2.5 The cost price of a stable

According to Bach (1982) a new-built stable with two rows of loose boxes, accommodating 10 to 15 horses (as described in 7.2.2.) costs ca. £ 1,875 per horse whilst the equipment (partitions, doors, troughs, water bowls etc.) costs ca. £ 625 per horse. The total cost price amounts then to not less than £ 2,500 per horse. Loose stables and the above-mentioned "Robusthaltung" of course cost less. Prefabricated stables are increasingly built in the U.S.A. and Canada (Langer, 1977) but are as yet not often found in Europe.

7.3 ZOOTECHNICAL AND LABOUR-TECHNICAL ASPECTS OF THE HOUSING OF HORSES

The box offers mainly ethological advantages by the larger freedom of movement it gives to the horses compared to the tie stall. Horses accommodated in boxes clearly show longer lying-down times and the latter form a larger part of the total sleeping time. Tethered horses show longer standing times and this often leads to an excessive load on the frog and the tendons resulting in damage to the front limbs.

Leg oedema and even slight lameness with older horses are also found with tethered animals. These lesions heal after moving the horses to a loose box. The incidence of accidents is also much higher in tie stalls than in boxes (often caused by neighbouring animals and parts of the stable equipment). Inflammations of the hoofs are more frequent in tie stalls and are due to a continuous contact of the hind limbs with the wet manure film on the floor. It is well known that the lack of movement with the horse leads to a number of digestive diseases (e.g. colics) and to the so-called "Monday-morning disease" (Myoglobinuria or Myoglobinaemia).

Moreover, feeding requires more time in a traditional tying stable because the feed has to be carried between the animals to the manger. Following Pirkelman and Wagner (1973) in the two-row tying stable with twenty horses, which he compared with boxes, a distance of 240 m had to be covered in the first stable versus 60 m in the house with boxes. It is however possible to install a feeding passage in front of each row of horses and the labour time requirement for the feeding of the horses is then practically equal for both types of stables. The time required for the daily care (feeding, mucking-out, littering, grooming, hoof care and cleaning of the passageways) of sport horses amounts to ca. 20 min per head. The time required for the periodic tasks (horse shoeing, assistance with veterinary treatment, care of tail and mane) viz. 3 min per head and per day and the daily walk viz. 7 min per head are to be added. The total labour time requirement both in the tying stable with a feeding passage as well as in the stable with boxes amounts in this way to ca. 30 min per horse and per day. This labour time requirement is very high compared to that for other domestic animals. On a normal working day (8 hours) a horseman is only able to care for 16 to 17 horses. In a loose stable the labour requirement for the care of a Haflinger horse amounts to a mere 8 min per day because the care is not individualized and the daily walk is omitted (Pirkelman and Wagner, 1973). The building costs of a tying stall of the traditional type are at least 50 % lower than those of box stalls. This difference however is smaller if the two-row tying stall is provided with feeding passages (1 m wide) and, in view of safety (against hoof-beats), with a wide passageway (3.5 m wide). The loose stable requires the lowest investment.

The consumption of straw is up to 50 % higher in box stalls than in a tying stall. In contrast to other domestic animals, the strawless housing of horses is only in the experimental stage and is rarely found in practice.

The following conclusions can be drawn from the preceding considerations. Draught horses, which perform labour practically every day, can for economic reasons be kept in tying stalls. Riding horses, which cannot be ridden every day by the owner and racehorses of which all-out efforts are wanted, must be housed in boxes where the animals can see each other. The temperature in the house must be around the comfort zone of the horse i.e. between 10 and 15°C. Too warm or too damp a house is disadvantageous for a horse, especially for its respiratory organs.

7.4 THE CONSTRUCTION AND EQUIPMENT OF MANEGES (RIDING SCHOOLS)

Riders of course wish to practise their hobby throughout the year and this is, owing to the climate, not always possible. An indoor manège is therefore practically indispensable (Schnitzer, 1972 ; Kondziella, 1973).

A covered manège is mostly an hangar of the portal frame type. It is normally orientated North-South (figs 7.8 and 7.9). The walls consist, from outside to inside, of either 1/2 a brick + cavity + 14 cm hollow blockwork or of wood. Prefabricated wall panels are also suitable. The walls are 4.2 m high to eaves'level. The roof is made of corrugated asbestos cement or metal sheets and has a pitch of 20° (36 cm/m). At the inside, along its entire perimeter, the manège is provided with a belt, which is a beam at a height of 1.50 m. Under this belt and sloping from the ground upwards (15° or 27 cm/m) are a series of tongued and grooved wooden planks, resting against horizontal wooden beams laid on the ground. The flooring is made of compacted earth covered with a mixture of wood shavings and sand. The dimensions are always given from belt to belt and are measured on the ground. The dimensions are : 15 m x 30 m for exercise-manèges ; 20 m x 40 m for tournament-manèges and extendible to 20 m x 60 m for dressage and jumping. The sloping entrances in the wooden belt are sited in its middle or at its extremes. One large entry is provided in the gable wall of the manège. This main entrance is 4.2 m high and 3.9 m wide and allows the entering and exit of a truck. The other entrances used by the horses and the riders are 3.9 m high and 3.3 m wide. A stand for spectators is sometimes installed along one longitudinal side or more often along one of the shorter sides (cheaper) of the hall. It is often a step-type structure with seats at the different levels.

The natural lighting of the hall is preferably through a number of translucent plastic panels in the roof. The artificial lighting is by fluorescent lamps, attached to the ceiling, and their number is based on a light intensity of 100 lux (5 W/m^2). The ventilation consists of adjustable inlets below the eaves on both longitudinal walls and of an adjustable open ridge which acts as an outlet.

The stables can be sited against one of the longitudinal walls of the manège and this is probably the cheapest way of construction. A number of drawbacks viz. concerning the lighting and ventilation of the manège itself make this building type less attractive.

Fig. 7.8 The manège for horses.

Fig. 7.9 shows the ground-plan of a more suitable layout which consists of :

- a manège including a stand for the spectators ; beneath the stand an office and a reception- and meeting room are installed ;
- a stable with 20 boxes, a feed storage, a room for the saddles and 5 tying stalls for saddling and tending of the horses ;
- an (open) hangar for the storage of hay and straw and large enough for a years supply ;
- an apron for the storage of manure. It must be large enough for a 4-week storage of the manure. It is surrounded by rush-mats of 3 m high.

Besides the indoor manège, for riders to practise, there should be an outdoor grassed area for tournaments, which measures 70 m x 110 m, and an exercise area which is normally 3,000 m^2 and also acts as an outside run for the horses. A sand circuit, at least 3 m wide, is necessary for race horses and is installed around the buildings.

Good access, ample parking space and a nice plantation will contribute towards a manège of high standing.

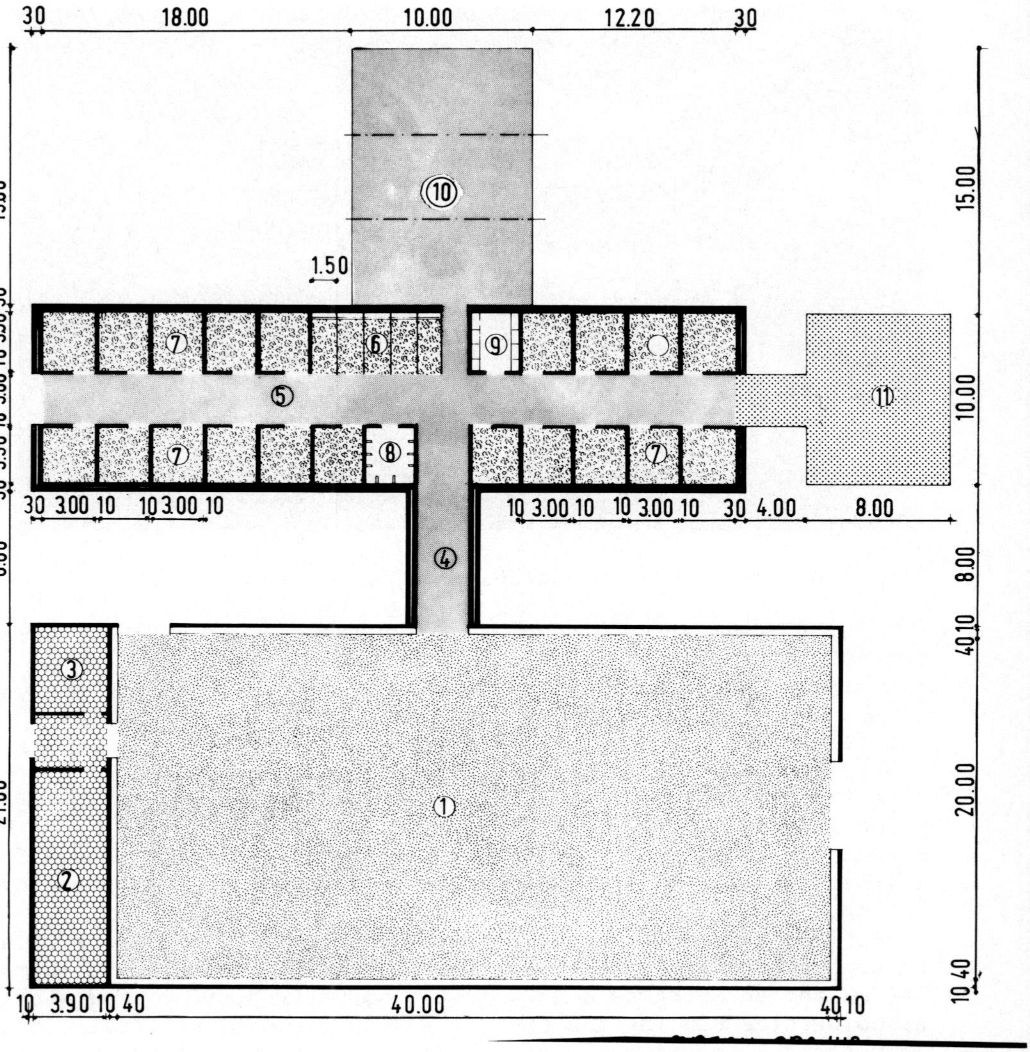

Fig. 7.9 Ground-plan of a manège with stables.
Legend : 1 = manège or riding school ; 2 = reception-room ; 3 = office 4 = corridor ; 5 = horse stable ; 6 = tying stalls ; 7 = individual loose boxes ; 8 = saddle room ; 9 = feed room ; 10 = storage for hay and straw ; 11 = manure stack.

REFERENCES

Anon., 1960-1982. Landbouw- en tuinbouwtelling, Nationaal Instituut voor de Statistiek, Brussel, Belgium.

Anon., 1983. Yearbook of Agricultural Statistics, Statistical Office of the European Communities - Eurostat, Brussel, Belgium, 286 pp.

Bach P., 1982. Wirtschaftliche Kenndaten der Pferdehaltung, Arbeiten der Bayer Landesanstalt für Betriebswirtschaft und Agrarstruktur, Heft 16, München, W. Germany.

Debruyckere M. and Neukermans G., 1973. Algemene richtlijnen in verband met de klimaatregeling in gesloten stallen, Landbouwtijdschrift, 26 : 251-282.

Franke H. and Nicolay W., 1973. Pferdeställe, Informationsbericht, nr. 13, A.L.B., Hessen, W. Germany, 75 pp.

Hardmuth R., 1972. Pferdeboxen, Bauen auf dem Lande, 23 : 165-168.

Kondziella W., 1973. Planung und Bau von Reitanlagen, Bauen auf dem Lande, 24 : 65-66.

Langer L., 1977. L'écurie moderne, Edisem Inc., Quebec, Canada, 64 pp.

Marten J., 1978. Robusthaltung von Pferden, Der Tierzüchter, 30 : 31-33.

Marten J., 1980. Haltung von Föhlen und Jungpferden, Der Tierzüchter, 32 : 77-79.

Marten J., 1982. Einsparungsmöglichkeiten beim Bau von Pferdeställen, Documentation der CIGR Arbeitstagung Braunschweig, W. Germany, pp. 237-249.

Pirkelman H. and Wagner M., 1973. Technische und arbeitstechnische Probleme in der Pferdehaltung, Bauen auf dem Lande, 24 : 61-64.

Schnitzer U., 1971. Anbinde- und Boxenstall bei der Reitpferdehaltung, Der Tierzüchter, 23 : 569-571.

Schnitzer U., 1972. Untersuchungen zur Planung von Reitanlagen, K.T.B.L. Bauschrift, nr. 6.

Chapter 8

THE HOUSING OF SHEEP

Chapter 8

THE HOUSING OF SHEEP

8.1 GENERALITIES

Sheep were among the first animals to be domesticated. The sheep and the lamb have played an important role in the cultural and religious history of mankind. The sheep is a ruminant belonging to the bovine family and is closely related to cattle, goats and antelopes. They are characterized by woolly fleece varying in length, fineness and colour, and by curving horns found in males, or rams, of many species. Wild sheep are typically inhabitants of mountain or plateau regions. Since sheep vary more widely than most animals and can quickly adapt to the environment, humans have succeeded in developing types for specific needs : meat, wool or milk. The sheep is one of the hardiest domestic animals. Large flocks of sheep are still found in North-Africa, the Middle East, Australia, New Zealand and South-America where they are an important trade. Large flocks are also found in some European countries (Great Britain, France, Italy, Greece) and especially in some desolate landscapes where they are probably the only possible branch of agriculture.

Table 8.1 shows the number of sheep and goats found in the different E.E.C.-countries in 1981 (Anon., 1983).

TABLE 8.1 Number of sheep and goats in the different countries of the E.E.C. in 1981 (x1,000) (Anon., 1983).

Country	Number of sheep (x 1,000)	%
W.-Germany	1,289	2.8
France	9,633	21.0
Italy	7,691	16.7
The Netherlands	646	1.4
BLEU	243	0.5
United Kingdom	13,352	29.1
Ireland	1,719	3.7
Denmark	16	0.3
Greece	11,274	24.5
Eur. 10	45,864	100.0

The breeding of sheep is one of the few branches of animal husbandry which escaped industrialization, although some changes are taking place. A change-over from cattle husbandry to sheep keeping in the E.E.C. is not likely. Since the surplus of dairy products forced the E.E.C.-authorities to take measures in 1984 and since large quantities of sheep's meat have to be imported in many E.E.C.-countries (ca. 270,000 tonnes) such a change-over seems to be an alternative from the macro-economic point of view. This transformation would however result in a substantial decrease in the income of the farmer and in a surplus of manpower (Calus, 1981). The present-day E.E.C.-policy does not stimulate sheep keeping since large import-quota's for sheep meat are fixed (Van Gelder, 1980).

Sheep can survive outside throughout the year : they tolerate large fluctuations of temperature as they are protected against the cold by their fleece and against the heat by their ability to intensify their respiration (Comberg and Hinrichsen, 1974). They mainly live on grass even in the winter but it is customary to provide them with supplementary concentrates and hay or silage in the hardest period of the year and especially from ca. 6 weeks prior to lambing. In this extensive sheep farming system the accommodation of the animals is limited to a minimum : the ewes are only brought into primitive shelters for lambing. The shelters are provided with partitions and the floor is straw-bedded forming a number of lambing pens : the ewes and lambs return to the pasture a few days after lambing.

The housing of sheep can bring managerial advantages in the supervision and shepherding of the flock ; the protection against rain and especially against wind, to which the sheep are rather sensitive, can improve their condition and thus the production results. Primitive shelters, protecting the sheep against wind can be built with walls of straw bales. There is however also a tendency to accommodate the sheep in sound buildings at least during the hardest winter-months and to provide them with roughage (fodder beets, hay, straw, silage) and concentrates. Lambing takes place in October or December (sale of Paschal Lambs) but mainly in March or April. Pregnant ewes and suckling lambs (pregnancy : 5 months ; suckling : 3 to 4 months) are better off, if provided with suitable accommodation while also better working conditions are created for the shepherd.

The breeding of sheep is mainly directed to the production of meat although its importance in the production of milk should not be under-estimated, especially in South-Europe and the Near East. Some cheeses, based on ewe's milk (e.g. Roquefort) are highly esteemed. The production of wool remains another important branch of sheep keeping. A combination of the three above-mentioned specialities is possible to a certain extent.

8.2 THE CONSTRUCTION AND EQUIPMENT OF A SHEEP HOUSE

The sheep house consists mainly of a low hangar covered, for instance, with corrugated asbestos cement or metal sheets and to a certain extent with translucent sheets which provides for the lighting of the house.

New sheep houses are constructed with an eaves height of at least 3 m. The higher the building, the greater the volume of air in the building, improving the ventilation which in turn prevents the build-up of a damp atmosphere within the shelter. The ventilation is carried out by means of an adjustable and continuous open ridge and by air inlets just below the eaves (15 cm opening for each 10 m of house-width, both for the inlets and the outlet).

The roof is supported by a number of intermediate posts and the walls are therefore merely a filling between the supports. The lower part of the wall (ca. 1 m high) consists of hollow bricks in order to be sufficiently strong to withstand the contact with livestock. The upper part of the wall can then be made of planking, treated with a suitable preservative to prevent penetration of moisture, corrugated asbestos cement sheets or plastic sheets fixed to a wooden frame. The whole wall can also be made of wood. The lower part of the wall must be heavily creosoted and rests upon a concrete foundation. Since the building doesn't need to be insulated, (prefab) concrete plates can also be used for the walls.

Sheep are prone to disease affecting their feet, problems which are made worse by damp or wet floors. Several types of floors are suitable, some with straw litter, others strawless : we will describe and compare them below.

The floor of the sheep house is traditionally made of compacted earth covered with litter. The manure of the sheep is consistent and the quantity of urine is rather limited so that both are easily absorbed by the litter thereby reducing the risk of infiltration into the soil. The straw consumption for a ewe is rated at ca. 75 kg for a 12 week housing period (Robinson, 1982). In some cases a layer of rubble is applied beneath the well-compacted earth which must prevent the rise of groundwater. Sheep buildings must always be built on a dry and preferably raised site.

Foot troubles of sheep led to the development of slatted floors, first reportedly used in Iceland around 1760 (Noton, 1982). Slatted floors became increasingly popular in the last few years. Timber slatted floors with a slat width of 4 to 5 cm and a slot width of ca. 2 cm are therefore used (Robinson, 1982 ; Burgkart and Mittrach, 1972). A slurry cellar with a depth of ca. 75 cm is situated beneath the slatted floor. The storage capacity of the slurry cellar generally exceeds one year and emptying is therefore not required annually. Slats made of woven wire mesh (galvanized iron) can also be used, meshes of for instance 2 cm x 2 cm or 7.5 cm x 1.2 cm give complete satisfaction (fig. 8.1). Flattened expanded metal slats may also be used. Concrete slats with an upper slat width of 7.5 cm and a lower slat width of 3.8 cm and with a slot width of 2.5 cm can also be considered, although sheep tend to befoul themselves more on concrete slats (Anon., 1971).

The application of slats results in a considerable reduction of the frequency of footrot compared to littered houses (Robinson, 1982) but the cost-price of a slatted house is much higher than that of a littered house.

Before mucking-out the slats have to be removed and this takes a lot of time. Mucking-out is usually accomplished with either a front loader or a manure crane. Furthermore feeding of hay can lead to the blockage of the slats. An all-mash feed could solve this problem but has as yet not given complete satisfaction.

Fig. 8.1 Sheep house with wire mesh floor and separate boxes for the lambs.

Sheep are normally kept in batches(batches of up to 50 sheep are feasible) in large pens. Larger batches may lead to problems in handling and supervision. Some farmers prefer to leave the ewe in the group at lambing rather than transferring it to an individual lambing pen (Robertson, 1978). Many others bring the ewes at lambing into removable lambing boxes where they remain with the lambs until the latter are accepted by the ewe. The individual lambing boxes measure 2 m x 1.5 m. Ewes with their lambs are then accommodated in groups in a pen. A separate box for lambs may be provided adjacent to the pen. Lambs can reach the ewes through an opening in the partition wall between the box for lambs and the pen for ewes. This opening, only passable by the lambs, can be locked and is 40 cm high and 25 cm wide. It is desirable that the boxes housing the lambs, can be split in two in a later stage to shed the ram lambs from the ewe lambs and hence avoiding the moving of lambs. Rams can later be accommodated with the ewes or can be kept in separate, individual boxes.

The following points must be taken into consideration when designing a new sheep house or sheep shelter :

- ca. 25 % of the ewes are annually replaced ;
- one ewe yeans on average 1.5 lambs per year ;
- the recommended floor space per large ewe (ca. 70 kg) in a group pen amounts to 1.2 - 1.4 m^2 on littered floors and to 0.95 - 1.1 m^2 on slatted floors (Robertson, 1978) ;
- the space allotment for ewes with lambs in a pen where no separate

box is provided for the lambs, amounts to 1.4 - 1.85 m^2 per large ewe (ca. 70 kg) for littered floors, whereas 1.2 - 1.7 m^2 is recommended for fully slatted floors (Robertson, 1978) ;

- the required space per lamb (of less than one year) is about 0.5 m^2 ;
- the area of the lambing pen is 2.5 - 3 m^2 ;
- the minimum length of the hay rack amounts to 0.4 to 0.5 m per sheep ;
- the minimum trough length is 0.4 to 0.5 m per sheep ;
- the minimum length of the hay rack for a lamb of 4 months is 0.20 to 0.25 m ;
- the minimum length of the hay rack for a lamb of 1 year is 0.30 to 0.35 m ;
- the self-feed silage length amounts to 0.1 m per sheep.

The necessary storage space for a three month winter housing and expressed per ewe with 1.5 lambs is rated at 0.5 m^3 for pressed hay, 0.6 m^3 for silage feed, 0.25 m^3 for concentrates and 0.5 m^3 for straw (Lovelidge, 1973). A frequently used design for housing sheep is the transverse division of the sheep house into a number of pens, and normally one pen per bay (fig. 8.2).

Troughs and racks can be arranged along the walls or placed in the middle of the pen. Pen area, trough length and rack length should be in accordance with the above-mentioned figures. Other facilities to be provided are : lambing pens (fig. 8.2), boxes for the rams, a pen for sick animals, feed storages and perhaps silos. A well-designed enterprise is completed with the addition of the following facilities which will contribute to an efficient management (Shepherd, 1974) (fig. 8.3):

- a collecting or gathering pen : where sheep are gathered, its area should be sufficient to hold at least 250 ewes with their lambs, if a larger flock is to be handled the sheep can be directed to additional flanking side pens ;
- the forwarding pen : this pen is optional and located between the gathering pen exit and the entrance to the shedder. Its purpose is to allow the handling of a small number of sheep before entering the shedding race. Marking of the lambs can be done in this pen ;
- the shedder : its position and length are critical and its design should receive attention. The sides are parallel and its width must be sufficient to allow most breeds to pass without being hindered or being able to turn (for instance 65 cm for large breeds). The sides must be smoothly finished to avoid injuries to the sheep and snagging of wool. The animals are forced into a single file. They can be weighed. Some farmers also install the footbath in the shedding race ;
- the footbath : which the animals pass through for the disinfection of their claws which are particularly sensitive to infection (e.g. footrot). The length of the footbath must be sufficient to allow the liquid to penetrate to all parts of the sheep's feet while it walks the length of the bath (ca. 7 m) ;
- the catching or forcing pen : consists of a square or circular pen, which must assure a suitable flow of sheep to the man at the

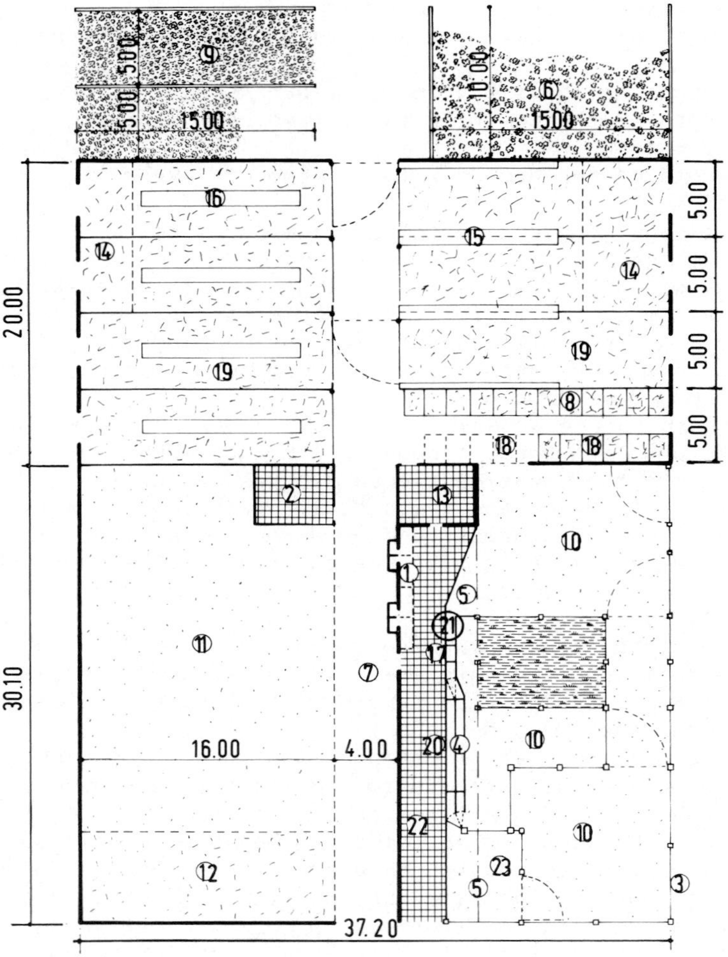

Fig. 8.2 A typical sheep farm.
Legend : 1 = the kennel for dogs ; 2 = the shearing room ; 3 = fence ; 4 = sheep dipper ; 5 = edge of roof overhang ; 6 = dung compound ; 7 = passageway ; 8 = section for rams ; 9 = silo ; 10 = gathering pen ; 11 = storage for feed ; 12 = sick bay ; 13 = room for shepherd ; 14 = section for lambs ; 15 = trough and rack against the wall of the pen ; 16 = trough and hay rack in the centre of the pen ; 17 = weigher ; 18 = "removable" lambing pens ; 19 = pens accommodating ca. 50 sheep ; 20 = footbath ; 21 = shedder ; 22 = passage ; 23 = dripping pen.

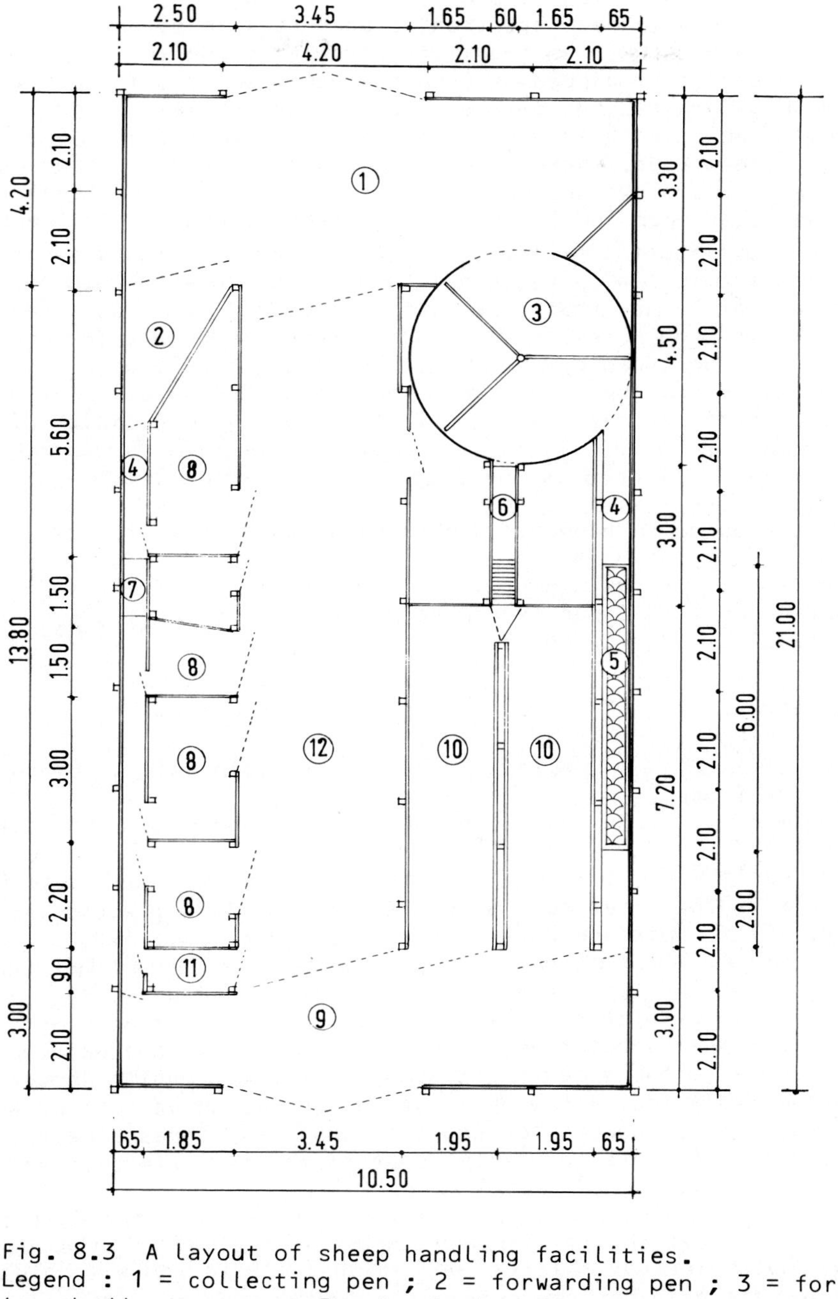

Fig. 8.3 A layout of sheep handling facilities.
Legend : 1 = collecting pen ; 2 = forwarding pen ; 3 = forcing pen ; 4 = shedder or race ; 5 = footbath ; 6 = dipper ; 7 = weigher ; 8 = side pens ; 9 = dispersal pen ; 10 = drip pen ; 11 = working pen ; 12 = central alley.

entrance to the dipper or the footbath and allow a convenient positioning of the sheep for dipping or footbathing ;

- the dipper : where sheep are treated against skin parasites. These parasites infest their host-animals throughout the year unless the sheep are correctly dipped in properly equipped handling facilities. The known skin-parasites of the sheep are classified as specific parasites which live their whole life on the sheep and are generally known as sheep keds, sheep lice and sheep scabmites and non-specific parasites (e.g. ticks and blow-flies) which are not completely dependent upon the sheep and of which the effective control is therefore much more difficult (Shepherd, 1974). Dipping is compulsory in many countries and is really essential. The sheep dipper should be designed in such a way that it minimizes labour as well as stress on sheep and it should result in a sufficient throughput. The choice of dipper is governed by the number of sheep to be dipped and its design determines the labour involved. Table 8.2 serves as a guide to economic dipper size (Watson et al., 1981). The capacity of the bath should be ca. 2.5 litres per head of the total flock and it must be at least 1.2 m deep as the sheep have to swim through it.

TABLE 8.2 Guide to economic dipper size (Watson et al., 1981).

Flock size (ewes)	Capacity of the dipper expressed in litres	Type of dipper
small flock (up to 500)	900 - 1,300	short swim
large flock (500 -1,000)	1,200 - 2,700	long swim or circular
very large flock (over 1,000)	up to 4,500	long swim or circular

A large variety of well-designed sheep dippers have been described in detail in the literature ad hoc (Shepherd, 1974 ; Watson and Speedy, 1977 ; Peck and Kelly, 1978 ; Watson et al., 1981). Dippers can be homemade but a large number of prefabricated dippers of most sizes and types are now commercially available. They are made of glass reinforced plastic, glass reinforced cement or galvanized steel sheet. The dippers should have an adequate water supply and a suitable drainage. Some farmers may prefer showering, jetting or spraying. Although these designs may have some advantages (portable, cheaper for small flocks, less stress for the sheep, etc.) they are generally regarded as less effective and sometimes inadequate (Watson et al., 1981) ;

- the drip or draining pens. Drip pens are provided at the exit of the dipper. They must ensure a speedy drainage of liquid to the dipper. They normally include a filter box which collects and withholds dung and wool. The drip pen can eventually be followed by a dispersal pen .
- a room for the shepherd ;

- a room where shearing of the sheep can be carried out ;
- a kennel for the sheep-dogs ;
- a manure compound if necessary.

The transverse arrangement of the pens and hence of the rack and the trough might result in a number of problems in view of the rational execution of the feeding and the mucking-out. Indeed, the mechanization of the feeding is thwarted by the location of the hay bunker and the trough. They are therefore nowadays installed along the feeding passage in the lengthwise direction of the house. This makes supervision of the house easier. Fig. 8.4 shows the ground-plan of such a sheep house consisting of pens abutting on a feeding passage and accommodating up to 25 sheep. The installation of racks and troughs along a central feeding passage results in an important labour saving and allows a mechanized feeding. The space per sheep amounts to 1.7 m² and this allows the accommodation of the sheep together with the lambs. Special feeders must be installed for the lambs to which the full-grown sheep have no access e.g. by choosing suitable feeder dimensions.

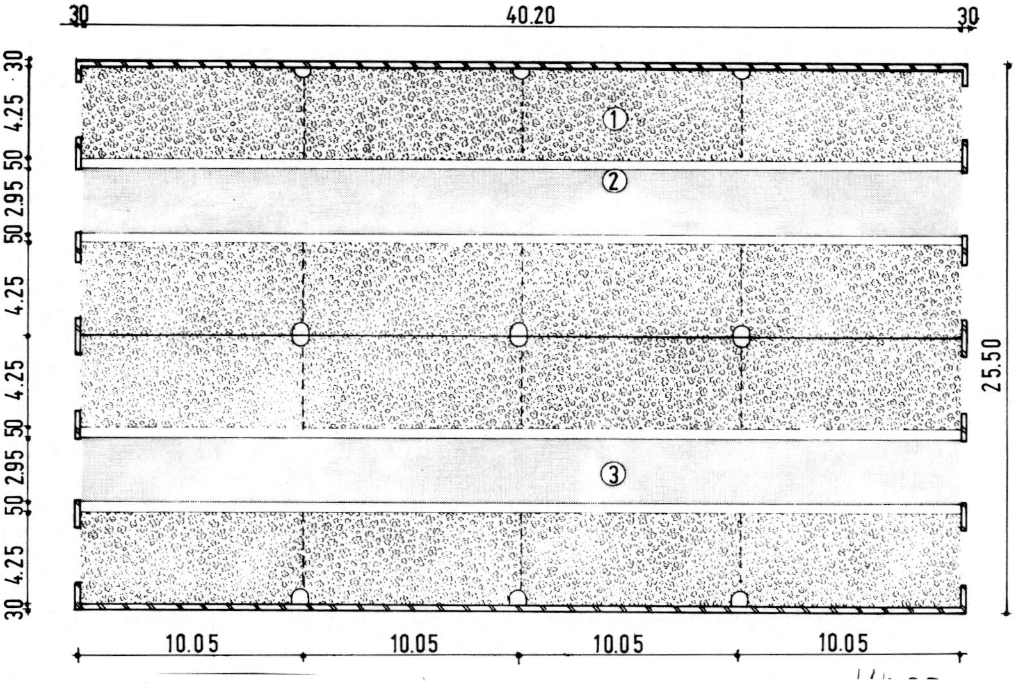

Fig. 8.4 The sheep house with pens along the feeding passages.
Legend : 1 = pen ; 2 = trough ; 3 = passage.

The construction of such a concentrate feeder box specifically intended for lambs is shown in fig. 8.5.

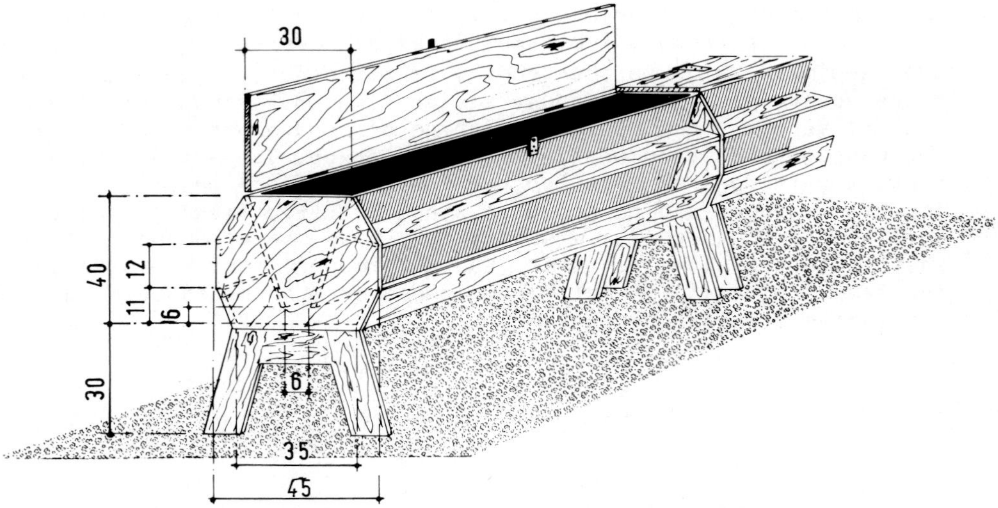

Fig. 8.5 Concentrate feeder box specifically designed for lambs.

Feeding of sheep is to be carried out rationally. Cauliformed roughage such as hay and straw are generally given in a rack while concentrates and perhaps silage are administered in a trough. A combined construction is commonly chosen where rack and trough form one unit, such a construction is not only cheaper but also requires less space. Fig. 8.6 shows a two-sided, combined trough and rack unit, for feeding roughage and concentrate, placed on the straw in the pen as is usually done in a transverse arrangement. If longitudinal pens are chosen they are generally situated along the feeding passage. The combined trough is then part of the pen partition and can be reached from the feeding passage (fig. 8.7).

The presence of a feeding passage undoubtedly increases the construction costs but enables rational feeding and even partial or full mechanization of the latter (feed trolley for concentrates or side-delivery forage wagon for roughage). The hay can also be fed in a Scandinavian trough and in this case the hay is covered with a wire mesh to avoid spillage (fig. 8.8).

The lambs are fed by means of a feed hopper or a trough in a separate box, adjacent to the pen of the ewes, and which, through a suitable opening, is not accessible to the ewes (fig. 8.1). If the lambs remain permanently with the ewes, a trough of a special design is employed whereby full-grown sheep have no access to the feed intended for the lambs (figs 8.5 and 8.9).

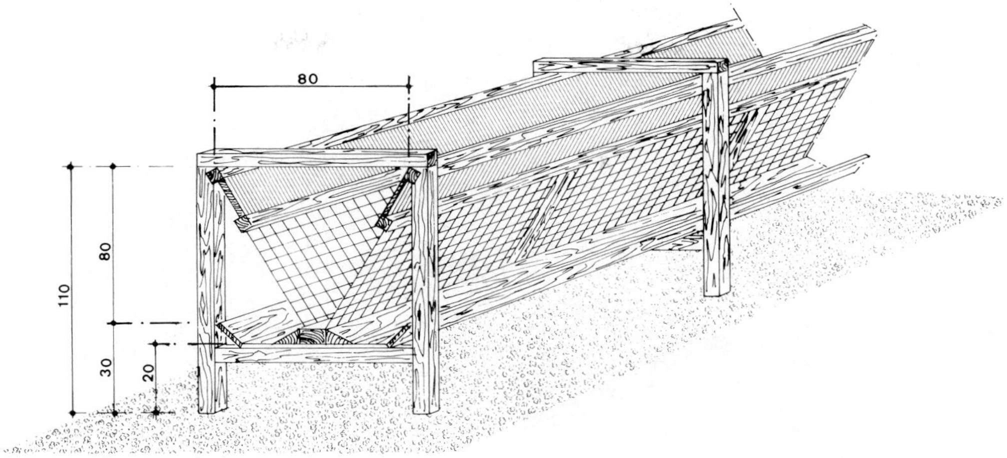

Fig. 8.6 Double-sided combined trough and rack for installation in the centre of a pen.

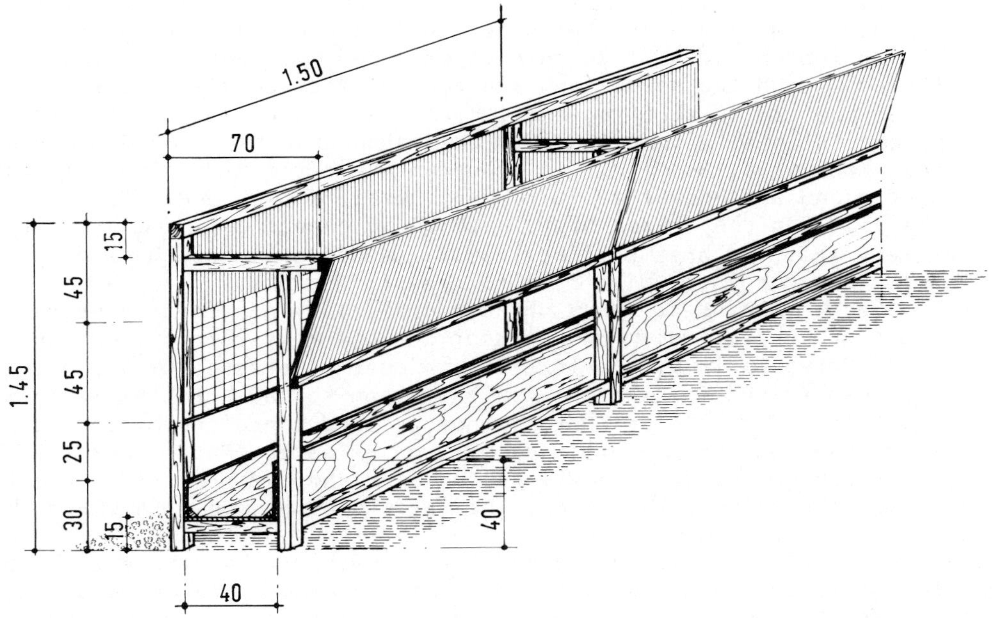

Fig. 8.7 The construction of a trough-and-rack combination along the feeding passage.

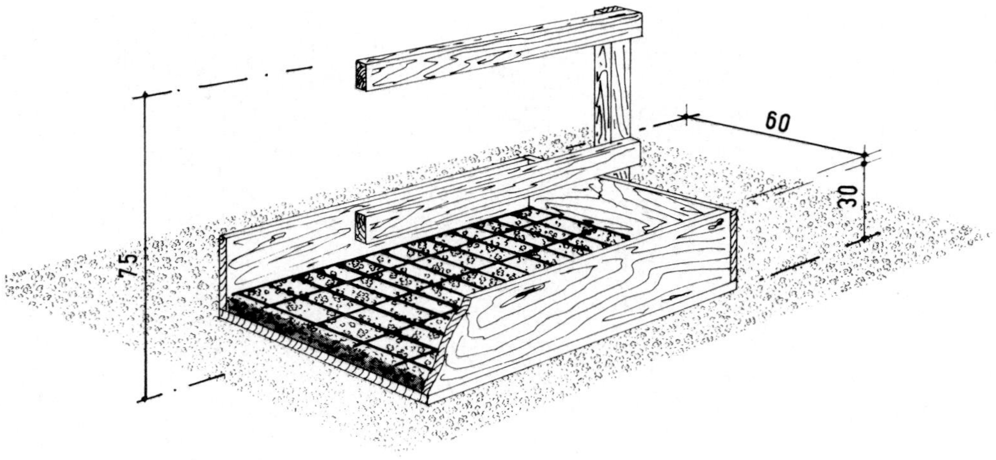

Fig. 8.8 The Scandinavian trough allowing the feeding of hay without the need for a rack.

Nowadays it is even possible to omit the trough and rack and to replace them with a synthetic belt which forms the bottom of a trough having wooden sides of 20 cm high extended by a vertical feeding fence through which the sheep can take concentrates or hay (Blanken, 1974).

Silage can be self-fed and for this a trench silo is ideally suited. A movable feed fence is then placed in front of the pitched-down feed (fig. 8.10). It is important that the dimension A is respected and a frontage of 10 cm per animal is provided. The floor of the silo and the access route have to be concreted, otherwise it is likely that the area will be transformed into a mire which can, among other things, give rise to claw infections.

The *watering facilities* are preferably provided by means of a piping system fed from a pressurized water system or from the main. Water bowls with a lever which must be pressed by the animals are less suited for sheep since they seem to have problems with activating it. Nipple drinkers cannot be used since sheep tend to play with them and this results in spillage of water (MacCormack, 1975). Float controlled water bowls can be used and their rim is normally placed 40 cm above the floor. The water consumption of a sheep amounts to ca. 3 litres per day.

Mucking-out is done periodically, either every three weeks or a few times during the winter and either by hand or by means of a front loader. A much longer period before mucking-out is allowed (even less than once a year) when slatted floors are used and when the slurry cellar is more than 60 cm deep.

The *doors and the partitions* respectively in the front and between the different sections of the sheep house are mostly wooden

Fig. 8.9 Feed hopper - with filling pipe - for lambs.

gates fixed to timber posts in the ground (fig. 8.11). The doors are also gates on hinges. All gates and catches should be easily operated ; the simplest catches work best and should be designed for quick operation with one hand e.g. simple ring on hook fastener or a hoop on the gate which drops over the gate post (Watson et al., 1981).

The above-mentioned sheep buildings are intended for the traditional breeding of sheep whereby the lambs are reared by the ewe. These lambs are then either slaughtered as Paschal Lambs at the age of 3 to 4 months when they have attained a weight of ca. 35 kg or they are kept on pasture or in the sheep house until they have reached a weight of ca. 50 kg.

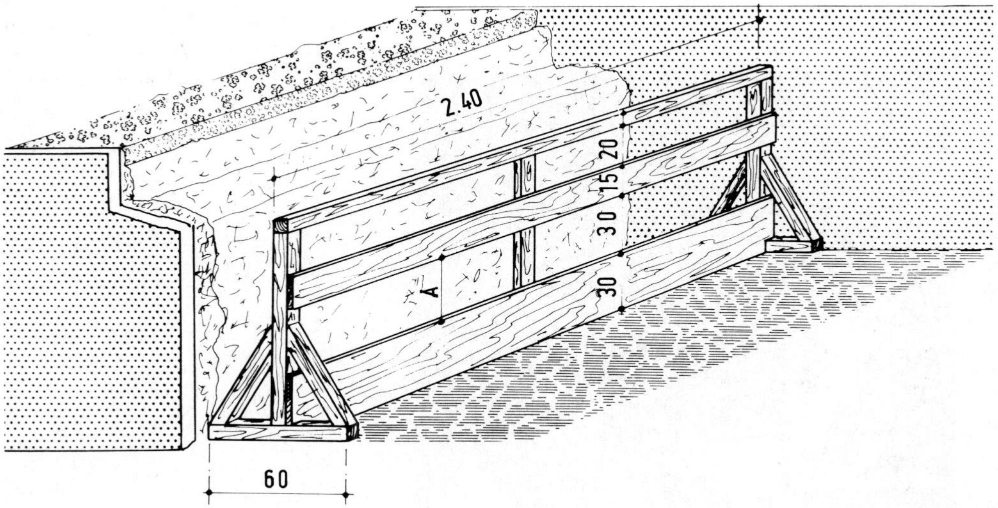

Fig. 8.10 A movable feeding fence for self-feeding of silage by sheep.

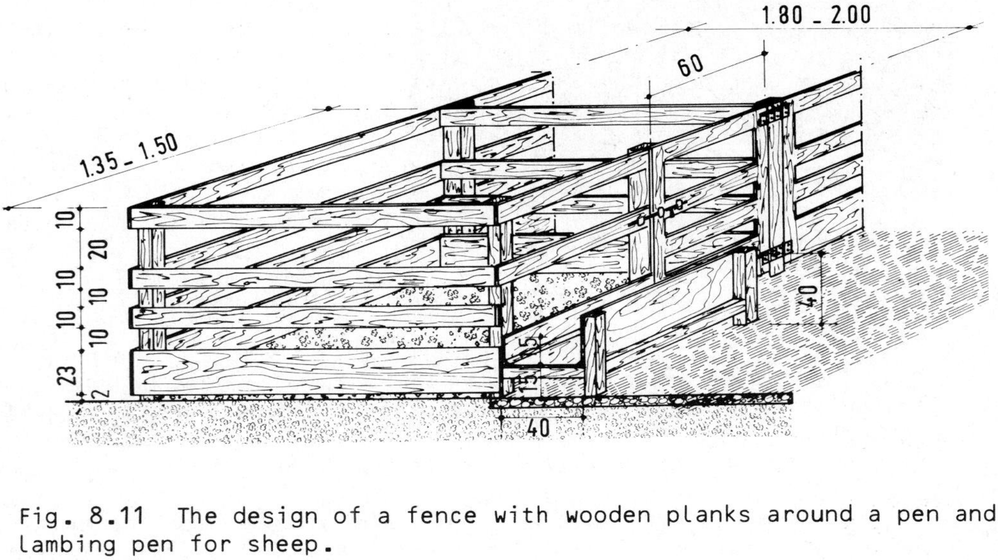

Fig. 8.11 The design of a fence with wooden planks around a pen and lambing pen for sheep.

The *ventilation* of the sheep house is always by natural ventilation as described above.

The *labour time requirement* in sheep keeping is rather low (Blamlen et al., 1970). The care of a flock of 250 ewes, accommodated in a sheep house of the traditional type, requires 1.10 min per sheep and per day if half of the ewes suckle lambs. The manual distribution of feed requires ca. 55 % of this time while littering amounts to ca. 15 %. Periodic and miscellaneous tasks require about 2 h per day. One person is thus able to care for 320 ewes. During the lambing period it is desirable to have the assistance of a second person. The manual distribution of feed from a tractor pulled forage wagon allows a reduction of the labour time requirement from 1.10 min to 0.55 min per ewe with lambs and per day.

The building costs of a sheep house in Scotland, according to Robinson (1982), amount to ca. £ 100 per m² including the fully slatted floor, troughs, fences, etc.

The *zootechnical and veterinary aspects* concerning the housing of sheep and the construction and equipment of the buildings have already been emphasized.

8.3 THE MILKING OF EWES

The production of milk in some regions and countries is an important activity in the keeping of sheep and is mainly intended for the production of cheese. In the E.E.C. for instance, only France, Greece and Italy produce sheep milk - in 1981, 1 million, 0.8 million and 0.6 million litres of sheep milk, respectively.

The housing of dairy ewes is similar to the housing of breeding and meat sheep. Manual milking is still widely applied. Mechanical milking is only practised in France and Israel (Labussiere, 1974). There are large differences in the milking ability of the ewes : 1 hour is sufficient for the milking of 75 Stara Zagora ewes or 65 Sardinian ewes but would hardly suffice for the milking of 20 Lacaune ewes. These differences are attributed to the nature of the milk secretion and the distribution of the milk in the udder of the ewe. The milk produced by the secretory tissues is partly retained in the lumen or the central cavity of the alveolus (alveolar milk) where it is stored while the other part is drained into the teat cisterns situated immediately above the teat (cisternal milk). The ratio of cisternal milk to the total milk quantity is important for Sardinian sheep but this is not the case with Lacaune sheep and this explains why milking of the latter is more difficult. On the other hand, the alveolar milk is not always ejected by the Lacaune sheep and this is mainly attributed to nervous impulses of pain and fear. These factors form a serious impediment for machine milking. The high incidence of mastitis, the low yield obtained with machine milking and the high investment required have contributed to the limited use of machine milking with sheep. According to Spanish research (Labussiere, 1974), machine milking would only be economically justified from 400 ewes onwards. Intensive research has been carried out over the last few years to improve the machine milking of sheep : the very high pulsation rate used until now (180 pulsations per minute and a vacuum and massage ratio of 2:1) is to a certain extent responsible for the incomplete milking and also makes manual stripping indispensable. A milking machine with a lower pulsation rate (e.g. 60 pulsations per minute) and a higher vacuum and massage ratio (e.g. 3:1) seem to have potential.

A Casse milking machine is currently used for the milking of sheep. The Casse milking system (fig. 8.12) consists of a herringbone parlour adapted to sheep. The sheep enter the platform in groups of 12 animals or a multiple of 12 and are positioned perpendicularly to the pit in which the milker is standing. By manipulating a movable fence in front of the animals the sheep are pushed backwards to the edge of the pit holding the milker. The milker then places the cluster through the hind limbs onto the udder of the ewe. With an improved model of this milking machine one person can easily milk up to 170 ewes per hour.

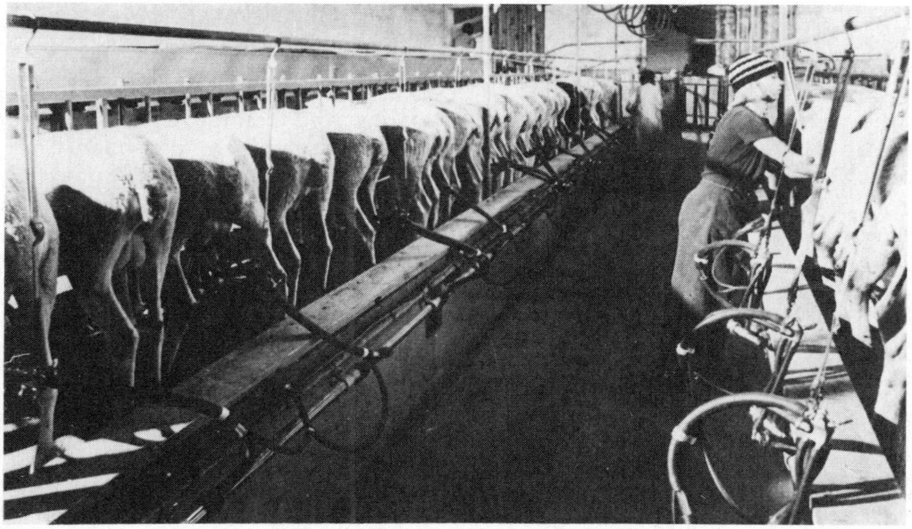

Fig. 8.12 The Casse milking system for sheep (2 x 24) with "low" milking line.

The milk let-down is fast with machine milking viz. in less than 75 seconds and this allows the use of half the number of clusters as would normally be required by the number of places.

This technique is gradually becoming popular although a number of obstacles are still present with the machine milking of sheep. Hand milking of sheep twice a day would normally take about 60 % of the daily labour time and is nowadays economically nor socially justified in developed countries. This probably explains the already large number of milking machines for sheep currently in use in France (ca. 2,000).

8.4 THE HOUSING OF GOATS

The keeping of goats is rather limited and they are seldom kept for economic reasons. In some countries and for example in France, this is an exception since goat milk is important for the preparation of some well-known cheeses. The housing of goats is similar to that of sheep. In accordance with the goat's jumping nature as in its original environment (mountainous areas), the feed stance is often raised by some 50 cm with respect to the floor and this makes the goat to jump onto the feed stance. This raised feed stance which is ca. 80 cm long is also used to milk the animals either by hand or by means of a machine. The goats are tethered to the feed fence, which abuts against the trough, during the milking (Toussaint and Grillot, 1976). Goats, like sheep, can be milked mechanically in a parlour but this is seldom done in view of its high cost price.

REFERENCES

Anon., 1971. Constructions rurales, Logement des ovins, Bulletin documentaire du Ministère de l'Agriculture, Paris, France, pp. 9-12.

Anon., 1979. Sheep handbook : housing and equipment, 3rd ed. Iowa State University, U.S.A., July 1979.

Anon., 1983. Yearbook of Agricultural Statistics, Statistical Office of the European Communities - Eurostat, Brussel, Belgium, 286 pp.

Biamlen G., Kuhner H. and Sebastian D., 1970. Arbeitsverfahren in der Schafhaltung, Neureuter Verlag, K.T.B.L., Flugschrift nr. 20.

Blanken G., 1974. Einrichtungen für die Schafhaltung, Landtechnik, 29 : 539-541.

Burgkart M. and Mittrach B., 1972. Schafställe, Bauen auf dem Lande, 23 : 4.

Calus A.C., 1981. Intensieve schapenhouderij als gedeeltelijke oplossing voor de E.E.G.-zuivelproblemen, Landbouwtijdschrift, 34 : 1181-1192.

Comberg G. and Hinrichsen K., 1974. Tierhaltungslehre, Ulmer Verlag, 464 pp.

Dickson I.A. and Stevenson D.E., 1979. The housing of ewes, Farm Buildings Association Journal, 24.

Labussiere J., 1974. La brebis laitière et la traite mécanique, L'élevage, Paris, France, nr. 33, pp. 103-109.

Lovelidge B., 1973. La maîtrise de l'agnelage par l'éclairement artificiel, L'élevage, Paris, France, nr. 23, pp. 105-107.

MacCormack J., 1975. House for 300 ewes, Farm Building Progress, nr. 41, pp. 7-8.

Noton N.H., 1982. Farm Buildings 1st Ed., College of Estate Management, Reading, G. Britain, 359 pp.

Peck D.N. and Kelly M., 1978. Round Swim sheep dipper at Craibstone, Farm Building Progress, nr. 52, pp. 7-8.

Robertson A.M., 1978. Sheep housing, Farm Building Progress, nr. 51, pp. 1-4.

Robinson T.W., 1982. Designing for the confinement of sheep, Farm Building Progress, nr. 69, pp. 9-14.

Shepherd C.S., 1974. Design and layout for sheep handling pens, The West of Scotland Agricultural College, Bull. 159, 50 pp.

Toussaint M. and Grillot M., 1976. Aménagement des Bâtiments pour la traite mécanique des caprins, Compte-rendu de la Journée d'études sur équipements et bâtiments au SIMA, L'élevage, Paris, France, 50 pp.

Van Gelder H., 1980. De schapenfokkerij in België, Agricontact, nr. 111, 7 pp.

Watson G.A.L. and Speedy A.W., 1977. Sheep dipping facilities, Farm Building Progress, nr. 47, pp. 15-18.

Watson G.A.L., Gerrie W.A.G. and MacCormack J.A.D., 1981. Sheep handling pens and dippers, Farm Building Progress, nr. 63, pp. 21-28.

Chapter 9

THE HOUSING OF RABBITS

Chapter 9

THE HOUSING OF RABBITS

9.1 GENERALITIES

Rabbits are mainly kept for the production of meat. There is nevertheless a great demand for rabbit hair, especially that of white rabbits ; on the contrary the demand for the pelt of rabbits, which was once an important source of revenue, has greatly decreased in recent years (Schlolaut, 1978).

Rabbits are very efficient in converting feed into edible meat and this efficiency is superior to most other domesticated species (Walsingham, 1972). Rabbits are rodents (now belonging to the order of the Lagamorphs) so that materials used for their housing must meet specific demands.

Rabbit meat is mainly consumed in West Europe : its consumption is the highest in France where it amounted to 5 kg per person in 1979.

The production of rabbit mainly takes place in Western Europe with France being the most important producer. In most of the European countries a decline or at least a stagnation can be noticed with the exception of Spain and the Netherlands where a steady growth in the production can be observed. In the Netherlands in 1983 8,500 tonnes of rabbit meat were produced against 4,600 tonnes in 1974 (Anon., 1984). Table 9.1 summarizes the production figures of a number of European countries (Stigter, 1982). The decline is largely attributed to increasing urbanization whereby the number of amateur producers of rabbits decreases.

TABLE 9.1 The production of rabbit meat in W. Europe in tonnes of slaughtered weight.

Country	1974	1977	1979
France	270,000	200,000	180,000
Italy	200,000	140,000	135,000
Spain	80,000	110,000	125,000
W. Germany	25,000	-	15,000
Great Britain	15,000	17,000	17,000
Belgium	6,500	5,000	5,000
The Netherlands	4,600	5,900	7,000

Since the demand for rabbit meat is greater than its production in all Western European countries, imports are necessary and China is by far the largest source of rabbit meat : about 80 % of the import of rabbit meat into the E.E.C. is attributed to China. The degree of self-sufficiency has dropped in most of the European countries (e.g. in Belgium from 85 % in 1968 to 50 % in 1979). The rabbit meat imported from China is however not so well adapted to the W. European market. A few European countries have therefore taken the opportunity to increase their exports through an important expansion of the production of meat rabbits (Stigter, 1982). Preference is nowadays given to a meat rabbit with a liveweight of 2 kg and which is slaughtered at the age of ca. 75 days giving ca. 1.25 kg of meat. A final liveweight of 2.5 kg is not uncommon. The meat must be pink-white, tasty and not thready (broiler rabbit).

Rabbit production by amateurs is still important (mainly by labourers, pensioners) although there is an increasing tendency towards industrial rabbit production. Advanced industrialization as is the case with layers or broilers is however not likely in the near future. The high risks involved indeed form a serious impediment for this industrialization. The rabbit appears to be particularly sensitive to a great number of diseases caused by bacteria (e.g. colibacillosis, pasteurellosis), viruses (e.g. myxomatosis, rotadiarrhoea) and amoeba (coccidiosis). An overall mortality of 20 % and more is not uncommon. The reduction of these excessive losses is therefore a major task and is tackled by breeding hybrid strains, by feed restriction with the kittens after weaning and by a rigorous medicinal prophylaxis.

Extensive breeding of rabbits is increasingly being abandoned in favour of intensive and semi-intensive breeding methods. The intensive breeding method implies post partum mating (mating at the same day as parturition or within a few days after it) and in this way the doe is capable of producing up to 10 litters per annum. This breeding method however leads to a lower conception rate, a smaller litter size and a premature culling of the does (Delage et al., 1975). The semi-intensive production of rabbits is therefore recommended : the doe is mated 10 to 20 days after partum, the pups are weaned 28 to 35 days after kindling and 7 to 9 litters are obtained per annum (pregnancy of 30 - 32 days). The semi-intensive breeding method has an expected output of up to 50 weaned pups per annum.

9.2 THE CONSTRUCTION AND EQUIPMENT OF A RABBITRY

The rabbitry is a *level hangar* of which the walls, roof and floor have a k-value lower than 0.8 W/(m² .K). The housing can be traditionally built or can consist of a so-called "rabbit-shelter". In the former case, which is most common, the building consists of walls made of bricks + cavity + cellular concrete or of wood + insulation material (see Chapter 3) ; the roof consists of corrugated asbestos cement or metal sheets with insulation boards beneath the sheeting.

The so-called rabbit shelter or rabbit tunnel (fig. 9.1) has a foundation of stone upon which a number of D-hoops of galvanized iron are installed. The latter are covered by light-tight and impervious boards of synthetic material beneath which an insulation layer is installed (e.g. glass-wool) held in place by a second board of synthetic material. This construction requires a smaller investment

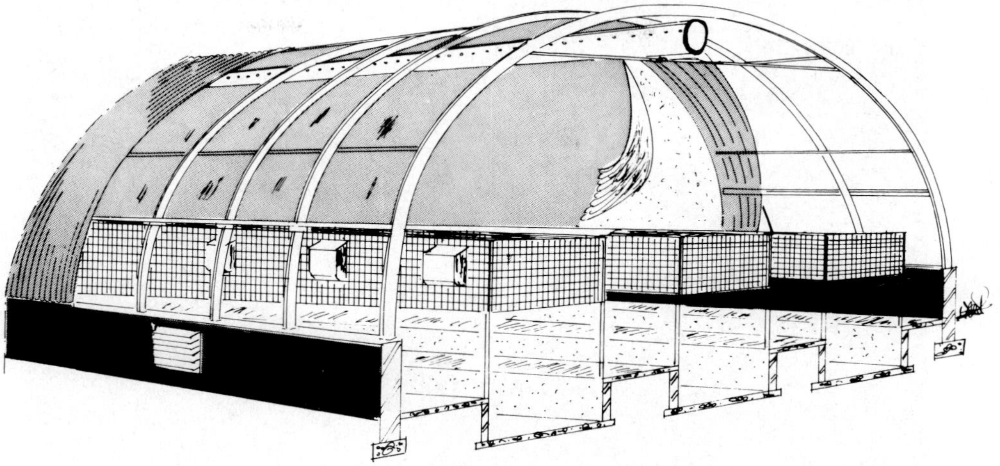

Fig. 9.1 The rabbit shelter or rabbit tunnel.

and therefore offers some potential for the future (Heylen, 1984).

The floor of the rabbitry is concreted. With the semi-intensive and intensive breeding of rabbits, the animals are housed in cages and litter is no longer used.

The *batteries* are composed of all-wire cages. The wire is double galvanized viz. once prior to welding and a second time after it. The dimensions are discussed later. The battery is mounted on a frame of thick-walled tubing (2 mm) of ca. 50 mm x 30 mm. Three types of batteries are in use : the single tier or flat deck batteries, two or three-tiered ordinary and "Californian" batteries.

The first and last type of batteries enable a good survey of all the cages but the stock density is the greatest with the other type of battery (fig. 9.2) although the two-tier battery is also satisfactory and moreover results in a lower price per rabbit. Three-tier ordinary and "Californian" cage systems are increasingly used for the production of table rabbits (fig. 9.3).

The floor of the cage is manufactured from 3 mm thick wires welded in the form of meshes measuring 25 mm x 13 mm or 19 mm square. The sides and the top can be made from 2.5 mm thick wires welded in the form of 25 mm square meshes. The floor is sometimes made of tough plastic slats of 30 mm width with 15 mm slots. The sides are made of galvanized iron sheets and although this contributes to a more restful environment it is disadvantageous to the overall view over the cages. A feed box filled with 2 to 3 kg of concentrates and perhaps a hay rack with 35 mm square meshes (Anon., 1978a) are attached to the front of the cages (figs 9.3 and 9.4). A combined feed trough and hay rack can also be used.

Fig. 9.2 The single tier flat-deck battery for breeding rabbits. Note the nest box and feed hopper with concentrates in the front.

Fig. 9.3 A three-tier cage system for table rabbits.

Fig. 9.4 A cage of a battery, with feed hopper and nipple drinker.

The cages are assembled above and next to each other forming one, two or three-tier batteries. Faeces and urine are deposited on a manure tray made of wire-glass or plastic which slopes to the rear and which is installed beneath each tier of cages. These trays deflect the faeces and urine into a collection gutter behind the battery and often between two batteries installed back-to-back. The wastes are brushed or scraped into a manure cellar.

With the "Californian" tiered system and the single tier flat-deck cage system, the manure is collected on the floor below the cages (fig. 9.2) and is removed daily by means of a scraper or a moving belt.

The wire cages offer a number of advantages : easy and thorough cleaning and disinfection are possible, littering is no longer required thus enabling a substantial saving on the costs of straw and labour and thereby eliminating a potential source of infection, a maximum number of animals can be put per unit area.

Caged rabbits will normally experience no special problems although injuries to the feet can occur, especially with newly bought rabbits which have previously been kept on litter. In this case a plank of 20 cm x 30 cm, temporarily laid on the wire floor, might cure the problem. Drinking water is supplied by an automatic watering system. Such a system generally consists of a break pressure tank equipped with a float valve and fed from the main. Via this tank, water, sometimes supplemented with medicines, can be distributed through the waterers.

Rabbits are preferably accommodated in *different compartments* according to the breeding purpose and the stage in their production cycle. A number of guidelines have to be respected in this connection. On one hand the frequency of production of successive offspring is high with the semi-intensive breeding system. On the other hand weaned rabbits are kept for another forty days and replacement rabbits are selected from these around the age of 10 weeks. The remaining animals are destined for slaughter. Different housing compartments are therefore provided viz. one or more for does and bucks (depending on their number), one or more for meat rabbits and one or more for replacement rabbits.

The kittens can either remain in the cage of the does for fattening or they can be removed and fattened in another compartment. The former possibility induces less stress. The latter method, which is mostly used, sets less stringent demands for the heating and ventilation of the house. The young rabbits are then in fact fattened in groups of 8 to 10 in a separate, unheated fattening compartment.

The compartment for breeding stock

The *does* are housed individually in cages with an area of 0.6 to 0.7 m² (1 m x 0.6 m or 0.9 m x 0.8 m or 0.8 m x 0.8 m) which are 35 to 50 cm high. About 5 days prior to parturition the pregnant does are provided with a nest box which can resemble the wooden construction illustrated in fig. 9.5.

Injury to young litters by the doe jumping into or out of the nest box has to be prevented : the upper side of a suitable nest box should therefore be covered while its front side must be 15 cm high.

The floor of the nest box is covered with clean straw. A trap door at the rear or a slide or lid forming the upper side of the nest box allows the rabbit keeper to inspect the nest box. Cylindrical nest boxes made of P.V.C. with a wooden floor or box-type nest boxes of P.V.C. or galvanized iron (fig. 9.6) with a lid of the same material are nowadays also in use. With these types of nest boxes the entrance for the doe is also placed in such a way as to sufficiently protect the offspring. These types of nest boxes are also littered. Open nest boxes made of P.V.C. are sometimes used because they are cheaper, but they offer no protection to the pups and require a higher house temperature (ca. 2°C higher) and hence higher heating costs. The closed nest box therefore can be recommended. A feeder and an automatic waterer (water trough or nipple drinker) are of course provided in the cage. If a combined hay-and-concentrate feeder is used, the latter should be installed out of reach of the young rabbits since they should not be supplied with hay ; this would induce too strong a development of the intestines which has an unfavourable influence on the dressing percentage. Before or at weaning the nest box is removed and thoroughly cleaned and disinfected.

The *bucks* are accommodated in the same house as the does, in individual cages which can measure for instance 50 cm x 60 cm and which are 40 cm high. The cages are identical to the ones described higher. One buck is provided for every ten does. The mating takes place in the cage of the buck. A day length regime of 16 hours is preferably maintained in the compartment for breeding stock as this

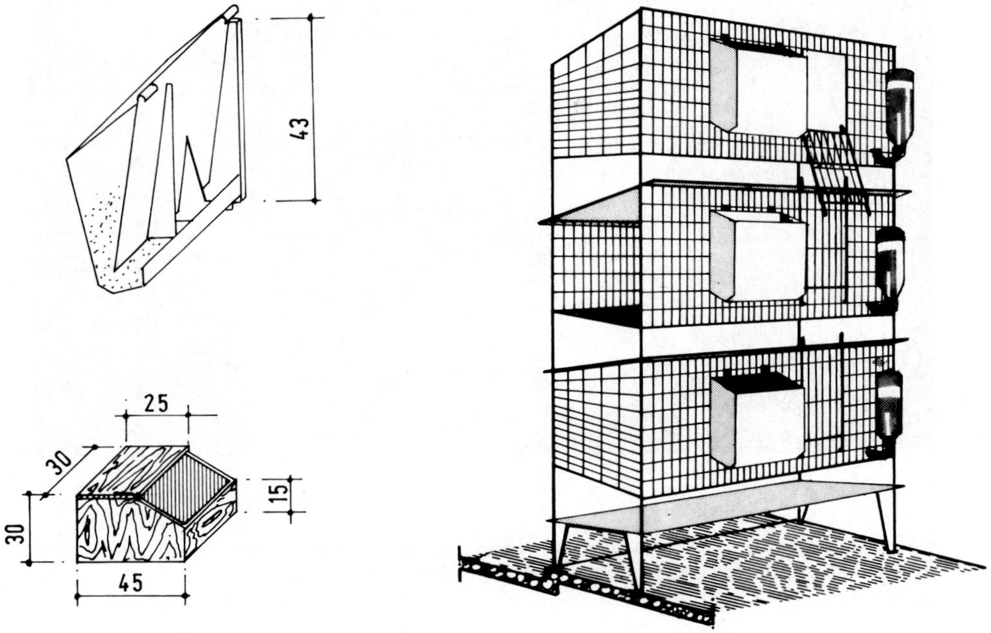

Fig. 9.5 A cage system for rabbits with a detailed design of a nest box.

leads to the most regular heat rhythm of the does (Delage et al., 1975).

The compartment for meat rabbits

Weaned rabbits are transferred to another compartment of the house. They are housed in groups of ca. 8 animals in community cages which have an area of ca. 0.5 m^2 and a height of 30 cm. The feeding and watering facilities are similar to those described earlier. The best performers are selected at the age of 10 weeks and are kept as replacement stock. The others are destined as *meat rabbits* and are ready for slaughtering at the age of ca. 75 days.

The compartment for replacement stock

The *replacement breeding rabbits* are transferred to the third compartment. They are all accommodated in individual cages measuring ca. 50 cm x 60 cm and 35 cm high. Bucks have to be kept separately, (certainly from 12 weeks of age) as they will otherwise engage in bloody fights often resulting in the castration of the

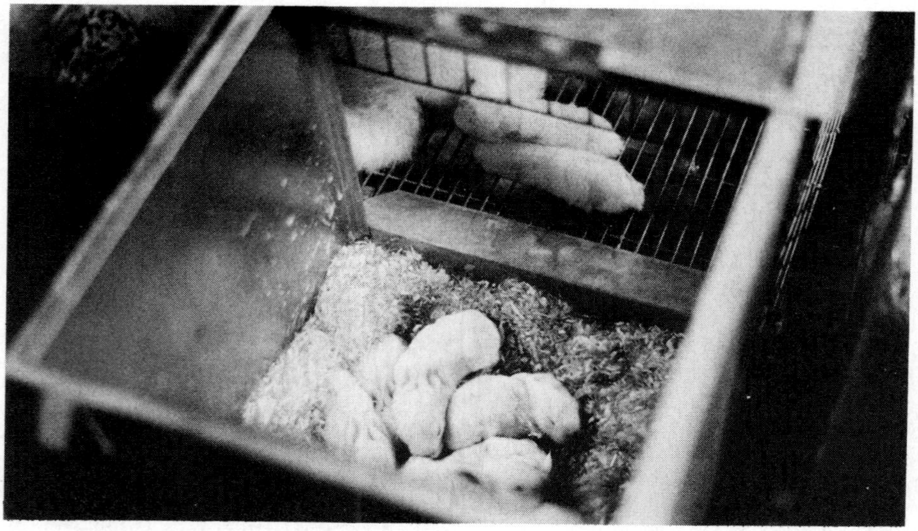

Fig. 9.6 A nest box made of galvanized iron.

weakest. Group housed young does tend to mount and pseudopregnancy or ovulation results from one doe riding another. This would inevitably lead to production losses. The bucks can be moved from the age of 5 months and the does from the age of 4.5 months to the breeding stock compartment where they can be put into production. There is an increasing tendency towards the application of the all-in all-out system in commercial rabbit production whereby the animals are kept in *compartments*. Breeding stock for instance can be kept in units of up to 120 animals, the meat rabbits derived from this breeding stock can also be kept in a separate compartment.

The *climatic environment* is of great importance for the rabbits. A windowless house is preferable because an adjustment of the climate can then be carried out more easily. Windows (1/15th of the floor area) are however sometimes provided in the compartments for breeding stock and for replacement stock to lower the lighting costs.

A lighting programme is applied in each compartment. The compartment for young breeding stock is subjected to a day length of 12 hours with either natural or artificial lighting and with a light intensity of 1.5 to 2 W/m^2 (striplight). The compartment for breeding stock has a day length of 16 hours.

The temperature in the rabbitry should be maintained between 15 and 18°C (Okerman, 1979), its relative humidity between 65 and 70 % and the air velocity at the level of the animals at 0.1 to 0.2 m/s (Anon., 1978b).

Pressure ventilation is often employed in rabbit houses. A fan blows air (sometimes filtered) through a tube with a large number of holes. This tube is installed under the ridge of the house (fig. 9.1). Foul air either leaves the house through a number of openings in the lower part of the side walls or is extracted by means of an extraction fan connected to a duct which is installed along both side walls (Claeys et al., 1982). The inlet air can also be heated by e.g. an heat exchanger which is thermostatically controlled. Extraction ventilation, especially in combination with cross ventilation, is also often used. The foul air is thereby extracted by fans placed as low as possible in one wall while the air inlets are distributed over the entire opposite wall. The fan speed is thermostatically controlled and should allow a ventilation volume of 5 m^3/h/kg liveweight in the summer and 1 m^3/h/kg liveweight in the winter.

The *zootechnical and veterinary aspects* of the housing of rabbits should be given due consideration. Unfavourable housing conditions and mainly too high a stocking density with meat rabbits, inadequate ventilation or draught, noise and the presence of strangers, rats, dogs and cats will lead to pathological reactions with the rabbit. They can give rise to cannibalism whereby the doe might injure or eat her pups or where meat rabbits will bite each others ears or tail or will proceed to auto-castration. In this connection philophagy can also occur. It consists of fur eating, where rabbits eat their own fur or the fur of other rabbits. It is most likely due to a faulty ration whereby the rabbits are not receiving the proper amount of feed or where the latter does not contain sufficient proteins or fibre. Finally disorders of the metabolism may occur. The rabbit excretes two types of faeces in a normal behavioural pattern, called caecotrophy. The soft faeces produced in the caecum (blind gut), are eaten directly from the anus (caecotrophy) and will, after passing a second time through the digestive tract, be definitively excreted as hard faeces. The hard faeces are the faecal pellets commonly seen and which are produced by the large intestine. Perturbations in the caecotrophy can lead to digestive troubles.

A thorough sanitation policy is very important in controlling disease (Delage et al., 1975) and is easily applicable with the all-in all-out system and with compartmentalizing. At regular intervals a compartment of the rabbitry is emptied and thoroughly disinfected with a commercial 40 % formaldehyde solution followed by steam cleaning viz. with superheated steam under high pressure (Blanken, 1975). A thorough cleaning of the house completes the treatment after which the house is left empty for an additional two weeks. The equipment of the house e.g. troughs, cages, racks etc. are also disinfected with steam and successively cleaned. All compartments of the rabbitry are in this way disinfected and cleaned.

The *organization of the rabbit farm* is on one hand based on the precise calculation of the number of cages required and on the other hand on the updating of suitable zootechnical record-keeping (Delage et al., 1975). As an example we calculate the required number of cages for the exploitation of a 100-doe rabbitry according to the semi-intensive breeding system :

- number of cages for does, in which they remain permanently = 100 e.g. divided over two rows of two-tier batteries (cages of 100 cm x 60 cm x 45 cm) ;
- number of cages for bucks = 10 to 12 (50 cm x 60 cm x 40 cm) ;
- number of cages for young breeding stock (cages : 50 cm x 60 cm x 35 cm ;rearing from the age of 10 weeks to the age of 20 weeks or for 70 days ; 60 replacement animals per annum ; ca. 10 % spare cages) :

$$\frac{70 \times 60 \times 1.1}{365} = 12.6 \text{ or } 12 \text{ for constructional reasons ;}$$

- number of cages for meat rabbits (each litter in one community cage of 1 m x 0.6 m ; 8 litters per annum ; finishing from weaning at ca. 30 days to slaughtering at 75 days or 45 days in total ; ca. 10 % spare cages) :

$$\frac{8 \times 100 \times 45 \times 1.1}{365} = 108.5 \text{ or } 108 \text{ for constructional reasons.}$$

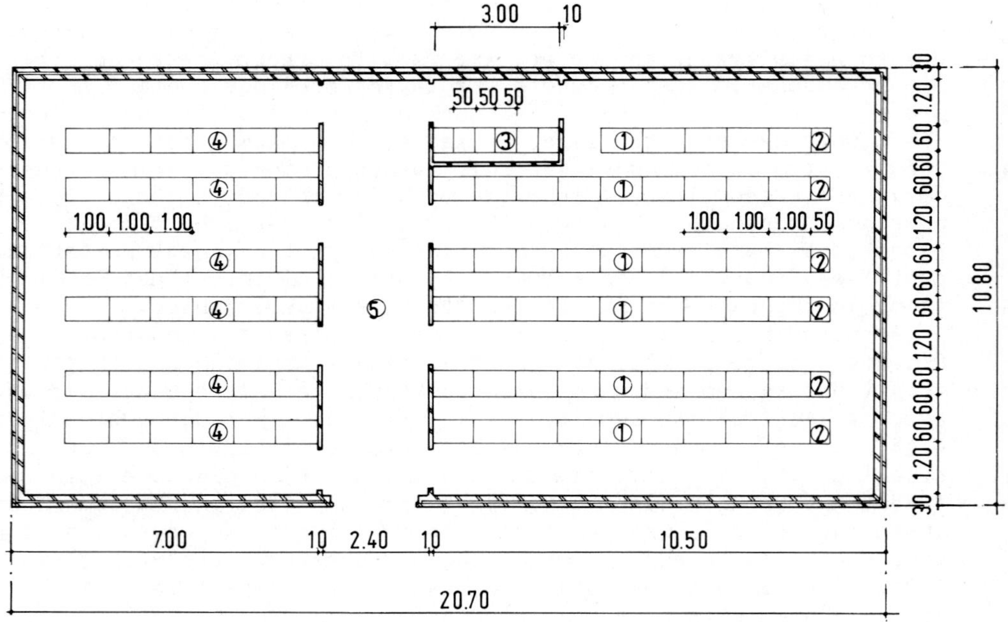

Fig. 9.7 Ground-plan of a closed 100-doe rabbitry with three compartments.
Legend : 1 = does (2-tier batteries) ; 2 = bucks (2-tier batteries) ; 3 = breeding stock (2-tier batteries) ; 4 = meat rabbits (3-tier batteries) ; 5 = feed store.

A total of 232 cages are thus required spread over three different compartments. Fig. 9.7 shows the ground-plan of a closed rabbitry for 100 productive does of which the offspring are grown to meat rabbits on the farm itself. It consists of three compartments. Larger rabbitries will have more units, each consisting of three compartments and housing ca. 100 does. There is however nowadays a tendency to reduce the population per unit to ca. 80 does.

The *labour productivity* amounts to 400 does per stockman. An exploitation at the level of a family enterprise can consist of 600 does. The investment required for such an enterprise (turn-key and incl. land) is calculated at £ 100 to £ 140 per doe for 600 does, which is indeed a high investment. This together with the risks mentioned higher explain why the industrialization of the rabbit production proceeds so slowly.

REFERENCES

Anon., 1978a. De huisvesting van konijnen, Ministerie van Landbouw, Direktie voor Landbouwtechniek, Brussels, Belgium, 10 pp.

Anon., 1978b. Bulletin d'Information, nr. 3 de la Station Expérimentale d'Agriculture, Ploufragan, France.

Anon., 1984. Konijnenhouderij krijgt meer aandacht, Bedrijfsontwikkeling, 15 : 736.

Blanken G., 1975. Geräte zur Stalldesinfektion, Der Tierzüchter, 5 : 28-30.

Claeys N., Hens J. and Sierens G., 1982. Succesvol konijnen houden, 4e herwerkte uitgave Belgische Boerenbond, Leuven, Belgium, 72 pp.

Cousin J.F., 1975. La saine gestion d'un élevage de lapins de chair exclut le pilotage à vue, L'élevage, Paris, Numéro Special, pp. 119-126.

Delage et al., 1975. Le Lapin, L'élevage, Paris, numéro spécial, 154 pp.

Heylen J., 1984. Italiaans batterijtype, Pluimvee, 19 : 305-306.

Okerman F., 1979. Klimaatregeling in konijnenstallen, Pluimvee, 14 : 394-396 and 419-423.

Schlolaut W., 1978. Entwicklungstendenzen in der Kaninchenproduktion und deren produktionstechnische Probleme, Der Tierzüchter, 30 : 436-438.

Stigter E., 1982. De konijnenhouderij in Nederland, Bedrijfsontwikkeling, 13 : 743-745.

Walsingham J.M., 1972. Ecological efficiency studies : Meat Production from rabbits, Grassland Research Institute, Hurley, G. Britain, Technical Report nr. 12.

Chapter 10

THE HOUSING OF FURRED ANIMALS

Chapter 10

THE HOUSING OF FURRED ANIMALS

10.1 GENERALITIES

Furs have been used for millenniums for warmth and adornment. The Chinese, the Assyrians and the Greeks made use of them. In the Middle-Ages the wearing of fur became so widespread that its use was forbidden to all but nobility. Ermine is nowadays still used for regal coronation robes. The principal fur-producing countries are the United States of America, Russia, Scandinavia and Canada. The demand for fur in the welfare-countries has gradually increased since World War II : this higher demand together with the relative scarcity of the fur-bearing animals have led to more difficulties in providing furs from animals living in the wild. The chinchilla's for example, which lived on the slopes of the Andes in South-America have practically been exterminated by local tribes of Indians. As a result, many countries have started to raise some furred animals in confinement e.g. minks and chinchilla's. In the Sixties minks were often raised by amateurs on a very limited scale but over the last two decades a number of larger and specialized enterprises made their appearance. Farms having several ten thousands of minks are nowadays not uncommon.

Selected female minks are always employed for the breeding of minks (Seutin, 1976 ; Claerhoudt, 1983). The bitches come on heat three times during the month of March and are then to be mated. Bitches are always taken to the male where mating will take place. The two animals are separated immediately after service.
A second mating is organized the next day followed by an interval of seven days. A third and fourth service are then performed on two successive days. This service schedule is not the most common : often a first service is followed by an interval of 8 days after which two services are organized on the two successive days. It is not always that easy : often the bitch does not accept service and afterwards more services have to be planned. It is not uncommon that the bitch refuses a first service and that even five repeats, carried out on every other day, are still unsuccessful. After a new interval of 1 day two matings may be organized at two successive days. Older bitches often accept a first service but tend to refuse any further services after an interval of 8 days. Needless to say this period is very labour-intensive for the mink-keeper.

A litter is normally delivered around May 1, (20th April - 15th May) after a gestation of 45 to 60 days. Minks produce only one litter per annum and the average litter size is 4 young (1 to 10). The young are weaned after 6 weeks.

Minks intended for breeding, are sexually mature at the age of 7 months and are accommodated in breeding sheds for about 3 years.

One male is kept for every four bitches. The young minks, born in May and intended for fur production are, after weaning, housed in so-called fur sheds until November-December of the same year. They are then killed either by dislocating their neck or by suffocation. The males then weigh ca. 2 kg while the bitches have a weight of ca. 1 kg. Female minks have a finer fur while the males have a thicker and nicer fur.

Minks are very sensitive to a number of diseases (e.g. : distemper, botulism, virus-enteritis, pseudo-rabies) and they can be vaccinated against some of the diseases at weaning.

The killed minks are brought into a rotating mill to remove impurities from their fur. The animals are then stripped and the pelts are stored in a fridge. After a few days the fat is scraped from the pelts and they are then ready for marketing.

Some farms produce, process and sell the pelts but most of the pelts are handled through the four international markets viz. New-York, London, Oslo and Leipzig. The fat from minks is also processed, viz. in cosmetic products.

Improved breeds of bitches are also used for the breeding of *chinchilla's*. One male is kept for every 6 to 10 bitches and the young are sexually mature at the age of 7 months. Gestation takes 111 days and an average litter contains 2 young (1 to 6) which are weaned at the age of 6 weeks. They are housed together up to the age of 3 months and are then separated. They can be killed at the age of 6 months. Chinchilla's normally produce two to three litters per year. The suckling bitch is mated shortly after parturition. Male animals are kept for ca. 5 years while the bitches are kept up to 10 years. Finally it must be mentioned that the amateur keepers of chinchilla's have often been abused by unfair merchants.

10.2 THE HOUSING OF MINKS

Minks can withstand the cold but should be protected against the rain. They are therefore kept in open-sided sheds of which the roof is made of corrugated metal or asbestos cement sheeting. Small ventilation stacks may be provided in the ridge to assist in the ventilation during warm and windless weather conditions. Important in the housing of minks is an adequate supply of light. The production of those hormones inducing oestrus is initiated by the increase of the day length. Ample sized and rather high hangars are therefore required which will guarantee sufficient light. Figs 10.1, 10.2 and 10.3 show the construction and equipment of a modern mink house and more in particular the construction of a breeding shed.

Animals are kept in cages, arranged in a row and on a single tier of which the floor is ca. 60 cm above the ground. Each open-sided shed normally holds two rows of cages separated from each other by a service passage with a width of ca. 1.4 m. Installing more rows of cages in a hangar has to be avoided since the cages in the middle will not receive enough light.

This aspect is of lesser importance in the fur sheds where there are no objections to the installation of several rows of cages.

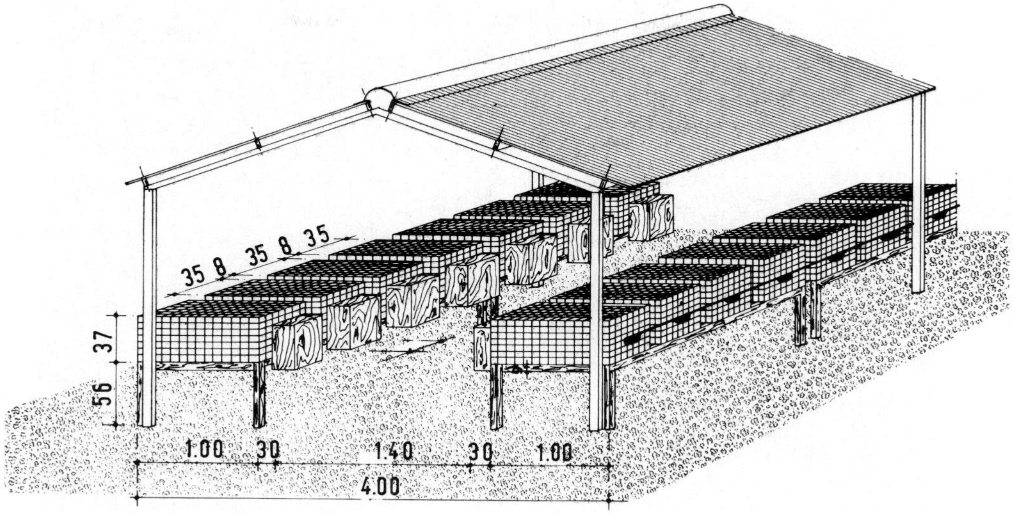

Fig. 10.1 A mink house.

Fig. 10.2 Outside view on a mink house.

Fig. 10.3 Inside view of a modern mink house.

The cages are made of wire mesh. The droppings fall through the wire floor practically always on the same place viz. away from the nest. Cage dimensions for bitches are (breeding cages) :
- for the cage : 0.35 m (W) x 1 m (L) x 0.37 m (H) ;
- for the nest box : 0.30 m (W) x 0.25 m (L) x 0.37 m (H).

The meshes of the floor are 1" x 1" while those of the sides are 1" x 3/2". With such large meshes the cages are normally installed 8 cm apart to prevent the animals biting each other. If only one partition is used, thus without an intermediate space between two cages, the animals tend to bite each others tongue or tail, even if the meshes are small. The top is equipped with a hinged door which is part of the wirework.

The nest box consists of solid wooden sides, while the floor is made of wire mesh. The floor of the nest box is normally 9 cm lower than the floor of the cage if front-loading nest boxes are used, but it is not uncommon to find nest boxes which are installed above the cages.

A pop hole with a diameter of 9 cm is provided in the wooden side of the nest box and forms a connection between the nest box and the cage. The latter is accessible through a hinged lid which forms part of the top of the nest box. Beneath this lid a supplementary wire mesh is provided.

The floor of the nest box is covered with wood shavings shortly before parturition. A special wire cage is sometimes placed in the nest box : it is a box-like cage (width 19 cm, height 19 cm, length 20 cm) with meshes of 1.2 cm x 2.4 cm and surrounded with straw. This tray-shaped nest with limited area forces the young to stay in the nest. In this way they cannot hide and they can then hardly be neglected by the bitch.

Bitches are always housed individually and in their own cage ; they are only taken to the male's cage for mating and are then returned to their cage. Weaned minks intended for the production of pelts are usually housed individually. They can sometimes be housed in groups provided that they belong to one and the same litter and that they have always remained with each other in the previous period.

The dimensions of the cages for minks intended for pelt-production (pelt cage) are ca. 1.20 m (length) x 0.18 m (width) x 0.37 m (height). The cages are 8 cm apart. The meshes are similar to the ones used in the breeding cages.

The installation for the distribution of drinking water is fitted to the rear of the cages. Nipple drinkers can be applied although small drinking dishes can also be used, which are then all installed at the same height and filled to the same level by means of a filling tube (and a float). The feed for minks consists of raw, minced meat, minced offal of poultry slaughterhouses and of fish supplemented with vitaminized meal. The feed is placed daily on top of the wire cage. Suckling minks are fed three times a day. Young minks are fed solid feed after ca. 3 weeks. The bitches together with their young receive their solid feed on the wire mesh at the top of the nest box.

Mechanized feed distribution is sometimes practised on the larger mink farms. The reservoir of the feed trolley is filled with ground and homogenized semi-liquid feed (fig. 10.4). The feed trolley consists of a converted mini-tractor equipped with a reservoir which can hold ca. 600 kg of feed and a pedal switch which activates the pump for the delivery of the feed. The mink keeper drives his feed trolley through the service passage. Each time he touches the pedal switch, the pump sucks a certain amount of feed (ca. 200 g) from the reservoir and pushes it into a flexible rubber tubing provided with a stainless steel mouthpiece. The driver uses his left foot for the gas pedal and activates the pump by pressing the pedal switch with his right foot. He uses his left hand for the steering wheel while with his right hand guiding the flexible rubber tubing he distributes the required amount of feed on the top of the cages or the nest boxes. In this way he is able to distribute the feed for one row of cages in one travel. A reservoir with a content of 600 kg suffices for the distribution of the ration to ca. 1,800 cages without the need for re-filling.

Such a mechanized feed distribution system allows a serious reduction of the labour demand viz. a mere 1.5 man-hours per day compared to ca. 8 man-hours per day for the manual feed distribution to 3,700 breeding bitches. Five labourers then suffice for the daily care of a breeding stock of 3,700 bitches with their young compared to 12 persons without mechanization.

A "mink-tight" fence is installed around the mink farm, in order to prevent the escape of minks. A wind-break is also necessary to protect the animals against draught.

Fig. 10.4 The mechanized feed distribution in a mink house.

The housing of minks in completely closed and windowless buildings where a light programme is applied is still in the experimental stage. According to Norwegian research (Skrede and Reiten, 1975) mink-pelts are ready for processing 1.5 months earlier when the minks have been subjected to a constant light programme of 6 hours of light and 18 hours of darkness. Their weight is then lower than that of minks grown in traditional buildings but the pelts are normally larger while their hair is shorter. The feed consumption and the labour time requirement are also lower in a house where such a light programme has been applied since the animals are ready for killing earlier. The electricity consumption, however is significant since no electricity is used in a traditional house.

10.3 THE HOUSING OF CHINCHILLA'S

Chinchilla's are accommodated in low level hangars, which are completely closed ($k < 1.1$ W/(m^2.K)), well-lit and ventilated. Each animal is individually housed in a wire cage of ca. 45 cm x 45 cm x 45 cm (fig. 10.5).

Flat deck batteries as well as 2, 3 or 4 tier batteries are used. The upper part of the cage front is equipped with a hinged door of wirework. Each cage is provided with a dish or a basin filled with fine, unsharp sand in which the animals can roll themselves and which has a beneficial influence on the quality of the pelt. A manure tray made of asbestos cement is placed underneath each removable cage and is covered with a layer of wood shavings which absorb both the urine and the faeces. A combined trough for concentrates and hay, is used and is fitted to the front of the cage. The bottle may be held

Fig. 10.5 Individual cage housing of chinchilla's.

in position by a leather, plastic or metal strap. A central wire passage runs at the top rear between two rows of cages installed back-to-back. The male can travel freely in this passage and has access through pop-holes to all the cages of the bitches. He has to find his feed and water in the cages of the bitches. The bitches wear a collar around their neck the diameter of which is slightly greater than the diameter of the pop-holes : thus they have no access to the passage (= male's cage). The collar is removed shortly before the parturition. Indeed, during the parturition the bitch pulls the young with her muzzle out of her vagina and such a collar might form an obstacle during the delivery. Nest boxes are permanently fitted to the cages and resemble those used by the minks. Weaned young are transferred per litter to a cage with the same dimensions (45 cm x 45 cm x 45 cm). The female chinchilla's, intended as breeding stock, are housed in individual cages from the age of about 3 months. The others, together with the males, are killed at the age of 6 months.

A service passage is provided in front of a row of cages or between two rows of cages and has a width of ca. 1.25 m. The labour time requirement for a farm with 20 bitches, males, replacement animals and pelt animals amounts to ca. 20 hours per annum (Bek, 1975). The daily tasks consist of the distribution of feed, the replacement of drinking water and bathing sand, the removal and the cleaning of the manure trays. Chinchilla's are night-animals and they normally sleep during daytime. The daily tasks must therefore be carried out in the early morning or the late afternoon. Thorough hygiene is very important in the raising of chinchilla's.

REFERENCES

Bek A., 1975. Die Wirtschaftlichkeit der Chinchillahaltung, Der Tierzüchter, 27 : 25-27.

Claerhoudt H., 1983. Private, unpublished communication concerning the breeding of minks, Sijsele, Belgium.

Seutin A., 1976. Private, unpublished communication concerning the breeding of minks, Heist-op-den-Berg, Belgium.

Skrede A. and Reiten J., 1975. Effect of light regulation on fur development and growth of mink, Scientific reports of the Agricultural University of Norway, Jg. 54, nr. 24, 21 pp.

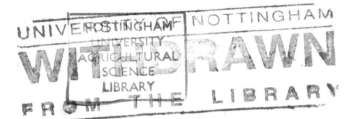